Three Facets of Public Health and Paths to Improvements

Three Facets of Public Health and Paths to Improvements

Behavior, Culture, and Environment

Edited by

Beth Ann Fiedler

Independent Research Analyst, Jacksonville, FL, United States

Academic Press is an imprint of Elsevier
125 London Wall, London EC2Y 5AS, United Kingdom
525 B Street, Suite 1650, San Diego, CA 92101, United States
50 Hampshire Street, 5th Floor, Cambridge, MA 02139, United States
The Boulevard, Langford Lane, Kidlington, Oxford OX5 1GB, United Kingdom

Library of Congress Cataloging-in-Publication Data
A catalog record for this book is available from the Library of Congress

British Library Cataloguing-in-Publication Data
A catalogue record for this book is available from the British Library

ISBN 978-0-12-819008-1

For information on all Academic Press publications
visit our website at https://www.elsevier.com/books-and-journals

Publisher: Andre Gerhard Wolff
Acquisitions Editor: Kattie Washington
Editorial Project Manager: Sara Painavilla
Production Project Manager: Maria Bernard
Cover Designer: Christian Bilbow

Typeset by SPi Global, India

Transferred to Digital Printing in 2020

The path is always easier when you are not walking alone.

Contents

Section C Environmental impact on public health

Contributors

Zaccheus James Ahonle University of Florida, Department of Occupational Therapy; Veterans Rural Health Resource Center-GNV (VRHRC-GNV), North Florida/South Georgia Veterans Health System, Malcom Randall VAMC, Gainesville, FL, United States

Samantha Angel University of Central Florida, Orlando, FL, United States

Yara Asi Department of Health Management and Informatics, University of Central Florida, Orlando, FL, United States

Matt Thomas Bagwell Public Administration Division, Tarleton State University, RELLIS Campus, Bryan, TX, United States

Carmen Besselli U.S. Navy Fleet & Family Support Center, Monterey, CA, United States

Eve Bleyhl Nebraska Family Support Network, Omaha, NE, United States

Colleen L. Campbell NFL/SG VHA Home-Based Primary Care, The Villages, FL, United States

Diane C. Cowper Ripley Veterans Health Administration, North Florida/South Georgia Veterans Health System, Malcom Randall VAMC, Gainesville, FL, United States

Michael E. Dunn Department of Psychology, University of Central Florida, Orlando, FL, United States

Robert D. Dvorak Department of Psychology, University of Central Florida, Orlando, FL, United States

Lynette Feder University of Central Florida, Department of Criminal Justice, Orlando, FL, United States

Beth Ann Fiedler Independent Research Analyst, Jacksonville, FL, United States

Jennifer Hale Gallardo Veterans Health Administration, North Florida/South Georgia Veterans Health System, Research Service, Malcom Randall VAMC, Gainesville, FL, United States

Shanti P. Ganesh Department of Physical Medicine and Rehabilitation, North Florida/South Georgia Veterans Health System, Malcom Randall VAMC, Gainesville, FL, United States

Huanguang Jia Veterans Health Administration, North Florida/South Georgia Veterans Health System, Research Service, Malcom Randall VAMC, Gainesville, FL, United States

Matthew P. Kramer Department of Psychology, University of Central Florida, Orlando, FL, United States

Angelina Leary Department of Psychology, University of Central Florida, Orlando, FL, United States

Mi Jung Lee University of Florida, Department of Occupational Therapy; Veterans Rural Health Resource Center-GNV (VRHRC-GNV), North Florida/South Georgia Veterans Health System, Malcom Randall VAMC, Gainesville, FL, United States

Charles E. Levy Physical Medicine and Rehabilitation Service, North Florida/South Georgia Veterans Health System, Malcom Randall VAMC, Gainesville, FL, United States

Vanessa Lopez-Littleton California State University, Monterey Bay, Seaside, CA, United States

Dawood Ahmed Mahdi Faculty of Languages and Translation, King Khalid University, Abha, Saudi Arabia

Amit Mukherjee School of Business, Stockton University, Galloway, NJ, United States

Dmitry Nikolaenko Environmental Epidemiology, Kiev, Ukraine

Andrés Cubillos Novella Institute of Public Health, Pontificia Javeriana University, Bogotá, Colombia

Naz Onel School of Business, Stockton University, Galloway, NJ, United States

Roselyn Peterson Department of Psychology, University of Central Florida, Orlando, FL, United States

Daniel Pinto Department of Psychology, University of Central Florida, Orlando, FL, United States

Carla Jackie Sampson Fox School of Business, Temple University, Department of Risk, Insurance, and Healthcare Management, Philadelphia, PA, United States

Ivan G. Savchuk Institute of Geography of National Academy of Sciences, Kyiv, Ukraine

Luz Mairena Semeah Health Services Research and Development, Veterans Health Administration, North Florida/South Georgia Veterans Health System, Research Service, Malcom Randall VAMC, Gainesville, FL, United States

Kenan Sualp College of Community Innovation and Education, University of Central Florida, Orlando, FL, United States

Xinping Wang Veterans Health Administration, North Florida/South Georgia Veterans Health System, Research Service, Malcom Randall VAMC, Gainesville, FL, United States

Tracy Wharton School of Social Work, College of Medicine: Clinical Sciences, University of Central Florida, Orlando, FL, United States

Cynthia Williams Brooks College of Health, University of North Florida, Jacksonville, FL, United States

Preface

We are shaped by our environment, learn within our culture, and are judged by our behavior. While there is a constant debate as to whether we are destined in the womb, cultivated by our family, and/or molded by our environment, the cumulative impact can lead to a high quality of life or an abbreviated life with a sentence of diminished capacity. On the other hand, some will say that we shape our own environment and generate culture by our behavior. I cannot say for certain if these perspectives are completely right, but I can say for certain that none of them is completely wrong.

My prior works in public policy, regulation, and environmental and public health were foundational in my desire to further delineate relevant underlying concepts. With this in mind, I researched a variety of fundamental parameters within the topics of behavior, culture, and environment in order to cultivate a network of contributing authors best able to address some important problems associated with improving public health. I asked myself, "What are some behaviors that we can change and what will leaders, parents, and scholars have to do in order to challenge the impact of negative cultural trends while retaining positive cultural heritage?" Then I asked, "How does our environment enhance or reduce the human capacity to elicit the positive?" Thus, the impetus for "*Three Facets of Public Health and Paths to Improvement*" was born.

The process of fundamental research revealed that the nature of the concepts of behavior, culture, and environment was important to our quality of life. For example, how we process information and why we come to certain conclusions (culture), what challenges we or our caregivers face on a daily basis (environment), and how we react to difficult circumstances or certain conditions (behavior) can all lead to different outcomes—some better, some worse. Facets of public health clearly emerged in the desire to demonstrate the diversity in local problems and yet were found repeatedly in the global commonality of fundamental problems. Thus, it is easy to recognize that no single book can encompass the many characteristics of environment, culture, and behavior that impact public health. The challenge in this quest then is to both distinguish the three factors and to demonstrate their wide span of application. We accomplish this by presenting layers of behavioral, cultural, and environmental impact on public health to further illuminate, educate, and promulgate solutions that resolve real world problems.

I have found the conduct of a fundamental and systematic literature review to form the book proposal to be helpful in selecting emergent and topical subjects, potential contributing authors, and the formation of a knowledge base of relevant research to supplement literature searches conducted by individual lead authors. In turn, I utilize this foundation gleaned from a variety of sources, such as the Cochrane Library, EBSCO host, ProQuest ABI, PubMed, relevant multilevel government agencies, nonprofits, and businesses, to prompt subject matter experts to interpret high level topics to produce quality material. Chapter contributions are an important aspect of learning from professionals using a combination of academic writing, organizational perspective, existing policy, volunteerism, and life experience.

The first section, on the general topic of behavioral impact on public health, views behavior from the perspective of those actions that require change and the manner in which the effort of organizations, collaborative relationships, and educators facilitate that change. Topics range from the problem of sedentary behavior to how business education must embrace sustainable development. The second section, cultural impact on public health, demonstrates how cultural awareness can benefit public health by understanding the relationship of care, history, religion, and law. The final section three, environmental impact on public health, may provide the widest span of perspectives even when they are viewed at ground level. Important environmental topics include the growing rate of maternal deaths in concentrated areas of minorities, the impact of infectious disease and energy choices, and Veteran health.

I believe the book and inclusive chapters represent the threefold perspective on public health by providing the reader with two overarching pieces of vital information: (1) a deeper depth of understanding of each facet of behavior, culture, and environment, and (2) practical solutions to resolve long-standing global problems. I hope that general readers, educators, students, concerned citizens, and policymakers find that we have achieved the goal to unleash a greater good by increasing understanding of mankind's current condition and potential solutions. Thank you for joining us on this journey. P.S.: I would like to thank Elsevier representatives for their help along the way: Kattie Washington, Sara Pianavilla, and Swapna Praveen.

Section A

Behavioral impact on public health

Intersecting global health metrics with personal change

Beth Ann Fiedler
Independent Research Analyst, Jacksonville, FL, United States

Abstract

Changing poor health behaviors is a persistent challenge to both practitioner and patient. We rely on aggregated data on health behaviors and metabolic risk indicators to measure health status and point to likely mortality. However, the scale of empirical evidence focusing on high-profile global disease indicators may lead to less than optimal care when local populations are experiencing indicators with less widespread concern. The multilevel scale of data in case studies here demonstrates variations in important trends, giving rise to methods that merge the institutional and social responsibility of data collection and analysis with healthcare and individual accountability. Repeatable analysis incorporating global trends with individual characteristics is one way to translate aggregated health data to predict mortality susceptibility in individual patients. We suggest that statistical modeling of health behaviors intersecting global health metrics with personal change objectives requires new methods of collecting, analyzing, and applying local knowledge from patient history.

Keywords: Behavior modification, Metabolic risk, Burden of disease, Data collection.

Chapter outline

Three Facets of Public Health and Paths to Improvements. https://doi.org/10.1016/B978-0-12-819008-1.00001-8

Focusing global data and risk behaviors on individual patients

What I am is what I am are you what you are or what?

Brickell and Withrow (1988)

If we can be anything, do anything, and believe in ourselves or some higher being enough to become strong, resilient, and fierce, why are we stuck in old habits that are killing us? Seemingly, we are information-saturated, but not information-practical application because there is a profound gap in the awareness of needing change and the difficulties associated with achieving change. But, perhaps unraveling why we hold onto old habits is too broad in scope because there really isn't a one-size-fits-all answer in the same way that aggregated data do not always address local needs. This is consistent with the premise that global or national indicators may not be the focus of trending diseases in your local practitioner's office, and thus, may require other metrics and methodologies to identify and address them. In this context, there emerges the problem of transitioning quantifiable, large-scale data to the average Joe, Josephine, and their children to informed decision-making for individual patient care. This change of focus impacts the fundamental question of why we choose dangerous behaviors to, "How do we facilitate relevant data to address the collective circumstances that lead to improvements across patient populations when local mortality disease projections are not the focus of trending policy, funding, or research?" This concept is illustrated in the persistent global funding for HIV/AIDS (Institute for Health Metrics and Evaluation (IHME), 2019b) in the face of other diseases (e.g., chronic obstructive pulmonary disease (COPD), reemergence of sexually transmitted diseases, and previously eliminated items such as polio) that are not always the primary target for morbidity or mortality reduction in a given area.

Research has repeatedly confirmed that "risk-modification strategies might be the best targets to improve health," (GBD, 2018, p. 1924); therefore, understanding the metrics of global change through this lens is one key to identify unhealthy behaviors that can facilitate positive changes in personal health. Prominent research concurs that "targeting population health interventions and indicators" is important but that "applying these approaches to primary health care" has little direction (Ford-Gilboe et al., 2018, p. 635). Their research at four primary health clinics in Canada where $N = 395$ demonstrates an important connection between global metrics and personal health metrics that is not always clear at the level of care. While they focus on an equity-oriented health care (EOHC) approach, they envelope trauma-related indicators including "interpersonal (e.g., child maltreatment, intimate partner violence) and structural (e.g., poverty, racism)," (Ford-Gilboe et al., 2018, p. 640) factors that they demonstrate can positively impact patient health outcomes. Thus, the introduction of an analytic approach to consider trends against personal factors is plausible to bridge the gap in the present methods of prescribing optimal health measures in primary care.

We understand that the mechanism to facilitate change must be a product of the intersection of several inputs that meet the patient head-on (Farrer, Marinetti, Cavaco, & Costongs, 2015). Therefore, resolution must include multiple factors accumulated by a variety of disciplines such as social science population statistical information and trends, business assessments that reduce costs and streamline processes, and health care professionals generating personal medical interventions that cumulatively lead the patient down the path to improvements. Advancements in the required path to change dictate that there must be a convergence of concepts directly focused on understanding patient and items that influence their daily choices to improve patient outcomes by changing behavior. Thus, unraveling optimal treatment regimens for behavioral problems leading to disease is informed by a multidisciplinary process that is inclusive of the individual patient to achieve long-term compliance beginning with how, who, what, when, and where we collect and analyze patient data.

While there is compelling literature discussing the negative aspects of the pitfalls of the business marketing approach to healthcare and lack of policy support (Kelly & Barker, 2016), other researchers have presented the flip side of this position demonstrating the positive advantages of incorporating sound business and clinical practice into patient care (Serxner, 2013; Young, 2014). These processes rely on social science theory-based research, data collection and analysis, and government-financed data policy support that permits the analysis to project multisector trends (Centers for Disease Control and Prevention (CDC), 2016; Global Burden of Diseases, Injuries, and Risk Factors Study, 2018; World Health Organization (WHO), 2018a) and costs (Holt et al., 2015; Vigo, Kestel, Pendakur, Thornicroft, & Atun, 2018), leading to successful models that change behavior. In turn, awareness becomes a practical intervention with general guidelines on a national scale, and thus, the ongoing closed cycle of demonstrated gaps in healthcare, research, data collection/analysis/validation followed by dissemination in practice can culminate in the reduction of harmful risk factors. But there are also inherent difficulties associated with achieving health equity through policy because of the challenges in policy outcome evaluation (Farrer et al., 2015; Lee et al., 2018) and the impact on local health when key global or national indicators of poor health differ in local physician offices.

Notable is the CDC efforts to localize data in their 500 cities program (CDC, 2018a), but the number of cities must expand and the data focus must continue to narrow. We concur that understanding historical patterns of change are easily demonstrated on a large scale and predictable based on recognized social determinants of health (SDOH). However, general policy application derived from this data to individuals from a variety of subpopulations with a number of comorbidities can be cumbersome and fuzzy. A product of the CDC, the Healthy People 2020 (n.d.) objectives incorporate a "place-based" framework where community context is embedded in the neighborhood environment in relation to SDOH. Thus, the impetus for refined data is due, in part, to overcome the difficulty in translating information from big population data sets focusing on SDOH to individual care. While SDOH can narrow patient targets, health behaviors are a complex web of interactive components impacted by social inequalities, structure of society for social support, institutional design, and prevailing culture (Short & Mollborn, 2015).

Following along these lines is the syndemics approach that is also strongly emerging in literature (Hart & Horton, 2017; Mendenhall, 2017; Mendenhall, Kohrt, Norris, Ndetei, & Prabhakaran, 2017; Singer, Bulled, Ostrach, & Mendenhall, 2017, Tsai,

Mendenhall, Trostle, & Kawachi, 2017). "Syndemics are defined as the aggregation of two or more diseases or other health conditions in a population in which there is some level of deleterious biological or behaviour interface that exacerbates the negative health effects of any or all of the diseases involved," (Singer et al., 2017, p. 941). The health model establishes a variety of disease patterns designated as disease clusters that recognize the synergistic interaction span of effects on individuals, groups, and entire populations. Vulnerabilities (e.g., physical and behavioral) and health inequalities (e.g., poverty, stress, or violence) increase the likelihood of emergence of disease. Cumulatively, this suggests that general guidelines based on large data sets require further refinement and methods of local evaluation.

In this chapter, we familiarize readers with the metrics that establish the risk behavior with adverse health outcomes. We demonstrate key aspects of multilevel health data from the Global Burden of Disease Study 2017, United States, and Duval County, Florida, which coincides with city of Jacksonville, Florida. We present how changing behavior along positive paths to improvement can be accomplished through optimizing global metrics in conjunction with gathering new information on standard medical records to support a method of repeatable analysis, pinpoint different aspects of risk behavior, and target individual care without overburdening practitioners.

Where business, health care professionals, and social science data can converge

Risky behavior is associated with vulnerability in certain populations which is often characterized by a variety of health inequalities based on the social determinants of health (Baker et al., 2017; Braveman & Gottlieb, 2014; Hicken, Kraivtz-Wirtzb, Durkee, & Jackson, 2018; Working Group for Monitoring Action on the Social Determinants of Health, 2018). Cultural factors such as race, ethnicity, and gender, as well as environmental factors which include birth location, place of residence, individual access to care, and income also play a role in pinpointing disease (Marmot & Allen, 2014; Short & Mollborn, 2015).

Several items contribute to the problematic nature of addressing known health inequalities. These include notable gaps in health information (Farrer et al., 2015; Morse, 2019) and data sharing (Fiedler, 2014, 2015); poor policy development (Lee et al., 2018); and inconsistent metrics not conducive to advanced statistical analysis (Cook & Fiedler, 2018; Fiedler & Cook, 2017; Fiedler & Ortiz-Baerga, 2017). Nevertheless, there is reasonable consensus that poor health outcomes are evident in vulnerable populations in a disproportionate amount based on individual behavioral factors (Hicken et al., 2018; Marteau, Hollands, & Kelly, 2015; Mendenhall et al., 2017; Willen, Knipper, Abadía-Barrero, & Davidovitch, 2017), often resulting from the interplay of the contextual impact of cultural and environmental conditions (GBD 2016 Healthcare Access and Quality Collaborators, 2018; GBD 2017 SDG Collaborators, 2018; World Health Organization (WHO), 2018a, 2018b). These data points are important to understand syndromic surveillance linked to individual attributes, behavioral risk factors, and metabolic risks in the assessment of global health (GBD 2017 Risk Factor Collaborators, 2018), national health (CDC, 2016, 2018b; Davies, Paltoglou, & Foxcroft, 2016; Heron, 2018; Holt et al., 2015; Nichols, 2017; Vigo et al., 2018), regional health (EU Member States and European Commission, 2016; European Commission, n.d.), and localized data (CDC, 2018a; FDOH, 2019). More importantly, they are necessary for the proper evaluation, diagnosis, and treatment of progressively smaller subpopulations (e.g., regional, city, individual) in order to properly address a wide spectrum of current and emerging health needs.

While there are many theoretical models and social science theories to promote successful behavioral interventions, we focus on some that serve as an example of a health organization perspective and clinical perspective on enhancing data to determine optimal patient health interventions. Kaiser Permanente's Behavior Change Pathway is one method to identify behavioral factors that incorporate data from outside influences, while a health care professional seeks to engage, connect, and apply information to specific patient care (Stulberg, 2014; Young, 2014). Interesting components of the Behavior Change Pathway include a focus on preparing the patient for change and to gather information on the amount of exercise the patient engages in as a vital sign. The overarching mindset is to prescribe success versus a medical intervention. The phenomenon of availability and access becomes key contributors to decision-making that greatly impacts patient choices where poor decisions have led to the onset of lifestyle diseases influenced by environmental context (Stulberg, 2014). In short, a person's action can be greatly influenced by the actual context in which they are asked to make decisions. "The core principal of implementing healthy behavior changes is making the healthy choice the easy choice," (Young, 2014, p. 89). The same principals should apply to the diagnostic and treatment methods utilized by physicians.

Another example is the Optum ASM (awareness, skill-building, and maintenance) behavior change model (Serxner, 2013). The Optum model envelopes several social science theoretical premises (Health Belief, Balance Theory, Reasoned Action, Self-Efficacy, Goal Setting, and Adult Learning Theory) to enhance patient success at behavioral modification. These various premises rely on patient participation in prevention, preparing a path for individual change, understanding how our personal behavior is influenced by external factors, and to what extent an individual can participate in their own path to change. A definitive book on health behavior edited by Glanz, Rimer, and Viswanath (2015) can provide further details on theory and practice. In addition, the National Institute of Health (NIH) (n.d.) offers online resources to unravel key constructs of behavioral and social science theoretical research.

Other potential interventions originate from health care professions, such as clinical care nurses, who advocate a theoretical approach to behavior modification using concepts embedded in the Integrated Theory of Health Behavior Change (Ryan, 2009). This approach emphasizes enhanced health care professional/patient communication which gathers data on patient knowledge and beliefs. The culmination of this assessment is instrumental in determining the best-practice mechanism for behavioral modification intervention.

An important element of both the Kaiser Permanente's Behavior Change Pathway and the Optum ASM is the inclusion of a variety of multisector factors. Many suggest the prominent goal of simplifying the information provided to the patient towards a successful behavioral modification. Ironically, the process is not simple for even a highly skilled health care practitioner with advanced statistical skills who spends a great deal of time disseminating various data. The person-centered intervention is the best approach for the patient, but will likely consume too much of health care professional's time unless we provide "ease of use" mechanisms to the practitioner. Otherwise, we open the door back up for generalized population solutions which is the exact opposite of clinical objectives. Next, we look at how and what data are being provided under current processes.

Associated behaviors, metabolic risks, and your health

The overall concept of who we are, what we consume, how often and where we move about is embraced in multilevel metrics. Why we do things is debatable and often intangible, but how this impacts public health is the subject of tangible international concern. Global metabolic risks indicating global burden of disease (GBD 2017 SDG Collaborators, 2018; GBD 2016 Healthcare Access and Quality Collaborators, 2018; GBD 2016 Alcohol Collaborators, 2018), US national chronic disease indicators (CDI) (CDC, 2018a) and other statistics linked to cause of death classification (Heron, 2018), and European core health indicators (ECHI) (EU Member States and European Commission, 2016) embrace socioeconomic factors in relation to demography. They also indicate general population health status and generally consider how several items (e.g., determinants of health, existing health services, health promotion, and policy) impact health status. Generally, all levels (e.g., global, national, and local) target the presence of existing diseases, access to care, social phenomena such as aging and socioeconomic status, and the important nature of health in all policies. However, the overall review of data will also demonstrate some informative differences. Understanding the metrics is an important component of clinical applications and research limitations.

Aggregated global behavior and metabolic risk factors on mortality

The relationship between risk and injury/disease is well-documented through the Global Burden of Disease studies spanning the last 10 years. The comparative risk assessment (CRA) is based on a common factor of level of development so that countries with similar development status are compared to one another. Together with the ratio of observed risks versus exposure level (O/E ratio) and the Socio-demographic Index (SDI), researchers can estimate trends in morbidity and mortality based on risk associated with three Level 1 risk categories. The Level 1 risk factors are behavioral, environmental and occupational, and metabolic risks. These three risk factors comprise half of the components used to calculate disability-adjusted life years (DALYs) that represent a significant loss in quality of life (GBD 2017 Risk Factor Collaborators, 2018).

GBD studies clearly define and understand the parameters of their data by implicitly stating, "We have not quantified the contribution of distal social, cultural, and economic risk factors" (GBD 2017 Risk Factor Collaborators, 2018, p. 1926).

However, the calculation of the SDI and association with SDOH factors have accounted for economic risks and some social and cultural components to some extent.

From a global disease perspective without regard to gender but ranked by number of DALYs, "the five leading risks in 2017 were high systolic blood pressure, smoking, high fasting plasma glucose, high body-mass index, and short gestation for birthweight," (GBD 2017 Risk Factor Collaborators, 2018, p. 1925). "Consequently, four of the five leading risks were behavioral risks in 1990, whereas three of the five leading risks were metabolic risks in 2017." Despite a decline in some global behavioral risks in 2017, indicating that assessments have led to successful strategic health planning to moderate progression of certain risks, the proposition that behavior matters is evident in continued research reported in the Global Burden of Diseases, Injuries, and Risk Factors Study (Fig. 1.1). Males struggle with Level 2 risk factors or negative behaviors such as smoking and alcohol use, while metabolic risks for women are a larger predictor of cause of death

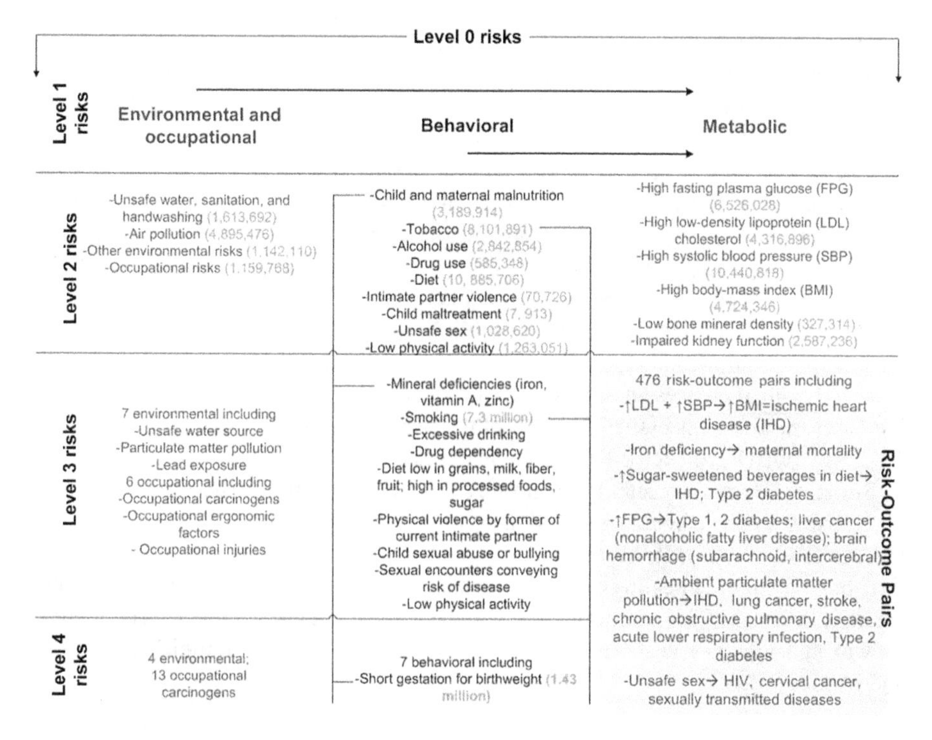

Fig. 1.1 Sampling of global, nongender-specific clusters of mortality risks by the number of attributable deaths (shown in *red*—light gray in print version) (Global Burden of Disease Collaborative Network, 2018 [database]; Institute for Health Metrics and Evaluation, 2019a, [databases]); example of the global burden of disease nested structure of five hierarchical data Levels 0–4 where Level 0 = all factors 1–4; table highlights two specific Level 4 behavioral risks associated with leading causes of death and risk-outcome pairs (GBD 2017 Risk Factor Collaborators, 2018, pp. 1928–1930, 1937–1938, 1964).

(GBD 2017 Risk Factor Collaborators, 2018). Overall, other prominent Level 2 risk factors include behavioral dietary risk (19.46% of deaths), environmental air pollution (8.75%), and metabolic risk of high systolic blood pressure (18.86%) (Global Burden of Disease Collaborative Network, 2018 [database]; Institute for Health Metrics and Evaluation, 2019a, [database]).

> *The majority of deaths worldwide are due to four noncommunicable diseases: cancer, cardiovascular disease, type 2 diabetes, and respiratory diseases; most of these deaths are potentially preventable. Four sets of behaviors contribute to the high and growing burden of noncommunicable disease: consumption of tobacco, alcohol, and highly processed foods, as well as physical inactivity.*
>
> *Marteau et al. (2015,* para *1)*

The World Health Organization weighs in on unacceptable levels of noncommunicable diseases (NCDs). "In 2016, NCDs were responsible for 41 million of the world's 57 million deaths (71%). 15 million of these deaths were premature (30 to 70 years)," (WHO, 2018b, p. 8). The NCD contribution to global death was validated by Global Burden of Disease studies at 64% (GBD 2017 Risk Factor Collaborators, 2018, p. 1939). Certain demographic locations and income categories proved to be more hazardous. The WHO South-East Asia Region reports the highest proportion of NCD deaths at 50% for those who are 30–69 years old, while the WHO European Regions is the lowest proportion at about 28% (WHO, 2018b, p. 12).

Income disparities in the same age span of 30–69 years of age identify low middle income nations having the highest proportion of NCD deaths at about 48%, while higher income nations predictably have the lowest proportion around 26% (WHO, 2018b, p. 11). The rise of noncommunicable disease risks is attributed to epidemiological transition defined as the relationship between "dramatic declines in communicable, maternal, neonatal, and nutritional diseases (CMNNDs) that have generally occurred with increases in socioeconomic development as well as the subsequent increases in life expectancy and absolute burden of noncommunicable diseases (NCDs)," (GBD 2017 Risk Factor Collaborators, 2018, p. 1925).

Smoking, alcohol consumption, unsafe sex, and poor eating habits remain in the top 20 global risks and can contribute to the problems measured in the metabolic risks in various GBD studies. Recognizing risky behavior is the first step to informing and directing new public health policy and personal accountability. Thus, tracking global changes in cause of death, development, and demographic factors, such as population growth and aging, reveals and projects risks associated with diminished capacity and death.

United States national chronic disease indicators and mortality

In the United States, the Centers for Disease Control and Prevention (CDC) aggregate data on measures spanning multiple levels of data including national, state, county, and in some cases, at the city level. They report the prevalence of chronic disease indicators (CDIs) that point to specific metrics including physical activity and nutrition, tobacco use, and cardiovascular disease. The CDC CDI indicators underwent a major upgrade that includes the addition of 22 systems and environmental change indicators (Holt et al., 2015) recognizing the important nature of these factors to public health. Systems and environmental change indicators appear in four groups: (1) alcohol; (2) nutrition, physical activity, and weight status; (3) oral health, and (4) tobacco. This section reports the similarities and differences between various levels of public health data. Overall, the leading causes of death are not reported in global terms of risks (e.g., environmental and occupations, behavior, or metabolic), but, in general, focus on disease-specific categories supplemented with information related to patient adherence to medically prescribed methods of preventive medicine and health indicators. In this way, bad and good health behaviors are rated. Dissimilarity also occurs because US national metrics are reporting in various statistical measures including rates per 100,000, mean, and others.

The national scale of leading causes of death varies by age group in the United States (Fig. 1.2) just as age is a factor in global health. However, two diseases, heart disease and malignant neoplasm (cancer), were responsible for taking more than 1.2 million US lives in 2017 or roughly 45% of death in 2017 (CDC, 2017; Murphy, Xu, Kochanek, & Arias, 2018; Nichols, 2017). Injuries from accidents, COPD and other chronic respiratory diseases such as asthma, and cerebrovascular diseases (i.e., stroke) round out the top 5 (CDC, 2017).

In line with global metrics, the CDC tracks subjective reporting of a variety of items including fruit and vegetable consumption and obesity (dietary risks), alcohol and tobacco use among youth and adults, high blood pressure awareness (high systolic blood pressure), and self-monitoring for glucose (high fasting plasma glucose). Information extends to reported patterns of preventive medicine (foot examinations for diabetic patients), inoculations, and hospitalization for cardiac events. Other risk factors include asthma and COPD, dental cleaning or tooth loss, hospitalization for hip fracture, and renal disease as a problem associated with diabetes. We also review reporting changes in overarching conditions and variance in the prevalence of key indicators such as poverty and lack of health insurance. Heart health and individual behaviors (e.g., tobacco and alcohol use, unprotected sex) remain problematic at the national level.

As in the case for global data, poverty and lack of insurance coverage are important for the overall well-being of national health. In 2009, the prevalence of US residents aged 18–64 years old without insurance was 16.9% and 23.5% of Florida residents compared to 12.2% and 21.9% in 2016. Crude prevalence poverty rates in US/Florida have risen from 13.4% and 13.2% in 2007 to 14.0 and 14.7% in 2016, respectively (CDC, 2016 [database]). Lower socioeconomic status is a predictor of growing prevalence of noncommunicable diseases.

As recently as 10 years ago, the CDC reported asthma and chronic obstructive pulmonary disease (COPD) in the general category of "other diseases and risk factors." Today, respiratory health is a major concern for the United States and the global population. In 2014, the COPD age-adjusted mortality rate per 100,000 people 45 years of age or older was 111.7 which is higher than the 105.1 reported in the State of Florida (CDC, 2016 [database]). Chronic low respiratory disease was rated as the 4th leading cause of death in the United States across all age spans (CDC, 2017).

Rank	Age groups										Total
	<1	1–4	5–9	10–14	15–24	25–34	35–44	45–54	55–54	65+	
1	Congential anomalies 4580	Unintentional injury 1267	Unintentional injury 718	Unintentional injury 860	Unintentional injury 13,441	Unintentional injury 25,669	Unintentional injury 22,828	Malignant neoplasms 39,266	Malignant neoplasms 114,810	Heart disease 519,052	Heart disease 647,457
2	Short gestation 3749	Congential anomalies 424	Malignant neoplasms 418	Suicide 517	Suicide 6252	Suicide 7948	Malignant neoplasms 10,900	Heart disease 32,658	Heart disease 80,102	Malignant neoplasms 427,896	Malignant neoplasms 599,108
3	Maternal pregnancy Comp. 1432	Malignant neoplasms 325	Congential anomalies 188	Malignant neoplasms 437	Homicide 4905	Homicide 5488	Heart disease 10,401	Unintentional injury 24,461	Unintentional injury 24,408	Chronic low respiratory disease 136,139	Unintentional injury 169,936
4	SIDS 1363	Homicide 303	Homicide 154	Congential anomalies 191	Malignant neoplasms 1374	Heart disease 3681	Suicide 7335	Suicide 3561	Chronic low respiratory disease 18,667	Cerebro-vascular 125,653	Chronic low respiratory disease 160,201
5	Unintentional injury 1317	Heart disease 127	Heart disease 75	Homicide 178	Heart disease 913	Malignant neoplasms 3616	Homicide 3351	Liver disease 13,737	Diabetes mellitus 14,904	Alzheimer's disease 120,107	Cerebro-vascular 146,383
6	Placentra cord. membranes 843	Influenza and pneumonia 104	Influenza and pneumonia 62	Heart Disease 104	Congential anomalies 355	Liver disease 918	Liver disease 3000	Diabetes mellitus 6409	Liver disease 13,737	Diabetes mellitus 59,020	Alzheimer's disease 121,404
7	Bacterial spesis 592	Cerebro-vascular 66	Chronic low respiratory disease 59	Chronic low respiratory disease 75	Diabetes mellitus 248	Diabetes mellitus 823	Diabetes mellitus 2118	Cerebro-vascular 5198	Cerebro-vascular 12,708	Unintentional injury 55,951	Diabetes mellitus 83,564
8	Circulatory system disease 449	Septicemia 48	Cerebro-vascular 41	Cerebro-vascular 56	Influenza and pneumonia 190	Cerebro-vascular 593	Cerebro-vascular 1811	Chronic low respiratory disease 3975	Suicide 7982	Influenza and pneumonica 46,862	Influenza and pneumonica 55,672
9	Respiratory distress 440	Benign neoplasms 44	Septicemia 33	Influenza and pneumonia 51	Chronic low respiratory disease 188	HIV 513	Septicemia 854	Septicemia 2447	Septicemia 5838	Nephritis 41,670	Nephritis 50,633
10	Neonatal hemorrhage 379	Perinatal period 42	Benign neoplasms 31	Benign neoplasms 31	Complicated pregnancy 168	Complicated pregnancy 512	HIV 831	Homicide 2275	Nephritis 5671	Parkinson's disease 31,177	Suicide 47,173

Fig. 1.2 Ten leading causes of death by age group, United States—2017 (CDC, 2017).
Database information provided courtesy of the National Vital Statistics System, National Center for Health Statistics, CDC; produced by the National Center for Injury Prevention and Control, CDC using WISQARS™.

Less than 1/3 of US citizens and Florida residents (29.5% age-adjusted prevalence) in 2015 report awareness of high blood pressure for those 18 years of age or older, despite the concern for cardiovascular health as a global leader in mortality. Screening for cholesterol is about 80% age-adjusted prevalence for adults 18 years or older in 2015, revealing high cholesterol in slightly over 30% age-adjusted prevalence in adults in the United States and Florida. The good news is that adherence to blood pressure medication taken by adults over 18 years of age with age-adjusted prevalence is at 59% and 60% for the United States and Florida, respectively. Overall, the US/Florida figures on mortality for diseases of the heart have dropped in actual numbers from 2007 (616,067, 42,254) to 2014 (614,348/44,511) with age-adjusted rates per 100,000 dropping in the United States (190.9–167) and in Florida (162.4–151.3) (CDC, 2016 [database]).

In 2009, cigarette smoking in the United States for adults 18+ years of age was reported at 17.9% prevalence, while Florida rates were somewhat lower at 17.1%. Figures in 2016 remain about the same in age-adjusted prevalence in the United States at 17.8%, but have lowered to 16.2% in Florida. Alternative to tobacco, such as *E*-cigarettes, is likely contributing to lower teen smoking rates reported in some areas in the United States. However, the status of vaping has simultaneously contributed to improving health outcomes for adults who use the produce in place of tobacco (CDC, 2019) and in introducing new health hazards, especially for teens, whose use of e-cigarettes is on the rise (FDA, 2019).

Binge drinking among adults 18 years or older has an age-adjusted mean of 7.1% prevalence in the United States, while heavy drinking among the same age category has an age-adjusted prevalence of 6.7% for the United States and 6.9% in the State of Florida (CDC, 2016 [database]). In 2014, the age-adjusted rate per 100,000 of 10.4 in the United States and 11.8 in the State of Florida (with crude rates of 12.0 and 15.3, respectively) for chronic liver disease mortality is not surprising (CDC, 2016, [database]).

Sexually transmitted diseases (STDs) previously declined to record lows or were close to being completely wiped out in the United States about 10 years ago. Now, there is a resurgence of STDs in adolescents and young adults which is causing concern due to the adverse impact of STDs on reproductive health leading to infertility and susceptibility to HIV/AIDS, among others. Chlamydia is a reported disease and screening has been implemented for persons 15–24 years of age. "In 2017, almost two-thirds of all reported chlamydia [1,708,569] cases were among persons aged 15–24 years," (CDC, 2018b, p. 1) representing "3635.3 cases per 100,000 females, an increase of 4.9% from 2016 and of 8.8% from 2013." Other STDs, such as gonorrhea and syphilis, related to unprotected sex are on the rise. The "rates of reported gonorrhea increased 75.2% since the historic low in 2009 and increased 18.6% since 2016" (CDC, 2018b, p.2). Two problems are associated with an increase in reported gonorrhea: (1) the disease is antimicrobial resistance limiting treatment to ceftriaxone and azithromycin, and (2) the nonspecific distribution of the disease making attempts to generate awareness to the general public difficult to target and disseminate.

The national data inform about the diseases' profile closer to home and, to some extent, replicate concerns for heart health, prompt to decrease use of harmful substances, and monitor vital statistics and other health indicators. However, the national data also measure patient participation in prescription adherence, preventive screenings, and self-monitoring. But, there are some missing clues in the national perspective in order for local physicians to assess and treat various health concerns of subpopulations in their practice.

Local syndromic surveillance in Duval County, Florida, USA

The Florida Department of Health collects data for the state using the CDC survey methods embedded in the Behavioral Risk Factor Surveillance System (BRFSS). The survey output is presented at county level to identify localized variance in disease trends, funding needs, and other policy targets to support health. The 2016 Florida BRFSS Data Report measures various diseases according to several controls including gender, race, ethnicity, age group, education, annual income, and marital status. The data assists in targeting specific patient profiles across the county. For the most part, data is reported in terms of percentages permitting data points to be noted for statistical significance in reported data at different data levels; specifically, statistical analysis points out significant changes between the county subpopulation and state level data in 2016 as well as between county data for 2013 and 2016 (FDOH, 2017, p. 9).

The top reportable diseases were tallied by age and vary greatly in Duval County. For instance, there were 347 cases of Salmonellosis making this diagnosis number 1 in three age brackets (< 1, 1–4, and 5–9 years of age). Chlamydia, a STD with 6, 928 reported cases, was at the top of the list for five age groups (10–14, 15–19, 20–24, 25–34, and 35–44). Nearly 4000 incidents of other STDs, Gonorrhea (3241) and syphilis (552), were also reported. Four age groups in Duval County struggle with Chronic Viral Hepatitis C including perinatal (45–54, 55–64, 65–74, and 75–84 years of age). (FDOH, 2017, pp. 183, 186). Unsafe sex was a leading risk factor from a local perspective.

Duval was the 7th leading county of 67 in the State of Florida in terms of new cancer diagnosis following Orange County (5123). Duval accounted for 4, 683/112,503 or about 4% of new cases in 2015 (FDOH, 2017, p. 154). Five other counties (e.g., Miami-Dade—12,216; Broward—9271; Palm Beach—8852; Hillsborough—6408, and Pinellas—6283) accounted for more than a third of new cancer diagnosis (38%) and cancer deaths (36%) across all Florida counties in 2015. "For all cancers combined, the Florida age-adjusted rate of occurrence for new cancer cases was 419.0 per 100,000 population and 152.1 per 100,000 population for cancer-related deaths," (FDOH, 2017, p. 152). All cancers (e.g., lung and bronchus, prostate, breast, colorectal, etc.) disproportionately impact the white race for both genders in the state.

Cardiovascular disease in Duval County, Florida, in the Northeast section of the state has seen declines in this disease in men across the county data (5.0 in 2013; 4.4 in 2016) and in comparison to state levels (5.7 in 2016). Women, on the other hand, have marked increases across the county data (2.8 in 2013; 5.3 in 2016) and state levels (3.9 in 2016). The highest measures for cardiovascular disease are among non-Hispanic White women at the county level (6.3 in 2016) compared to 4.4 in 2016 state percentages. Several other factors further inform this data. They include the highest proportion of individuals 45–64 years of age with less than a high school education, married, and earning less than $25,000 annual income. However, these changes were not reported as statistically significant. There is, however, a statistically significant measure between 2016 county data and 2016 state data showing a decline in men and, specifically, non-Hispanic White men who have experienced a heart attack (FDOH, 2017, p. 7) and non-Hispanic White men who have experienced a stroke (FDOH, 2017, p. 8).

Respiratory disease is another obstacle for non-Hispanic White women. They lead the percentage of adults with chronic obstructive pulmonary disease or other respiratory disease, such as emphysema or chronic bronchitis, showing increases in county level percentages from 10.8 in 2013 to 11.3 in 2016; both county measures are higher than 9.2 state levels in 2016 (FDOH, 2017, p. 10). While not considered statistically significant, other factors with the highest percentages in this category further quantify this category. They include age over 65+, < a high school education, <$25,000 annual income, and unmarried (FDOH, 2017, p. 10).

Women were engaged in binge drinking at 15.9 in 2016 at the county level which is statistically significant against 9.3 county level measures in 2013. Statistical significance was established for non-Hispanic White women who engaged in binge drinking at 19.6 in 2016 at the county level compared to 9.8 county level measures in 2013. Other demographic factors were not statistically significant, but demonstrated high percentages in the age category of 18–44, high school education, and being unmarried with an income >$50,000 annually (FDOH, 2017, p. 23).

Other race/ethnicities are experiencing disease trends. The 2016 state and county level decline in any type of cancer in men and specifically non-Hispanic White men is statistically significant as is the control variable of education at the level of high school graduate (FDOH, 2017, p. 5). While the statistically significant percentage of Hispanic women with diabetes has declined to 3.8 county levels in 2016 compared to 11.9 state levels in 2016, the largest percentage of diabetes is found in non-Hispanic, Black men (18.6 in 2016 county), outpacing both 2013 county (15.9) and 2016 state percentages (15.1) (FDOH, 2017, p. 12). Around ¾ of Duval county residents who are non-Hispanic Black men (76.1) and women (74.2) are overweight or obese, but both show a decline compared to 2013 measures from 81.3 and 78.5, respectively (FDOH, 2017, p. 18). While we recognize the value of maintaining a healthy weight, changes in being overweight or having a high body-mass index are not statistically significant in Duval County according to the Florida Department of Health Report (2017), but are quite prominent in global data. However, obesity in the age category of 45–64 is a statistically significant percentage at 40.5 for 2016 county levels compared to 32.1 state levels in 2016 (FDOH, 2017, p. 21). Finally, current smokers who are non-Hispanic Black men account for 30.5 (2016 county) smokers compared to 17.2 (2016 state) percentages. Important demographics indicating <high school education are also statistically significant in this category (FDOH, 2017, p. 25).

What stands out in the 2016 Florida BRFSS Data Report is the statistical significance of the percentage of people <65 years of age who have ever been tested for HIV (Table 1.1). Testing between 2013 and 2016 county data and between 2016 county

Table 1.1 Statistically significant HIV/AIDS screening participation at least once in Duval County, Florida, 2016 for adults <65 years of age compared to 2016 Florida data[a] and/or to Duval County 2013 measures[b] indicating percentage of adults have participated in testing at least once and/or have done so in the past 12 months of survey

General category	Specific demographic	Duval County measure 2016	Florida state measure 2016	Duval County measure 2013
All	Overall	64.1[a,b]	55.3	53.9
Gender	Men	60.9[a]	51.9	50.5
	Women	67.0[a]	58.7	57.2
Race/ethnicity	Non-Hispanic White	58.1[a,b]	49.6	41.0
Gender, race/ ethnicity	Non-Hispanic White men	56.9[a,b]	46.9	37.4
Age group	45–64	62.0[a,b]	50.5	44.1
Education level	High school/GED	65.6[a]	51.2	55.7
	>High school	65.2[a]	57.2	55.1
Annual income	$25,000–$49,999	73.1[a,b]	55.0	50.3
Marital status	Not married/couple	69.2[a]	56.0	58.0

(compiled from the Florida Department of Health, 2019, p. 42).

data and 2016 state data was dotted with statistical significance. Overall, the percentage of people <65 years of age who have ever been tested for HIV is statistically significant for the 2013 and 2016 county comparisons and the 2016 county percentages compared to the 2016 state information. The same holds true for demographics such as non-Hispanic White men, 45–64 age group, and annual income between $25,000 and $49,999. Statistical significance is also achieved for several measures between 2016 county and 2016 state data including both genders, high school graduates, >high school graduates, and unmarried. Finally, the percentage of non-Hispanic White women (21.2 in 2016 county data) is statistically significant compared to 2013 county data at 8.6. A similar pattern of numerous data points achieving statistical significance holds true for the percentage of adults who have been screened within the past 12 months or adults who have had ever been tested for HIV with some obvious variants in percentage (FDOH, 2019, p. 42–43).

This measure is of particular interest when combined with the Florida Department of Health HIV epidemiology presentation by the HIV AIDS Section in 2018. The report covers Area 4 which includes the counties of Duval, Clay, Nassau, St. John's, and Baker, but does not include correctional facilities in these areas. While there has been an 18% decrease in HIV from 2008 (435) to 2017 (358) (FDOH, 2018, slide 8), with 307/358 or about 86% of the cases in Duval County (FDOH, 2018, slide 9), recent changes are trending upwards which could be attributed to high rates of recent HIV/ AIDS screening. From 2016 (1,640,747) to 2017 (1,667,767), there has been a 1.6% population increase, but a 10.5% increase in diagnosed HIV cases from 324 in 2016 to 358 in 2017 (FDOH, 2018, slide 6). The recent upward trend in local HIV cases coupled with increasing local risk behaviors (e.g., unsafe sex, transmission of STDs) associated with HIV spotlights an urgent community health problem not necessarily reflected in decreasing global or national cases in most areas.

"Nationally, the problem has fallen under the radar, particularly amongst groups that are not disproportionately affected by the disease. Yet, approximately 40,000 people are diagnosed every year in the United States with HIV, and over 1 million people are living with the disease," said Samantha Kwiatkowski, Education & Prevention Manager, CAN Community Health in Jacksonville, Florida. "For those who are experiencing persistent rates of HIV, such as in Duval County, we need to find ways to keep funding so that patients will not experience a break in urgent health services."

As with national data collected by the CDC, tracking patient prevention activity through participation in office visits, vaccinations, and screening is a prominent theme in local data. But there are inconsistencies in actual budget allocation for prevention and public health professional advocacy which values prevention versus the higher costs of treatment. For example, HIV prevention funding constitutes only 3% of the $34.8 billion in FY 2019 federal budget, while 71% pays for domestic care and treatment (62%) and domestic cash and housing assistance (9%) (Kaiser Family Foundation, 2019, p. 1).

Another critical theme is that of patient screening. We have already noted the lack of statistical significance placed on local changes in global items such as BMI. In many cases, however, demographic information such as age, education, income, and race/ethnicity are statistically significant and relevant control points in global, national, and local data. Measuring change has some advantages in an overall population view, but can lack statistical relevance or links to mortality risk, as shown in the Duval County level data, where changes that are statistically relevant stand out against general trends that did not achieve any significance. However, while we are obtaining some personal demographics to identify and aid physician decision-making, the level of information is still missing important factors about our dietary risk, mental health, inoculations, and other relevant information in relation to our actual disease classifications. Frequency counts and percentages give a general understanding of the problem, but not a clear path from health indicators in relation to personal health outcomes that individuals are presently experiencing and/or are trending towards that disease.

Discussion

We have noted within the presentation of the general parameters in various levels of health data (e.g., global, national, and county) how the multilevel data approach contributes to the body of knowledge for practitioners and policy makers. The inclusion of specific age groups, race, and location across all levels is certainly informative. So too is the growing burden of NCDs (e.g., cancer, cardiovascular disease, diabetes, and respiratory ailments). Also, we have pointed out how the important nature of changes in data points represents new trends in global data, but does not necessarily translate to the interpretation of local data. We have shown how local data can bring forth statistical significance to atypical trends far removed from the focus of problems at the national or global level. Despite the positive information, we note an opportunity to enhance current data collection and presentation methods that focus on frequency counts of incidence of disease, death, and behaviors. Moving towards the addition of data on behavior or life circumstances in direct relation to existing medical conditions, demographics, and other control variables for smaller populations where treatment decisions must be made. Logically, we suggest taking another step from data aggregation to application of data at the point of care because there is a tangible separation from global or national data to individual care. We also recognize that the capacity to remain steadfast in pointing out or addressing personal flaws is more difficult close-up. Is who we are still what we are and why?

> *Smoking, eating, drinking and the amount of physical activity people do are ingrained in people's everyday lives and their routines and habits. These things to a very important extent help people define who and what they are: their sense of self is in part derived from these activities. Thus...behaviours that persist tend to be functional for people.*
>
> *Kelly & Barker, 2016, pp. 5–6*

So, the answer is yes. We are still what we are and we are only beginning to understand why. The solution is managing or eliminating specific risks that can improve health outcomes by applying data to patient care. This requires a greater level of knowledge about a patient population including the "uniquely personal context of an individual's knowledge, beliefs, attitudes, values, cultural norms and social environment," (Serxner, 2013, p. 2). Without this, a transition from limited data points and frequency counts to detailed information that could be assessed in relation to existing health conditions and associated behaviors change will continue to be difficult to achieve.

The data we collect can be driven by the physician/patient relationship. Are we collecting the behavioral data by asking the right questions and planning data types for multiple types of comparison to support decision-making? The answer would be "not quite."

Arguably, we cannot conduct advanced statistical analysis on each and every patient. But we can utilize existing healthcare frameworks already used in medical records, such as the International Classification of Diseases, 10th Revision (ICD-10) codes, to identify diseases in conjunction with additional items obtained from patients. In 2018, US National Vital Statistics Report, the use of ICD-10 codes, is demonstrated and the limits of cause of death statistics (i.e., not all causes of death meet eligible criteria for ranking; rankings denote most frequently occurring eligible causes) including "rankings do not illustrate cause-specific mortality risks as depicted by mortality rates" (Heron, 2018, p. 1). Thus, the conduct of analysis at the organization or practice level of a physician is plausible and rests on expanding existing prompts to include risky behaviors that will extend information on electronic medical records. The proximity of data can help to assess and reveal underlying causalities in poor decision-making, such as depression, or other underdiagnosed ailments that impact ability to move away from debilitating behaviors. The following demonstrates how data points can become personal motivators.

Recommendations: A place for new data models at the organization and practice level

Collecting patient information is not about the data, but about the patient. How we handle data from people is important to unlock valuable assistance in behavior modification. Ironically, much of the required information is already gathered in local medical records in the United States, reported to state health departments, and then aggregated to national data (Fiedler, 2015). The process entails the collection of data at local medical facilities. This premise of optimizing data collected using the foundation of the International Classification of Diseases, 9th Revision (ICD-9) and now the 10th Revision (ICD-10), was discussed in relation to data sharing health records to decrease time to identify bioterrorist activity or epidemics (Fiedler, 2014, 2015). Further analysis revealed that existing data points in many national, state, and local databases were problematic due to the inconsistencies of data types; behaviors (predictor variables) were different from outcome variables, and therefore, not conducive to advanced statistical analysis (Fiedler & Cook, 2017).

The concept of aligning predictors with actual ICD-10 disease classifications was introduced to the CDC as a potential solution to their Phase 1 Health Behavior Data Challenge (Fiedler and Ortiz-Baerga, 2017). While the CDC competition emphasized the incorporation of digital tracking devices and wearable technology, such as Fitbit, versus estimates of exercise and other behaviors currently reported during telephone interviews, Fiedler and Ortiz-Baerga (2017, p. 5) building on previous research (Cook & Fiedler, 2018; Fiedler & Cook, 2017) suggested corrective action to questions to subscribe to consistent data types for advanced causal analysis and the incorporation of new data into medical records.

For example, instead of asking, “Amount of Moderate-to-Vigorous Physical Activity (MVPA) time per day,” we suggest “Do you exercise for at least twenty minutes three times a week?” If yes, “How many calories do you estimate that you burn?” Dichotomous (Yes/No) questions lend themselves to statistical analysis, while the element of calories burned can further validate and quantify the activity. In the original question, frequency counts were added for each similar response, but little information was gleaned in how this relates to patient health. Thus, incorporating this type of question and the ICD-10 condition of the patient could lead to analysis where certain behavior could be statistically defined as contributing to the patient’s poor health status.

In other existing questions relating to sedentary behavior, “Amount of sedentary time accrued while at work, at home and/or in transit” can be changed to understandable questions with quantifiable answers. Instead, “Do you sit about eight hours per day?” or “Do you spend less than two hours recreation time on a computer/screen including watching TV, videos, playing computer games, emailing or using the internet per day?” These examples and all questions can also use other data types to permit greater levels of analysis by using a 5 scale ordinal response range where 1 is strongly disagree, 2 disagree, 3 neither disagree or agree, 4 agree, and 5 strongly agree. The ability to select from controlled responses in the ordinal data type method incorporates the ability to test the statistical reliability of the survey instrument.

In terms of nutrition, we often ask if an individual consumes dairy products, but do not ask if they are lactose-intolerant which would significantly skew findings. At some point, this information can be collected on personal data records in the same fashion that EMRs contain information on family history, previous surgeries, or other relevant data.

Since information is routed up from local practitioners through state agencies to the CDC to compile national data, including key local attributes is vital to increase the depth and strength of information. Additionally, enhancing electronic medical records with pertinent information would enrich findings to improve response to trending conditions locally where the problems are being faced.

Utilizing electronic health records (EHRs) more efficiently and effectively has recently garnered support as a new method of data collection. "EHRs should include information on housing, food, transportation, and other needs. Systems must transform their thinking, create a new strategy, empower multidisciplinary teams, educate health professionals, invest in research, and 'raise our voices to drive change,'" says Claire Pomeroy, CEO and president of the Albert and Mary Lasker Foundation (Morse, 2019, *para* 17).

Ultimately, the individual nuances that can propagate improved individual solutions are lost in the translation of current health data. It is not enough to know how many people participate in a behavior, but to understand how that behavior in your local environment impacts your own health. In existing analysis, we have an isolated series of predictors (Xs) and an isolated series of aggregated outcomes such as actual existing conditions (Ys) that are currently not measured against one another. Instead, we should be looking at behavioral factors linked directly to outcomes commonly expressed in scientific analysis as $X \rightarrow Y$.

More importantly, we bring attention to how subgroups or sectors of national populations provide additional data to support health services at the local level. We consider that the potential solution of repeatable analysis for diagnosis/treatment optimizing data collected at the local level at the organization or practice-level has merit based on the increasing level of specific demographic and other control factors that could facilitate targeted diagnosis and treatment. To change behavior, we have to better understand individual behavior and the health problems that are associated with them.

Thus, we proposed the relationship of individual factors (e.g., sleep, physical activity, alcohol use, and nutrition) measured against actual patient health status identified by ICD-10 codes in the formation of a structural equation model (Cook & Fiedler, 2018, p. 124; Fiedler, 2014, 2015; Fiedler & Ortiz-Baerga, 2017). The model, based on artifacts and data collected from the U.S. Center for Disease Control (CDC) and Prevention and additions recommended by the researchers, suggests the modified utilization of existing data collected by the CDC and compiled as the Behavioral Risk Factor Surveillance System (BRFSS at www.cdc.gov/brfss) can be more effective, efficient, and equitable in the collection of data and targeting dissemination of funding in specific areas of health need by associating known patient disease status with their behaviors and personal attributes known as control variables (Fig. 1.3). Thus, developing methods that enhance analysis in conjunction with existing data collection

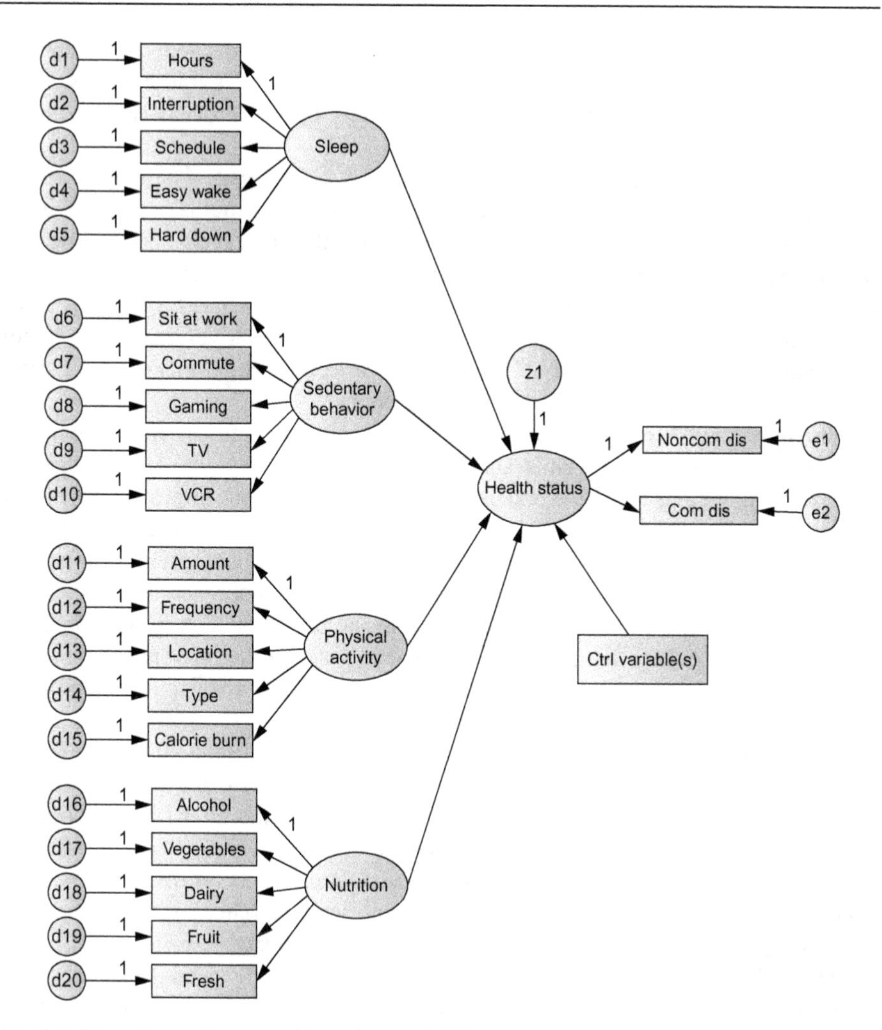

Fig. 1.3 The advanced structural equation model depicts how existing United States Centers for Disease Control and Prevention Behavioral Survey Strategies, existing patient conditions as identified by the International Classification of Disease, and direct communication with patients to form an "Ease of Use" option for health care providers to optimize wide scale data for customized treatment targeting key patient factors in local practices.
Courtesy of Cook & Fiedler, 2018; Fiedler & Cook, 2017; Fiedler & Ortiz-Baerga, 2017; Fiedler, 2015; Fiedler, 2014.

leading to improved patient treatment objectives can address existing gaps in analysis due to the lack of local predictive data gathered in relation to actual patient outcomes and extenuating factors. Additionally, the information can be utilized directly by practitioners to elicit improvements on patient understanding, adherence, and action in pursuit of improved quality of life. Data will still be moved to the state and onto the

CDC for national aggregation. However, local data will have the added advantage of advanced analysis to direct physician care. Notable is that incorporation of new questions and data types can still preserve the continuity of data previously collected in frequency counts.

The need for data that reveal individual choices as opposed to national trends is apparent because how one individual reacts to an external stimulus or medical intervention is unpredictable (Kelly & Barker, 2016). On the other hand, larger scale data can vividly depict observed patterns of common behaviors in relation to the outcomes such as the impact of alcohol or tobacco use on population health over prescribed periods of time and location (Kelly & Barker, 2016). Cumulatively, targeting trends with individual information at a local level is a key to targeting optimal patient outcomes in conjunction with large data predictions.

The introduction of the concept of clinical microsystems, "small groups of professionals who work together on a regular basis to provide care to discrete populations of patients," is a method in which health systems can improve the delivery of care (Likosky, 2014, p. 33). The concept of the multilevel engagement in clinical microsystems in Likosky's approach is a derivative of individuals best known for their academic contribution (James Brian Quinn) and statistical approach to consulting in business management (W. Edwards Deming) in the late 20th century with practical applications today. Together, they provide a sound basis for the convergence of social science, business, and clinical application in patient quality care also prevalent in the new data model presented here.

Understanding various dynamics involved in the investigation of public health and social phenomena is vital in terms of how these factors increase and decrease the risk of poor health outcomes in different group populations, communities, and individuals. Without distinct efforts to close the loop of information to practitioners as a conduit to patients, there will almost assuredly be a communication gap in the efforts to translate data to patient care and long-term behavioral modifications. This method can solidify the communication link from big data to practitioner/patient.

"Health behaviors are increasingly recognized as multi-dimensional and embedded in health lifestyles, varying over the life course and across place and reflecting dialectic between structure and agency that necessitates situating individuals in context," (Short & Mollborn, 2015, p. 78). The assessment of multilevels of data (e.g., global, national, and local) and model supports the proposition that care must be a derivative of aggregated trends and individual behaviors that are supplemented by contextual information in new data collection methods and analysis.

Conclusion

In this chapter, we illustrate that the intersection of multilevel health data metrics with personal characteristics of individual patients requires new methods of collecting, analyzing, and applying local knowledge from patient history and other data gathered in medical records. The case study analysis of global, US and Duval County, Florida health data demonstrates how narrowing the population focus reveals subpopulation schisms and identifies root causes of manifestations of poor health that can require a drastically different approach to patient care. Operationalizing data and analysis at the organizational or practice level provides optimized individual patient diagnosis and treatment in addition to contributing better data to state, national, and potentially global data. Recommendations include (1) linking patient behaviors to physical diseases following scientific protocols to review what factors may be significant to their present outcomes otherwise known as existing ($X \rightarrow Y$); (2) instituting improvements in the type of questions asked to acquire data types that are compatible for advanced statistical analysis; and (3) enable physicians to implement automated repeatable analysis on their patient populations from data in medical records.

Acknowledgments

The author would like to thank the researchers dealing with the Global Burden of Diseases, Injuries, and Risk Factors Study, the Centers for Disease Control and Prevention, the Institute for Health Metrics and Evaluation, the World Health Organization, and the Florida Department of Health for access to their timely information and data. The Florida Behavioral Risk Factor Surveillance System data used in this report were collected by the Florida Department of Health. The author is solely responsible for views and interpretation of all data source content and use does not necessarily reflect the opinions of the various agencies.

References

Baker, P., Friel, S., Kay, A., Baum, F., Strazdins, L., & Mackean, T. (2017). What enables and constrains the inclusion of the social determinants of health inequities in government policy agendas? A narrative review. *International Journal of Health Policy and Management*, *7*(2), 101–111. https://doi.org/10.15171/ijhpm.2017.130.

Braveman, P., & Gottlieb, L. (2014). The social determinants of health: It's time to consider the causes of the causes. *Public Health Reports (Washington, DC: 1974)*, *129*(Suppl. 2), 19–31.

Brickell, E., & Withrow, K. (1988). *What I am [Recorded by Edie Brickell and New Bohemians]. On Shooting rubberbands at the stars [album]*. USA: Geffen Records.

Centers for Disease Control and Prevention (CDC). (2018a). *500 cities: Local data for better health, 2018 release. Retrieved August 11, 2019 from https://chronicdata.cdc.gov/500-Cities/500-Cities-Local-Data-for-Better-Health-2018-relea/6vp6-wxuq*.

Centers for Disease Control and Prevention (CDC). (2018b). *Sexually transmitted disease surveillance 2017*. Atlanta: U.S. Department of Health and Human Services. Retrieved January 9, 2019 from *https://www.cdc.gov/std/stats*.

Centers for Disease Control and Prevention (CDC). (2019, March 11). *Electronic cigarettes. Retrieved August 12, 2019 from https://www.cdc.gov/tobacco/basic_information/e-cigarettes/index.htm*.

Centers for Disease Control and Prevention (CDC), National Center for Injury Prevention and Control. (2017). *Ten leading causes of death by age group in the United States 2017. Retrieved August 12, 2019 from https://www.cdc.gov/injury/wisqars/LeadingCauses.html (Last review 10 April 2019)*.

Centers for Disease Control and Prevention, National Center for Chronic Disease Prevention and Health Promotion, Division of Population Health. (2016). *Chronic disease indicators (CDI) data (online). Retrieved August 11, 2019 from https://nccd.cdc.gov/cdi.*

Cook, K., & Fiedler, B. A. (2018). Foundations of community health: Planning access to public facilities. In B. A. Fiedler's (Ed.), *Translating national policy to improve environmental conditions impacting public health through community planning* (pp. 107–130). Springer International Publishing.

Davies, E. L., Paltoglou, A. E., & Foxcroft, D. R. (2016). Implicit alcohol attitudes predict drinking behaviour over and above intentions and willingness in young adults but willingness is more important in adolescents: Implications for the prototype willingness model [Epub]. *British Journal of Psychology*, *22*(2), 238–253. https://doi.org/10.1111/bjhp.12225.

EU Member States and European Commission. (2016, September 7). *The European core health indicators (factsheet). Retrieved August 11, 2019 from https://www.amplexor.com/en/resources/news/the-european-core-health-indicators–echi–factsheet.html.*

European Commission (n.d.). ECHI—European core health indicators. Retrieved May 14, 2019 from https://ec.europa.eu/health/indicators/echi/list_en.

Farrer, L., Marinetti, C., Cavaco, Y. K., & Costongs, C. (2015). Advocacy for health equity: A synthesis review. *The Milbank Quarterly*, *93*(2), 392–437.

Fiedler, B. A. (2014). Constructing legal authority to facilitate multi-level interagency health data sharing in the United States. In: *American society of public administration conference in Washington, DC, March 14–18. ASPA 2014 Founders' Fellow.*

Fiedler, B. A. (2015). Constructing legal authority to facilitate multi-level interagency health data sharing in the United States. *International Journal of Pharmaceutical and Healthcare Marketing*, *9*(2), 175–194.

Fiedler, B. A., & Cook, K. (2017). Foundations of community health: Planning access to public facilities. In: *Southeast conference of public administration (SECoPA), defending public administration in a time of uncertainty*, Hollywood, FL: Education, Health, and Social Welfare Policy. October 4–7.

Fiedler, B. A., & Ortiz-Baerga, J. (2017). *Extending public health surveillance reporting through digital data collection of behavior patterns and existing conditions.* US Centers for Disease Control and Prevention, CDC Data Challenge, Phase I. August 4.

Florida Department of Health (FDOH), Division of Community Health Promotion. (2019, July 23). *Behavioral risk factor surveillance system (BRFSS). Retrieved August 12, 2019 from http://www.floridahealth.gov/statistics-and-data/survey-data/behavioral-risk-factor-surveillance-system/index.html.*

Florida Department of Health (FDOH), Division of Disease Control and Health Protection. (2017). *Florida annual morbidity statistics report, 2017. Retrieved August 13, 2019 from http://www.floridahealth.gov/diseases-and-conditions/disease-reporting-and-management/disease-reporting-and-surveillance/data-and-publications/fl-amsr1.html.*

Florida Department of Health (FDOH), HIV AIDS Section (2018, June). *HIV epidemiology, area 4. Retrieved August 13, 2019 from http://www.floridahealth.gov/diseases-and-conditions/aids/surveillance/area-slide-sets.html.*

Ford-Gilboe, M., Wathen, C. N., Varcoe, C., Herbert, C., Jackson, B. E., et al. (2018). How equity-oriented health care affects health: Key mechanisms and implications for primary health care practice and policy. *The Milbank Quarterly*, *96*(4), 635–671.

GBD 2016 Alcohol Collaborators. (2018). Alcohol use and burden for 195 countries and territories, 1990–2016: a systematic analysis for the Global Burden of Disease Study 2016. *Lancet*, *392*, 1015–1035. https://doi.org/10.1016/S0140-6736(18)31310-2.

GBD 2016 Healthcare Access and Quality Collaborators. (2018). Measuring performance on the healthcare access and quality index for 195 countries and territories and selected subnational locations: A systematic analysis from the Global Burden of Disease Study 2016 (online). *Lancet*. https://doi.org/10.1016/S0140-6736(18)30994-2.

GBD 2017 Risk Factor Collaborators. (2018). Global, regional, and national comparative risk assessment of 84 behavioural, environmental and occupational, and metabolic risks or clusters of risks for 195 countries and territories, 1990–2017: A systematic analysis for the Global Burden of Disease Study 2017. *Lancet*, *392*, 1923–1994.

GBD 2017 SDG Collaborators. (2018). Measuring progress from 1990 to 2017 and projecting attainment to 2030 of the health-related Sustainable Development Goals for 195 countries and territories: A systematic analysis for the Global Burden of Disease Study 2017. *Lancet*, *392*, 2091–2138.

Glanz, K., Rimer, B. K., & Viswanath, K. (Eds.), (2015). *Health behavior: Theory, research, and practice*. (5th ed.). John Wiley & Sons, Inc..

Global Burden of Disease Collaborative Network. Global Burden of Disease Study 2017 (GBD 2017) Results. Seattle, United States: Institute for Health Metrics and Evaluation (IHME). (2018). Retrieved August 23, 2019 from http://ghdx.healthdata.org/gbd-results-tool.

Hart, L., & Horton, R. (2017). Syndemics: Committing to a healthier future. *Lancet*, *389* (10072), 888–889.

Healthy People 2020. (n.d.). Social determinants of health. Retrieved August 19, 2019 from https://www.healthypeople.gov/2020/topics-objectives/topic/social-determinants-of-health (Last update 14 August 2019).

Henry J. Kaiser Family Foundation. (2019, March). *U.S. federal funding for HIV/AIDS: Trends over time (Fact Sheet). Retrieved August 15, 2019 from http://files.kff.org/attachment/Fact-Sheet-US-Federal-Funding-for-HIVAIDS-Trends-Over-Time*.

Heron, M. (2018). *Deaths: Leading causes for 2016. National Vital Statistics Reports. Vol. 67(6)*. Hyattsville, MD: National Center for Health Statistics.

Hicken, M. T., Kraivtz-Wirtzb, N., Durkee, M., & Jackson, J. S. (2018). Racial inequalities in health: Framing future research. *Social Science & Medicine*, *199*, 11–18.

Holt, J. B., Huston, S. L., Heidari, K., Schwartz, R., Gollmar, C. W., et al. (2015). Indicators for chronic disease surveillance—United States, 2013. *Morbidity and Mortality Weekly Report*, *64*(RR-1), 1–252.

Institute for Health Metrics and Evaluation (IHME). (2019a). *Global health data exchange (database). http://ghdx.healthdata.org/*.

Institute for Health Metrics and Evaluation (IHME). (2019b). *Global HIV/AIDS spending 2000–2016*. Seattle, United States: Institute for Health Metrics and Evaluation (IHME).

Kelly, M. P., & Barker, M. (2016). Why is changing health-related behaviour so difficult? *Public Health*, *136*, 109–116. https://doi.org/10.1016/j.puhe.2016.03.030.

Lee, J., Schram, A., Riley, E., Harris, P., Baum, F., et al. (2018). Addressing health equity through action on the social determinants of health: A global review of policy outcome evaluation methods. *International Journal of Health Policy and Management*, *7*(7), 581–592. https://doi.org/10.15171/ijhpm.2018.04.

Likosky, D. S. (2014). Clinical microsystems: A critical framework for crossing the quality chasm. *Journal of Extra-Corporeal Technology*, *46*(1), 33–37.

Marmot, M., & Allen, J. J. (2014). Social determinants of health equity. *American Journal of Public Health*, *104*(Suppl. 4), S517–S519.

Marteau, T. M., Hollands, G. J., & Kelly, M. P. (2015). *Changing population behavior and reducing health disparities: Exploring the potential of "Choice Architecture" interventions (content last reviewed July 2015)*. Rockville, MD: Agency for Healthcare Research and Quality. http://www.ahrq.gov/professionals/education/curriculum-tools/population-health/marteau.html.

Mendenhall, E. (2017). Syndemics: A new path for global health research. *Lancet, 389*, 889–891.

Mendenhall, E., Kohrt, B. A., Norris, S. A., Ndetei, D., & Prabhakaran, D. (2017). Syndemics 2. Non-communicable disease syndemics: Poverty, depression, and diabetes among low-income populations. *Lancet, 389*, 951–963.

Morse, S. (2019, February 12). *Spending money on the social determinants is an investment: Value-based care demands the switch to wellcare to raise outcomes and decrease costs, Claire Pomeroy says.* Available from https://www.healthcarefinancenews.com/news/spending-money-social-determinants-investment?mkt_tok=eyJpIjoiTldZNFptRmxaVEprTURJNCIsInQiOiJrbGM4aWRxZlhkbjNNNm1cL3ZwVVNLd1BlZWVwWXNmZE9iN1BiZlBKZnJINWNTQW12QlRvVkl4S2tPdW56b1QrU1dibmNQV2syT3RvSjEzXC9rb1orMXJjeG41UzdyaVZ0NFh1UW0yZENVQTNod2VTTjMweERoQjB4eWhZUWkxYTg3In0%3D. (Accessed 25 July 2019).

Murphy, S. L., Xu, J. Q., Kochanek, K. D., & Arias, E. (2018). Mortality in the United States, 2017. In *NCHS Data Brief, no 328*. Hyattsville, MD: National Center for Health Statistics.

National Institute of Health (NIH), Office of Behavioral and Social Science Research (n.d.). eSource, social and behavioral theories, chapter 5. Interventions to change health behavior. Retrieved August 4, 2019 from http://www.esourceresearch.org/eSourceBook/SocialandBehavioralTheories/5InterventionstoChangeHealthBehavior/tabid/737/Default.aspx.

Nichols, H. (2017, February 3). *What are the leading causes of death in the US? [Medical News Today] [Last update July 4, 2019]. Retrieved July 23, 2019 from https://www.medicalnewstoday.com/articles/282929.php.*

Ryan, P. (2009). Integrated theory of health behavior change: Background and intervention development. *Clinical Nurse Specialist, 23*(3), 161–172. https://doi.org/10.1097/NUR.0b013e3181a42373.

Serxner, S. (2013). *Optum modeling behavior change for better health [White paper]. Retrieved May 14, 2019 from www.optum.com/resources/library/behavior-change.html.*

Short, S. E., & Mollborn, S. (2015). Social determinants and health behaviors: Conceptual frames and empirical advances. *Current Opinion in Psychology, 5*, 78–84. https://doi.org/10.1016/j.copsyc.2015.05.002.

Singer, M., Bulled, N., Ostrach, B., & Mendenhall, E. (2017). Syndemics. Syndemics and the biosocial conception of health. *Lancet, 389*, 941–950.

Stulberg, B. (2014). *The key to changing individual health behaviors: Change the environments that give rise to them. Retrieved May 14, 2019 from http://harvardpublichealthreview.org/the-key-to-changing-individual-health-behaviors-change-the-environments-that-give-rise-to-them/.*

Tsai, A. C., Mendenhall, E., Trostle, J. A., & Kawachi, I. (2017). Co-occurring epidemics, syndemics, and population health. *Lancet, 389*, 978–982.

US Food & Drug Administration (FDA) (2019, February 6). *2018 NYTS data: A startling rise in youth E-cigarette use. Retrieved August 12, 2019 from https://www.fda.gov/tobacco-products/youth-and-tobacco/2018-nyts-data-startling-rise-youth-e-cigarette-use.*

Vigo, D. V., Kestel, D., Pendakur, K., Thornicroft, G., & Atun, R. (2018). Disease burden and government spending on mental, neurological, and substance use disorders, and self-harm: Cross-sectional, ecological study of health system response in the Americas [online]. *The Lancet Public Health*. https://doi.org/10.1016/S2468-2667(18)30203-2.

Willen, S. S., Knipper, M., Abadía-Barrero, C. E., & Davidovitch, N. (2017). Syndemics 3. Syndemic vulnerability and the right to health. *Lancet, 389*, 964–977.

Working Group for Monitoring Action on the Social Determinants of Health (2018). Towards a global monitoring system for implementing the Rio political declaration on social determinants of health: Developing a core set of indicators for government action on the social determinants of health to improve health equity. *International Journal for Equity in Health, 17*(1), 136. https://doi.org/10.1186/s12939-018-0836-7.

World Health Organization (WHO). (2018a). *Global status report on alcohol and health 2018.* Geneva: WHO License: CC BY-NC-SA 3.0 IGO.

World Health Organization (WHO). (2018b). *Noncommunicable diseases country profile 2018. Open Access under CC by 4.0 license. Retrieved July 19, 2019 from https://www.who.int/nmh/publications/ncd-profiles-2018/en/.*

Young, S. (2014). Healthy behavior change in practical settings. *The Permanente Journal, 18*(4), 89–92. https://doi.org/10.7812/TPP/14-018.

Supplementary reading recommendations

Carey, R. N., Connell, L. E., Johnston, M., Rothman, A. J., De Bruin, M., Kelly, M. P., et al. (2018). Behavior change techniques and their mechanisms of action: A synthesis of links described in published intervention literature. *Annals of Behavioral Medicine, 53*(8), 693–707. https://doi.org/10.1093/abm/kay078.

European Health Interview Survey (EHIS). Reference Metadata in Euro SDMX Metadata Structure (ESMS) (Last update April 12, 2019; Certified July 5, 2018). Compiling agency: Eurostat, the Statistical office of the European Union. Retrieve August 11, 2019 from https://ec.europa.eu/eurostat/cache/metadata/en/hlth_det_esms.htm.

Farrer, L., Marinetti, C., Cavaco, Y. K., & Costongs, C. (2015). Advocacy for health equity: A synthesis review. *The Milbank Quarterly, 93*(2), 392–437.

Michie, S., Carey, R. N., Johnston, M., Rothman, A. J., deBruin, M., Kelly, M. P., et al. (2016). From theory-inspired to theory-based interventions: A protocol for developing and testing a methodology for linking behaviour change techniques to theoretical mechanisms of action. *Annals of Behavioral Medicine*. https://doi.org/10.1007/s12160-016-9816-6 (online).

National Association for Public Health Statistics and Information Systems. https://www.naphsis.org/.

National Association of Health Data Organizations. https://www.nahdo.org.

National Institute of Health (NIH), National Heart, Lung, and Blood Institute (NHLBI) (n.d.). Guide to behavior change: Your weight is important. Retrieved May 14, 2019 from https://www.nhlbi.nih.gov/health/educational/lose_wt/behavior.htm.

Oregon State University (2015, November 10). Changing habits to improve health: New study indicates behavior changes work. *ScienceDaily*, Retrieved May 13, 2019 from www.sciencedaily.com/releases/2015/11/151110170916.htm.

Pedrana, L., Pamponet, M., Walker, R., Costa, F., & Rasella, D. (2016). Scoping review: National monitoring frameworks for social determinants of health and health equity. *Global Health Action, 9*, 28831. https://doi.org/10.3402/gha.v9.28831.

Roberto, C., & Kawachi, I. (Eds.), Behavioral economics and public health. Oxford, UK: Oxford University Press. Retrieved May 20, 2019 from https://oxfordmedicine.com/view/10.1093/med/9780199398331.001.0001/med-9780199398331.

Waller Wellness Center. (2019). *Healthy behaviors for a healthier lifestyle. Retrieved May 14, 2019 from https://www.wallerwellness.com/healthy-behaviors-for-a-healthier-lifestyle.*

Definitions

Adjusted mean "Mean…standardized to the age distribution of a specific population, usually the U.S. 2000 standard population" (CDC, 2016).

Age-adjusted prevalence "Prevalence…standardized to the age distribution of a specific population, usually the U.S. 2000 standard population" (CDC, 2016).

Age-adjusted rate "Crude rate…standardized to the age distribution of a specific population, usually the U.S. 2000 standard population" (CDC, 2016).

Biosocial complex term associated with the syndemics model of health; emerging synergistic approach to health in which diseases, social, and environmental factors redefine the progression of disease as a complex social condition and biological interaction resulting in quantifiable pathways for specific individuals or populations.

Comparative risk assessment (CRA) framework/decision-making tool for measuring, ranking, and synthesizing evidence on risk and associated outcomes to comprehensively quantify exposure to risks impacts human health; the assessment formulates new risks and risk-outcome pairs as the basis for new data structure to further assess risk.

Contextually tailored care (CTC) "moves beyond culturally sensitive approaches to explicitly address inequitable power relations, racism, discrimination, and ongoing effects of historical and current inequities within health care encounters" (Ford-Gilboe et al., 2018, p. 640).

Culturally safe care (CSC) "expands the individually focused concept of patient-centered care to include offering services tailored to the specific health care organization, the populations served, and the local and wider social contexts" (Ford-Gilboe et al., 2018, p. 640).

Crude rate "the measure number of deaths, causes of conditions, diseases or hospitalizations during a specific year—specified as rates per 1000, per 10,000, per 100,000 or rates per 1,000,000 persons" (CDC, 2016).

Disability-adjusted life years (DALYS) comprised of exposure to three major factors (e.g., environmental and occupational risks, behavior, and metabolic risks) combined with population growth, changes in the age structure, and changes due to all other factors.

Disease clusters term associated with the syndemics model of health, indicating an established pattern of multiple diseases affecting individuals, groups, and large populations.

Epidemiological transition when socioeconomic development increases, there is an expected increase in life span and the emergence of noncommunicable diseases versus communicable, nutritional, and other diseases normally associated with low socioeconomic status.

Mean "the measured or estimated mean (average or central value of a discrete set of numbers)—weighted to the population characteristics for an attribute or disease during a specific year" (CDC, 2016).

Prevalence "the measured or estimated number of people—weighted to population characteristics—with an attribute or disease during a specific year" (CDC, 2016).

Socio-demographic index (SDI) relationship between level of economic development and the exposure prevalence of risk associated with socioeconomic status. Increasing SDI has an inverse relationship with several decreasing risks such as unsafe water sources, intimate partner violence, and underweight children while other risks increase directly with increasing SDI such as high red meat consumption, alcohol use, smoking, and discontinued breastfeeding.

Syndemics model of health the study of the proliferation of chronic disease by examining the biosocial complex resulting in predictable disease clusters as it relates to social inequalities and injustices in certain populations.

Syndromic surveillance a form of public health surveillance achieved by monitoring health indicators (e.g., signs, symptoms) to aid in the identification of trending diseases and to inform policy to limit adverse effects on public health.

Trauma- and violence-informed care (TVIC) extends beyond trauma-informed practice to explicitly acknowledge and address the intersection and cumulative effects of interpersonal (e.g., child maltreatment, intimate partner violence) and structural (e.g., poverty, racism) forms of violence on people's lives and health (Ford-Gilboe et al., 2018, p. 640).

Don't sit this one out: Moderating the negative health impact of sedentary behavior at work

2

Matt Thomas Bagwell[a], *Beth Ann Fiedler*[b]
[a]Public Administration Division, Tarleton State University, RELLIS Campus, Bryan, TX, United States, [b]Independent Research Analyst, Jacksonville, FL, United States

Abstract

Sedentary behavior is a global problem contributing to a variety of noncommunicable diseases leading to premature mortality in nearly 1 out of 10 people in the world population. This qualitative content analysis reviewed recent research to investigate the health effects of sitting and sedentary work. This study recommends an adaptive strategy for behavioral change toward an optimal and active work-life balance, using the frameworks of Self-Efficacy Theory and Goal Setting Theory, and also prescribes intervention approach recommendations from synthesized findings. Major findings include the identification of six optimal behavior change approaches to modify and/or decrease sedentary inactivity, including (1) redesigning the physical environment and adding devices, objects, or tools to the environment; (2) outcome goal setting; (3) behavioral self-monitoring; (4) information on health consequences of sedentary behavior with directions on how to perform behaviors; (5) having social support; and (6) receiving advice or feedback on behavior.

Keywords: Behavior modification strategies, Occupational health, Sedentary behavior, Goal setting theory, Thematic content analysis

Chapter outline

Three Facets of Public Health and Paths to Improvements. https://doi.org/10.1016/B978-0-12-819008-1.00002-X

Sitting to death

Be the change you want to see in the world.

Ghandi

An object at rest tends to stay at rest. An object in motion tends to stay in motion. Apply Newton's first physical law of motion to human behavior and this idea emerges in the global offices in industrialized countries as a silent epidemic endangering life expectancy at alarming rates. Sedentary behavior was the cause of approximately 9% of premature deaths, or more than 5.3 million of the 57 million deaths worldwide in 2008 (Lee et al., 2012, p. 227; World Health Organization (WHO), 2011 [database]). WHO (2018) also reports that increased levels of sedentary inactivity is one of the leading global risk factors for death. Consequentially, people who are highly engaged in immobile, sedentary behaviors, occupations, and lifestyles have a 20%–30% higher likelihood in their risk of death compared to people who are engaged in moderate physical activity (WHO, 2018, *para* 6). The global scope of the problem is important to the content of this chapter in which we define sedentary behavior and adverse health consequences, demonstrate the impact of sedentary behavior on select populations, and conduct a qualitative analysis of literature to summarize notable recommendations from findings to offset this debilitating but modifiable negative behavior pattern.

Sedentary behavior has been defined as any waking behavior characterized by an energy expenditure of 1.5 metabolic equivalents or less, undertaken while sitting or lying down (Sedentary Behavior Research Network (SBRN), n.d.; SBRN, 2012; Thivel et al., 2018; Tremblay et al., 2017). This standard definition has been widely adopted by those in the field of public health and physiology, and who study sedentary behavior whether in or outside of an office work environment. This definition is operationalized for the purpose of this study to include sedentary work that is a growing occupational hazard. Thus, sedentary occupational work is an example of a waking behavior, such as sitting or lying down.

Excessive sitting is a ubiquitous problem impacting many industrialized nations such as the United Kingdom with an estimated 70,000 preventable deaths in 2016 (Heron, O'Neill, McAneney, Kee, & Tully, 2019, *para* 3) and the United States where over 8.3% of deaths were attributable to inadequate levels of physical activity (Carlson, Adams, Yang, & Fulton, 2018, *para* 3). However, prevention may be achieved by decreasing sedentary behavior by increasing the proportion of physical activity. Research within the US population indicates that the potential gains in life expectancy could be around 2 years if sedentary behavior can be reduced to less than 3 hours per day (Katzmarzyk & Lee, 2012).

In addition to premature mortality, "negative effects such as an increased risk for type 2 diabetes, heart disease and certain cancers (breast and colon) are all found to be associated with the sitting disease," (LifeSpan, 2016, *para* 1). Excessive sitting has also been linked to a wide variety of adverse conditions such as chronic obstructive pulmonary disease (COPD) (Cavalheri, Straker, Gucciardi, Gardiner, & Hill, 2016), colon and rectal cancer (Mahmood, MacInnis, English, Karahalios, & Lynch, 2017), ovarian cancer (Patel, Rodriguez, Pavluck, Thun, & Calle, 2006), kidney disease (American Society of Nephrology, 2015), fatty liver disease (Kwak & Kim, 2018), and other deleterious health effects such as obesity (Brown, Miller, & Miller, 2003). Amassing physio-medical evidence and empirical research support that inactive sedentary occupational work, while interfacing with a computer screen in a sitting position, negatively affects health outcomes of many modern communities in industrialized countries where office work is an upward trending occupational requirement.

Those with sedentary work occupations are at increased risk because at least two-thirds of their working days are spent sitting and engaging in low-energy expending, sedentary tasks, and behaviors (Parry & Straker, 2013; Ryan, Grant, Dall, & Granat, 2011). Chau et al. (2013, p. 9–10) report that "associations between total sitting and all-cause mortality found that each additional hour of daily sitting is associated with an overall 2% increased risk of all-cause mortality" and can increase "up to 8% per hour, when consecutive total time spent sitting is above 8 hours per day," (Thivel et al., 2018, p. 2). Another study by Biswas et al. (2015) also supports these findings in adults that increased sedentary time and the associated risk factors for disease incidence, mortality, and hospitalization are likewise increased.

Work occupations with excessive sedentary behavior can contribute to spinal-related issues as well as low back and neck pain (Hallman, Ekman, & Lyskov, 2014; Hallman, Gupta, Mathiassen, & Holtermann, 2015; Lis, Black, Korn, & Nordin, 2007; Sitthipornvorakul, Janwantanakul, & Lohsoonthorn, 2015; Steffens et al., 2016). Moreover, musculoskeletal problems throughout the body are commonplace in workers who are engaged in excessive sedentary behavioral occupations; however, desk and office environment modifications may help to abate these common problems (Foley, Engelen, Gale, Bauman, & Mackey, 2016; Gao, Nevala, Cronin, & Finni, 2016; Karol & Robertson, 2015; Parry, Straker, Gilson, & Smith, 2013; Sethi, Sandhu, & Imbanathan, 2011). Corrective action through smarter, human-friendly office environments and work stations is possible in order to improve working conditions and promote increased physical activity levels for more productive and satisfied employees.

Literature supports that sedentary behavior has a negative effect on mental health in terms of the psychological effects. Multiple studies support that decreased sedentary behavior and increased physical activity can mitigate depression (Helgadóttir et al., 2017; Mammen & Faulkner, 2013; Stanton, Happell, & Reaburn, 2014). Moreover, bio-psychological cognitive research studies also suggest that there is decreased brain function and mental aptitude due to low physical activity and sedentary behavior (Falck, Davis, & Liu-Ambrose, 2017; Mikkelsen, Stojanovska, Polenakovic, Bosevski, & Apostolopoulos, 2017; Singh, Neva, & Staines, 2014; Smith & Ainslie, 2017). These findings also speak about a potential decrease in work productivity and how productivity in the workplace may be negatively impacted by increased sedentary behavior and physiological inactivity, along with psychological slumps, sluggishness, and impairment in cognitive brain function and mental aptitude of those working in office environments where sedentary behavior is the norm. These chronic comorbid conditions, psychological malaises, related mortalities, and medical reasons highlight the need for behavioral change to ameliorate these harmful health effects on public health populations and communities who work in sedentary occupations.

A major behavioral modification in the modern and industrial office work environment is to decrease overall sedentary behavior, specifically to reduce time spent sitting for office-work employees while on the job, which is the focus of this paper. Both progressive companies and employees alike are seeking healthier, balanced alternatives to the normative work behaviors and practices of the past for typical office occupations and employment where productivity primarily occurs when the employee is sitting. Activity levels are considered "at rest." Noteworthy is the reemergence of scholarly attention in this subject who study generational differences and attribute this increasing trend to Generation X actively searching for a more balanced work-life state of being (Gibson, Greenwood, & Murphy, 2009). The quest for improved health is timely as many researchers, like office workers, spend a majority of time sitting. Together, they face their personal desire to make healthier choices to offset family medical history and other lifestyle choices by seeking a work-life balance. While links between sedentary inactivity and health have been well-established in the literature (Bauman et al., 2011; Martin et al., 2015; Parry & Straker, 2013; Rezende et al., 2016), some areas that lack investigation include: (1) strategic best practices to improve sedentary behavior, and (2) behavioral approaches that model adaptive change toward improvement in healthy activity by optimizing work-life balances for people in sedentary occupations.

Generating awareness about the ill effects of sedentary behavior is a key public health problem that can be improved through healthy behavioral modification intervention approaches. This chapter investigates strategic best practices and synthesizes behavioral change approaches in research literature. The research objectives are to provide: (1) an understanding of the integral components of a healthier life style (increasing physical activity, decreasing sedentary behavior) in occupational work practice; (2) actionable behavioral change intervention approaches to improve health outcomes for sedentary work occupations; and (3) strategic best practices to reduce the negative health impact of sedentary occupational work and patterns of behavior. We begin by establishing a foundation for sedentary behavior and proceed to provide the research methodology. General findings from the content analysis of important literature on changing or modifying sedentary behavior suggest that goal setting, social support, and environmental adaptation are three major thematic strategies for altering sedentary behavior.

General overview of applicable behavior change theories

The premise of behavioral change theories endeavors to capture the underlying problem of why behavior change seems to be somewhat difficult for humans to adopt. Are people simply creatures of habit or could there be more behind the mechanisms between the binominal states of doing and not doing? More appropriately, can one adopt a standard of goal setting in terms of marking outcomes in a way which increases the likelihood of having goals attained? This section highlights two prominent behavioral change theories: (1) Goal Setting Theory, and (2) Self-Efficacy Theory, as the foundation leading to adaptive strategies for behavioral change toward an optimal work-life balance.

A predominant psychological theory related to behavioral change is Goal Setting Theory (Locke, 1968). This theory asserts that being specific in setting goals which are relatively larger or more ambitious tends to lead to increased goal performance or execution of these predetermined goals. Moreover, these goals should be specific and have a relatively high degree of difficult and predefined parameters in order to assess if goals were either achieved or not achieved (Locke & Latham, 2006). In terms of applying this theory to sedentary inactivity, one needs to understand in specificity what her or his preestablished goals are with a relatively high degree of difficulty. In addition to clarity, challenge, and task complexity, two other principles emerged to include commitment and feedback (Locke & Latham, 2006). Accountability and long-term dedication are important to sustain behavioral changes.

Bandura (1977) devised a theory of self-efficacy which conceptualizes one's confidence or capacity to exert control and execute behaviors needed to produce or attain a specific performance or action. Four keys are important to unlock the capacity for self-efficacy. They are (1) vicarious experience or achieving personal awareness through observation, (2) verbal persuasion which can occur in positive encouragement or negative criticism, (3) physiological feedback or the body's reaction to inputs, and (4) performance outcomes or the ability to repeat experiences that promote desired outcomes. This type of theoretical grounding is fundamental in establishing meaningful and specifically defined parameters of behavioral change in many capacities and relevant in terms of behavior change for sedentary inactivity. Together, these theoretical constructs inform the research methodology that follows.

Research methodology

This study uses the qualitative research method of thematic content analysis to investigate and understand behavioral change approaches and intervention strategies to curtail sedentary behavior among adults in sedentary work or labor occupations. This methodology has been tested and utilized for several decades within the social sciences in qualitative research and focuses on examining themes within narrative or transcribed data (Daly, Kellehear, & Gliksman, 1997; Polit & Beck, 2010). Content analysis emphasizes structured organization and in-depth description of thematic or narrative data or informative content (Guest, MacQueen, & Namey, 2012). Rather than simply counting words or phrases within text, the method identifies both explicit and implicit ideas, meanings, and concepts within thematic content or narrative data (Guest et al., 2012). The primary process for extracting related patterns and developing trends within raw data is by coding narrative information and identifying major, minor, and/or emerging themes, meanings, ideas, and concepts (Boyatzis, 1998). As expected, the study reveals various levels of themes related to sedentary behavior change approaches.

As one of the most prevalent methods of qualitative inquiry, thematic content analysis emphasizes pinpointing, examining, and reporting narrative patterns and thematic meanings or key ideas among data (Braun & Clarke, 2006; Guest et al., 2012). Using this method, this chapter study explores and presents Targeted Behavior Change (TBC), major goals, Invention Prescriptions (IPs), and Behavioral Change Approaches (BCAs) from the reviewed research on the TBC related to sedentary behavior.

Overview of methods

Keyword or term searches of the current published literature were conducted and presented in the next section.

(1) Search literature for keywords: "changing sedentary behavior," "modifying sedentary behavior," "changing sedentary behavior pre post intervention outcomes" studies with intervention arms; reviewing only adults over age 18 and also older adults;
(2) Intervention review of key studies over approximately the last 10 years (2009–19);
(3) Compile studies that encapsulate behavioral change interventions or approaches;
(4) Analyze and categorize the major thematic content on behavior change qualitatively;
(5) Identify what strategies move or progress toward optimal work-life balance.

Inclusion and exclusion criteria for the articles were established for this study and are provided below.

Inclusion criteria:

- Published articles that investigated sedentary behavior of persons 18 and older, or adults.
- Articles published between January 1, 2009 and April 30, 2019 (approximately the last 10 years from when this research inquiry was conducted, May and June of 2019.
- Articles whereby clear behavioral change approaches or interventions were presented and examined in the study research of the article reviewed.

Exclusion criteria:

- Published articles that investigated sedentary behavior of persons under 18, children.
- Articles published prior to 2009.
- Articles whereby clear behavioral change approaches or interventions were not presented, specified, or examined in the study research of the article reviewed.

This thematic content analysis examined published literature on sedentary occupational work, health, and behavior change outcomes among adults age 18 and older. The conduct of the qualitative analysis reviewed published literature on sedentary occupational work, health, and behavior change which provide empirical results on sedentary behavioral change intervention approaches and outcomes. The publication dates span over a 10-year period from approximately 2009–19 following the guidelines set forth in inclusion and exclusion criteria.

Search protocol

Three research inquiry strategies were implemented to cull relevant and timely published studies from the literature. First, germane sources were identified from an existing review of sedentary behavior reduction interventions conducted by Gardner, Smith, Lorencatto, Hamer, and Biddle (2016). Second, an electronic search of three databases (MEDLINE, PsycInfo, and Web of Science) was conducted in June of 2019. For each database, the query parameters or search filters were set or fixed to specifically retrieve articles related to these key terms or phrases: "sedentary behavior and health," "occupational work and health," "sedentary occupational work," and "sedentary behavioral change intervention approaches with outcomes." The publication date range was limited to January 1, 2009 through April 30, 2019. Third, a search of the same key terms and phrases was also conducted via Google Scholar. Data saturation was achieved in the review of these research articles with regard to thematic content for subsequent analysis (Polit & Beck, 2010).

The systematic literature search yielded 30 published meta-analysis review items using key terms or phrases on listed databases and alternative search engines. Two notable and recently published studies, Ekelund et al. (2016) and Patterson et al. (2018), among others were excluded from this qualitative investigation upon review because behavioral change approaches or prescriptive interventions were not clearly presented or specified within the articles, or could not be determined by the reviewer. Seemingly, there were no specific behavioral change recommendations or prescriptive interventions provided in terms of decreasing or modifying sedentary behavior, rather data related to the impact of sedentary, inactive behavior and the link to all-cause morbidity, cancer, and other chronic comorbid disease conditions.

Findings, raw data, and behavior change implications

Thirty research articles were reviewed for specific behavioral change approaches and prescriptive intervention recommendations. Data saturation, the balance between obtaining enough data for research replication and meaningful findings without being redundant, was achieved. The qualitative data was then coded for major goals, targeted behavior change approaches, and intervention prescriptions.

Table 2.1 presents the studies that were reviewed for thematic content analysis in terms of sedentary behavioral change intervention approaches and outcomes. The TBCs identify problematic behavior, BCAs suggest methods to modify behavior,

Table 2.1 Content analysis of prescriptive findings for behavioral change interventions in current literature.

Lead author, year	Targeted behavior change (TBC)	Major goal	Intervention prescription (IP)
Alkhajah et al., 2012	SB, PA	Decreasing SB	Environmental change and adaptation
BCAs: Directions on how to perform behavior; information about social and emotional costs/consequences; changing/adapting the physical environment; adding devices, objects, and/or tools to the environment.			
Barwais, Cuddihy, & Tomson, 2013	SB, PA	Decreasing SB	Empower/enable, educate
BCAs: Outcome goal setting; behavioral advice/feedback; social support; adding devices, objects, and/or tools to the environment.			
Burke et al., 2013	Diet, PA	Increasing PA, improving diet	Empower/enable
BCAs: Outcome goal setting; behavioral advice/feedback; behavioral self-monitoring; social support; adding devices, objects, and/or tools to the environment.			
Burke et al., 2010	Diet, PA	Increasing PA, improving diet	Empower/enable
BCAs: Outcome goal setting; behavioral advice/feedback; behavioral self-monitoring; social support; adding devices, objects, and/or tools to the environment.			
Cavalheri et al., 2016	SB, PA	Decreasing SB, increasing PA engagement/participation (light intensity)	Incentivizing, empower/enable, environmental change, education, modeling behavior; persuasion, motivating
BCAs: Outcome goal setting; behavioral advice/feedback; PA data collecting/measurable monitoring; social support; adding devices, objects, and/or tools to the environment; frequent contact with health care provider for motivation, advice, support.			
Chang, Fritschi, & Kim, 2013	SB, PA	Decreasing SB, increasing PA	Education/training, empowering/enabling
BCAs: Outcome/goal setting; activity/action planning; peer or social support behavior monitoring, self-monitoring, social support, instruction/direction on how to perform behavior(s), information on health consequences of SB, behavior practice/rehearsal, task evaluation from healthcare provider or other knowledge expert, adding devices, objects, and/or tools to the environment.			
De Cocker, Spittaels, Cardon, De Bourdeaudhuij, & Vandelanotte, 2012	PA	Increasing PA	Pedometer only; pedometer with advice/feedback; educating, empowering/enabling, persuading
BCAs: Outcome goal setting, problem solving, action, advice/feedback on behavior, adding devices, objects, and/or tools to the environment.			

Table 2.1 Continued

Lead author, year	Targeted behavior change (TBC)	Major goal	Intervention prescription (IP)
Dewa, deRuiter, Chau, & Karioja, 2009	SB, PA	Decreasing SB, increasing PA	Empowering/enabling
BCAs: Behavior self-monitoring, adding devices, objects, and/or tools to the environment.			
Dunstan et al., 2013	SB, PA	Decreasing SB	Empowering/enabling, environmental change/adaptation, educating/training
BCAs: Outcome goal setting; problem solving; action planning; goal review; advice/feedback on behavior; self-monitoring; social support; directions/instructions how to perform behavior(s); information on health consequences of SB; cues/prompts; behavior substitutes; behavior practice/rehearsal; habit formation/routine-making; task/behavior evaluation; physical environmental adapting/redesign; social/peer environment change/restructuring; adding devices, objects, and/or tools to the environment.			
Ellegast, Weber, & Mahlberg, 2012	PA	Increasing PA	Persuading, incentivizing, environmental change/adaptation, behavior modeling, empowering/enabling
BCAs: Outcome goal setting, looking at the disparity between current behavior and goal behavior; advice/feedback on behavior; self-monitoring; social support; directions/instructions how to perform behavior(s); peer comparison; behavior practice/rehearsal; behavior substitute; material incentives/rewards; environmental adapting/redesign; social/peer change; adding devices, objects, and/or tools to the environment.			
Evans et al., 2012	SB	Decreasing SB	Educating, empowering/enabling, environmental change/adaptation
BCAs: Outcome goal setting; active action planning; directions/instructions how to perform behavior(s); information on health consequences of SB; cues/prompts; behavior substitutes.			
Fitzsimons, Baker, Gray, Nimmo, & Mutrie, 2012	PA	Increasing PA	Empowering/enabling
BCAs: Outcome goal setting; action planning; problem solving; advice/feedback on behavior; self-monitoring; social support; directions/instructions how to perform behavior(s); task/behavior evaluation; risks and rewards (pros/cons); adding devices, objects, and/or tools to the environment.			

Continued

Table 2.1 Continued

Lead author, year	Targeted behavior change (TBC)	Major goal	Intervention prescription (IP)
Fitzsimons et al., 2013	SB	Decreasing SB	Empowering/enabling, education/training
BCAs: Outcome goal setting; action planning; problem solving; commitment; advice/feedback on behavior; directions/instructions how to perform behavior(s); information on health consequences of being inactive; behavior practice/rehearsal; habit formation/routine-making; habit reversal/routine-breaking of inactivity; risks and rewards (pros/cons).			
Gardner et al., 2016	SB	Decreasing SB	Educating/training, persuading, environmental change/adaptation
BCAs: Outcome goal setting; problem solving; goal review; looking at the disparity between current behavior and goal behavior; self-monitoring; advice/feedback on behavior; social support; information on health consequences of SB; cues/prompts; behavior substitutes; social/peer environment change/restructuring; physical environmental adapting/redesign; risks and rewards (pros/cons); habit formation/routine-making; behavior practice/rehearsal.			
Gilson et al., 2009	PA	Increasing PA	Empowering/enabling
BCAs: Outcome goal setting; self-monitoring; directions/instructions how to perform behavior(s); task/behavior evaluation; behavior substitutes; adding devices, objects, and/or tools to the environment.			
Hansen et al., 2012	PA	Increasing PA	Empowering/enabling
BCAs: Outcome goal setting; problem solving; advice/feedback on behavior; self-monitoring; biofeedback; social support; information on social and emotional health consequences of inactivity; advice from knowledge expert; social/peer environment change/restructuring.			
Healy et al., 2013	SB, PA	Decreasing SB	Empowering/enabling, environmental change/adaptation, educating/training
BCAs: Outcome goal setting; problem solving; action planning; goal review; advice/feedback on behavior; self-monitoring; social support; directions/instructions how to perform behavior(s); information on health consequences of SB; cues/prompts; behavior substitutes; behavior practice/rehearsal; habit formation/routine-making; task/behavior evaluation; physical environmental adapting/redesign; social/peer environment change/restructuring; adding devices, objects, and/or tools to the environment.			
John et al., 2011	PA	Increasing PA	Environmental change/adaptation
BCAs: Physical environmental adapting/redesign; adding devices, objects, and/or tools to the environment.			

Table 2.1 Continued

Lead author, year	Targeted behavior change (TBC)	Major goal	Intervention prescription (IP)
Kozey-Keadle, Libertine, Lyden, Staudenmayer, & Freedson, 2011	SB	Decreasing SB	Educating, empowering/ enabling
BCAs: Outcome goal setting; action planning; self-monitoring; instructions how to perform behavior(s); information on health consequences of SB; cues/prompts; behavior substitutes.			
MacMillan et al., 2011	PA	Increasing PA	Empowering/enabling
BCAs: Outcome goal setting; self-monitoring; problem solving; social support; directions/ instructions how to perform behavior(s); information on health consequences of physical inactivity; task/behavior evaluation; risks and rewards (pros/cons); adding devices, objects, and/or tools to the environment.			
Martin et al., 2015	SB	Decreasing SB	Educating, empowering/ enabling, environmental change/adaptation
BCAs: Outcome goal setting; self-monitoring; information on health consequences of SB; risks and rewards (pros/cons); adding devices, objects, and/or tools to the environment; data collecting/measurable monitoring.			
Mutrie et al., 2012	PA	Increasing PA	Empowering/enabling
BCAs: Outcome goal setting; self-monitoring; problem solving; social support; directions/ instructions how to perform behavior(s); information on health consequences of physical inactivity; task/behavior evaluation; risks and rewards (pros/cons); adding devices, objects, and/or tools to the environment.			
Neuhaus et al., 2014	SB, PA	Decreasing SB	Empowering/enabling, environmental change/ adaptation, educating/ training
BCAs: Outcome goal setting; problem solving; action planning; goal review; advice/ feedback on behavior; self-monitoring; social support; directions/instructions how to perform behavior(s); information on health consequences of SB; cues/prompts; behavior substitutes; behavior practice/rehearsal; habit formation/routine-making; task/behavior evaluation; physical environmental adapting/redesign; social/peer environment change/restructuring; adding devices, objects, and/or tools to the environment.			
O'Donoghue et al., 2016	SB	Decreasing SB	Environmental change/ adaptation, educating/ training
BCAs: Outcome goal setting; social support; physical environmental adapting/redesign; social/peer environment change/restructuring.			

Continued

Table 2.1 Continued

Lead author, year	Targeted behavior change (TBC)	Major goal	Intervention prescription (IP)
Prince, Saunders, Gresty, & Reid, 2014	SB	Decreasing SB	Empowering/educating/enabling change, environmental change/adaptation
BCAs: Outcome goal setting; information on health consequences of SB; habit formation/routine-making; physical environmental adapting/redesign; social/peer environment change/restructuring; adding devices, objects, and/or tools to the environment.			
Pronk, Katz, Lowry, & Payfer, 2012	SB	Decreasing SB	Environmental change/adaptation, empowering/enabling
BCAs: Physical environmental adapting/redesign; social/peer environment change/restructuring; adding devices, objects, and/or tools to the environment.			
Schwartz, Kapellusch, Baca, & Wessner, 2019	SB	Decreasing SB	Environmental change/adaptation
BCAs: Physical environmental adapting/redesign; adding devices, objects, and/or tools to the environment.			
Shrestha, Ijaz, Kukkonen-Harjula, Kumar, & Nwankwo, 2015	SB, PA	Decreasing SB, increasing PA	Empowering/educating/enabling change, environmental change/adaptation
BCAs: Outcome goal setting; information on health consequences of SB; habit formation/routine-making; physical environmental adapting/redesign; social/peer environment change/restructuring; adding devices, objects, and/or tools to the environment.			
Shrestha et al., 2018	SB, PA	Decreasing SB, increasing PA	Empowering/educating/enabling change, environmental change/adaptation
BCAs: Outcome goal setting; information on health consequences of SB; habit formation/routine-making; physical environmental adapting/redesign; social/peer environment change/restructuring; adding devices, objects, and/or tools to the environment.			
Verweij, Proper, Weel, Hulshof, & van Mechelen, 2012	SB, PA, diet	Decreasing SB, increasing PA, improving Diet	Empowering/enabling
BCAs: Outcome goal setting; action planning; goal review; self-monitoring; social support; risks and rewards (pros/cons); adding devices, objects, and/or tools to the environment.			

Abbreviations: *BCA*, behavioral change approaches; *IP*, intervention prescription; *PA*, physical activity; *SB*, sedentary behavior; *TBC*, targeted behavior change.

and the IPs also suggest ways to curb sedentary inactivity through various forms of social and physical interaction. A majority of the studies focused on behavior needing change in terms of decreasing sedentary behavior (SB) and increasing physical activity (PA). Three distinct levels of information emerge from the analysis and are presented herein. They are major goals or TBCs related to sedentary inactivity, behavior change approaches related to sedentary inactivity, and intervention prescriptions for sedentary behavior.

Major goals and targeted behavior change (TBC) related to sedentary inactivity

In terms of comparing the major goals and TBCs, the raw data produced synonymously indistinguishable or homogenously equivalent results. Nine studies (30% of the total reviewed) reported the major goal and TBC as seeking to decrease sedentary behavior. Eight studies (27% of the total reviewed) reported the major goal and TBC as seeking to increase physical activity. Ten studies (33% of the total reviewed) reported dual major goals and TBCs as seeking both to decrease sedentary behavior and increase physical activity. Two studies (7% of the total reviewed) reported dual major goals and TBCs as seeking both to increase physical activity and optimize or improve diet. One study (3% of the total reviewed) reported three major goals and TBCs as seeking to (1) decrease sedentary behavior, (2) increase physical activity, and (3) optimize or improve diet.

Intervention prescriptions (IPs) for sedentary behavior and interpretation

The raw data of IPs for sedentary behavior indicated the following themes within the reviewed studies. Twenty-five studies reported IPs in term of empowering and enabling sedentary behavioral change and decreasing inactivity. Sixteen studies reported IPs of changing or adapting the environment and educating. Seven studies reported an IP of training. Four studies reported an IP of persuading. Two studies reported IPs in terms of modeling behavior and incentivizing. At least one study reported IPs in terms of motivating, using a pedometer along with receiving advice and/or feedback, and using a pedometer only (Table 2.2).

These intervention prescriptions (IPs) suggest ways to curb sedentary inactivity by offering actionable ways in which to facilitate and promote positive behavior changes related to sedentary inactivity at work—moving from thought to action. The primary intervention prescriptions in this category included these five IPs: empowering, enabling, changing or adapting the environment, educating, and training that unlock human will power and the resolve to action. Engaging in motion, such as reaching out for Newton's apple, is one way to combat sedentary inactivity to enable meaningful change. Six secondary, or less frequently cited, IPs were persuading, modeling behavior, incentivizing, motivating, using a pedometer with providing advice or feedback,

Table 2.2 Identified intervention prescriptions (IPs) stratified or categorized by thematic frequency among the reviewed studies.

Intervention prescriptions (IPs)
Empowering = 25
Enabling = 25
Changing/adapting environment = 16
Educating = 16
Training = 7
Persuading = 4
Modeling behavior = 2
Incentivizing = 2
Motivating = 1
Using pedometer with providing advice/feedback = 1
Using pedometer only = 1

and using a pedometer only. Prescriptive intervention ideas can be strong catalysts toward achieving better health and leading change in people from inaction to action.

While one prescription may not offer a panacea, in aggregation, these recommendations provide meaningful conceptual and theoretical insights in an attempt to spur movement, improve overall occupational or work-related activity, and promote better health outcomes in adults who work in offices. Together, these keys provide the motivation, knowledge, and change agency to increase activity.

Behavior change approaches (BCAs) related to sedentary inactivity and interpretation

The overall raw data for BCAs related to sedentary inactivity indicate the following themes within the reviewed studies: major BCA themes, minor BCA themes, outlier BCA themes, and emerging BCA themes, which were identified from this thematic content analysis research review. The categories are designated by meeting a specific threshold in the number of studies where the concept was found. Specifically, major BCA themes are defined as being identified in 13 or more studies, minor BCA themes between 11 and 5 studies, and emerging BCA themes in 2 or 3 studies. Outlier BCAs are defined as being identified in at least one study (Table 2.3).

Outcome goal setting was the leading major theme in behavior change approaches (BCAs) with 25 studies reporting this element, while the leading minor theme was changing and/or restructuring the social or peer environment in 11 studies. Two studies reported emerging BCAs of data collecting and/or measurable monitoring and looking at or examining the disparity between current behavior and goal behavior. Several outlier themes emerged including some that engage major concepts of peer support and methods of receiving/applying health information.

Table 2.3 Identified behavioral change approaches (BCAs) stratified or categorized by thematic frequency in reviewed studies.

Behavior change approaches (BCAs)
Major themes
Outcome goal setting; 25
Adding devices, objects, and/or tools to the environment; 24
(Behavioral) self-monitoring; 17
Information on health consequences of sedentary behavior; 16
Social support; 15
Changing/adapting/redesigning the physical environment; 14
Directions/instructions on how to perform behavior(s); 13
Advice/feedback on behavior; 13
Minor themes
Social/peer environment change/restructuring; 11
Activity/action planning; 10
Problem solving; 10
Behavior substitutes; 9
Habit formation/routine-making; 8
Task/behavior evaluation; 7
Behavior practice/rehearsal; 7
Risks and rewards (pros/cons); 7
Cues/prompts; 6
Goal review; 5
Emerging themes
Data collecting/measurable monitoring; 2
Looking at the disparity between current behavior and goal behavior; 2
Outlier themes
Information about social and emotional costs/consequences; 1
Peer or social support behavior monitoring; 1
Peer comparison; 1
Frequent contact with health care provider for motivation, advice, support; 1
Task evaluation from healthcare provider or other knowledge expert; 1
Advice from knowledge expert; 1
Material incentives/rewards; 1
Habit reversal/routine-breaking of inactivity; 1
Commitment; 1
Biofeedback; 1

Overall, eight major themes related to behavior change approaches (BCAs) were identified, along with 10 minor, 2 emerging, and 10 outlier themes. Promoting sedentary behavioral change is vital to improving the overall health of those who sit and work in offices around the world. The next section transfers the presentation of major, minor, emerging, and outlier themes found within the reviewed literature data from tables to narrative form.

Major themes related to sedentary behavior change

Eight major behavioral change approaches (BCAs) were discovered from this thematic content analysis. Much of the literature on changing or modifying sedentary behavior suggests that three goals are essential: (1) outcome goal setting, (2) environmental adaptation (encompassing two major subthemes described as adding devices or tools, and redesigning the physical environment), and (3) behavioral self-monitoring. Four other major BCAs include receiving information on health consequences of sedentary behavior, social support, receiving directions on how to perform behavior(s), and receiving advice or feedback on behavior. Moreover, the major themes of "information on health consequences" and "social support" as well as "directions on how to perform behavior(s)" and "advice or feedback on behavior" are combined in the same subsection so that all eight BCAs are discussed in the following five categories.

Outcome goal setting

The most predominant theme in relation to sedentary behavior change was outcome goal setting. In terms of theoretical importance, this behavior change approach is critical. Locke's (1968) initial prescription for individuals to elicit change is to predefine or set outcome goals that are specific, somewhat difficult, have predefined parameters, and are measureable. Measurable goals enable monitoring and the ability to determine if goals are being met. Outcome goal setting, according to the findings of this study, is of vital importance in achieving behavioral change related to decreasing sedentary activity and increasing physical activity. The likelihood of decreasing sedentary behavior and increasing physical activity is possible when approached with a positive outlook believing positive outcomes are possible.

Environmental adaptation

The second most predominant theme involved components of adaptation or redesign of the individual's environment. First, the most predominant subtheme discussed within this section was adding devices, objects, or tools to the environment. Ergonomic modifications, such as sit-stand desks, of both the home and work office spaces are becoming more common. Other office desks are being outfitted with devices that enable workers to break free of the bonds of the everyday office chair at their computer terminals. This is also true for those workers who telecommute at their home office. Moreover, the second major subtheme was changing, adapting, or redesigning the physical environment. There has been discussion throughout the articles within this study about the critical impact of personal environment in relation to their overall outlook.

Self-monitoring of behavior

The third major theme in this study was the idea or concept of behavioral self-monitoring, due to the association of this activity enabling the development of an awareness of both healthy habits and detrimental hazards. Self-monitoring establishes a baseline for behavior from which people can set outcome goals that eventually may move them into the goal achievement of increased physical activity, which decreases overall sedentary behavior in the home office or work environment. Awareness is a key concept in leading or enabling healthy behavioral change which mitigates harmful habits. One should strive to be keenly aware of periodic break times in order to plan light to moderate exercise in between performing stationary work.

Information on health consequences and social support

Another predominant theme for promoting behavioral change was receiving information on health consequences of sedentary inactivity. Having knowledge and information is important in terms of producing meaningful and healthy behavior change. Indeed, this may be the first step in changing or adapting behavior, receiving and heeding knowledgeable expert information, and understanding the consequences of inaction upon quality of life and longevity. Social support is important when considering that health information can be provided through medical professionals, family members, or requirements at work requesting that employees live healthier life styles. Having social support is also vital toward achieving meaningful and healthy behavior change. Coworkers, friends, and family can be important sources of encouragement in becoming more active and decreasing sedentary behaviors. Employer support could also be a factor in office culture. Managers who encourage active breaks for their employees can improve health and wellness as well as productivity. Also, employees may benefit from rewards from employers when funds used to pay high-risk premiums can be diverted to pay increases, bonus incentives, or additional benefits.

Directions on how to perform behavior(s), and advice or feedback on behavior

Other predominant themes to elicit healthy behavior changes revealed in this study include receiving directions or instructions on how to perform behaviors and how to operationalize feedback on behaviors. Proper knowledge increases the likelihood of executing the replacement behavior or exercise with the correct physical motion in the most optimal manner. Proper execution of an exercise maximizes the activities' effect on the human body for improving health. Finally, the last theme for promoting behavioral change was receiving advice or feedback on behavior. This is important in achieving meaningful and healthy behavior change because feedback facilitates the enforcement of positive behavior that allows them to take root, while harmful behaviors are potentially diminished or mitigated.

Minor, emerging, and outlier themes related to sedentary behavior change

The minor themes that arose from this review suggest that modifying sedentary behavior can be achieved by changing or restructuring the social/peer environment, activity or action planning, problem solving, behavior substitutes, habit formation or routine-making, task or behavior evaluation, behavior practice or rehearsal, reviewing risks and rewards (pros and cons), setting reminder cues or prompts, and goal review. They represent other approaches to drive behavior change related to sedentary inactivity which appear to a lesser extent in the literature review in this research.

Social/peer environment change, activity or action planning, and problem solving

Birds of a feather flock together. Although categorized as a minor theme, the idea of changing one's peer group or social environment for initiating sedentary behavior change was referenced somewhat often throughout the reviewed literature. Social norms and peer pressure play a relatively large role in increasing or decreasing the likelihood of the successful achievement of behavioral change according to the studies reviewed. Even slight modifications to a work peer group may permit an increase in motivation to be more actively engaged, thus removing the existing fetters that keep most employees chained to their desks and computer screens.

Activity or action planning to succeed was another theme that ties into the goal setting approach. Sometimes people do not intentionally plan to fail, rather they fail to actively plan for sedentary behavior change. Identifying ways to remove barriers to physical activity in the office can go far in improving health as well. According to the study findings, active planning and problem solving can play a relatively large role in increasing the likelihood of successfully achieving sedentary behavior change.

Behavior substitutes through habit formation and routine-making

Changing it up. Substituting behaviors, or replacing one behavior for another healthier behavior, can also drive behavior change in the office. Opting to grab an apple and go for a walk instead of sitting and eating more sugar could do wonders toward improving health. This idea also ties to following up on this more optimal behavior and forming habits that are healthier, rather than more hazardous. Making healthier, new habits part of the daily office routine can have a positive impact toward mitigating the negative effects of sedentary inactivity.

Behavior evaluation and rehearsal

Behavior epiphanies! The concept of behavior evaluation was also captured in this study. Evaluating self-behavior can lead to awareness of existing healthy habits and those that are more detrimental to health. Insightfully taking stock throughout the day of personal activity and inactivity can lead to shifting behavior patterns overtime for health improvements during working hours and overall. Behavior evaluation can lead individuals to taking the next step to then routinely practice or rehearse the activities of living actively every day at the office. People are creatures of habit. Practice makes perfect the routines that we adhere to in our daily lives.

Risks and rewards, cues and prompts, and goal review

Weighing your options. Examining the risks and rewards of personal behavior can also lead to discovering how to minimize sedentary behavior. Knowledge is power, and in this respect, having the knowledge that sitting too long is literally killing people can be an empowering motivator to spur activity. Light to moderate exercise, such as walking throughout the day, can be highly beneficial to physical and mental health while at work. Establishing prompts and cues in daily calendars is an excellent way to set timers and alarms to get and stay active as well. This also provides an easy way to be accountable to planned goals and determine if existing goals are being interrupted by outside circumstances. Seek ways to overcome obstacles by reviewing goals and having alternatives to planned activities. Always review lack of adherence as another way to adjust goals that will work.

Assessing whether or not an individual is sufficiently maintaining a physical activity level that promotes good health and wellness is a necessary dimension of behavior change. The important process of reviewing personal goals is similar to task evaluation. Assessing initial goals by determining whether or not the goals were accomplished is imperative in almost every process of understanding a behavioral phenomenon. Goal review can play a relatively large role in increasing the likelihood of the successful achievement of changing sedentary behavior, according to these study findings.

Emerging and outlier themes

Two emerging themes were identified as behavior change approaches from the thematic content review. They are (1) data collecting and measurable monitoring, and (2) looking at the disparity between current behavior and goal behavior. The first can provide beneficial information so that individuals who are employed in an office can identify patterns of behavior and subsequently work to correct or improve their physical activity throughout the work day. The second theme is closely related to the concepts of goal setting and goal review discussed earlier in this chapter.

Ten BCA outliers were revealed in this review including: information about social and emotional costs/consequences; peer or social support behavior monitoring; peer comparison; frequent contact with health care provider for motivation, advice, and support; task evaluation from healthcare provider or other knowledge expert; receiving advice from a knowledge expert; material incentives or rewards; habit reversal or routine-breaking of inactivity; commitment; and biofeedback. Interestingly, the research here identified breaking the bonds of the office chair similar to that of quitting smoking.

Summation: Application of findings

The transformative process, moving from inactivity to activity, is critical to better health and begins with breaking the habit of sedentary behavior. In comparing major goals and targeted behavior change (TBC) approaches, we surmise that the overall content analysis results may have some important practical implications in terms of goal setting, goal framing, and goal prioritization. The decision to choose a course(s) of action is important in terms of personal behavior. A leading number of studies reviewed (33%) emphasized setting or prioritizing the dual goals of decreasing sedentary behavior and increasing physical activity, followed closely by studies emphasizing setting or prioritizing only one goal, either decreasing sedentary behavior (30%) or increasing physical activity (27%). Two studies (7%) emphasized setting or prioritizing dual goals of increasing physical activity and optimizing or improving diet. However, one study (3%) emphasized setting or prioritizing three (3) goals of decreasing sedentary behavior, increasing physical activity, and optimizing or improving diet. Selecting the most comfortable personal option is preferable to remaining at rest for long periods of time.

Developing an intervention strategy is not an easy proposition, but rests on important questions such as, "which goal setting approaches are most effective as innovative strategic interventions?" And conversely, "which goal setting approaches are least effective as intervention strategies?" In order to address these questions of effectiveness, future research should develop and test intervention strategies beginning with the major themes discussed in this thematic content analysis. The results of testing intervention strategies and goal setting approaches in an ethical, randomized control trial (RCT) experimental design would have both pragmatic and sound implications in terms of which intervention or interventions in combination are empirically most effective in improving overall health outcomes. From these subsequent findings, public health recommendations on behavioral change intervention strategies and goal setting approaches could be made more optimally and prescribed to the communities which in the field of public health itself seeks to serve, guide, and advise with practical, evidenced-based information.

Recommendations and tools for change

Several recommendations were identified from this qualitative research by deploying thematic content analysis on the frequency of behavioral change approaches (BCAs) that were found in the articles reviewed in this study and previously summarized. Summarily, they include (1) outcome goal setting, which was identified predominantly as the major theme or idea that leads to behavioral change; followed secondarily by (2) adding devices, objects, or tools to the environment; (3) self-monitoring of one's own behavior; (4) receiving and heeding information on health consequences of sedentary behavior; (5) having social support (e.g., team up with a positive, physical activity peer group, a work walking club); (6) changing, adapting, or redesigning the physical environment; (7) receiving directions or instructions on how to properly

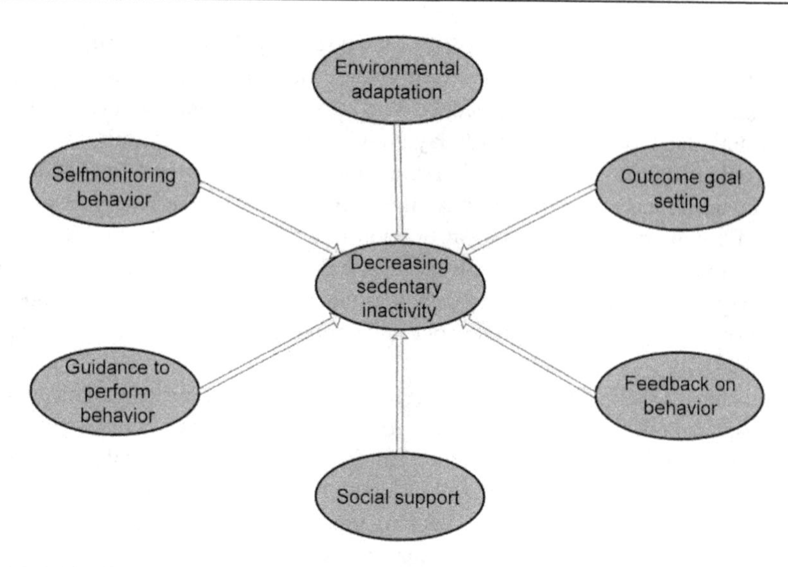

Fig. 2.1 Major behavior change approaches (BCAs) in decreasing sedentary inactivity.

perform other active exercise behaviors; and (8) receiving advice or feedback on one's own behavior, which is similar to goal review and task evaluation.

Fig. 2.1 illustrates the eight major behavior change approach (BCA) themes into six key thematic BCAs or conceptual tools for change. We present these as vital strategies toward improving health and reducing sedentary inactivity at work with extended application to home and other environments. "Environmental adaptation" and "guidance to perform the behavior" (correctly) combine concepts previously discussed. Cumulatively, these six key BCAs can effectively mitigate and decrease sedentary inactivity.

Other adaptive strategies for decreasing sedentary inactivity at work include designing human-friendly work environments; leveraging innovations in technology to promote physical activity in lieu of sedentariness; becoming more keenly self-aware of personal behavior patterns, and resolving to optimize them for health outcomes of wellness, rather than chronic illness. Also, leaning on peer groups for increased social support as well as receiving and heeding information, directions, and feedback on staying physically active and decreasing sedentary behavior.

In line with the behavior change approach of social support, employers can help improve physical activity by encouraging their employees to develop a more active office environment and work culture every day. Collectively, it is not enough to watch the apple fall to the ground, sometimes you have to reach for it. This is also true of setting individual goals for better health outcomes. Get moving and stay moving; pair up for walks; and drink plenty of water.

We suggest several simple ways to facilitate positive motivation, adapt free activity tracking applications, and engage others in the work place to enlist the aid of peer support based on findings. Several apps from Pedometer (https://play.google.com/store/

apps/details?id=com.tayu.tau.pedometer&hl=en), MyFitnessPal (https://play.google.com/store/apps/details?id=com.myfitnesspal.android), Leap Fitness Step Counter (https://play.google.com/store/apps/details?id=pedometer.steptracker.calorieburner.stepcounter), and Google Fit (https://play.google.com/store/apps/details?id=com.google.android.apps.fitness&hl=en_US) are available for little or no cost. Monitoring steps can be an easy way to see how much or how little exercise you get every day. Find a coworker to join you on this journey to improved health. Anyone can be a great external source to get you moving and keep you accountable. Remember to take steps that make the journey easy and not burdensome. The idea is not to be bogged down by another set of rules, but to be mindful of ways to increase longevity and productivity.

If you haven't had a recent physical, get one! If you have a record of a recent physical, please read the information carefully to determine what you should focus on during this change to improve your health. If you find that your cholesterol is a bit high or your blood pressure is creeping up, follow-up with your physician as you begin a new approach to maintaining your health. Seeking guidance from health professionals who can monitor progress is important to your success.

Other points to consider that span the eight BCAs include engaging in active outcome goal setting, such as predetermining that you will ask peers to "walk and talk" instead of those ad hoc meetings in your cubicle, purchase/bring healthy snacks that you must obtain by physically moving to the company refrigerator (not in your desk drawers), or by planning to walk the perimeter of the building during break times. Overcoming the pitfalls of a body at rest can be as easy as standing up and stretching or walking to the break room to fill your water bottle each hour or during periodic breaks.

As we alluded to Newton's Law at the beginning of this chapter, an object at rest tends to stay at rest and an object in motion tends to stay in motion. Breaking out of ruts at work can be challenging, but not impossible. Getting the ball rolling at work can be as simple as, well, getting the ball rolling during planned activity breaks or incorporating simple, but varied exercises that promote light physical activity. With a bit of effort, you can get moving building healthier work day routines. Take the time to schedule activities on your calendar or electronic gadget until you have built that habit into a normal part of your work schedule. Consider maintaining health and wellness as an important contribution to consistent office productivity and personal longevity.

Conclusion

In this chapter, a thematic content analysis was instrumental in providing key recommendations on strategic best practices to guide individuals engaged in sedentary office work to decrease sedentary inactivity and embrace new behavioral change approaches. Pursuing this path of behavioral modification can mitigate or reduce the harmful and sometimes deadly health impacts of sitting and sedentary inactivity, which often contributes to premature death in many industrial nations.

The goals of modifying, adapting, and decreasing sedentary behavior have long-term implications in improving employee health for those who must sit for extended periods of time. The objective of reducing sedentary behavior has many benefits to the employee, such as improvements in physical and mental health, and to the employer who benefits from increases in productivity when employees are healthy. Thus, improving working conditions and promoting the mental and physical health of their employees should be a vital mission of organizations around the world to improve overall public health. The top three recommendations from the findings of this study include: outcome goal setting, environmental redesign, and behavioral self-monitoring.

Acknowledgment

The authors would like to thank the World Health Organization for access to timely information and research. The authors are responsible for analysis and interpretation of WHO content in this manuscript.

Disclosure

The authors do not have any financial conflict of interest regarding general or specific recommendations to utilize online software applications to improve health.

References

Alkhajah, T. A., Reeves, M. M., Eakin, E. G., Winkler, E. A. H., Owen, N., & Healy, G. N. (2012). Sit-stand workstations. A pilot intervention to reduce office sitting time. *American Journal of Preventive Medicine*, *43*, 298–303. https://doi.org/10.1016/j.amepre.2012.05.027.

American Society of Nephrology (ASN). (2015, October 20). People with sedentary lifestyles are at increased risk of developing kidney disease. *ScienceDaily*. Retrieved July 26, 2019 from www.sciencedaily.com/releases/2015/10/151020094824.htm.

Bandura, A. (1977). Self-efficacy: Toward a unifying theory of behavioral change. *Psychological Review*, *84*(2), 191. https://doi.org/10.1037/0033-295X.84.2.191.

Barwais, F. A., Cuddihy, T. F., & Tomson, L. M. (2013). Physical activity, sedentary behavior and total wellness changes among sedentary adults: A 4-week randomized controlled trial. *Health and Quality of Life Outcomes*, *11*, 183. https://doi.org/10.1186/1477-7525-11-183.

Bauman, A., Ainsworth, B. E., Sallis, J. F., Hagströmer, M., Craig, C. L., et al. (2011). The descriptive epidemiology of sitting: A 20-country comparison using the international physical activity questionnaire (IPAQ). *American Journal of Preventive Medicine*, *41*(2), 228–235. https://doi.org/10.1016/j.amepre.2011.05.003.

Biswas, A., Oh, P. I., Faulkner, G. E., Bajaj, R. R., Silver, M. A., et al. (2015). Sedentary time and its association with risk for disease incidence, mortality, and hospitalization in adults: A systematic review and meta-analysis. *Annals of Internal Medicine*, *162*(2), 123–132.

Boyatzis, R. E. (1998). *Transforming qualitative information: Thematic analysis and code development*. Sage.

Braun, V., & Clarke, V. (2006). Using thematic analysis in psychology. *Qualitative Research in Ppsychology*, *3*(2), 77–101. https://doi.org/10.1191/1478088706qp063oa.

Brown, W. J., Miller, Y. D., & Miller, R. (2003). Sitting time and work patterns as indicators of overweight and obesity in Australian adults. *International Journal of Obesity*, *27*(11), 1340.

Burke, L., Jancey, J., Howat, P., Lee, A., Kerr, D., Shilton, T., et al. (2010). Physical activity and nutrition program for seniors (PANS): Protocol of a randomized controlled trial. *BMC Public Health*, *10*, 751. https://doi.org/10.1186/1471-2458-10-751.

Burke, L., Lee, A. H., Jancey, J., Xiang, L., Kerr, D. A., Howat, P. A., et al. (2013). Physical activity and nutrition behavioural outcomes of a home-based intervention program for seniors: A randomized controlled trial. *International Journal of Behavioral Nutrition and Physical Activity*, *10*, 14. https://doi.org/10.1186/1479-5868-10-14.

Carlson, S. A., Adams, E. K., Yang, Z., & Fulton, J. E. (2018). Percentage of deaths associated with inadequate physical activity in the United States. *Preventing Chronic Disease*, *15*. https://doi.org/10.5888/pcd18.170354.

Cavalheri, V., Straker, L., Gucciardi, D. F., Gardiner, P. A., & Hill, K. (2016). Changing physical activity and sedentary behaviour in people with COPD. *Respirology*, *21*(3), 419–426. https://doi.org/10.1111/resp.12680.

Chang, A. K., Fritschi, C., & Kim, M. J. (2013). Sedentary behavior, physical activity, and psychological health of Korean older adults with hypertension. Effect of an empowerment intervention. *Research in Gerontological Nursing*, *6*, 81–88. https://doi.org/10.3928/19404921-20121219-01.

Chau, J. Y., Grunseit, A. C., Chey, T., Stamatakis, E., Brown, W. J., et al. (2013). Daily sitting time and all-cause mortality: A meta-analysis. *PLoS ONE*, *8*(11), 1–14. https://doi.org/10.1371/journal.pone.0080000.

Daly, J., Kellehear, A., & Gliksman, M. (1997). *The public health researcher: A methodological guide*. Oxford University Press.

De Cocker, K., Spittaels, H., Cardon, G., De Bourdeaudhuij, I., & Vandelanotte, C. (2012). Web-based, computer-tailored, pedometer-based physical activity advice: Development, dissemination through general practice, acceptability, and preliminary efficacy in a randomized controlled trial. *Journal of Medical Internet Research*, *14*, e53. https://doi.org/10.2196/jmir.1959.

Dewa, C. S., deRuiter, W., Chau, N., & Karioja, K. (2009). Walking for wellness: Using pedometers to decrease sedentary behaviour and promote mental health. *International Journal of Mental Health Promotion*, *11*, 24–28. https://doi.org/10.1080/14623730.2009.9721784.

Dunstan, D. W., Wiesner, G., Eakin, E. G., Neuhaus, M., Owen, N., et al. (2013). Reducing office workers' sitting time: Rationale and study design for the Stand Up Victoria cluster randomized trial. *BMC Public Health*, *13*, 1057. https://doi.org/10.1186/1471-2458-13-1057.

Ekelund, U., Steene-Johannessen, J., Brown, W. J., Fagerland, M. W., Owen, N., et al. (2016). Does physical activity attenuate, or even eliminate, the detrimental association of sitting time with mortality? A harmonised meta-analysis of data from more than 1 million men and women. *The Lancet*, *388*(10051), 1302–1310. https://doi.org/10.1016/S0140-6736(16)30370-1.

Ellegast, R., Weber, B., & Mahlberg, R. (2012). Method inventory for assessment of physical activity at VDU workplaces. *Work*, *41*, 2355–2359. https://doi.org/10.3233/WOR-2012-0464-2355.

Evans, R. E., Fawole, H. O., Sheriff, S. A., Dall, P. M., Grant, P. M., & Ryan, C. G. (2012). Point-of-choice prompts to reduce sitting time at work: A randomized trial. *American Journal of Preventive Medicine*, *43*(3), 293–297. https://doi.org/10.1016/j.amepre.2012.05.010.

Falck, R. S., Davis, J. C., & Liu-Ambrose, T. (2017). What is the association between sedentary behaviour and cognitive function? A systematic review. *British Journal of Sports Medicine*, *51*(10), 800–811.

Fitzsimons, C. F., Baker, G., Gray, S. R., Nimmo, M. A., & Mutrie, N. (2012). Does physical activity counseling enhance the effects of a pedometer-based intervention over the long-term: 12-month findings from the Walking for Wellbeing in the West study. *BMC Public Health*, *12*, 206. https://doi.org/10.1186/1471-2458-12-206.

Fitzsimons, C. F., Kirk, A., Baker, G., Michie, F., Kane, C., & Mutrie, N. (2013). Using an individualised consultation and activPAL feedback to reduce sedentary time in older Scottish adults: Results of a feasibility and pilot study. *Preventive Medicine*, *57*, 718–720. https://doi.org/10.1016/j.ypmed.2013.07.017.

Foley, B., Engelen, L., Gale, J., Bauman, A., & Mackey, M. (2016). Sedentary behavior and musculoskeletal discomfort are reduced when office workers trial an activity-based work environment. *Journal of Occupational and Environmental Medicine*, *58*(9), 924–931.

Gao, Y., Nevala, N., Cronin, N. J., & Finni, T. (2016). Effects of environmental intervention on sedentary time, musculoskeletal comfort and work ability in office workers. *European Journal of Sport Science*, *16*(6), 747–754.

Gardner, B., Smith, L., Lorencatto, F., Hamer, M., & Biddle, S. J. H. (2016). How to reduce sitting time? A review of behaviour change strategies used in sedentary behaviour reduction interventions among adults. *Health Psychology Review*, *10*(1), 89–112. https://doi.org/10.1080/17437199.2015.1082146.

Gibson, J. W., Greenwood, R. A., & Murphy, E. F., Jr. (2009). Generational differences in the workplace: Personal values, behaviors, and popular beliefs. *Journal of Diversity Management*, *4*(3).

Gilson, N. D., Puig-Ribera, A., McKenna, J., Brown, W. J., Burton, N. W., & Cooke, C. B. (2009). Do walking strategies to increase physical activity reduce reported sitting in workplaces: A randomized control trial. *International Journal of Behavioral Nutrition and Physical Activity*, *6*, 43. https://doi.org/10.1186/1479-5868-6-43.

Guest, G., MacQueen, K. M., & Namey, E. E. (2012). Introduction to applied thematic analysis. *Applied Thematic Analysis*, *3*, 20.

Hallman, D. M., Ekman, A. H., & Lyskov, E. (2014). Changes in physical activity and heart rate variability in chronic neck–shoulder pain: Monitoring during work and leisure time. *International Archives of Occupational and Environmental Health*, *87*(7), 735–744.

Hallman, D. M., Gupta, N., Mathiassen, S. E., & Holtermann, A. (2015). Association between objectively measured sitting time and neck–shoulder pain among blue-collar workers. *International Archives of Occupational and Environmental Health*, *88*(8), 1031–1042.

Hansen, A. W., Grønbæk, M., Wulff Helge, J., Severin, M., Curtis, T., & Schurmann Tolstrup, J. (2012). Effect of a web-based intervention to promote physical activity and improve health among physically inactive adults: A population-based randomized controlled trial. *Journal of Medical Internet Research*, *14*, e145. https://doi.org/10.2196/jmir.2109.

Healy, G. N., Eakin, E. G., LaMontagne, A. D., Owen, N., Winkler, E. A. H., et al. (2013). Reducing sitting time in office workers: Short-term efficacy of a multicomponent intervention. *Preventive Medicine*, *57*, 43–48. https://doi.org/10.1016/j.ypmed.2013.04.004.

Helgadóttir, B., Owen, N., Dunstan, D. W., Ekblom, Ö., Hallgren, M., & Forsell, Y. (2017). Changes in physical activity and sedentary behavior associated with an exercise intervention in depressed adults. *Psychology of Sport and Exercise*, *30*, 10–18.

Heron, L., O'Neill, C., McAneney, H., Kee, F., & Tully, M. A. (2019). Direct healthcare costs of sedentary behaviour in the UK. *Journal of Epidemiology and Community Health*, *73*(7), 625–629. https://doi.org/10.1136/jech-2018-211758.

Jette, M., Sidney, K., & Blümchen, G. (1990). Metabolic equivalents (METS) in exercise testing, exercise prescription, and evaluation of functional capacity. *Clinical Cardiology*, *13*(8), 555–565. https://doi.org/10.1002/clc.4960130809.

John, D., Thompson, D. L., Raynor, H., Bielak, K., Rider, B., & Bassett, D. R. (2011). Treadmill workstations: A worksite physical activity intervention in overweight and obese office workers. *Journal of Physical Activity and Health*, *8*, 1034–1043.

Karol, S., & Robertson, M. M. (2015). Implications of sit-stand and active workstations to counteract the adverse effects of sedentary work: A comprehensive review. *Work*, *52*(2), 255–267.

Katzmarzyk, P. T., & Lee, I. M. (2012). Sedentary behaviour and life expectancy in the USA: A cause-deleted life table analysis. *BMJ Open*, *2*(4), e000828. https://doi.org/10.1136/bmjopen-2012-000828.

Kozey-Keadle, S., Libertine, A., Lyden, K., Staudenmayer, J., & Freedson, P. S. (2011). Validation of wearable monitors for assessing sedentary behavior. *Medicine & Science in Sports and Exercise*, *43*, 1561–1567. https://doi.org/10.1249/MSS.0b013e31820ce174.

Kwak, M. S., & Kim, D. (2018). Non-alcoholic fatty liver disease and lifestyle modifications, focusing on physical activity. *The Korean Journal of Internal Medicine*, *33*(1), 64–74. https://doi.org/10.3904/kjim.2017.343.

Lee, I. M., Shiroma, E. J., Lobelo, F., Puska, P., Blair, S. N., Katzmarzyk, P. T., et al. (2012). Effect of physical inactivity on major non-communicable diseases worldwide: An analysis of burden of disease and life expectancy. *The Lancet*, *380*(9838), 219–229. https://doi.org/10.1016/S0140-6736(12)61031-9.

LifeSpan. (2016, April 4). *Sitting disease is taking a toll on your body*. Retrieved July 26, 2019 from https://www.lifespanfitness.com/workplace/resources/articles/sitting-all-day-is-taking-a-toll-on-your-body.

Lis, A. M., Black, K. M., Korn, H., & Nordin, M. (2007). Association between sitting and occupational LBP. *European Spine Journal*, *16*(2), 283–298.

Locke, E. A. (1968). Toward a theory of task motivation and incentives. *Organizational Behavior and Human Performance*, *3*(M), 157–189. https://doi.org/10.1016/0030-5073(68)90004-4.

Locke, E. A., & Latham, G. P. (2006). New directions in goal-setting theory. *Current Directions in Psychological Science*, *15*(5), 265–268. https://doi.org/10.1111/j.1467-8721.2006.00449.x.

MacMillan, F., Fitzsimons, C., Black, K., Granat, M. J., Grant, M. P., et al. (2011). West End Walkers 65+: A randomised controlled trial of a primary care-based walking intervention for older adults: Study rationale and design. *BMC Public Health*, *11*, 120. https://doi.org/10.1186/1471-2458-11-120.

Mahmood, S., MacInnis, R. J., English, D. R., Karahalios, A., & Lynch, B. M. (2017). Domain-specific physical activity and sedentary behaviour in relation to colon and rectal cancer risk: A systematic review and meta-analysis. *International Journal of Epidemiology*, *46*(6), 1797–1813.

Mammen, G., & Faulkner, G. (2013). Physical activity and the prevention of depression: A systematic review of prospective studies. *American Journal of Preventive Medicine*, *45*(5), 649–657.

Martin, A., Fitzsimons, C., Jepson, R., Saunders, D. H., Van Der Ploeg, H. P., et al. (2015). Interventions with potential to reduce sedentary time in adults: Systematic review and meta-analysis. *British Journal of Sports Medicine*, *49*(16), 1056–1063. https://doi.org/10.1136/bjsports-2014-094524.

Mikkelsen, K., Stojanovska, L., Polenakovic, M., Bosevski, M., & Apostolopoulos, V. (2017). Exercise and mental health. *Maturitas*, *106*, 48–56.

Mutrie, N., Doolin, O., Fitzsimons, C. F., Grant, P. M., Granat, M., et al. (2012). Increasing older adults' walking through primary care: Results of a pilot randomized controlled trial. *Family Practice*, *29*, 633–642. https://doi.org/10.1093/fampra/cms038.

Neuhaus, M., Healy, G. N., Fjeldsoe, B. S., Lawler, S., Owen, N., et al. (2014). Iterative development of stand up Australia: A multi-component intervention to reduce workplace sitting. *International Journal of Behavioral Nutrition and Physical Activity*, *11*, 21. https://doi.org/10.1186/1479-5868-11-21.

O'Donoghue, G., Perchoux, C., Mensah, K., Lakerveld, J., Van Der Ploeg, H., et al. (2016). A systematic review of correlates of sedentary behaviour in adults aged 18-65 years: A socio-ecological approach. *BMC Public Health*, *16*(1). https://doi.org/10.1186/s12889-016-2841-3.

Parry, S., & Straker, L. (2013). The contribution of office work to sedentary behaviour associated risk. *BMC Public Health*, *13*(1). https://doi.org/10.1186/1471-2458-13-296.

Parry, S., Straker, L., Gilson, N. D., & Smith, A. J. (2013). Participatory workplace interventions can reduce sedentary time for office workers—A randomised controlled trial. *PLoS One*, *8*(11), e78957.

Patel, A. V., Rodriguez, C., Pavluck, A. L., Thun, M. J., & Calle, E. E. (2006). Recreational physical activity and sedentary behavior in relation to ovarian cancer risk in a large cohort of US women. *American Journal of Epidemiology*, *163*(8), 709–716.

Patterson, R., McNamara, E., Tainio, M., de Sá, T. H., Smith, A. D., et al. (2018). Sedentary behaviour and risk of all-cause, cardiovascular and cancer mortality, and incident type 2 diabetes: A systematic review and dose response meta-analysis. *European Journal of Epidemiology*, *33*(9), 811–829. https://doi.org/10.1007/s10654-018-0380-1.

Polit, D., & Beck, C. (2010). Qualitative designs and approaches. In D. Polit & C. T. Beck (Eds.), *Essentials of nursing research: Appraising evidence for nursing practice* (7th ed., pp. 258–283). Philadelphia, PA: Wolters Kluwer.

Prince, S. A., Saunders, T. J., Gresty, K., & Reid, R. D. (2014). A comparison of the effectiveness of physical activity and sedentary behaviour interventions in reducing sedentary time in adults: A systematic review and meta-analysis of controlled trials. *Obesity Reviews*, *15*(11), 905–919. https://doi.org/10.1111/obr.12215.

Pronk, N. P., Katz, A. S., Lowry, M., & Payfer, J. R. (2012). Reducing occupational sitting time and improving worker health: The Take-a-Stand Project, 2011. *Preventing Chronic Disease*, *9*, 110323. https://doi.org/10.5888/pcd9.110323.

Rezende, L. F. M., Sá, T. H., Mielke, G. I., Viscondi, J. Y. K., Rey-López, J. P., & Garcia, L. M. T. (2016). All-cause mortality attributable to sitting time: Analysis of 54 countries worldwide. *American Journal of Preventive Medicine*, *51*(2), 253–263. https://doi.org/10.1016/j.amepre.2016.01.022.

Ryan, C. G., Grant, P. M., Dall, P. M., & Granat, M. H. (2011). Sitting patterns at work: Objective measurement of adherence to current recommendations. *Ergonomics*, *54*(6), 531–538. https://doi.org/10.1080/00140139.2011.570458.

Schwartz, B., Kapellusch, J. M., Baca, A., & Wessner, B. (2019). Medium-term effects of a two-desk sit/stand workstation on cognitive performance and workload for healthy people performing sedentary work: A secondary analysis of a randomised controlled trial. *Ergonomics*, 1–17. https://doi.org/10.1080/00140139.2019.1577497.

Sedentary Behavior Research Network (SBRN). (2012). Standardized use of the terms "sedentary" and "sedentary behaviours". *Applied Physiology, Nutrition, and Metabolism, 37*, 540–542. https://doi.org/10.1139/H2012-024.

Sedentary Behavior Research Network (SBRN). (n.d.). What is sedentary behaviour? Retrieved June 2, 2019 from https://www.sedentarybehaviour.org/what-is-sedentary-behaviour/.

Sethi, J., Sandhu, J. S., & Imbanathan, V. (2011). Effect of body mass index on work related musculoskeletal discomfort and occupational stress of computer workers in a developed ergonomic setup. *Sports Medicine, Arthroscopy, Rehabilitation, Therapy & Technology, 3*(1)22.

Shrestha, N., Ijaz, S., Kukkonen-Harjula, K. T., Kumar, S., & Nwankwo, C. P. (2015). Workplace interventions for reducing sitting at work. *Cochrane Database of Systematic Reviews, 1*, CD010912. https://doi.org/10.1002/14651858.CD010912.pub2.

Shrestha, N., Kukkonen-Harjula, K. T., Verbeek, J. H., Ijaz, S., Hermans, V., & Pedisic, Z. (2018). Workplace interventions for reducing sitting at work. *Cochrane Database of Systematic Reviews, 6*, CD010912. https://doi.org/10.1002/14651858.CD010912.pub4.

Singh, A. M., Neva, J. L., & Staines, W. R. (2014). Acute exercise enhances the response to paired associative stimulation-induced plasticity in the primary motor cortex. *Experimental Brain Research, 232*(11), 3675–3685.

Sitthipornvorakul, E., Janwantanakul, P., & Lohsoonthorn, V. (2015). The effect of daily walking steps on preventing neck and low back pain in sedentary workers: A 1-year prospective cohort study. *European Spine Journal, 24*(3), 417–424.

Smith, K. J., & Ainslie, P. N. (2017). Regulation of cerebral blood flow and metabolism during exercise. *Experimental Physiology, 102*(11), 1356–1371.

Stanton, R., Happell, B., & Reaburn, P. (2014). The mental health benefits of regular physical activity, and its role in preventing future depressive illness. *Nursing: Research and Reviews, 4*, 45–53.

Steffens, D., Maher, C. G., Pereira, L. S. M., Stevens, M. L., Oliveira, V. C., et al. (2016). Prevention of low back pain a systematic review and meta-analysis. *JAMA Internal Medicine, 176*(2), 199–208. https://doi.org/10.1001/jamainternmed.2015.7431.

Thivel, D., Tremblay, A., Genin, P. M., Panahi, S., Rivière, D., & Duclos, M. (2018). Physical activity, inactivity, and sedentary behaviors: Definitions and implications in occupational health. *Frontiers in Public Health, 6*(October), 1–5. https://doi.org/10.3389/fpubh.2018.00288.

Tremblay, M. S., Aubert, S., Barnes, J. D., Saunders, T. J., Carson, V., et al. (2017). Sedentary behavior research network (SBRN)–terminology consensus project process and outcome. *International Journal of Behavioral Nutrition and Physical Activity, 14*(1), 75.

Verweij, L. M., Proper, K. I., Weel, A. N. H., Hulshof, C. T. J., & van Mechelen, W. (2012). The application of an occupational health guideline reduces sedentary behaviour and increases fruit intake at work: Results from an RCT. *Occupational and Environmental Medicine, 69*, 500–507. https://doi.org/10.1136/oemed-2011-100377.

World Health Organization (WHO). (2011). *WHO. Global health observatory data repository.* Retrieved July 26, 2019 from http://apps.who.int/ghodata/.

World Health Organization (WHO) (2018, February 23). *Physical activity [Fact sheet].* Retrieved June 2, 2019 from https://www.who.int/en/news-room/fact-sheets/detail/physical-activity.

Supplementary reading recommendations

Buckley, J. P., Hedge, A., Yates, T., Copeland, R. J., Loosemore, M., et al. (2015). The sedentary office: An expert statement on the growing case for change toward better health and productivity. *British Journal of Sports Medicine*, *49*(21), 1357–1362.

Davis, R., Campbell, R., Hildon, Z., Hobbs, L., & Michie, S. (2015). Theories of behaviour and behaviour change across the social and behavioural sciences: A scoping review. *Health Psychology Review*, *9*(3), 323–344.

Physical Activity Guidelines Advisory Committee. (2018). *Physical activity guidelines advisory committee scientific report, part F. chapter 2. Sedentary behavior* (pp. 1–42). Washington, DC: U.S. Department of Health and Human Services. Retrieved June 2, 2019 from https://health.gov/paguidelines/second-edition/report/pdf/08_F-2_Sedentary_Behavior.pdf.

Rav-Marathe, K., Wan, T. H., & Marathe, S. (2016). A systematic review on the KAP-O framework for diabetes education and research. *Archives of Medical Research*, *4*, 1–22.

Song, Z., & Baicker, K. (2019). Effect of a workplace wellness program on employee health and economic outcomes: A randomized clinicaltTrial. *JAMA*, *321*(15), 1491–1501. https://doi.org/10.1001/jama.2019.3307.

Sullivan, A. N., & Lachman, M. E. (2017). Behavior change with fitness technology in sedentary adults: A review of the evidence for increasing physical activity. *Frontiers in Public Health*, *4*, 289.

Thyfault, J. P., Du, M., Kraus, W. E., Levine, J. A., & Booth, F. W. (2015). Physiology of sedentary behavior and its relationship to health outcomes. *Medicine and Science in Sports and Exercise*, *47*(6), 1301–1305. https://doi.org/10.1249/MSS.0000000000000518.

Van der Ploeg, H. P., Chey, T., Korda, R. J., Banks, E., & Bauman, A. (2012). Sitting time and all-cause mortality risk in 222 497 Australian adults. *Archives of Internal Medicine*, *172*(6), 494–500.

Wan, T. T. (2014). A transdisciplinary approach to health policy research and evaluation. *International Journal of Public Policy*, *10*(4–5), 161.

World Bank. (2010). *Theories of behavior change (English). Communication for governance and accountability program (CommGAP).* Washington, DC: World Bank. Retrieved June 3, 2019 from http://documents.worldbank.org/curated/en/456261468164982535/Theories-of-behavior-change.

Definitions

Generation X people generally born between the years ranging from approximately the early-to-mid 1960s to the early-1980s; inclusivity parameters of this generation as well as other generations may vary across sources, but this span is generally widely accepted.

Metabolic equivalents (MET) generally the amount of oxygen consumed while sitting at rest and conventionally is set or equal to 3.5 mL of oxygen (O_2) per kg body weight per minute (Jette, Sidney, & Blümchen, 1990); MET is the ratio of a person's working metabolic rate relative to the resting metabolic rate used to evaluate function capacity where one MET is scientifically defined as the energy cost of sitting quietly and is equivalent to the caloric consumption of 1 kcal/kg/h.

Sedentary behavior any waking behavior characterized by an energy expenditure that is ≤ 1.5 metabolic equivalents (METs), while in a sitting, reclining, or lying posture (SBRN, 2012).

Sedentary occupational work any waking behavior characterized by an energy expenditure that is ≤ 1.5 metabolic equivalents (METs), while in a sitting, reclining, or lying posture, even while at work or in the office (SBRN, 2012).

ments

related harm among students: Past issues and future directions

3

Robert D. Dvorak, Angelina Leary, Roselyn Peterson, Matthew P. Kramer, Daniel Pinto, Michael E. Dunn
Department of Psychology, University of Central Florida, Orlando, FL, United States

Abstract

College student drinking continues to be a significant public health concern because they consume alcohol at rates higher than their same-age, noncollege, peers. In 2002, the National Institute of Alcohol Abuse and Alcoholism's Task Force on college campus drinking identified mechanisms to address this epidemic as the basis for prevention programs projected to reduce drinking. Subsequent policy implemented the program across US college campuses. However, recent analysis of these prevention programs has shown that researchers overestimated the magnitude of effect and longevity of these programs. The current chapter discusses some of the issues that have arisen due to nationwide policy adoption for these programs in the absence of large-scale evidence. We then provide a more nuanced approach to understanding and targeting college student drinking which addresses all three levels of analysis identified by the NIAAA Task Force. Finally, we provide a roadmap for future research aimed at reducing this significant problem.

Keywords: College students, Alcohol use, Binge drinking, Alcohol harm, Prevention, Intervention

Chapter outline

Three Facets of Public Health and Paths to Improvements. https://doi.org/10.1016/B978-0-12-819008-1.00003-1

College student drinking: Alcohol use and alcohol-related harm

Alcohol consumption and the number of deadly incidents associated with the recreational activity remain on college campuses despite efforts by government leaders, agency researchers, and appointed task members to address the systemic problem. The National Institute of Alcohol Abuse and Alcoholism's (NIAAA) implemented a multilevel (e.g., individual, student body, and community) approach to reduce college-age campus drinking. While the initial NIAAA implementation did not yield an expected reduction in consumption or related injuries and fatalities, this chapter demonstrates how the basic *3-1 framework* can be utilized as the backbone for stable policy for prevention focusing on the individual. The suggested approach, with some modifications based on peer-reviewed research with sound design, yields four categories of alternative prevention approaches each with two strategies: (1) feedback approaches for alternative targets, (2) changing cognitive factors linked to alcohol use, (3) interventions grounded in behavioral economics, and (4) increasing safe drinking strategies. Application of these four main alternative paths to college-age drinking prevention requires an understanding of the historical attempts to address the problem since the initial NIAAA implementation across college campuses in the United States beginning near the millennium.

In 2002, the surgeon general issued a "Call to Action" in order to address the high prevalence, severity, and persistence of alcohol use and alcohol-related harm on college campuses (NIAAA, 2002). Despite decades of research, college student alcohol use remains a significant public health issue. Research consistently shows that college students consume alcohol at higher rates than their noncollege peers (Linden-Carmichael & Lanza, 2018; Slutske, 2005). Furthermore, college student drinking is linked to approximately 2000 deaths annually, including deaths from heavy consumption as well as engagement in high-risk behaviors while intoxicated such as drinking and driving (Hingson, Heeren, Winter, & Wechsler, 2005). More than 600,000 students are physically assaulted each year as a direct result of alcohol consumption (White & Hingson, 2013).

Alcohol is linked to over 70,000 sexual assaults each year, with approximately 1 in 4 women experiencing an alcohol-related sexual assault during their years in college (Hingson, 2010b; Lawyer, Resnick, Bakanic, Burkett, & Kilpatrick, 2010; White & Hingson, 2013). Among college men who have perpetrated sexual aggressive acts, including sexual assault, 74% report consuming alcohol just prior to the acts (Koss, Dinero, Seibel, & Cox, 1988). Annually, over 400,000 students report engaging in high-risk sexual activities while drinking, with a sizable number (greater than 100,000) reporting that they have been too intoxicated to know if they even consented to sexual activities.

Furthermore, alcohol use has led to significant legal issues. Over 2 million students report drinking and driving annually (Hingson, Zha, & Weitzman, 2009, p. 15), 5% of students are involved with law enforcement due to their drinking (Hingson et al., 2005, p. 266), and approximately 1825 deaths from alcohol use occurs each year in college campuses (Hingson et al., 2009, p. 15–18). With 3 out of 5 college students reporting alcohol use over the past month and 2 out of 5 college students endorsing at least one heavy episodic drinking occasion (i.e., binge drinking episode) in the past month (SAMHSA, 2017 [database]) identifying factors that promote heavy consumption rates, targeting these factors with effective interventions remains vitally important for public health.

Risk factors of alcohol-related harm among college students

The NIAAA Task Force on college student drinking (2002) highlighted a multivariate approach to tackle the college student drinking epidemic (NIAAA, 2002). The overarching framework emphasized three basic levels of analysis for comprehensive integrated programs to address college student alcohol use and alcohol-related consequences. This approach, defined as the *3-in-1 framework*, suggested that campus prevention/intervention programs simultaneously target each level of analysis. The *3-in-1 framework* was designed to target risk factors at each level which interact to promote higher consumption rates and increase alcohol-related problems. The identified levels are: (1) the individual, (2) the student body, and (3) the college and surrounding community. Based on the 2002 report, the NIAAA also compiled a tool, the Alcohol Intervention Matrix, to assist colleges with the development and implementation of a comprehensive *3-in-1 framework* approach that could be tailored to the specific needs of individual campuses. At each level, there are identified risk factors that may be potential intervention targets. It is important to note that the factors in each level do not necessarily warrant intervention at the same level. Indeed, some individual factors are targeted at the student body and/or community levels. The levels, associated risk factors within each level, and recommended prevention/intervention strategies at each level are outlined below.

Individual level

The NIAAA Task Force highlighted the fact that intervention efforts are needed for individuals who meet diagnostic criteria for an alcohol use disorder as well as individuals at risk for developing an alcohol use disorder (NIAAA, 2002). Indeed, the vast majority of severe alcohol-related consequences among college students (i.e., deaths, disability, and damages) occur among students without an alcohol use disorder diagnosis (Knight et al., 2002). Thus, targeting *at-risk* drinkers across the spectrum is extremely important. There are numerous individual-level factors that place college students at an increased risk across a variety of domains. These include a variety of factors beginning with perceptions of drinking norms such as the base-rates of drinking among peers and the perceived acceptability of drinking among peers (Prentice & Miller, 1993); motivations for drinking (e.g., drinking to cope with negative mood, drinking to enhance positive mood, drinking for social reasons, and/or drinking to "fit in" with peers, Cooper, 1994); and expectations of the pharmacological effects of alcohol to reduce tension or provide liquid courage to increase confidence (Aarons, Goldman, Greenbaum, & Coovert, 2003; Baer, 2002). Other factors include individual temperament such as emotional stability, impulsivity, affective functioning, and personality traits (Dvorak, Pearson, & Day, 2014; Dvorak, Pearson, Sargent, Stevenson, & Mfon, 2016; Dvorak, Simons, & Wray, 2011); cognition (e.g., response inhibition, implicit associations, attentional biases) (Wiers et al., 2002); and personal values, goals, and beliefs (e.g., being a responsible person, maintaining good grades) (Monahan, Bracken-Minor, McCausland, McDevitt-Murphy, & Murphy, 2012; White & Hingson, 2013). These domains vary both in malleability as well as in the extent to which they promote alcohol use (i.e., consumption levels) versus alcohol-related consequences (i.e., harm).

There have been a number of approaches targeting alcohol use and problems at the individual-level. Among individuals identified as *at risk*, there has been good evidence supporting a multifaceted approach that incorporates aspects of feedback about an individual's own drinking, coupled with motivational approaches that try to leverage individual values and/or goals using a cognitive-behavioral approach (Dimeff, Baer, Kivlahan, & Marlatt, 1999; Kulesza, McVay, Larimer, & Copeland, 2013). This has resulted in the development of several comprehensive, individual-level, Brief Alcohol Interventions (BAIs) and the adoption of BAIs on virtually every campus in the country (Hingson, 2010a; Jernigan, Shields, Mitchell, & Arria, 2019). One of the most supported BAIs is a program known as BASICS: Brief Alcohol Screening and Intervention for College Students (Dimeff et al., 1999). BASICS is a multifaceted individual-level BAI, utilized with *at risk* drinkers, meant to curtail alcohol-related consequences. The program targets several core features at the individual level including feedback about base-rates of drinking on campus relative to the individual's consumption rates, clarification of goals and the impact of drinking on goal attainment, psychoeducation about alcohol and its effects, and ways to reduce harm in the context of drinking. A recent meta-analysis of college student alcohol interventions found BASICS (and similar BAIs) to be the most effective approach to reduce adverse alcohol-related outcomes among college students (Huh et al., 2015).

Student-body level

At the student-body level, the NIAAA Task Force identified factors that (1) promote high-risk drinking and (2) reduce safeguards against college students engaging in risky behaviors (NIAAA, 2002). The Task Force noted five areas that encourage or assist in high-risk drinking: (1) widespread availability of alcohol for college students (including underage students), (2) promotion of alcohol in social and commercial spheres, (3) extended periods of unstructured time, (4) inconsistent enforcement of rules and laws on campuses (e.g., officials sometimes "looking the other way" regarding drinking on a dry campus or occasionally giving verbal warnings instead of formal sanctions for underage drinking), and (5) student perceptions of normative alcohol use (Jones-Webb et al., 1997; Larimer et al., 2011; Lenk, Toomey, Wagenaar, Bosma, & Vessey, 2002; Nelson, Toomey, Lenk, Erickson, & Winters, 2010; Toomey & Wagenaar, 2002; Wagenaar & Toomey, 2002). Since the NIAAA Task Force's report, research has continued to find that the above areas increase high-risk alcohol consumption (Hingson, 2010a). For example, Knight et al. (2002, p. 267) found that students' reports of consistent enforcement on campus vary greatly, with approximately 24% of students endorsing that campus policy regarding alcohol is strongly enforced and that heavy drinking rates among college students were inversely associated with consistent, strict policy enforcement. Another study supported the assertion that unstructured time is a risk for increased alcohol consumption and consequences (Patrick, Maggs, & Osgood, 2010).

Perhaps the greatest risk is students' normative beliefs. Over the last several decades, two findings have been consistently reported. First, perceptions of how much other students drink (drinking norms) are strongly associated with one's own drinking. Second, there is a consistent discrepancy between perceived and actual drinking norms (Borsari & Carey, 2001; Cox et al., 2019; Lewis & Neighbors, 2006a, 2006b; Perkins, Haines, & Rice, 2005). In particular, students tend to overestimate the consumption rates of other college students. These findings have provided a rationale for broad-based social norms alcohol intervention strategies. Furthermore, social norms information has been incorporated into almost all college alcohol prevention programs that provide some form of multicomponent personalized feedback (Hingson, 2010a; Miller & Leffingwell, 2013).

The "gold standard" prevention approach for college student drinking exists at the student-body level. This approach involves campus campaigns and campus-wide programming aimed at changing perceptions of drinking norms on campus. Currently, there are campus-wide prevention programs aimed at (1) modifying the perception that drinking is common and acceptable on campus and (2) correcting individual-level misperceptions about drinking base-rates by the broader college community. The evidence for normative feedback interventions has been shown time and again throughout the literature (Carey, Scott-Sheldon, Carey, & DeMartini, 2007; LaBrie et al., 2013; Lewis & Neighbors, 2006a; Miller et al., 2013; Neighbors et al., 2010). The integration of normative feedback began at the *individual level* with the use of feedback during individual and group sessions that applied brief motivational techniques for more problematic college student drinkers. However, as the evidence base of personalized normative feedback grew, so too did the appeal of utilizing this approach as a means to target broad swaths of the college student community.

In one of the first large-scale tests of a computer-delivered personalized normative feedback intervention, neighbors and colleagues (2004) showed that this approach had small-to-moderate effects on college student drinking and that these effects were mediated via changes in perceived norms (Neighbors, Larimer, & Lewis, 2004). A number of studies have subsequently replicated these early results and extended the effects to identify more effective ways to enhance the normative feedback approach (LaBrie et al., 2013; Lewis & Neighbors, 2006a; Miller et al., 2013). However, research has also shown that computer-/web-based personalized normative feedback does not result in effects as robust as those observed when feedback is delivered face-to-face in some sort of brief motivational intervention (Carey et al., 2007; Neighbors et al., 2010). Nonetheless, the ability to effect small changes in college student drinking, but on a very large scale via mass delivery modes (e.g., web-based approaches), presents the potential to have a very broad public health impact. This latter point has been the driving force behind the continued push to incorporate technology into college drinking prevention programs.

As noted above, norm-based interventions, especially personalized normative feedback, have become a staple of virtually every campus-wide drinking prevention program. For nearly two decades, researchers and policy makers have accepted the evidence that these interventions are effective. Indeed, one could ask, "Why invest in new college drinking interventions when we already have several that work?" The answer lies in a series of recent studies that have taken a microscope to the *personalized normative feedback/brief motivational interventions* data. In the last several years, significant efforts have been made to better understand the true effects of brief drinking prevention/interventions. This effort, known as Project INTEGRATE (Huh et al., 2015; Mun et al., 2015), utilized integrative data analysis to combine outcomes across 24 large-scale college student drinking interventions. This resulted in combined person-level data on 12,630 college student drinkers who had participated in major RCTs aimed at reducing deleterious college student alcohol involvement. The research team then conducted subject-level meta-analyses using sophisticated analytics (i.e., Bayesian multilevel over-dispersed Poisson hurdle models). The results of this project indicated that (a) many of the "robust" effects previously reported in the literature were significantly overestimated, and (b) for most interventions, the effects are no different than zero. In fact, Project INTEGRATE found that in-person brief motivational approaches were the only approaches to show consistent effects on college student drinking. This may help explain why we have barely moved the needle on drinking-related harms despite decades of research on personalized normative feedback and the implementation of some form of technology-delivered alcohol personalized normative feedback on virtually every campus in the country.

College and community level

The NIAAA Task Force highlights the following college-, university-, and community-level factors as contributing to excessive levels of student drinking: residing in college-affiliated housing (e.g., dorms), dominant Greek systems (i.e., fraternities and sororities), prominent athletic department, being labeled a "wet" campus, schools located in the northeast region of the United States, selling liquor near campus (especially to minors), and failure on the part of the legal system to properly penalize inappropriate use, among others (Hingson, 2010b; NIAAA, 2002; Wechsler, Davenport, Dowdall, & Moeykens, 1994; Wechsler, Dowdall, Davenport, & Castillo, 1995; Wechsler, Kelley, Weitzman, San Giovanni, & Seibring, 2000;

Wechsler & Nelson, 2008; Werner & Greene, 1992). Unfortunately, colleges and universities often feel embarrassed to reveal that a problem exists on campus with regard to alcohol (Wechsler & Nelson, 2008). Thus, the NIAAA Task Force suggests working with the surrounding community to aid in the decrease of alcohol-related harm among college students.

Perhaps the most common and well-known community approach to reducing alcohol-related use and consequences is the minimum legal drinking age (MLDA) (Toomey & Wagenaar, 2002). Toomey and Wagenaar (2002) found higher legal drinking ages reduce alcohol consumption, possibly due to reducing alcohol sales to minors. Stricter enforcement of MLDA laws has better outcomes in reducing alcohol use (Montgomery, Foley, & Wolfson, 2006). Furthermore, enforcing laws related to driving under the influence of drugs (DUIs), lowering legal blood alcohol limits to 0.08% for adults over the age of 21, and 0.02% for drivers under the age of 21 have been found to reduce injuries and deaths related to alcohol-impaired driving (Dee & Evans, 2001; Hingson, Heeren, & Winter, 1994, 1996, 2000; Schultz et al., 2001; Voas, Tippetts, & Fell, 2000). DUI laws have been enforced, and alcohol-impaired driving has been reduced with the use of sobriety checkpoints (Schultz et al., 2001) and administrative license revocation laws (Dee & Evans, 2001). Reducing access to alcohol at the community may also reduce alcohol use and consequences. Taxes on alcoholic beverages may decrease many alcohol-related problems, and price appears to affect all drinker types (Cook & Moore, 1993). Also, individuals selling alcohol, such as bartenders and waiters, who receive training on standard drink sizes, cutting off alcohol to intoxicated patrons, checking age identification, detecting falsified identifications, and refusing service have been found to reduce alcohol consumption and alcohol-related problems (Gliksman et al., 1993; Lang, Stockwell, Rydon, & Beel, 1998; Russ & Geller, 1987; Saltz, 1989).

What doesn't work to reduce college student drinking and related problems

Although a number of strategies have been successful in reducing alcohol-related use and consequences on college campuses nationwide, there still remain a number of unsuccessful strategies that need to be considered. For example, a review of the literature by Larimer and Cronce (2002) found little to no evidence for the utility of educational or awareness programs. Numerous studies' findings have supported the ineffectiveness of educational programs among college student populations (Pan & Bai, 2009; Thombs, 2000; West & O'Neal, 2004); yet, undeterred by these (and similar) findings, informational and educational programs continue to be the most frequently used prevention methods on college campuses. Nelson et al. (2010, p. 1689) found that nearly all of the colleges in their sample (98%) reported the use of educational methods to teach their students of the risks associated with alcohol use and abuse. Project INTEGRATE researchers have also shown that non-personalized, multicomponent, interventions may actually be linked to an *increase* in alcohol use and problems proportional with the amount of content (Ray et al.,

2014). According to the NIAAA Task Force, another ineffective strategy is to provide blood alcohol content (BAC) feedback to students. Some researchers hold that this strategy may lead to harmful and counterproductive results (Johnson, Voas, Kelley-Baker, & Furr-Holden, 2008).

An additional study evaluated the effectiveness of 35 official campus alcohol policies (CAPs) and 13 individual sanctions from 15 diverse college campuses (Jernigan et al., 2019). They found that the least effective policies were (1) requiring students to register their kegs if used on campus, (2) no mention of local police force collaboration with the institution, and (3) the use of student funds in order to purchase alcohol over the phone. Notable is that two additional policies: (1) bans on drinking paraphernalia and (2) recovery houses on campus were categorized as "not scored." The authors explained that while these strategies are important both symbolically and for the treatment of addiction overall, they are also impractical and unlikely to have an impact on the student body as a whole.

Changes in college student drinking since the "Call to Action" report

According to data from the National Survey on Drug Use and Health (2017 [database]), monthly college student drinking declined from 64.3% in 2002 to 58.0% in 2016, a reduction of 6.3%. Similar change was observed among same-age, noncollege peers, who showed a reduction from 54.2% in 2002 to 46.8% in 2016 (SAMHSA, 2017 [database]). Thus, while more college students consume alcohol than noncollege peers, they have also shown a trend toward lower rates which is consistent with the general population. Despite changes in the number of students consuming alcohol each month, data from the Core Institute show that campus drinking norms have remained relatively stable from 2006, when 91% reported a belief that peers drink weekly, to 2016, when 88% reported a belief that peers drink weekly (Core Institute, 2017 [database]). Furthermore, although college campuses have observed declines in consumption across the last 15 years, data from the Core institute also suggest very little change in alcohol-related consequences from 2006 to 2016 (Core Institute, 2017 [database]). This may be due to differences in the topography of drinking. For example, in a recent latent class analysis of $n = 2213$ drinkers between the ages of 18 and 22 from the National Epidemiologic Survey on Alcohol and Related Conditions-III (NESARC-III), Linden-Carmichael and Lanza (2018, p. 2161) found that 38.9% of college students engage in high-intensity drinking, while only 27.2% of noncollege peers engage in drinking at simlar levels.

The most problematic type of drinking topography involves consuming large quantities of alcohol in a short period of time. This type of drinking is referred to as "Heavy Episodic Drinking" (HED; i.e., binge drinking) and entails consuming 4+ (for women) or 5+ (for men) drinks per drinking occasion (NIAAA, 2004). According to data from the 2016 National Survey on Drug Use and Health (SAMHSA, 2016 [database]), ~39% of college students report engaging in HED in the past month, compared to ~32% of noncollege peers. Furthermore, trends in this data show that from 2002 to 2011 (the decade following the NIAAA "Call to Action" report), college students showed a decline in heavy episodic drinking from 44.4% to 39.6%, a reduction of 4.8%. This trend was on pace with that observed in the general public of 38.9% to 34.7% (a reduction of 4.2%) during the same time period. Over the next 5 years (2011–16), the noncollege population continued this trend, declining from 34.7% to 31.7%, a decline of 3.0%. In contrast, college student binge drinking showed a leveling-off, moving from 39.6% in 2011 to 38.9% in 2016, a decline of just 0.7%. Finally, heavy alcohol use (i.e., engaging in binge drinking on five or more days over the last 30 days) was reported by 10.2% of college students, but only 8.1% of their noncollege peers in 2016 (SAMHSA, 2016 [database]), again supporting the notion that college student alcohol use patterns are more dangerous than their noncollege peers.

In summary, since the "Call to Action" report, we have observed declines in the proportion of college students who drink on a monthly basis, and this decline has kept pace with larger population trends. However, we have seen little change in campus-wide perceptions of drinking norms despite significant efforts to change this perception. Furthermore, we have observed a leveling-off of the most problematic forms of drinking (heavy episodic drinking) despite continued declines in this type of drinking among the general population. This may explain why alcohol-related harm on campuses has shown little change over the last decade.

Alternative prevention approaches: Targeting individual risk factors

Evidence suggests that college level drinking interventions implemented at the individual level can have meaningful and lasting effects by changing campus culture (Dimeff et al., 1999; Huh et al., 2015; Kulesza et al., 2013; Ray et al., 2014). Additionally, approaches which utilize a partnership between campuses and the broader noncollege community can have important impact (Jernigan et al., 2019). This approach also entails moving away from the embedded social norms-based approach about which more recent research has suggested that the magnitude of effects of these programs has been overestimated (Huh et al., 2015; Mun et al., 2015).

Why norm-based prevention programs have not produced long-term effects on college student drinking is unclear. We propose three primary reasons. First, despite support for social norms as a basis for heavy drinking, the operationalization has remained relatively narrow. Social norms interventions have generally not taken broader theoretical perspectives into account. Often the theoretical model for these sorts of interventions is simply called "social norms theory." However, social norms are an umbrella construct useful in understanding a variety of other theories of normative behavior, but are not identified as a theory of behavior with specific testable hypotheses (aside from the basic hypothesis of higher norms equals more behavior). Second, these interventions have not had much application beyond the exclusive focus on quantity and frequency of drinks consumed. Perhaps these shortcomings may explain the lack of efficacy across these types of interventions as they do not address "nonconsumption" factors known to be linked to consequences. Finally, these interventions do not lead to changes in underlying personal characteristics (e.g., values, beliefs, motivations, etc.) that may influence safe/responsible drinking behaviors. We turn now to four approaches that can address these shortfalls and show promise for future prevention/intervention efforts.

Feedback approaches for alternative targets

An overabundance of research examining the effects of personalized normative feedback on college student drinking has spawned a web of studies trying to apply this approach to a variety of other alcohol-related outcomes. While some of these have shown no effects, others have indicated substantial promise. Below we discuss two feedback approaches that were selected based on either the explicit or implicit ways that they address the previously identified shortfalls.

Drinking motivation feedback

Drinking motivation has long been a stable of alcohol-related research (Cooper, 1994). Originally conceptualized in drive reduction theory (Hull, 1943), motivation to drink has been seen as a primary etiological factor for alcohol use and alcohol-related consequences at both the within- and between-subjects levels (Cook, Newins, Dvorak, & Stevenson, in press; Dvorak, Pearson, & Day, 2014; Littlefield, Vergés, Rosinski, Steinley, & Sher, 2013; O'Donnell et al., 2019; Stevenson et al., 2019). Research has shown that drinking motives mediate the links between many risk factors (e.g., emotion regulation, for alcohol-related outcomes) (Gaher, Simons, Jacobs, Meyer, & Johnson-Jimenez, 2006; Simons, Gaher, Correia, Hansen, & Christopher, 2005). Perhaps most importantly, given the current state of the literature, drinking motives appear to mediate the association between drinking norms and alcohol-related outcomes (Halim, Hasking, & Allen, 2012; Neighbors, Lee, Lewis, Fossos, & Larimer, 2007; Simons, Hahn, Simons, & Murase, 2017). In general, there are four basic motivational factors linked to alcohol-related outcomes: enhancement, coping, socializing, and conformity. These are divided into internal motivations and external motivations that carry either a positive or negative valence (Kuntsche, Knibbe, Gmel, & Engels, 2005). For the purposes of this chapter, we focus on the two most researched motives, the positive and negative valanced internal motives. Positively valanced, internally driven, motivations are often called "enhancement" motives. These sorts of drinking motives tend to be directly associated with consumption and indirectly associated with alcohol-related consequences and harm via consumption. Negatively valanced, internally driven, motivations are often called "coping" motives. These sorts of drinking motives tend to be directly associated with alcohol-related consequences and harm and are often not associated with use at all (Kuntsche et al., 2005). This latter point is particularly relevant, given that use is only one factor linked to alcohol-related consequences and harm. As noted above, targeting frequency and quantity of use may produce some effects on consumption rates, but the effects on more clinically relevant outcomes (e.g., consequences, harm) are less pronounced, if at all. Thus, targeting drinking motivation allows for an approach nested in drive reduction theory to address core underlying factors linked to both consumption and consequences which also extends the normative approach beyond that of frequency/quantity of alcohol consumption.

In the only attempt to date, Blevins and Stephens (2016) developed a protocol to compare a standard normative feedback approach, targeting alcohol consumption quantity/frequency, to a normative feedback approach that included feedback on "drinking to cope" motivation. They randomly assigned $n = 170$ college student drinkers, who endorsed weekly alcohol consumption, to receive one of the two conditions. After the initial assessment and randomization, participants received personalized normative feedback with or without personalized motives feedback. At the 2-month follow-up, both conditions showed reductions in consumption and in alcohol-related consequences, and these effects did not differ significantly across conditions. These effects are in-line with those observed in other personalized normative feedback interventions. Most important, however, were the effects on drinking to cope with anxiety and drinking to cope with depression. In both cases, there were significant declines in drinking to cope motivation in the motives feedback condition, but not in the normative feedback condition. In addition, an examination of indirect effects showed that the effect of the motives feedback intervention on alcohol-related outcomes was mediated via changes in drinking to cope. Given the lack of long-term effects for alcohol quantity/frequency normative feedback approaches, this change in motives and indirect effects through changes in motives may suggest a more sustainable approach that targets a known, and prominent, risk factor for alcohol-related harm. While promising, there are several caveats that must be noted. This study lacked a true control condition; however, testing this approach against the "gold standard" approach gives greater confidence in the results. Further, the follow-up was only 2 months, which may not have been long enough to observe the long-term efficacy of changes in motives. Nonetheless, noteworthy are that both conditions resulted in similar reductions in use and problems, but appeared to achieve reductions via different mechanisms. This may suggest that the effects of changes in motives are still latent and may manifest via long-term efficacy. While this remains to be seen, this approach does appear promising.

Injunctive norms feedback

Thus far, our discussion of norms has focused on a single aspect of drinking norms: descriptive norms. Descriptive drinking norms refer specifically to the base-rates of alcohol consumption. Thus, personalized normative feedback is generally an approach in which individuals report a base-rate of drinks consumed by a typical student on a drinking occasion or over a set drinking time. However, there is a different flavor of drinking norm that has been largely ignored in the drinking intervention literature: injunctive drinking norms. Injunctive norms refer to the perceived acceptability of consumption and not to the base-rate of consumption. In this regard, injunctive norms tap into a core assumption about the social acceptability of alcohol use. Previous research has shown that descriptive and injunctive norms are differentially associated with alcohol consumption and alcohol-related consequences (Neighbors et al., 2007).

To date, there have been four studies that have utilized some form of feedback in an attempt to modify injunctive norms (Barnett, Far, Mauss, & Miller, 1996; Prince & Carey, 2010; Prince, Maisto, Rice, & Carey, 2015; Schroeder & Prentice, 1998). In a combined descriptive and injunctive norms intervention, Barnett et al. (1996) observed a change in both norm types, as well as reduction in reported consumption. However, the intervention did not allow for the disambiguation of descriptive and injunctive norms. In another study, Schroeder and Prentice (1998) found that an intervention, which utilized a discussion about pluralistic ignorance, resulted in a change in descriptive (but not injunctive) norms among male students. Neither of these studies provided detail on intervention content. In a more recent study, Prince and Carey (2010) used a feedback approach to test a mechanism for modifying injunctive norms. The results of this study showed that a very brief message could be used to modify injunctive norms. Interestingly, this approach had a significant impact on descriptive norms as well. This highlights a complicated issue for normative feedback interventions. Specifically, that disambiguating descriptive and injunctive norms may be quite difficult to achieve in the context of prevention/intervention programs.

With this in mind, Prince et al. (2015) developed and tested the first personalized injunctive norms feedback intervention aimed at comparing personalized injunctive norms feedback alone to personalized descriptive norms feedback (the "gold standard" approach), a combined injunctive/descriptive norms approach, and an attention control. The results were quite intriguing. There were no effects of personalized descriptive norms feedback on alcohol use or alcohol consequences, an effect that is inconsistent with larger literature on personalized descriptive norms feedback interventions (Lewis & Neighbors, 2006a), but we have noted that recent research has started to indicate that the effects of these previous interventions may have been exaggerated (Huh et al., 2015; Mun et al., 2015). Prince and colleagues did find that personalized injunctive norms feedback was effective as both a stand-alone intervention and in combination with descriptive norms feedback. Interestingly, the combination approach yielded no better results than injunctive feedback alone. This too is just a single study, but it offers promise. Personalized injunctive norms feedback takes an effective approach, normative feedback, and extends this approach by leveraging personally relevant material (social desirability) as a mechanism for change. Replication of these findings in a larger, more diverse sample will be important for future research.

Changing cognitive factors linked to alcohol use

There are a number of cognitive factors that have been implicated in alcohol use and alcohol-related consequences/harm. These range from drug-specific factors (e.g., implicit attitudes about alcohol, attention bias to alcohol cues, memory networks of alcohol effects, impulse control deficits to alcohol-related cues, etc.) to broader drug irrelevant factors such as response inhibition, cognitive control, and working memory. While the etiological research has shown support for cognitive factors linked to alcohol use and alcohol-related consequences, the intervention research has been mixed. Below we introduce two different approaches to addressing cognitive factors linked to consumption and subsequent harm and discuss both shortcomings with the initial approaches as well as intriguing future approaches.

Cognitive retraining programs

The research on cognition and alcohol-related outcomes has consistently linked cognitive biases toward alcohol to alcohol-related outcomes. For example, implicit attitudes are underlying, subconscious, beliefs about some aspect of alcohol (e.g., acceptability, effects, personal liking/wanting, etc.). Research has shown that automatic cognitive processes are associated with higher alcohol craving (Marhe, Waters, van de Wetering, & Franken, 2013), more consumption (Houben & Wiers, 2006), and more severe alcohol-related consequences (Wiers et al., 2006). The link between automatic cognitive processes and alcohol-related outcomes led to an explosion of research on treatment approaches targeting aspects of cognition that could be *potentially* retrained. This retraining typically takes the form of some sort of cognitive task through repeated exposure over time that has the effect of changing the underlying cognitive and semantic architecture of alcohol-relevant information (Friese, Hofmann, & Wiers, 2011). While there are a number of promising results, there are also a host of failed replications and null effects. Below we discuss three primary approaches which may hold promise.

In 2002, researchers began attempting to modify attention bias to selective stimuli. The first of these studies utilized attention bias modification to retrain attention toward emotionally salient words as a way to reduce attentional avoidance (MacLeod, Rutherford, Campbell, Ebsworthy, & Holker, 2002). This was later adopted by alcohol researchers and adapted to push attention away from alcohol-related stimuli and toward more healthy stimuli (Wiers et al., 2015). This approach is most commonly implemented in a modified dot-probe task in which individuals are presented with an alcohol-related image on one side of a screen and a nonalcohol-related image on the opposing side of the screen. The images are briefly displayed and then replaced by a dot. The subject's task is to respond as quickly as possible to the dot. In the attention modification condition, the dot is always behind the nonalcohol stimuli, resulting in a gradual shift of attention toward the nonalcohol stimuli and away from the alcohol-relevant stimuli (Fig. 3.1, panel A). Early research on attention bias modification suggested this approach may have effects on alcohol-related outcomes (Schoenmakers et al., 2010). However, more recent research among clinical samples has called these effects into question (Boffo et al., 2019). Nonetheless, research with college students indicates there may be some effects on outcomes that could translate to reduced problematic use (Luehring-Jones, Louis, Dennis-Tiwary, & Erblich, 2017).

Building off of earlier research in attention bias modification, Wiers, Eberl, Rinck, Becker, and Lindenmeyer (2010) developed a new form of cognitive retraining that involved modifying implicit approach/avoidance motivation toward alcohol-relevant stimuli. In these tasks, subjects would push a joystick away from them or pull a joystick toward them in a simulated approach/avoidance environment. As images

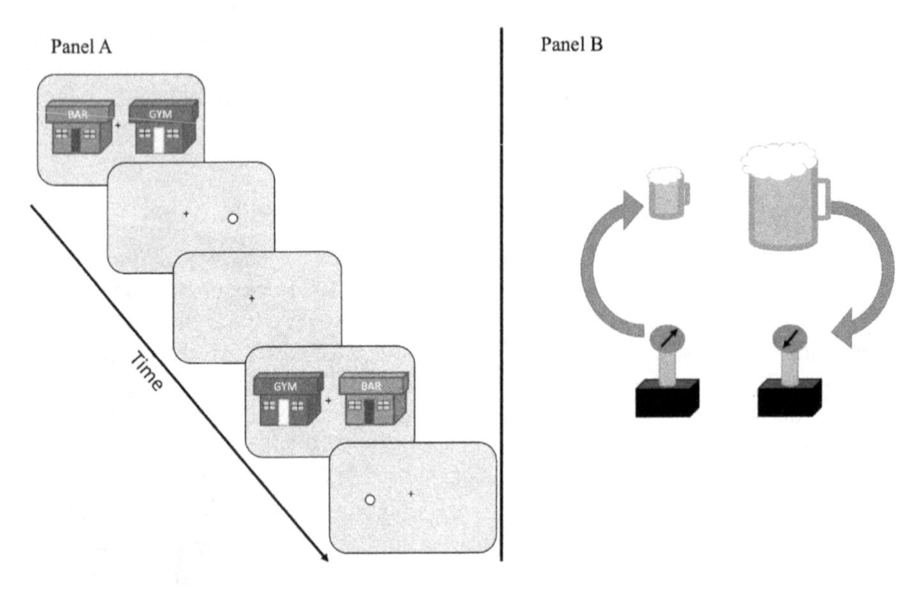

Fig. 3.1 Panel A depicts a dot-probe task example trial sequence of attention modification condition; Panel B depicts a joystick cognitive task example of approach/avoidance (pushing alcohol away, becomes smaller; pulling alcohol toward, becomes larger).
Credit: Robert Dvorak.

appeared on the screen, participants "pushed" away or "pulled" toward them, depending on the shape of the image (landscape vs. portrait). The image would then seem to move toward them on the screen (grow larger) or move away from them on the screen (shrink away) (Fig. 3.1, panel B). Subjects were, unbeknownst to them, trained to push alcohol-related images away (i.e., alcohol images were consistently presented in the "push" format) or trained to pull alcohol-related images toward them (i.e., alcohol images were consistently presented in the "pull" format). This resulted in subsequent changes in alcohol consumption in a pseudo-alcohol taste test performed after the intervention. In a follow-up to this study, Wiers, Rinck, Kordts, Houben, and Strack (2011) utilized this approach in a sample of treatment seeking individuals with an alcohol use disorder. The results suggested changes in implicit drinking motivation consistent with the retraining program. In addition, individuals who completed the training, relative to control, had a significantly lower likelihood of relapse 1 year after treatment. While there is considerable promise for this approach, there is also reason to warrant some restraint. In a recent replication and extension of this approach across two college student samples, Lindgren et al. (2015) found no effects of implicit approach/avoidance modification training. In fact, not only were effects on alcohol-relevant outcomes not observed, there were also no effects on underlying cognitive structures linked to implicit attitudes or motivations. It is possible that the earlier work was effective because it was being done with treatment seeking individuals, perhaps who wanted to change their drinking. In contrast, the latter studies utilized a very different sample. This seems to suggest that this approach may be limited to a specific subset of individuals trying to stop/limit consumption. Regardless, the strong theoretical underpinning coupled with a novel approach that has the potential to be widely disseminated makes this option a promising candidate worthy of additional research.

Changing alcohol expectancy networks

Alcohol expectancies represent a person's explicit and implicit beliefs regarding the subjective effects of alcohol (Aarons et al., 2003). Alcohol expectancies cover a wide variety of topics and are similar in many ways to the concept of alcohol use motives. For example, if an individual believes that alcohol will have sedating effects on anxiety and/or stress (i.e., tension-reduction expectancies), they may also be motivated to drink in order to cope with anxiety and/or stress. Alcohol expectancies are generally categorized across two broad domains: prosocial/antisocial expectancies (e.g., the belief that alcohol will make you more/less sociable and/or more/less obnoxious) and arousing/sedating (e.g., the belief that alcohol will make you feel more/less aroused and/or alcohol will make you feel more/less anxious). Early research attempted to construct hypothetical memory network models and plot the likely activation of alcohol expectancies through these networks. This research suggested that the heaviest/most problematic drinkers tended to have expectancies emphasizing the activation of arousing/sedating effects (Goldman, 1994; Goldman & Darkes, 2004; Magri et al., 2020).

What is perhaps most interesting is that many of the perceived subjective effects of alcohol are not actually true pharmacological effects of alcohol. Instead, many of these expectancies are a result of observational learning, anecdotal beliefs, and placebo effects (Goldman & Darkes, 2004). Consequently, interventions aimed at changing expectancies have generally used an "expectancy challenge" (EC)-based approach, in which individuals consume a nonalcoholic beverage, under the guise that they are consuming alcohol. They report the effects they are experiencing as a result of consuming "alcohol." They are then informed of the ruse. This is followed by a discussion of alcohol expectancies and the role of expectancies in alcohol use/problems (Darkes & Goldman, 1993). Individuals who receive this intervention show changes in the activation of alcohol expectancies in their semantic memory network. These changes activation result in subsequent changes in a number of consumption-relevant outcomes (Fig. 3.2). This *experiential* approach has shown modest effects across an array of settings and has been replicated numerous times throughout the literature (Dunn, Lau, & Cruz, 2000). However, this approach is fairly demanding, requires a very specific environment (typically a lab modified to resemble a bar), can only be conducted with a few individuals at a time, and requires the use of special materials (e.g., nonalcohol products). All of these factors severely limit the broad dissemination requisite of campus-wide alcohol prevention programs. Furthermore, the effects of this approach do not appear to extend much beyond 30 days (though, at least one study has shown effects were maintained up to 3 months; Scott-Sheldon, Carey, Elliott, Garey, & Carey, 2014).

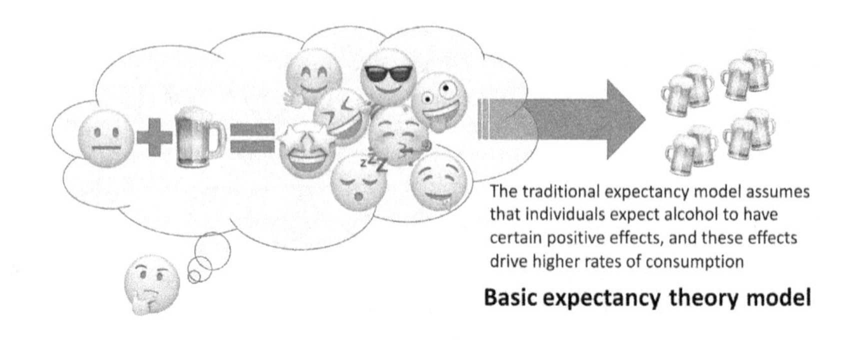

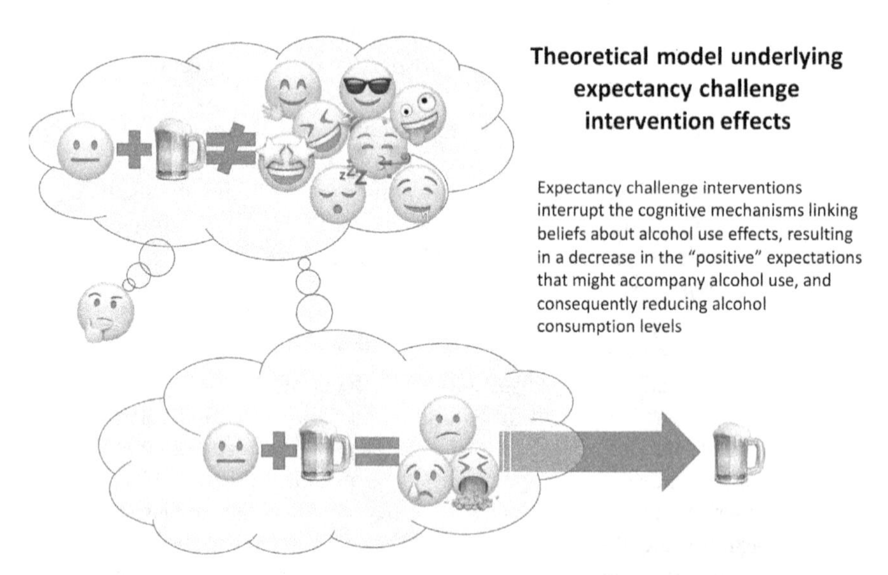

Fig. 3.2 Theoretical model of the expectancy challenge approach.
Credit: Robert Dvorak.

However, recently there have been modifications to this EC which utilize a *non-experiential* approach. The Expectancy Challenge Alcohol Literacy Curriculum (ECALC) has adapted the initial challenge approach into a series of educational modules and tasks that aim to facilitate information processing, maximize understanding, and retention. The research on this approach has been quite promising. Dunn and colleagues showed that ECALC was more effective than an in-person Brief Motivational Intervention at changing every aspect of alcohol expectancy (Dunn, Fried-Somerstein, Flori, Hall, & Dvorak, 2019; Fried & Dunn, 2012). This is especially promising, given that the individuals in ECALC never had their expectancies challenged in any *experiential* setting. Further, there were significant effects on every index of alcohol consumption, and these effects were mediated via changes in alcohol expectancies. While there is considerable promise for this approach, there is limited evidence for the effectiveness beyond 30 days. Furthermore, the evidence to date seems to indicate that this approach may be most effective for consumption-related outcomes. While this does suggest that this approach could yield significant impact on harmful consumption (e.g., heavy episodic drinking), the effects on other forms of alcohol-related consequences and harm are weak, at best. Nonetheless, of the cognitive factors, this approach remains the most promising for widespread implementation with significant public health impacts.

Interventions grounded in behavioral economics

In the last several decades, there has been a surge in the use of basic economic principles as a way to understand human behavior. Starting in the 1960s, researchers began examining models of behavioral choice, trying to understand the mechanisms that lead individuals to prefer short-term, often maladaptive, rewards over long-term adaptive rewards. Indeed, this lies at the heart of college student drinking, in which individuals often find themselves consuming larger amounts of alcohol, in the moment, at the expense of later cost. Examining the ratios of cost and reward has shown consistent links with more problematic patterns of use (Bickel, Mueller, MacKillop, & Yi, 2016; Stein & Madden, 2013). Recently, two different forms of intervention, grounded in behavioral economics theory, have been proposed and tested. Each offers an interesting potential approach to addressing the problem of behavioral choice in the context of college student drinking.

Changing delay discounting and reward value via working memory training

Delay discounting refers to the selection of smaller-sooner rewards over larger-later rewards. For example, individuals are asked "would you prefer to have $5 now or $50 one month from now?" The point at which a person's preference shifts from preferring the larger-later reward marks their discount rate (i.e., the ratio at which they discount future reward in favor of current reward). Individual differences in the delay discounting ratio are a fairly robust predictor of a wide range of maladaptive behaviors (McClure, Ericson, Laibson, Loewenstein, & Cohen, 2007) including problematic alcohol use (Murphy & Garavan, 2011). Furthermore, research has shown that certain aspects of cognitive control, specifically working memory, are correlated with lower delay discounting rates (i.e., a lower discount of future value) (Wesley & Bickel, 2014). This finding is linked to basic "dual process" model theories in which hot (i.e., impulsive) processing can be overcome/overridden by stronger cool (e.g., controlled) processing (Friese et al., 2011; Wiers et al., 2006). A host of research in the area of addiction and substance use supports the general dual process model of addiction. Based on this research, Bickel and colleagues have sought to enhance controlled cognitive processing by increasing general working memory via training tasks (Bickel, Moody, & Quisenberry, 2014; Bickel, Yi, Landes, Hill, & Baxter, 2011). To date, they have shown promising results. In a small sample of individuals,

Bickel et al. (2011) showed that a computerized working memory training task could be used to decrease delay discounting. Extending this work, Houben, Wiers, and Jansen (2011) found that working memory training resulted in decreased alcohol use and that this was maintained up to 1 month after training. Furthermore, this effect was strongest among individuals with more positive implicit drinking attitudes. Despite the early success of such approaches, more recent attempts to replicate these effects have fallen short (Wanmaker et al., 2018). However, this may be due to varying training effects as a function of initial delay discounting rates (Bickel, Quisenberry, Moody, & Wilson, 2015). Thus, while promising, future research is needed to identify the conditions under which this approach may be effective vs. those when it would not be warranted.

Increasing substance-free reinforcement

The interest in interventions that utilize basic behavioral economics has also spread from the traditional concept of monetary reward discounting to the environmental level. The basic premise of this approach assumes that substance use environments tend to be more reinforcing (e.g., opportunities for social interaction, positive/enhancing effects of the drug, etc.). Indeed, research has shown that substance use is higher among individuals who have environments devoid of substance-free reinforcement (Murphy, Correia, & Dennhardt, 2013). In a novel approach, some researchers have begun working to help individuals structure their environments to include positive/rewarding experiences in the absence of substance use (Murphy et al., 2019, 2012; Yurasek, Dennhardt, & Murphy, 2015). These interventions are generally added as a supplement to ongoing interventions and involve a session and/or discussion to identify substance-free activities a person could engage in and increase the frequency of those activities. While this approach does not necessarily outperform treatment as usual (Murphy et al., 2019), it does appear to reduce delay discounting while also increasing use of safe drinking strategies (Dennhardt, Yurasek, & Murphy, 2015). Thus, adding this brief adjunct may result in prolonged effects of other interventions by enhancing known protective factors linked to alcohol use and alcohol-related consequences.

Increasing safe drinking strategies

While many interventions place the focus on reducing alcohol consumption rates, there have also been a handful of studies that take a harm reduction approach. Harm reduction in the context of alcohol was pioneered by Marlatt and Witkiewitz (2002). The goal of the approach was to reduce problematic behavior rather than a strict focus on consumption. This approach was born out of the need to offer a form of intervention that did not require strict drinking abstinence. Given that many college students are drinking in excess, but do not have an alcohol use disorder, an approach that emphasizes reducing harm while still allowing them to engage in this social activity is appealing. Studies taking this approach have targeted a series of activities collectively known as "Protective Behavioral Strategies" for alcohol use. These strategies emphasize behaviors like avoiding shots of liquor, having a designated driver, setting limits on drinks and/or time drinking, and avoiding drinking games, just to name a few. Research has consistently shown an inverse association between the use of protective strategies and both alcohol consumption and alcohol-related consequences (Pearson, 2013). Below we present two different strategies that have shown promise for increasing safe drinking by targeting behaviors that protect against alcohol-related consequences.

Safe drinking strategies feedback

Despite a consistent negative relationship between protective behavioral strategy use and alcohol consumption/problems in both cross-sectional and longitudinal research (Barnett, Murphy, Colby, & Monti, 2007; Dvorak, Kramer, Stevenson, Sargent, & Kilwein, 2017; Larimer et al., 2007; Martens et al., 2007), the evidence for stand-alone protective behavioral strategy interventions has shown mixed support. For example, Martens, Smith, and Murphy (2013) implemented an intervention involving both personalized normative feedback for alcohol use and protective behavioral strategy feedback, meant to increase use of protective behavioral strategies by highlighting normative discrepancies in protective behavioral strategy use. Across three different consumption measures, personalized normative feedback (Cohen's $ds = -0.77$ to -1.16) was more effective at reducing alcohol consumption than protective behavioral strategy feedback (Cohen's $ds = -0.22$ to -0.68)—though, protective behavioral strategy feedback did result in a significant reduction in two of these outcomes. However, these findings must be taken with a grain of salt, given the results of the large-scale meta-analysis conducted in Project INTEGRATE. It is unclear if the personalized normative feedback effects on drinking are true effects, if these effects are actually as robust as indicated, if these effects had any lasting impact, and if the changes in drinking resulted in lasting changes in alcohol-related harms. More intriguing, however, was that protective behavioral strategy feedback showed the only significant increase in protective behavioral strategy use at the 6-month follow-up (personalized normative feedback: Cohen's $d = 0.18$; protective behavioral strategy feedback: Cohen's $d = 0.54$), as well the most robust (and again, only statistically significant) decrease in alcohol-related consequences at the 6-month follow-up (personalized normative feedback: Cohen's $d = -0.32$; protective behavioral strategy feedback: Cohen's $d = -0.64$). Recently, LaBrie, Napper, Grimaldi, Kenney, and Lac (2015) found evidence that a stand-alone protective behavioral strategy intervention, which utilized skills training and personal protective behavioral strategy feedback, increased protective behavioral strategy use. Further, there was a reduction in alcohol consumption, which mediated the association between group and alcohol use. However, both the personal protective behavioral strategy feedback and control group showed reductions in alcohol use, and while there appeared to be mediated effects via protective behavioral strategy use on alcohol use, simple group differences on indices of alcohol use were not observed (LaBrie et al., 2015). Similarly, Sugarman and Carey (2009) found that a simple intervention, instructing students to use more protective behavioral strategies, resulted in greater protective behavioral strategy use, but no change in alcohol use. In contrast, Kenney, Napper, LaBrie, and Martens (2014) found that a stand-alone PBS intervention, delivered in a group format to first-year college women, produced increased protective behavioral strategy use and subsequent decreases in both heavy alcohol use and alcohol-related consequences. In addition, use of protective behavioral strategies mediated intervention effects on alcohol consequences postintervention, at least among participants with heightened levels of anxiety (Kenney et al., 2014). Thus, while stand-alone protective behavioral strategy interventions appear to increase protective behavioral strategy use consistently, the effects on alcohol consumption are inconsistent. However, they may be more effective at reducing the most relevant outcome of alcohol-related consequences (harms).

Targeted health messaging using deviance regulation theory

A few recent studies have been harnessing a relatively new social psychology theory to increase safe drinking. This theory, known as Deviance Regulation, posits that individuals will strive to "fit in" to the larger peer group if they believe that standing out will affect their social status in a negative way. Alternatively, they will strive to "stand out" from the larger peer group if they believe that standing out will positively affect their social status. This approach allows for targeted messaging that interacts with perceived norms. For example, if a behavior is seen as uncommon, a positive message about the behavior (or individuals who engage in the behavior) should push a person to engage in that behavior as it allows them to stand out in positive ways among a small group of people. In contrast, if a behavior is seen as common, a negative message about NOT engaging in the behavior (or about individuals that do not engage in the behavior) should push people to engage in the behavior to ensure they DO NOT stand out in negative ways. The basic predictive model of the theory is shown in Fig. 3.3. Using this approach, Dvorak and colleagues have conducted a series of studies aimed at increasing the use of protective behavioral strategies among college student drinkers (Dvorak et al., 2017; Dvorak, Kramer, & Stevenson, 2018; Dvorak, Pearson, Neighbors, Martens, & Stevenson, 2016; Dvorak, Pearson, Neighbors, Martens, & Williams, 2015; Dvorak, Raeder, et al., 2018; Dvorak, Troop-Gordon, et al., 2018; Sargent et al., 2018). Thus far, the results are quite promising. They have shown that a positive message about individuals who use protective behavioral strategies results in more use of these strategies if individuals believe that the base-rate of these safe drinking behaviors is low among their peers; thus, allowing them to stand out in positive ways. In contrast, if an individual believes the base-rate of safe drinking strategies is high among their peers, a message emphasizing the negative qualities of people that DO NOT use these strategies results in an increase in individual strategy use. For both of these groups, the use of more protective behavioral strategies is linked to lower alcohol consumption and fewer alcohol-related consequences. This approach has, thus far, been supported in two samples of typical college student drinkers (Dvorak et al., 2015; Dvorak, Raeder, et al., 2018), a sample of incoming college freshman (Leary & Dvorak, 2019), and two different samples of college students going on Spring Break (Dvorak et al., 2017; Sargent et al., 2018). However, all

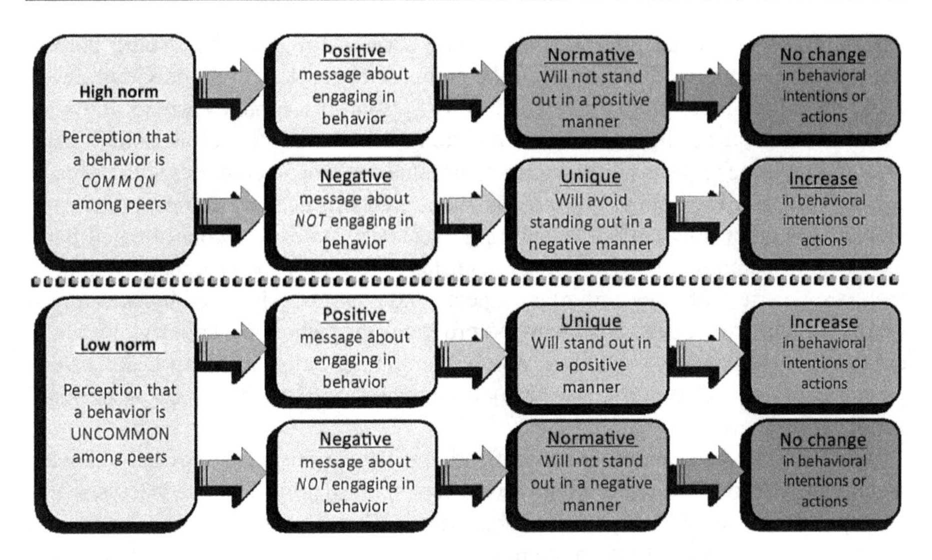

Fig. 3.3 Theoretical model of the deviance regulation theory approach.
Credit: Robert Dvorak.

of these studies suffer from limitations, the most glaring of which is the lack of long-term efficacy. Regardless, this approach shows promise and certainly warrants additional research.

Discussion and recommendations

Despite seeing a number of promising approaches implemented on virtually every campus in the country, alcohol use among college students remains a significant public health issue. As discussed, this is likely due to the fact that broad-based social norms prevention programs have not produced changes in the campus culture that we believed they would. We have not seen reductions in the most problematic forms of alcohol use on campus, nor have we seen meaningful changes in alcohol-related harm on campus over the last three decades. We propose that this lack of meaningful change may be the result of the nationwide adoption of ineffective campus-level prevention efforts. Current campus-level efforts combine some form of awareness education (already shown to be ineffective) coupled with mixed messages about campus drinking policies and ineffective social norms interventions that (a) are limited in scope, (b) are not theoretically grounded, and (c) do not affect variables known to be of importance/relevance to the individual. To address this issue, we present emerging research aimed at correcting these shortcomings. Based on this research, we recommend the following adaptions to the campus-level of the *3-in-1 framework* prevention approach.

First, we suggest that each campus establish a campus-wide drug and alcohol program that serves as a "go-to" source for all substance-related issues. There are currently several model programs that could serve as guides. This would help to ensure a single intermediary between the community-level and the individual-level. Furthermore, a single program or council for substance use policy would allow for the establishment of clear messaging regarding the role of alcohol and other drugs on campus. Some of the tasks for such a committee could be interactions with the campus police regarding policy, discussions about alcohol sales on campus, policies on alcohol and other drugs in dorms/sporting events/etc., sales of alcohol-related paraphernalia on campus, and advertising of substance-free events on campus. Of course, there are a variety of other important aspects that could be addressed by such a program. While there are certainly countless programs throughout the country, we would point to the President's Council on Alcohol and Other Drugs at North Dakota State University (www.ndsu.edu/alcoholinfo) as a model for universities seeking to establish new programs.

Second, we would encourage a modification to the current practice of campus-wide alcohol prevention. The use of broad-based personalized normative feedback remains an important aspect of the campus-level approach. This method provides a framework for the personalization of information at the individual-level as a means of changing perceptions at the campus-level. We would argue that this approach should also be expanded. Based on the current research, it appears that identifying risk factors at the individual level (like normative misperceptions) can be used to tailor strategies that would be most effective for individuals but implemented at the campus-level. For example, during the first week of classes, most universities require an initial alcohol education program. Rather than using this as an opportunity to provide a broad strokes approach to changing normative perceptions of drinking norms, which clearly has no effects, colleges could use this opportunity to identify specific risk factors that are amenable to change (e.g., normative misperceptions, high delay discounting rates, low use of safe drinking strategies, alcohol expectancy misperceptions, positive attitudes about alcohol, etc.) via personalized assessment of risk. Following this initial assessment, students could receive individually targeted prevention approaches designed to modify identified risk factors. One feature of every emerging prevention strategy listed above is the ability to leverage technology in some fashion to implement the program, thus making them appropriate for wide dissemination.

Third, we would recommend, where feasible, the establishment of an office for alcohol and drug treatment. This could be a combined office that provides (a) initial prevention for *at risk* students, (b) personalized counseling services for individuals with substance-related issues, and (c) support for individuals in recovery. This would ensure stability at the individual-level and allow for a triage of services based on risk. Further, this strategy could be used as a compliment for campus-level interventions that require more intensive approaches for *at risk* students. For example, if an individual is identified for intensive cognitive retraining, this office could serve as a point of contact for such services. Many universities already have alcohol and drug treatment programs as part of mandated state requirements. In many cases, these offices could be augmented or expanded to address this area of need.

Finally, we would recommend a more nuanced tracking program that examines the efficacy at each level of the individualized *3-in-1 framework* approach. This would mean longitudinal follow-ups of students who receive individualized prevention, as well as yearly assessment and intervention strategies based on identified risk factors. This would provide evidence for implemented approaches and allow for the removal of approaches as they are deemed ineffective/insufficient. This strategy could include a series of short assessments, perhaps tied to other aspects of student requirements, which would allow for the tracking of individual alcohol-related outcomes and broader campus level perceptions. This information could then be combined with conduct reports from law enforcement and de-identified treatment data from campus counseling centers.

Conclusion

The current chapter presented: (1) the current state of affairs regarding college student alcohol use and alcohol-related consequences, (2) the three levels of factors that promote alcohol-related issues on campus, (3) the *3-in-1 framework* approach to addressing alcohol-related consequences outlined by the NIAAA Task Force on college student drinking, (4) efforts thus far in addressing alcohol-related issues on campus, (5) shortcomings of the approaches to reduce alcohol-related issues on campus, and (6) promising interventions to address alcohol-related issues among college student drinkers. We have identified shortcomings in the implementation of efforts to reduce college student drinking and provide potential solutions to these programs through a personalized prevention program implemented at the campus-level. Finally, we offer a broad set of recommendations for implementing such a program.

References

Aarons, G. A., Goldman, M. S., Greenbaum, P. E., & Coovert, M. D. (2003). Alcohol expectancies: Integrating cognitive science and psychometric approaches. *Addictive Behaviors*, *28*(5), 947–961. https://doi.org/10.1016/S0306-4603(01)00282-9.

Baer, J. S. (2002). Student factors: Understanding individual variation in college drinking. *Journal of Studies on Alcohol*, 40–53. https://doi.org/10.15288/jsas.2002.s14.40.

Barnett, L. A., Far, J. M., Mauss, A. L., & Miller, J. A. (1996). Changing perceptions of peer norms as a drinking reduction program for college students. *Journal of Alcohol and Drug Education*, *41*(2), 39–62.

Barnett, N. P., Murphy, J. G., Colby, S. M., & Monti, P. M. (2007). Efficacy of counselor vs. computer-delivered intervention with mandated college students. *Addictive Behaviors*, *32*(11), 2529–2548. https://doi.org/10.1016/j.addbeh.2007.06.017.

Bickel, W. K., Moody, L., & Quisenberry, A. (2014). Computerized working-memory training as a candidate adjunctive treatment for addiction. *Alcohol Research: Current Reviews*, *36*(1), 123–126.

Bickel, W. K., Mueller, E. T., MacKillop, J., & Yi, R. (2016). Behavioral-economic and neuroeconomic perspectives on addiction. In K. J. Sher (Ed.), *Vol. 1. The Oxford handbook of substance use and substance use disorders* (pp. 422–446): Oxford University Press.

Bickel, W. K., Quisenberry, A. J., Moody, L., & Wilson, A. G. (2015). Therapeutic opportunities for self-control repair in addiction and related disorders: Change and the limits of

change in trans-disease processes. *Clinical Psychological Science*, *3*(1), 140–153. https://doi.org/10.1177/2167702614541260.
Bickel, W. K., Yi, R., Landes, R. D., Hill, P. F., & Baxter, C. (2011). Remember the future: Working memory training decreases delay discounting among stimulant addicts. *Biological Psychiatry*, *69*(3), 260–265. https://doi.org/10.1016/j.biopsych.2010.08.017.
Blevins, C. E., & Stephens, R. S. (2016). The impact of motives-related feedback on drinking to cope among college students. *Addictive Behaviors*, *58*, 68–73. https://doi.org/10.1016/j.addbeh.2016.02.024.
Boffo, M., Zerhouni, O., Gronau, Q. F., van Beek, R. J. J., Nikolaou, K., Marsman, M., et al. (2019). Cognitive bias modification for behavior change in alcohol and smoking addiction: Bayesian meta-analysis of individual participant data. *Neuropsychology Review*, *29*(1), 52–78. https://doi.org/10.1007/s11065-018-9386-4.
Borsari, B., & Carey, K. B. (2001). Peer influences on college drinking: A review of the research. *Journal of Substance Abuse*, *13*(4), 391–424. https://doi.org/10.1016/S0899-3289(01)00098-0.
Carey, K. B., Scott-Sheldon, L. A. J., Carey, M. P., & DeMartini, K. S. (2007). Individual-level interventions to reduce college student drinking: A meta-analytic review. *Addictive Behaviors*, *32*(11), 2469–2494. https://doi.org/10.1016/j.addbeh.2007.05.004.
Cook, P. J., & Moore, M. J. (1993). Violence reduction through restrictions on alcohol availability. *Alcohol Health & Research World*, *17*(2), 151–156.
Cook, M., Newins, A. R., Dvorak, R. D., & Stevenson, B. L. (in press). What about this time? Within- and between-person associations between drinking motives and alcohol outcomes. *Experimental and Clinical Psychopharmacology*. https://doi.org/10.1037/pha0000332.
Cooper, L. M. (1994). Motivations for alcohol use among adolescents: Development and validation of a four-factor model. *Psychological Assessment*, *6*(2), 117–128. https://doi.org/10.1037/1040-3590.6.2.117.
Core Institute. (2017). *Core alcohol and drug survey 2006–2016*. https://core.siu.edu/results/.
Cox, M. J., DiBello, A. M., Meisel, M. K., Ott, M. Q., Kenney, S. R., Clark, M. A., et al. (2019). Do misperceptions of peer drinking influence personal drinking behavior? Results from a complete social network of first-year college students. *Psychology of Addictive Behaviors*, *33*(3), 297–303. https://doi.org/10.1037/adb0000455.
Darkes, J., & Goldman, M. S. (1993). Expectancy challenge and drinking reduction: Experimental evidence for a mediational process. *Journal of Consulting and Clinical Psychology*, *61*(2), 344–353. https://doi.org/10.1037/0022-006X.61.2.344.
Dee, T. S., & Evans, W. N. (2001). Teens and traffic safety. In J. Gruber (Ed.), *Risky behavior among youths: An economic analysis* (pp. 121–165): University of Chicago Press Books.
Dennhardt, A. A., Yurasek, A. M., & Murphy, J. G. (2015). Change in delay discounting and substance reward value following a brief alcohol and drug use intervention. *Journal of the Experimental Analysis of Behavior*, *103*(1), 125–140. https://doi.org/10.1002/jeab.121.
Dimeff, L. A., Baer, J. S., Kivlahan, D. R., & Marlatt, G. A. (1999). *Brief alcohol screening and intervention for college students (BASICS): A harm reduction approach*. New York: Guilford Press.
Dunn, M. E., Fried-Somerstein, A., Flori, J. N., Hall, T. V., & Dvorak, R. D. (2019). Reducing alcohol use in mandated college students: A comparison of a brief motivational intervention (BMI) and the expectancy challenge alcohol literacy curriculum (ECALC). *Experimental and Clinical Psychopharmacology*. https://doi.org/10.1037/pha0000290.
Dunn, M. E., Lau, H. C., & Cruz, I. Y. (2000). Changes in activation of alcohol expectancies in memory in relation to changes in alcohol use after participation in an expectancy challenge program. Experimental and Clinical Psychopharmacology, 8(4), 566–575. doi: 10.0.4.13/1064-1297.8.4.566

Dvorak, R. D., Kramer, M. P., & Stevenson, B. L. (2018). An initial examination of the effects of deviance regulation theory on normative perceptions. *Journal of Substance Use*, *23*(6), 567–573. https://doi.org/10.1080/14659891.2018.1459900.

Dvorak, R. D., Kramer, M. P., Stevenson, B. L., Sargent, E. M., & Kilwein, T. M. (2017). An application of deviance regulation theory to reduce alcohol-related problems among college women during spring break. *Psychology of Addictive Behaviors*, *31*, 295–306. https://doi.org/10.1037/adb0000258.

Dvorak, R. D., Pearson, M. R., & Day, A. M. (2014). Ecological momentary assessment of acute alcohol use disorder symptoms: Associations with mood, motives, and use on planned drinking days. *Experimental and Clinical Psychopharmacology*, *22*(4), 285–297. https://doi.org/10.1037/a0037157.

Dvorak, R. D., Pearson, M. R., Neighbors, C., Martens, M., & Stevenson, B. L. (2016). A road paved with safe intentions: Increasing intentions to use alcohol protective behavioral strategies via deviance regulation theory. *Health Psychology*, *35*(6), 304–313. https://doi.org/10.1037/hea0000327.

Dvorak, R. D., Pearson, M. R., Neighbors, C., Martens, M. P., & Williams, T. J. (2015). Fitting in and standing out: Increasing the use of alcohol protective behavioral strategies with a deviance regulation intervention. *Journal of Consulting and Clinical Psychology*, *83*(3), 482–493. https://doi.org/10.1037/a0038902.

Dvorak, R. D., Pearson, M. R., Sargent, E. M., Stevenson, B. L., & Mfon, A. M. (2016). Daily associations between emotional functioning and alcohol involvement: Moderating effects of response inhibition and gender. *Drug and Alcohol Dependence*, *163*(Suppl. 1), S46–S53. https://doi.org/10.1016/j.drugalcdep.2015.09.034.

Dvorak, R. D., Raeder, C., Kramer, M., Sargent, E. M., Stevenson, B. L., & Helmy, M. (2018). Using deviance regulation theory to target marijuana use intentions. *Experimental and Clinical Psychopharmacology*, *26*(1), 29–35. https://doi.org/10.1037/pha0000159.

Dvorak, R. D., Simons, J. S., & Wray, T. B. (2011). Alcohol use and problem severity: Associations with dual systems of self-control. *Journal of Studies on Alcohol and Drugs*, *72*(4), 678–684. https://doi.org/10.15288/jsad.2011.72.678.

Dvorak, R. D., Troop-Gordon, W., Stevenson, B. L., Kramer, M., Wilborn, D., & Leary, A. (2018). A randomized control trial of a deviance regulation theory intervention to increase alcohol protective strategies. *Journal of Consulting and Clinical Psychology*, *86*(12), 1061–1075. https://doi.org/10.1037/ccp0000347.

Fried, A. B., & Dunn, M. E. (2012). The expectancy challenge alcohol literacy curriculum (ECALC): A single session group intervention to reduce alcohol use. *Psychology of Addictive Behaviors*, *26*(3), 615–620. https://doi.org/10.1037/a0027585.

Friese, M., Hofmann, W., & Wiers, R. W. (2011). On taming horses and strengthening riders: Recent developments in research on interventions to improve self-control in health behaviors. *Self and Identity*, *10*(3), 336–351. https://doi.org/10.1080/15298868.2010.536417.

Gaher, R. M., Simons, J. S., Jacobs, G. A., Meyer, D., & Johnson-Jimenez, E. (2006). Coping motives and trait negative affect: Testing mediation and moderation models of alcohol problems among American red Cross disaster workers who responded to the September 11, 2001 terrorist attacks. *Addictive Behaviors*, *31*(8), 1319–1330. https://doi.org/10.1016/j.addbeh.2005.10.006.

Gliksman, L., McKenzie, D., Single, E., Douglas, R., Brunet, S., & Moffatt, K. (1993). The role of alcohol providers in prevention: An evaluation of a server intervention programme. *Addiction*, *88*(9), 1195–1203. https://doi.org/10.1111/j.1360-0443.1993.tb02142.x.

Goldman, M. S. (1994). The alcohol expectancy concept: Applications to assessment, prevention, and treatment of alcohol abuse. *Applied & Preventive Psychology*, *3*(3), 131–144. https://doi.org/10.1016/S0962-1849(05)80066-6.

Goldman, M. S., & Darkes, J. (2004). Alcohol expectancy multiaxial assessment: A memory network-based approach. *Psychological Assessment*, *16*(1), 4–15. https://doi.org/10.1037/1040-3590.16.1.4.

Halim, A., Hasking, P., & Allen, F. (2012). The role of social drinking motives in the relationship between social norms and alcohol consumption. *Addictive Behaviors*, *37*(12), 1335–1341. https://doi.org/10.1016/j.addbeh.2012.07.004.

Hingson, R. W. (2010a). "Implementation of NIAAA college drinking task force recommendations: How are colleges doing 6 years later?": Commentary on Nelson, Toomey, Lenk, et al. (2010). *Alcoholism: Clinical and Experimental Research*, *34*(10), 1694–1698. https://doi.org/10.1111/j.1530-0277.2010.01315.x.

Hingson, R. W. (2010b). Magnitude and prevention of college drinking and related problems. *Alcohol Research & Health*, *33*(1–2), 45–54. PMC3887494 23579935.

Hingson, R., Heeren, T., & Winter, M. (1994). Lower legal blood alcohol limits for young drivers. *Public Health Reports*, *109*(6), 738.

Hingson, R., Heeren, T., & Winter, M. (1996). Lowering state legal blood alcohol limits to 0.08%: The effect on fatal motor vehicle crashes. *American Journal of Public Health*, *86*(9), 1297–1299. https://doi.org/10.2105/AJPH.86.9.1297.

Hingson, R., Heeren, T., & Winter, M. (2000). Effects of recent 0.08% legal blood alcohol limits on fatal crash involvement. *Injury Prevention*, *6*(2), 109–114. https://doi.org/10.1136/ip.6.2.109.

Hingson, R. W., Heeren, T., Winter, M., & Wechsler, H. (2005). Magnitude of alcohol-related mortality and morbidity among U.S. college students ages 18–24: Changes from 1998 to 2001. *Annual Review of Public Health*, *26*, 259–279. https://doi.org/10.1146/annurev.publhealth.26.021304.144652.

Hingson, R. W., Zha, W., & Weitzman, E. R. (2009). Magnitude of and trends in alcohol-related mortality and morbidity among U.S. college students ages 18–24, 1998–2005. *Journal of Studies on Alcohol and Drugs. Supplement*, (16), 12–20. https://doi.org/10.15288/jsads.2009.s16.12.

Houben, K., & Wiers, R. W. (2006). Assessing implicit alcohol associations with the implicit association test: Fact or artifact? *Addictive Behaviors*, *31*(8), 1346–1362. https://doi.org/10.1016/j.addbeh.2005.10.009.

Houben, K., Wiers, R. W., & Jansen, A. (2011). Getting a grip on drinking behavior: Training working memory to reduce alcohol abuse. *Psychological Science*, *22*(7), 968–975. https://doi.org/10.1177/0956797611412392.

Huh, D., Mun, E., Larimer, M. E., White, H. R., Ray, A. E., Rhew, I. C., et al. (2015). Brief motivational interventions for college student drinking may not be as powerful as we think: An individual participant-level data meta-analysis. *Alcoholism: Clinical and Experimental Research*, *39*(5), 919–931. https://doi.org/10.1111/acer.12714.

Hull, C. L. (1943). *Principles of behavior: An introduction to behavior theory*. Oxford, England: Appleton-Century.

Jernigan, D. H., Shields, K., Mitchell, M., & Arria, A. M. (2019). Assessing campus alcohol policies: Measuring accessibility, clarity, and effectiveness. *Alcoholism: Clinical and Experimental Research*, *43*(5), 1007–1015. https://doi.org/10.1111/acer.14017.

Johnson, M. B., Voas, R. B., Kelley-Baker, T., & Furr-Holden, C. D. M. (2008). The consequences of providing drinkers with blood alcohol concentration information on assessments of alcohol impairment and drunk-driving risk. *Journal of Studies on Alcohol and Drugs*, *69*(4), 539–549. https://doi.org/10.15288/jsad.2008.69.539.

Jones-Webb, R., Short, B., Wagenaar, A., Toomey, T., Murray, D., Wolfson, M., et al. (1997). Environmental predictors of drinking and drinking-related problems in young adults. *Journal of Drug Education*, *27*(1), 67–82. https://doi.org/10.2190/RJYG-D5C3-H2F0-GJ0L.

Kenney, S. R., Napper, L. E., LaBrie, J. W., & Martens, M. P. (2014). Examining the efficacy of a brief group protective behavioral strategies skills training alcohol intervention with college women. *Psychology of Addictive Behaviors*, *28*(4), 1041–1051. https://doi.org/10.1037/a0038173.

Knight, J. R., Wechsler, H., Kuo, M., Seibring, M., Weitzman, E. R., & Schuckit, M. A. (2002). Alcohol abuse and dependence among U.S. college students. *Journal of Studies on Alcohol*, *63*(3), 263–270. https://doi.org/10.15288/jsa.2002.63.263.

Koss, M. P., Dinero, T. E., Seibel, C. A., & Cox, S. L. (1988). Stranger and acquaintance rape: Are there differences in the victim's experience? *Psychology of Women Quarterly*, *12*(1), 1–24. https://doi.org/10.1111/j.1471-6402.1988.tb00924.x.

Kulesza, M., McVay, M. A., Larimer, M. E., & Copeland, A. L. (2013). A randomized clinical trial comparing the efficacy of two active conditions of a brief intervention for heavy college drinkers. *Addictive Behaviors*, *38*(4), 2094–2101. https://doi.org/10.1016/j.addbeh.2013.01.008.

Kuntsche, E., Knibbe, R., Gmel, G., & Engels, R. (2005). Why do young people drink? A review of drinking motives. *Clinical Psychology Review*, *25*(7), 841–861. https://doi.org/10.1016/j.cpr.2005.06.002.

LaBrie, J. W., Lewis, M. A., Atkins, D. C., Neighbors, C., Zheng, C., et al. (2013). RCT of web-based personalized normative feedback for college drinking prevention: Are typical student norms good enough? *Journal of Consulting and Clinical Psychology*, *81*(6), 1074–1086. https://doi.org/10.1037/a0034087.

LaBrie, J. W., Napper, L. E., Grimaldi, E. M., Kenney, S. R., & Lac, A. (2015). The efficacy of a standalone protective behavioral strategies intervention for students accessing mental health services. *Prevention Science*, *16*(5), 663–673. https://doi.org/10.1007/s11121-015-0549-8.

Lang, E., Stockwell, T., Rydon, P., & Beel, A. (1998). Can training bar staff in responsible serving practices reduce alcohol-related harm? *Drug and Alcohol Review*, *17*(1), 39–50. https://doi.org/10.1080/09595239800187581.

Larimer, M. E., & Cronce, J. M. (2002). Identification, prevention, and treatment: A review of individual-focused strategies to reduce problematic alcohol consumption by college students. *Journal of Studies on Alcohol*, *Suppl. 14*, 148–163. https://doi.org/10.15288/jsas.2002.s14.148.

Larimer, M. E., Lee, C. M., Kilmer, J. R., Fabiano, P. M., Stark, C. B., et al. (2007). Personalized mailed feedback for college drinking prevention: A randomized clinical trial. *Journal of Consulting and Clinical Psychology*, *75*(2), 285–293. https://doi.org/10.1037/0022-006X.75.2.285.

Larimer, M. E., Neighbors, C., LaBrie, J., Atkins, D. C., Lewis, M. A., et al. (2011). Descriptive drinking norms: For whom does reference group matter? *Journal of Studies on Alcohol and Drugs*, *72*(5), 833–843. https://doi.org/10.15288/jsad.2011.72.833.

Lawyer, S., Resnick, H., Bakanic, V., Burkett, T., & Kilpatrick, D. (2010). Forcible, drug-facilitated, and incapacitated rape and sexual assault among undergraduate women. *Journal of American College Health*, *58*(5), 453–460. https://doi.org/10.1080/07448480903540515.

Leary, A. V., & Dvorak, R. D. (2019). A deviance regulation theory intervention to reduce alcohol-related problems among first-year college students. In *Poster Presented at the International Convention of the Psychological Sciences (ICPS), Paris, France*.

Lenk, K. M., Toomey, T. L., Wagenaar, A. C., Bosma, L. M., & Vessey, J. (2002). Can neighborhood associations be allies in health policy efforts? Political activity among neighborhood associations. *Journal of Community Psychology*, *30*(1), 57–68. https://doi.org/10.1002/jcop.1050.

Lewis, M. A., & Neighbors, C. (2006a). Social norms approaches using descriptive drinking norms education: A review of the research on personalized normative feedback. *Journal of American College Health*, *54*(4), 213–218. https://doi.org/10.3200/JACH.54.4.213-218.

Lewis, M. A., & Neighbors, C. (2006b). Who is the typical college student? Implications for personalized normative feedback interventions. *Addictive Behaviors*, *31*(11), 2120–2126. https://doi.org/10.1016/j.addbeh.2006.01.011.

Linden-Carmichael, A. N., & Lanza, S. T. (2018). Drinking patterns of college- and non-college-attending young adults: Is high-intensity drinking only a college phenomenon? *Substance Use & Misuse*, *53*(13), 2157–2164. https://doi.org/10.1080/10826084.2018.1461224.

Lindgren, K. P., Wiers, R. W., Teachman, B. A., Gasser, M. L., Westgate, E. C., et al. (2015). Attempted training of alcohol approach and drinking identity associations in US undergraduate drinkers: Null results from two studies. *PLoS ONE*, *10*(8). https://doi.org/10.1371/journal.pone.0134642.

Littlefield, A. K., Vergés, A., Rosinski, J. M., Steinley, D., & Sher, K. J. (2013). Motivational typologies of drinkers: Do enhancement and coping drinkers form two distinct groups? *Addiction*, *108*(3), 497–503. https://doi.org/10.1111/j.1360-0443.2012.04090.x.

Luehring-Jones, P., Louis, C., Dennis-Tiwary, T. A., & Erblich, J. (2017). A single session of attentional bias modification reduces alcohol craving and implicit measures of alcohol bias in young adult drinkers. *Alcoholism: Clinical and Experimental Research*, *41*(12), 2207–2216. https://doi.org/10.1111/acer.13520.

MacLeod, C., Rutherford, E., Campbell, L., Ebsworthy, G., & Holker, L. (2002). Selective attention and emotional vulnerability: Assessing the causal basis of their association through the experimental manipulation of attentional bias. *Journal of Abnormal Psychology*, *111*(1), 107–123. https://doi.org/10.1037/0021-843X.111.1.107.

Magri, T. D., Leary, A. V., De Leon, A. N., Flori, J. N., Crisafulli, M. J., Dunn, M. E., & Dvorak, R. D. (2020). Organization and activation of alcohol expectancies across empirically derived profiles of college student drinkers. *Experimental and Clinical Psychopharmacology*. https://doi.org/10.1037/pha0000346.

Marhe, R., Waters, A. J., van de Wetering, B. J. M., & Franken, I. H. A. (2013). Implicit and explicit drug-related cognitions during detoxification treatment are associated with drug relapse: An ecological momentary assessment study. *Journal of Consulting and Clinical Psychology*, *81*(1), 1–12. https://doi.org/10.1037/a0030754.

Marlatt, G. A., & Witkiewitz, K. (2002). Harm reduction approaches to alcohol use: Health promotion, prevention, and treatment. *Addictive Behaviors*, *27*(6), 867–886. https://doi.org/10.1016/S0306-4603(02)00294-0.

Martens, M. P., Cimini, M. D., Barr, A. R., Rivero, E. M., Vellis, P. A., Desemone, G. A., et al. (2007). Implementing a screening and brief intervention for high-risk drinking in university-based health and mental health care settings: Reductions in alcohol use and correlates of success. *Addictive Behaviors*, *32*(11), 2563–2572. https://doi.org/10.1016/j.addbeh.2007.05.005.

Martens, M. P., Smith, A. E., & Murphy, J. G. (2013). The efficacy of single-component brief motivational interventions among at-risk college drinkers. *Journal of Consulting and Clinical Psychology*, *81*(4), 691–701. https://doi.org/10.1037/a0032235.

McClure, S. M., Ericson, K. M., Laibson, D. I., Loewenstein, G., & Cohen, J. D. (2007). Time discounting for primary rewards. *Journal of Neuroscience*, *27*(21), 5796–5804. https://doi.org/10.1523/JNEUROSCI.4246-06.2007.

Miller, M. B., & Leffingwell, T. R. (2013). What do college student drinkers want to know? Student perceptions of alcohol-related feedback. *Psychology of Addictive Behaviors*, *27*(1), 214–222. https://doi.org/10.1037/a0031380.

Miller, M. B., Leffingwell, T., Claborn, K., Meier, E., Walters, S., & Neighbors, C. (2013). Personalized feedback interventions for college alcohol misuse: An update of Walters & Neighbors (2005). *Psychology of Addictive Behaviors*, *27*(4), 909–920. https://doi.org/10.1037/a0031174.

Monahan, C. J., Bracken-Minor, K. L., McCausland, C. M., McDevitt-Murphy, M. E., & Murphy, J. G. (2012). Health-related quality of life among heavy-drinking college students. *American Journal of Health Behavior*, *36*(3), 289–299. https://doi.org/10.5993/ajhb.36.3.1.

Montgomery, J. L., Foley, K. L., & Wolfson, M. (2006). Enforcing the minimum drinking age: State, local and agency characteristics associated with compliance checks and Cops in Shops programs. *Addiction*, *101*(2), 223–231.

Mun, E.-Y., de la Torre, J., Atkins, D. C., White, H. R., Ray, A. E., et al. (2015). Project INTEGRATE: An integrative study of brief alcohol interventions for college students. *Psychology of Addictive Behaviors*, *29*(1), 34–48. https://doi.org/10.1037/adb000004710.1037/adb0000047.supp.

Murphy, J. G., Correia, C. J., & Dennhardt, A. A. (2013). Behavioral economic factors in addictive processes. In P. M. Miller, et al. (Eds.), *Comprehensive addictive behaviors and disorders, Vol. 1: Principles of addiction* (pp. 249–257). https://doi.org/10.1016/B978-0-12-398336-7.00026-7.

Murphy, J. G., Dennhardt, A. A., Martens, M. P., Borsari, B., Witkiewitz, K., & Meshesha, L. Z. (2019). A randomized clinical trial evaluating the efficacy of a brief alcohol intervention supplemented with a substance-free activity session or relaxation training. *Journal of Consulting and Clinical Psychology*, *87*(7), 657–669. https://doi.org/10.1037/ccp0000412.

Murphy, J. G., Dennhardt, A. A., Skidmore, J. R., Borsari, B., Barnett, N. P., Colby, S. M., et al. (2012). A randomized controlled trial of a behavioral economic supplement to brief motivational interventions for college drinking. *Journal of Consulting and Clinical Psychology*, *80*(5), 876–886. https://doi.org/10.1037/a0028763.

Murphy, P., & Garavan, H. (2011). Cognitive predictors of problem drinking and AUDIT scores among college students. *Drug and Alcohol Dependence*, *115*(1–2), 94–100. https://doi.org/10.1016/j.drugalcdep.2010.10.011.

National Institute of Alcohol Abuse and Alcoholism. (2002). *A call to action: Changing the culture of drinking at U.S. colleges*. Baltimore, MD: D. of H. and H. Services. https://www.collegedrinkingprevention.gov/media/taskforcereport.pdf.

National Institute of Alcohol Abuse and Alcoholism. (2004). NIAAA council approves definition of binge drinking. *NIAAA Newsletter*, *3*, 3. https://pubs.niaaa.nih.gov/publications/Newsletter/winter2004/Newsletter_Number3.pdf.

Neighbors, C., Larimer, M. E., & Lewis, M. A. (2004). Targeting misperceptions of descriptive drinking norms: Efficacy of a computer-delivered personalized normative feedback intervention. *Journal of Consulting and Clinical Psychology*, *72*(3), 434–447. https://doi.org/10.1037/0022-006X.72.3.434.

Neighbors, C., Lee, C. M., Lewis, M. A., Fossos, N., & Larimer, M. E. (2007). Are social norms the best predictor of outcomes among heavy-drinking college students? *Journal of Studies on Alcohol and Drugs*, *68*(4), 556–565. https://doi.org/10.15288/jsad.2007.68.556.

Neighbors, C., Lewis, M. A., Atkins, D. C., Jensen, M. M., Walter, T., Fossos, N., et al. (2010). Efficacy of web-based personalized normative feedback: A two-year randomized controlled trial. *Journal of Consulting and Clinical Psychology*, *78*(6), 898–911. https://doi.org/10.1037/a0020766.

Nelson, T. F., Toomey, T. L., Lenk, K. M., Erickson, D. J., & Winters, K. C. (2010). Implementation of NIAAA College Drinking Task Force recommendations: How are colleges doing 6 years later? *Alcoholism: Clinical and Experimental Research*, *34*(10), 1687–1693. https://doi.org/10.1111/j.1530-0277.2010.01268.x.

O'Donnell, R., Richardson, B., Fuller-Tyszkiewicz, M., Liknaitzky, M., Arulkadacham, L., Dvorak, R. D., et al. (2019). Ecological momentary assessment of drinking in young adults: An investigation into social context, affect, and motives. *Addictive Behaviors*, *98*. https://doi.org/10.1016/j.addbeh.2019.06.008.

Pan, W., & Bai, H. (2009). A multivariate approach to a meta-analytic review of the effectiveness of the D.a.R.E. program. *International Journal of Environmental Research and Public Health*, *6*(1), 267–277. https://doi.org/10.3390/ijerph6010267.

Patrick, M. E., Maggs, J. L., & Osgood, D. W. (2010). Latenight Penn State alcohol-free programming: Students drink less on days they participate. *Prevention Science*, *11*(2), 155–162. https://doi.org/10.1007/s11121-009-0160-y.

Pearson, M. R. (2013). Use of alcohol protective behavioral strategies among college students: A critical review. *Clinical Psychology Review*, *33*(8), 1025–1040. https://doi.org/10.1016/j.cpr.2013.08.006.

Perkins, H. W., Haines, M. P., & Rice, R. (2005). Misperceiving the college drinking norm and related problems: A nationwide study of exposure to prevention information, perceived norms and student alcohol misuse. *Journal of Studies on Alcohol*, *66*(4), 470–478. https://doi.org/10.15288/jsa.2005.66.470.

Prentice, D. A., & Miller, D. T. (1993). Pluralistic ignorance and alcohol use on campus: Some consequences of misperceiving the social norm. *Journal of Personality and Social Psychology*, *64*(2), 243–256. https://doi.org/10.1037/0022-3514.64.2.243.

Prince, M. A., & Carey, K. B. (2010). The malleability of injunctive norms among college students. *Addictive Behaviors*, *35*(11), 940–947. https://doi.org/10.1016/j.addbeh.2010.06.006.

Prince, M. A., Maisto, S. A., Rice, S. L., & Carey, K. B. (2015). Development of a face-to-face injunctive norms brief motivational intervention for college drinkers and preliminary outcomes. *Psychology of Addictive Behaviors*, *29*(4), 825–835. https://doi.org/10.1037/adb0000118.

Ray, A. E., Kim, S.-Y., White, H. R., Larimer, M. E., Mun, E.-Y., et al. (2014). When less is more and more is less in brief motivational interventions: Characteristics of intervention content and their associations with drinking outcomes. *Psychology of Addictive Behaviors*, *28*(4), 1026–1040. https://doi.org/10.1037/a0036593.

Russ, N. W., & Geller, E. S. (1987). Training bar personnel to prevent drunken driving: A field evaluation. *American Journal of Public Health*, *77*(8), 952–954. https://doi.org/10.2105/AJPH.77.8.952.

Saltz, R. F. (1989). Research needs and opportunities in server intervention programs. *Health Education Quarterly*, *16*(3), 429–438. https://doi.org/10.1177/109019818901600310.

Sargent, E. M., Kilwein, T. M., Dvorak, R. D., Looby, A., Stevenson, B. L., & Kramer, M. P. (2018). Deviance regulation theory and drinking outcomes among Greek-life students during spring break. *Experimental and Clinical Psychopharmacology*. https://doi.org/10.1037/pha0000204.

Schoenmakers, T. M., de Bruin, M., Lux, I. F. M., Goertz, A. G., Van Kerkhof, D. H. A. T., & Wiers, R. W. (2010). Clinical effectiveness of attentional bias modification training in abstinent alcoholic patients. *Drug and Alcohol Dependence*, *109*(1–3), 30–36. https://doi.org/10.1016/j.drugalcdep.2009.11.022.

Schroeder, C. M., & Prentice, D. A. (1998). Exposing pluralistic ignorance to reduce alcohol use among college students. *Journal of Applied Social Psychology*, *28*(23), 2150–2180. https://doi.org/10.1111/j.1559-1816.1998.tb01365.x.

Schultz, R., Elder, R., Sleet, D., Nichols, J., Alas, M., Grande-Kulis, V., et al. (2001). Reviews of evidence regarding interventions to reduce alcohol impaired driving. *American Journal of Preventive Medicine*, *48*, 66–88. https://doi.org/10.1016/S0749-3797(01)00381-6.

Scott-Sheldon, L. A. J., Carey, K. B., Elliott, J. C., Garey, L., & Carey, M. P. (2014). Efficacy of alcohol interventions for first-year college students: A meta-analytic review of randomized controlled trials. *Journal of Consulting and Clinical Psychology*, *82*(2), 177–188. https://doi.org/10.1037/a0035192.

Simons, J. S., Gaher, R. M., Correia, C. J., Hansen, C. L., & Christopher, M. S. (2005). An affective-motivational model of marijuana and alcohol problems among college students. *Psychology of Addictive Behaviors*, *19*(3), 326–334. https://doi.org/10.1037/0893-164X.19.3.326.

Simons, R. M., Hahn, A. M., Simons, J. S., & Murase, H. (2017). Emotion dysregulation and peer drinking norms uniquely predict alcohol-related problems via motives. *Drug and Alcohol Dependence*, *177*, 54–58. https://doi.org/10.1016/j.drugalcdep.2017.03.019.

Slutske, W. S. (2005). Alcohol use disorders among US college students and their non-college-attending peers. *Archives of General Psychiatry*, *62*(3), 321–327. https://doi.org/10.1001/archpsyc.62.3.321.

Stein, J. S., & Madden, G. J. (2013). Delay discounting and drug abuse: Empirical, conceptual, and methodological considerations. In J. MacKillop & H. de Wit (Eds.), *The Wiley-Blackwell handbook of addiction psychopharmacology* (pp. 165–208): Wiley-Blackwell. https://doi.org/10.1002/9781118384404.ch7.

Stevenson, B. L., Dvorak, R. D., Kramer, M. P., Peterson, R. P., Dunn, M. E., Leary, A. V., & Pinto, D. (2019). Within- and between-person associations from mood to alcohol consequences: The mediating role of enhancement and coping drinking motives. *Journal of Abnormal Psychology*, *128*(8), 813–822.

Substance Abuse and Mental Health Services Administration. (2016). *National survey on drug use and health, 2016 [database]*. Rockville, MD: Substance Abuse and Mental Health Services Administration. https://www.samhsa.gov/data/sites/default/files/NSDUH-DetTabs-2016/NSDUH-DetTabs-2016.pdf.

Substance Abuse and Mental Health Services Administration. (2017). *National survey on drug use and health, 2002–2016 [database]*. Rockville, MD: Substance Abuse and Mental Health Services Administration. https://www.samhsa.gov/data/sites/default/files/cbhsq-reports/NSDUHDetailedTabs2017/NSDUHDetailedTabs2017.pdf.

Sugarman, D. E., & Carey, K. B. (2009). Drink less or drink slower: The effects of instruction on alcohol consumption and drinking control strategy use. *Psychology of Addictive Behaviors*, *23*(4), 577–585. https://doi.org/10.1037/a0016580.

Thombs, D. L. (2000). A retrospective study of DARE: Substantive effects not detected in undergraduates. *Journal of Alcohol and Drug Education*, *46*(1), 27–40.

Toomey, T. L., & Wagenaar, A. C. (2002). Environmental policies to reduce college drinking: Options and research findings. *Journal of Studies on Alcohol*, *Suppl. 14*, 193–205. https://doi.org/10.15288/jsas.2002.s14.193.

Voas, R. B., Tippetts, A. S., & Fell, J. (2000). The relationship of alcohol safety laws to drinking drivers in fatal crashes. *Accident Analysis & Prevention*, *32*(4), 483–492. https://doi.org/10.1016/S0001-4575(99)00063-9.

Wagenaar, A. C., & Toomey, T. L. (2002). Effects of minimum drinking age laws: Review and analyses of the literature from 1960 to 2000. *Journal of Studies on Alcohol*, *Suppl. 14*, 206–225. https://doi.org/10.15288/jsas.2002.s14.206.

Wanmaker, S., Leijdesdorff, S. M. J., Geraerts, E., van de Wetering, B. J. M., Renkema, P. J., & Franken, I. H. A. (2018). The efficacy of a working memory training in substance use patients: A randomized double-blind placebo-controlled clinical trial. *Journal of Clinical and Experimental Neuropsychology*, *40*(5), 473–486. https://doi.org/10.1080/13803395.2017.1372367.

Wechsler, H., Davenport, A., Dowdall, G., & Moeykens, B. (1994). Health and behavioral consequences of binge drinking in college: A national survey of students at 140 campuses. *Journal of the American Medical Association*, *272*(21), 1672–1677. https://doi.org/10.1001/jama.1994.03520210056032.

Wechsler, H., Dowdall, G. W., Davenport, A., & Castillo, S. (1995). Correlates of college student binge drinking. *American Journal of Public Health*, *85*(7), 921–926. https://doi.org/10.2105/AJPH.85.7.921.

Wechsler, H., Kelley, K., Weitzman, E. R., San Giovanni, J. P., & Seibring, M. (2000). What colleges are doing about student binge drinking: A survey of college administrators. *Journal of American College Health*, *48*(5), 219–226. https://doi.org/10.1080/07448480009599308.

Wechsler, H., & Nelson, T. F. (2008). What we have learned from the Harvard School of Public Health College Alcohol Study: Focusing attention on college student alcohol consumption and the environmental conditions that promote it. *Journal of Studies on Alcohol and Drugs*, *69*(4), 481–490. https://doi.org/10.15288/jsad.2008.69.481.

Werner, M. J., & Greene, J. W. (1992). Problem drinking among college freshmen. *Journal of Adolescent Health*, *13*(6), 487–492. https://doi.org/10.1016/1054-139X(92)90012-Z.

Wesley, M. J., & Bickel, W. K. (2014). Remember the future II: Meta-analyses and functional overlap of working memory and delay discounting. *Biological Psychiatry*, *75*(6), 435–448. https://doi.org/10.1016/j.biopsych.2013.08.008.

West, S. L., & O'Neal, K. K. (2004). Project DARE outcome effectiveness revisited. *American Journal of Public Health*, *94*(6), 1027–1029. https://doi.org/10.2105/AJPH.94.6.1027.

White, A., & Hingson, R. (2013). The burden of alcohol use: Excessive alcohol consumption and related consequences among college students. *Alcohol Research: Current Reviews*, *35*(2), 201–218.

Wiers, R. W., Eberl, C., Rinck, M., Becker, E. S., & Lindenmeyer, J. (2010). Retraining automatic action tendencies changes alcoholic patients approach bias for alcohol and improves treatment outcome. *Psychological Science*, *22*(4), 490–497. https://doi.org/10.1177/0956797611400615.

Wiers, R. W., Houben, K., Fadardi, J. S., van Beek, P., Rhemtulla, M., & Cox, W. M. (2015). Alcohol cognitive bias modification training for problem drinkers over the web. *Addictive Behaviors*, *40*, 21–26. https://doi.org/10.1016/j.addbeh.2014.08.010.

Wiers, R. W., Houben, K., Smulders, F. T. Y., Conrod, P. J., Jones, B. T., & Stacy, A. W. (2006). To drink or not to drink: The role of automatic and controlled cognitive processes in the etiology of alcohol-related problems. In *Handbook of implicit cognition and addiction* (pp. 339–361). . https://doi.org/10.4135/9781412976237.n22.

Wiers, R. W., Rinck, M., Kordts, R., Houben, K., & Strack, F. (2011). Retraining automatic action-tendencies to approach alcohol in hazardous drinkers. *Addiction*, *105*(2), 279–287. https://doi.org/10.1111/j.1360-0443.2009.02775.x.

Wiers, R. W., Stacy, A. W., Ames, S. L., Noll, J. A., Sayette, M. A., Zack, M., et al. (2002). Implicit and explicit alcohol-related cognitions. *Alcoholism: Clinical and Experimental Research*, *26*(1), 129–137. https://doi.org/10.1111/j.1530-0277.2002.tb02441.x.

Yurasek, A. M., Dennhardt, A. A., & Murphy, J. G. (2015). A randomized controlled trial of a behavioral economic intervention for alcohol and marijuana use. *Experimental and Clinical Psychopharmacology*, *23*(5), 332–338. https://doi.org/10.1037/pha0000025.

Supplementary reading recommendations

Babor, T., Caetano, R., Casswell, S., Edwards, G., Giesbrecht, N., et al. (2003). Alcohol: No ordinary commodity. A summary of the book. *Addiction*, *98*(10), 1343–1350. https://doi.org/10.1046/j.1360-0443.2003.00520.x.

Cleveland, M. J., Mallett, K. A., White, H. R., Turrisi, R., & Favero, S. (2013). Patterns of alcohol use and related consequence in non-college-attending emerging adults. *Journal of Studies on Alcohol and Drugs*, *74*(1), 84–93. https://doi.org/10.15288/jsad.2013.74.84.

Cronce, J. M., Toomey, T. L., Lenk, K., Nelson, T. F., Kilmer, J. R., & Larimer, M. E. (2018). NIAAA's college alcohol intervention matrix: College AIM. *Alcohol Research: Current Reviews*, *39*(1), 43–47.

Dvorak, R. D., Sargent, E. M., Kilwein, T. M., Stevenson, B. L., Kuvaas, N. J., & Williams, T. J. (2014). Alcohol use and alcohol-related consequences: Associations with emotion regulation difficulties. *American Journal of Drug and Alcohol Abuse*, *40*(2), 125–130. https://doi.org/10.3109/00952990.2013.877920.

Reid, A. E., & Carey, K. B. (2015). Interventions to reduce college student drinking: State of the evidence for mechanisms of behavior change. *Clinical Psychology Review*, *40*, 213–224. https://doi.org/10.1016/j.cpr.2015.06.006.

Shupp, M. R., Brooks, F., & Schooley, D. (2015). Assessing effective alcohol and other drug interventions with the college-age population: A longitudinal review. *Alcoholism Treatment Quarterly*, *33*(4), 422–443. https://doi.org/10.1080/07347324.2015.1077630.

Definitions

Alcohol-related consequences negative outcomes due to alcohol consumption, ranging from mild (e.g., headache) to severe (e.g., hospitalization or death).

Behavioral economics psychology as it relates to decision making when an individual must decide between two seemingly beneficial outcomes, one an immediate, lesser gain and the other a delayed, greater gain.

Binge drinking consumption of 4+ (for women) or 5+ (for men) standard alcoholic drinks during one drinking period, typically within a two-hour period.

Cognitive control ability to inhibit or delay automatic responses in service of achieving a desired goal, often relying on an individual's working memory system.

Coping motivation motives for consuming alcohol which serve to decrease negative effects while drinking (e.g., reducing anxiety or depressive feelings).

Drive reduction theory a motivation theory in which a "need" leads an individual to experience a "drive," which in turn leads to behavior that is meant to satiate that need.

Enhancement motives motives for consuming alcohol which serve to increase the positive effects while drinking (e.g., being more social).

Intervention a treatment or program (ideally empirically supported) with the primary purpose of curbing undesirable outcomes or behaviors and providing alternative, desirable outcomes or behaviors.

Negative valence the extent to which an aspect of motivation or emotion indexes a negative, punishing, threatening, or aversive state.

Nonconsumption factors factors linked to alcohol-related consequences that do not involve the consumption of alcohol (e.g., personality, behavioral strategies to reduce problems, cognitive factors, etc.).

Pluralistic ignorance a social psychology phenomenon in which the perceived descriptive norms that inform an individual's beliefs are different from the actual descriptive norms (i.e., true rate of occurrence).

Positive valence the extent to which an aspect of motivation or emotion indexes a positive, reinforcing, or rewarding state.

Social norms the mainstream beliefs/behaviors of the socially "in group".

Impacting healthcare outcomes: Connecting digital health within the Veteran's Health Administration Home-Based Primary Care (HBPC) program

Colleen L. Campbell[a], *Carmen Besselli*[b]
[a]NFL/SG VHA Home-Based Primary Care, The Villages, FL, United States
[b]U.S. Navy Fleet & Family Support Center, Monterey, CA, United States

Abstract

The Veteran's Health Administration, with a goal of defining excellence in healthcare in the 21st century, is an innovative leader in designing healthcare systems serving challenging populations. This chapter highlights how the application of digital healthcare has generally improved Veteran access to healthcare and healthcare outcomes, specifically for Veterans enrolled in the HBPC program. The program addresses the needs of frail, elderly Veterans, and/or Veterans who have chronic illnesses/diseases and are primarily homebound. In this chapter, a systematic literature review and subsequent analysis informs readers regarding existing digital health programs and program utilization within Veteran populations with varying chronic health conditions. Case studies supplement this information by demonstrating how digital health is used effectively within the HBPC population. Findings suggest that the use of digital healthcare programs to improve access to care and improve healthcare outcomes are a viable and cost-effective healthcare resource with universal applications across many patient populations.

Keywords: Digital health, Veteran's Health Administration, Geriatrics, Home-based care, Healthcare innovations.

Chapter outline

Three Facets of Public Health and Paths to Improvements. https://doi.org/10.1016/B978-0-12-819008-1.00004-3

Building on existing programs to enhance health service delivery

Home-Based Primary Care (HBPC), enhanced by digital health technologies and online health portals launched shortly after the millennium, is one of many alternative collaborative efforts in service delivery in a long history of innovation by the Veteran's Health Administration (VHA). The VHA, operating almost 1500 clinics nationwide and serving 7 million Veterans annually (GAO, 2018a, 2018b), is one of the largest healthcare systems in the United States. Staying ahead of service needs is important for providing Veterans with a comprehensive array of medical services including primary care and specialty medical appointments. However, servicing an estimated 19 million Veterans in the United States, with 50% of this population being over the age of 65 (US Census Bureau, 2017), presents a number of obstacles in terms of engaging Veterans who are not in the VHA system or computer-literate. Behavioral modifications to support the transition to digital health technologies among Veterans are the foundation for increasing access and supporting the transmission of digital healthcare delivery leading to cost-effective and convenient healthcare services. Innovators at the forefront of change also face a great number of obstacles in administration, implementation, operation, and application of new technologies during transition with substantial research costs.

The good news is that, for over 90 years, the VHA has been up to the challenge by leading the healthcare industry in healthcare research (US Department of Veterans Affairs, 2017) and notably leading the field in the areas of implementing innovations in healthcare (Razak, 2018; Sharp, Pineros, Hsu, Starks, & Sales, 2004). Advances in Veteran services provided by the VHA have been accomplished through collaboration and support of the Office of Research and Development, Health Services Research and Development (HSR&D), Centers of Innovation (COIN), and Centers of Innovation on Disability and Rehabilitation Research (CINDRR). Research and healthcare innovations include the implementation of electronic medical records (Hynes, Whittier, & Owens, 2013), advances in medical interventions, and innovative approaches to providing psychosocial support. The VHA has notably advanced in the provision of special healthcare needs and services to Veterans (Moye et al., 2019) in contrast to the private sector discussed in the subsequent section of this chapter. To this point, the VA was cited in December 2018 as the largest provider of telehealth services in the United States supporting their innovative status in this arena.

The VA has also been touted as leading the industry in both research and provision of services to the geriatric population, which is the focus of this chapter. One example is the provision of healthcare in noninstitutional care settings (Edes, 2010; Hughes et al., 2000; Kramer, Creekmur, Cote, & Saliba, 2015; Moye et al., 2019). Home-Based Primary Care (HBPC), described as an "innovative expansion," is a model of long-term care whereby the interdisciplinary primary care clinic is taken into the Veteran's home for those Veterans who are homebound (Campbell, Freytes, & Hoffman, 2015; Edes, 2010; Kramer et al., 2015; US Department of Veterans Affairs, 2007). This program focuses on delivery of care for Veterans with complex and debilitating chronic conditions that make the delivery of care in a traditional outpatient setting ineffective (Edes et al., 2014; US Department of Veterans Affairs, 2019b; Weaver et al., 1995). As an innovative delivery system established in 1972, HBPC has the highest satisfaction of all VA services (Beales & Edes, 2009; Edes et al., 2014). The program has been found to reduce hospitalizations and improve quality of care for homebound Veterans and improve patient satisfaction (Chang, Jackson, Bullman, & Cobbs, 2009, Cooper, Granadillo, & Stacey, 2017; Edes et al., 2014; Edes & Tompkins, 2007; Nelson et al., 2014; Weaver et al., 1995). In addition, home-based healthcare services outside of traditional institutional healthcare settings have been noted as transformational in improving access to healthcare (Semeah, Campbell, Cowper, & Peet, 2017).

Despite these unique attributes and assets, the VA Office of the Inspector General (OIG) acknowledges that there have been concerns in recent years about Veteran's access to care and has made recommendations to address this concern with a focus on improving Veterans' access to timely medical care (GAO, 2018a; Griffin, 2015). The Government Accountability Office notes that "access to timely primary care medical appointments is critical to ensuring that Veteran's obtain needed medical care," (GAO, 2016, p. 1). One factor that has contributed to this problem is difficulty with recruitment and retention of clinical healthcare staff (GAO, 2017). Several federal initiatives to address this problem include implementation of the VA Choice program, expanding clinic hours, and improving areas of human resources recruitment in addition to the use of digital health which have been found to be some ways of increasing access to care (Dang et al., 2019; GAO, 2018b; Houston et al., 2013; Levine, Lipsitz, & Linder, 2018).

Rising health care costs and improvements in patient outcomes are formidable hurdles in healthcare. Of the 13 wealthiest countries in the world, the US ranks 1st in healthcare costs per person at $10,224, representing an expenditure of 50% more than the next wealthiest country, Switzerland. The majority of healthcare costs are spent across five top diseases; (1) ill-defined conditions, (2) circulatory, (3) musculoskeletal, (4) respiratory, and (5) endocrine. In 2019, healthcare costs represented 17.8% of the US's Gross Domestic Product (GDP) (Statista, 2019). Studies have found that when telehealth is integrated into routine service delivery of healthcare in the VA geriatric population, the cost of patient visits is reduced by as much as $310 per visit (Hale, Haverhals, Manheim, & Levy, 2018). In the American healthcare system as a whole, research by the Federal Communication Commission notes that digital healthcare has the potential to save $305 billion every year through the provision of virtual healthcare visits (Miliard, 2019). As such, the use of digital health is proving effective in mitigating rising healthcare costs that are a persistent barrier to provision of care. Additionally, the application of digital health has limitless potential to personalize healthcare delivery to maximize positive healthcare outcomes. Controlling healthcare costs is a necessity in today's world and the use of digital technologies in a digital health approach represents a cost-effective alternative to care that is poised to change the face of healthcare not only in the U.S., but worldwide, with the VHA clearly front and center.

This chapter will review how the VHA is leading change in the healthcare industry, by their innovative use of digital health with HBPC Veterans and their caregivers, resulting in easier access to care and improving healthcare outcomes for this unique population with specific healthcare needs. First, we begin with a population overview followed by a summation of recent research providing specific information about how digital health interventions have increased access to care. Then we highlight the current understanding of the critical factors in facilitating behavior changes to accommodate the use of digital health in HBPC. Next, we discuss the viability of expansion of this concept into other health systems and make recommendations from the VHA experience to implement those changes. Finally, we provide a summary of key aspects of this chapter in the concluding statement.

Veteran population overview

The Veteran population in the United States postdeployment is a vulnerable population differentiated by the potential health impact from their experiences during service to the country and/or complex ailments in conjunction with the general needs associated with aging. This section will first define the Veterans and their unique characteristics and then discuss the provision of special healthcare needs unique to this population. Finally, we provide a review of general concern for the problems the geriatric Veteran population is experiencing in relation to access to care.

Veterans

The Veteran population consists of men and women who have served in the US Army, Navy, Air Force, Marine Corps, or Coast Guard (US Census Bureau, 2017). Veterans who meet VHA eligibility requirements, which are based upon Veteran's status, disability, income, and other factors, have a unique source for healthcare through the Veteran's Health Administration; but not all Veterans have access to this care. In 1980, minimum eligibility requirements changed to include active duty service and receipt of an honorable discharge, but limited budgets and capacity require that care is rationed based on eight eligibility priority tiers (APHA, 2014). Once enrolled, Veterans can obtain care at any VA medical facility which presently includes over 151 medical centers, 300 Veterans' centers, and in excess of 820 community-based outpatient clinics (APHA, 2014).

Unique Veteran characteristics

The US Department of Veterans Affairs identifies Veterans as a vulnerable population (GAO, 2015) for a myriad of reasons including higher risk than the general US population for joblessness, homelessness, mental health ailments, and financial instability (APHA, 2014). US Veterans are a unique population and are found to vary as a subculture dependent upon branch of military service and military experiences (Olenick, Flowers, & Diaz, 2015).

In contrast to the non-Veteran population, 50% of the US Veteran population is over the age of 65, whereas less than 15% of the non-Veteran population is in this age cohort (US Census Bureau, 2017). The US Census Bureau notes that this percentage reflects two cohorts of Veterans including (1) those from the draft era, and (2) those composed of a volunteer force. Draftera Veterans are comprised of those of the wartime eras of World War II, the Korean War, and the Vietnam era. Initially, this group was 12 times larger than the current all-volunteer force; presently, however, the mortality rate of this group exceeds the military discharge rate of the current all-volunteer force. Additionally, the mortality rates for draft era Veterans are higher than their non-Veteran counterparts, which have resulted in a decreasing draft era Veteran population (US Census Bureau, 2017).

Veterans are significantly overrepresented among the homeless population in the US; while Veterans account for less than 9% of the US population, they account for at least 12% of the homeless population (Olenick et al., 2015); of the homeless male population, Veterans compose 20% of this subgroup (National Coalition for Homeless Veterans, n.d.). The strongest risk factors for homelessness in the Veteran population are mental illness and substance use disorders (Rosenheck & Koegel, 1993). Homeless Veterans were noted to be older, better educated, and more likely to be male, married/have been married, and to have health insurance coverage as compared to their non-Veteran counterparts; however, these did not appear to be protective factors to reduce their risks for homelessness (Fargo et al., 2012). Additional contributing factors, including low income and other income-related factors such as unemployment, are consistent in both the Veteran and non-Veteran population (Petrovich, Pollio, & North, 2014). Social isolation, adverse childhood experiences, and past incarceration are additional risk factors in this population (Fargo et al., 2012; Petrovich et al., 2014; Rosenheck & Koegel, 1993).

Table 4.1 demonstrates unique characteristic of this population. For example, Veterans are less racially diverse than their non-Veteran US counterparts (US Census Bureau, 2017). This is largely attributed to the historical discriminatory policies and practices of the military (Webb & Herrmann, 2002), impacting where and when minorities served and what duties they were assigned. Although Executive

Table 4.1 Distinguishing factors of the United States Veteran in comparison to the non-Veteran population.

Demographic characteristic	Veteran	Non-Veteran
	Percentage	
Population[a]	8.5%	91.5%
Caucasian non-Hispanic[b]	77.7%	62.1%
Educational attainment (some college)[b]	37.5%	29.9%
Employment status (employed)[c]	44.2%	61.2%
Disability status (having at least one disability)[b]	30.1%	14.8%
Over the age of 65[b]	49.9%	14.9%
Diagnosed with posttraumatic stress disorder (PTSD)[a,d]	36%	8%
Homeless[a,d,e]	12%–20%	80%–88%
	Median age	
Male population[b]	65	42
Female population[b]	51	47

[a] United States Department of Veterans Affairs (2018a, 2019e).
[b] United States Census Bureau (2017).
[c] United States Bureau of Labor Statistics (2017).
[d] Olenick et al. (2015).
[e] Pew Research Center (2017).

Order 9981 issued by US President Harry S. Truman ostensibly abolished racial segregation in the US military on July 26, 1948, actual integration was not implemented until the Korean War (Webb & Herrmann, 2002). As a result, Caucasian Veterans continue to represent the largest Veteran group (Saha, Freeman, Toure, Tippens, & Weeks, 2007). For example, Caucasians comprise 78% of the Korean War era Veterans, whereas Caucasian non-Veterans aged 75 or older currently living in the United States is 62% (US Census Bureau, 2015).

Additionally, the US Veteran population is shaped by their education. Almost 87% of the non-Veteran population has a high school diploma, while, in contrast, 93.5% of Veterans hold a high school diploma or higher (US Census Bureau, 2015). Veterans not only have unique demographic characteristics compared to the non-Veteran population, but also are unique in terms of their multifaceted psychosocial and healthcare needs including reestablishing emotional connections and vocational rehabilitation (Canfield & Weiss, 2015), chronic illnesses with multiple symptoms, military sexual trauma (MST), posttraumatic stress disorder (PTSD), and traumatic brain injuries (TBI) (Matthews, Vinson, DeWitt, McGuire, & Karel, 2019; Miltner et al., 2013).

In terms of the impact of military service on mental health, Veterans make up a disproportionate percentage of the US population suffering from mental health ailments including depression, posttraumatic stress, traumatic brain injuries, and substance use disorders (Olenick et al., 2015). In contrast to their male civilian counterparts, male Veterans are four times more likely to be diagnosed with PTSD (Olenick et al., 2015). Veterans are also two times more likely to commit suicide than the civilian population; while Veterans make up less than 9% of the US Population, they account for 14% of successful suicides (Fox, 2018).

Although Veterans have unique characteristics and needs, they share a significant similarity with their US counterparts in the area of healthcare disparity. Studies show that minority Veterans within the VHA system have difficulty accessing care, obtaining specialized medical care/procedures, and experience overall poorer health (US Department of Veterans Affairs, 2017). Many of these factors are attributed to cultural differences in communication styles, patient and provider expectations, lack of economic resources, and a myriad of other factors, all of which impact how care is delivered, the quality of care, and healthcare outcomes (Rickles, Dominguez, & Amaro, 2010).

Access to care

Many Veterans who are eligible for VHA care do not use the VHA for their healthcare (Blais & Renshaw, 2013; Turvey et al., 2014; US Department of Veterans Affairs, 2012). Use of the VHA varies greatly based upon many factors; however, some of the contributing factors include distance to a VA facility and socioeconomic status. Currently, there are 4.7 million rural and highly rural Veterans and only 2.8 million are enrolled in the VA (US Department of Veterans Affairs, 2019c.). Rural areas generally have a higher population of elderly residents whose overall health is worse (Ward, Cope, & Elmont, 2017; Ward, Solomon, & Stearmer, 2011). Additionally, rural areas have limited access to healthcare due to fewer physicians, hospitals, and other healthcare delivery resources such as public transportation (US Department of Veterans Affairs, 2019c).

Personal socioeconomic factors also contribute to whether or not Veterans opt for VHA for their healthcare. Studies indicate that Veterans who seek VHA care have lower incomes and poorer health insurance coverage than their counterparts who do not use the VHA as their primary source for healthcare (RAND Corporation, 2015). Veterans with good health insurance coverage, such as Medicare and supplemental healthcare plans, tend to seek care outside of the VHA (Berchick, Hood, & Barnett, 2018). In some US territories and states, such as Puerto Rico and South Carolina, respectively, as many as 43%–44% of Veterans cite the VHA as their primary source of healthcare (US Bureau of Labor Statistics, 2014). Some of the contributing factors to the high VHA utilization rate in Puerto Rico may include 49% of Puerto Ricans living below the poverty level, ten years of economic recession further limiting healthcare access, the priority health demands of the Zika virus in 2016, and infrastructure damage caused by recent hurricanes (Seervai, 2017). In the case of South Carolina, there is both a large rural and minority population; this state did not accept the Affordable Care Act (ACA) Medicaid expansion and has been ranked the 13th worse state for healthcare in the country (Levitt, Neuman, & Brody, 2019; Lewis, Coleman, Abrams, & Doty, 2019). In states such as Delaware and New Jersey where VHA utilization is much lower, under 20%, there is better healthcare access, active state healthcare initiatives, and both states accepted the ACA's Medicaid expansion (Lewis et al., 2019; Mueller, 2018).

Enhancing Veteran access through use of HBPC and digital health within the VHA

The VHA healthcare system is large, complex, and ever evolving. These characteristics can result in Veterans having difficulty accessing VHA healthcare resources. However, there are several programs in place that are designed to enhance access and improve healthcare outcomes. This section will address enhancing access to care for Veterans through HBPC and the use of digital health.

Access for the HBPC Veteran population

HBPC was established by the Department of Veterans Affairs with a goal of providing care to homebound Veterans that is "accessible, comprehensive, coordinated and continuous" (Weaver et al., 1995, p. 84). The program is designed for frail, elderly Veterans and/or Veterans who have chronic illnesses and are primarily homebound as a result (Campbell et al., 2015). HBPC participants often include Veterans that reside in rural areas, have lower socioeconomic status, and/or are exempt from VA medication and medical treatment copayments. The program can enhance access to care for this population when Veterans are unable to consistently go to their primary care appointments in a clinic setting because of their medical conditions. This population also tends to have multiple chronic illness/disease which include but are not limited to cerebral vascular attacks (CVA), amyotrophic lateral sclerosis (ALS), coronary artery disease (CAD), hypertension (HTN), Alzheimer's and dementia type illnesses, and an array of other debilitating conditions (US Department of Veterans Affairs, 2019a; Weaver et al., 1995).

A vast majority of the VHA HBPC program serves Veterans who are older males with the majority of the population being Caucasian (Weaver et al., 1995), reflecting the demographics illustrated in the Veteran population as noted earlier. Veterans in receipt of this service also overwhelmingly require assistance performing activities of daily living (ADLs), which include dressing, bathing, toileting, and transferring (Hughes et al., 2000; US Department of Veterans Affairs, 2007). This program has proven effective in the provision of healthcare services, specifically for reducing costs, increasing access, and improving delivery of quality healthcare, and also reducing hospital and emergency room visits for Veterans with chronic and complex conditions (Beales & Edes, 2009; Hicken & Plowhead, 2010; Kramer et al., 2015; North, Kehm, Bent, & Hartman, 2008; Wolf, 2006).

Defining digital health and key policy

Digital health interventions have been defined as those which incorporate digital technologies into the delivery of healthcare services (Blanford et al., 2018; Levine et al., 2018) (Fig. 4.1). The term digital health, interchangeable in literature with the terms telemedicine and telehealth, is defined as "the use of medicinal information exchanged from one site to another via electronic communications to improve a

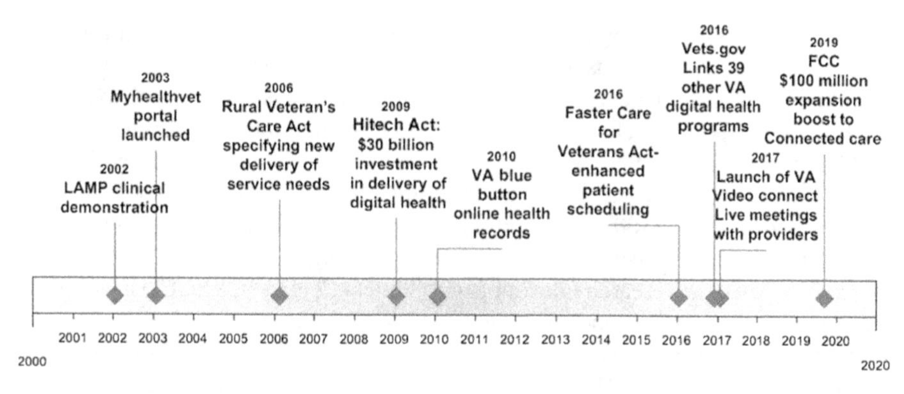

Fig. 4.1 Timeline of key digital health policy initiatives and program development to support Veteran's Health at the Veterans Health Administration.
Created by Beth Ann Fiedler.

patient's clinical health status," (Cahn, Akirov, & Raz, 2018, p. 12). Research has found that for this type of healthcare intervention to be successful, an investment of both time and resources is needed to incorporate digital health into the broader care system "in ways that work well for professionals, patients, and others involved in care," (Blanford et al., 2018, p. 8).

In general, the goal of digital health technologies is to gather and analyze healthcare data to improve patient health and healthcare delivery (Sharma et al., 2018). This is achieved by pairing technology with care coordination. Digital health has been found to have many benefits including improved communication with healthcare providers (Ahmadvand, Gatchel, Brownstein, & Nissen, 2018), prompt access to medical services (Liu et al., 2017), self-management of healthcare concerns (Sellers, Campbell, & Moore, 2018), increasing adherence to treatment plans (Walsh & Groarke, 2019), improving coordinated care, and overall improved health behaviors (Neter & Brainin, 2019). In cases where face to face support is not practical, digital health has been demonstrated as effective in management of chronic diseases among geriatric patients (Sweet et al., 2018). Bi-directional digital health programs, where patients and providers may interact with one another in real time, is an application of digital health that has been cited for improving patient-centered care through fostering relationships with healthcare providers and increasing patient participation in their care (Houston et al., 2013; Lafontaine et al., 2018). The Chief Officer of the VA Office of Connected Care asserts that "what matters are the relationships between the patient and their health care team that technology can secure and support," (Evans, 2017, *para* 3); thus, integration of bi-directional digital health platforms becomes a fundamental component of the provision of quality care for our nations' Veterans.

With implementation of the Federal Health Information Technology for Economic and Clinical Health (HITECH) Act of 2009, over $30 billion federal dollars have been invested in delivery of digital health technologies (Levine et al., 2018; Sharma et al., 2018). In part, this is due to digital health being viewed as a promising means of delivering services to Veterans (Hicken et al., 2017). The Rural Veterans Care Act of 2006 reinforced the notion that building more healthcare facilities is insufficient for meeting the needs of rural Veterans. Innovations in digital healthcare services for Veterans, including Veterans living in rural areas facing typical obstacles to care such as poverty and lack of transportation, have been found to address traditional barriers to receipt of healthcare services for Veterans (Luptak et al., 2010). Supporting the notion that digital health can have such an instrumental impact, in June of 2019, the FCC commissioner advanced a $100 million pilot program to support the expansion of digital telehealth to increase access to healthcare for rural patients (Miliard, 2019).

The VA Geriatric Research Education and Clinical Centers (GRECCs) have led the way towards developing innovative approaches for addressing the needs of older Veterans in hard to reach settings (Luptak et al., 2010). Many telehealth programs currently offered by the VHA operate under the service umbrella of geriatrics and extended care for this reason. Unlike the private sector, the VHA has a readily available study group which allows for the VA to develop and offer new innovations in healthcare. In general, the objective of the digital health programs is to pair technology with care coordination. Similarly, the goal is to provide the Veteran and caregiver with tools that will increase their knowledge enabling the Veteran to remain in their home and independent for as long as possible. Although each digital program has its own eligibility criteria specific to the disease or health condition, generally the primary criterion is internet or telephone access and a willingness to enter data as required. For example, the program that monitors blood pressure may require the Veteran to daily take their blood pressure either with the monitoring device that automatically updates the information or by a method that requires the Veteran to manually enter data. Patient compliance is also a common denominator of these programs. Consistently failing to use and/or record outputs from a monitoring device data may result in disenrollment from the prescribed digital health program (US Department of Veterans Affairs, 2019a).

Digital health programs in VHA

Several digital health technology tracking tools are used to monitor activities of daily living (ADL) and various indicators of health. Some digital health technologies in use within the VHA include LAMP (low activities of daily living (ADL) monitoring program), myhealthevet, video telehealth, and Tech care for Chronic Illnesses (Table 4.2). Many of these innovative digital health programs are currently being utilized by Veterans within the HBPC program (US Department of Veterans Affairs, 2019a).

LAMP is one of the early VHA adapted digital health programs that began as a clinical demonstration project in 2002 with the aim of attenuating functional decline of elderly Veterans through the use of (1) assistive technology, and (2) adaptive equipment. Eligibility for LAMP includes indication that the Veteran requires assistance with at least two ADLs, resides in their home, and has the infrastructure to support electronic internet connections. Should the Veteran agree to participate and have a need for remote monitoring of their condition, they are able to enroll in the program. Several HBPC Veterans currently utilize LAMP in many ways including

Table 4.2 Digital health technologies utilized by the Veterans Health Administration (VHA) to improve patient health.

Program	Purpose	Eligibility	Resource URL
Live/synchronous			
VA video connect	Permits synchronous digital face to face meetings with provider and auxiliary staff via a secured online portal	Enrollment in VHA care	https://mobile.va.gov/app/va-video-connect
Video telehealth	Patient initiated self-scheduling and real-time video conferencing with healthcare providers	Enrollment with VHA for receipt of care; requires video conferencing equipment (camera and microphone)	www.telehealth.va.gov
Asynchronous			
LAMP (low activities of daily living (ADL) monitoring program)	Low ADL monitoring program unique to VISN08 providing home-based Veterans remote monitoring and in-home messaging	Enrollment in VISN08 for healthcare and dependent on at least two activities of daily living; requires home phone and internet connectivity. Provider consult is needed	www.northflorida.va.gov/services/lamp.asp
Myhealthevet	Direct online portal access to patient health record allowing Veterans to refill prescriptions, email provider, track vital statistics, and monitor lab results	Available to any Veteran who is both eligible for and enrolled in VHA Care; Premium account is free and enables communication with provider(s)	www.myhealthevet.va.gov
Tech care for chronic illnesses	Coordination with PCP to promote safety and independence to remain at home by increasing access to VA services and decreasing unscheduled and emergency visits; alerts the Veterans' PCP about how the Veteran is managing between healthcare appointments	Eligibility is based upon the chronic illness condition (cardiac, diabetes management, mental health, congestive heart failure, palliative care, etc.). A consult from the VHA provider is needed	Specific to the VISN in which the Veteran is enrolled

Continued

Table 4.2 Continued

Program	Purpose	Eligibility	Resource URL
VA mobile apps	VA mobile apps store includes a plethora of smartphone and tablet applications to enhance access to care	Many do not have enrollment criteria unlike others who require enrollment in VHA Care to allow connection with providers	www.mobile.va.gov/appstore
Weight management	Move! is a national weight management program focusing on health and wellness. Many facilities have an in-person component, but some facilities offer a digital health program and all offer a supplemental phone app: The Move! Coach	Only available to Veterans receiving care in the VA; website content is accessible to Veterans who are not enrolled in the VA and the general public. A consult from the VHA provider is needed	www.move.va.gov/MOVE/MoveCoach.asp www.move.va.gov
VA prescription refill and tracking	A web-based service that allows Veterans to manage VA prescriptions online	Requires Veterans to create an account and have a verified ID.me or Premium Myhealthevet account	www.va.gov/health-care/refill-track-prescriptions/

Notes: Primary care physician (PCP); Veterans integrated service network (VISN) designations.

(1) encouragement of self-management of their chronic illnesses, (2) receiving continuous care coordination, and (3) to receive daily support from VHA providers (Bendixen, Levy, Olive, Kobb, & Mann, 2009). Two of the goals of the LAMP program are to prevent long-term institutional care and to promote safety and independence, both of which are in direct concert with the goals of the HBPC program. In essence, these programs combine unique characteristics of individual care coordination, technology, and traditional healthcare delivered in a home care setting to bring about the best healthcare outcomes for elderly homebound Veterans (Bendixen et al., 2009; US Department of Veterans Affairs, 2019d).

Another innovative use of technology is the Myhealthevet portal that was launched in 2003. This online portal allows Veterans to conveniently refill prescriptions, request copies of portions of their electronic health record (EHR), track their vitals, and monitor their test and laboratory results. More than 20 million prescriptions were refilled through this platform only two years after the program was launched (Haggstrom et al., 2011). Touted as technology that allows Veterans 24/7 access to their healthcare records, Veterans also have the ability to email their healthcare providers directly facilitating increased communication and information sharing (Naditz, 2008). To be eligible to access the Myhealthevet portal, the Veteran must be registered, receiving care at a VHA facility, and have the ability to access the internet. Blue Button access, a second registration requirement and security strategy, is necessary to obtain their sensitive healthcare information (Turvey et al., 2014). The process requires verification of the identity of the Veteran (or surrogate) at a primary care clinic to enable access to the Veteran's medical record information within a few days. Additionally, healthcare information following healthcare encounters is also readily available and can be viewed, downloaded, or printed. The ability to easily and conveniently access this information allows Veterans to coordinate their care with any providers they may have outside of the VHA system. The use of this technology has several advantages including (1) 24/7 access; (2) increased opportunities to communicate between the Veteran and their healthcare team, (3) timely follow-up when the Veteran reports any healthcare incidents or other healthcare concerns, and (4) improvements in the quality and accuracy of reporting (Turvey et al., 2014). In 2016 alone, myhealthevet had 55.8 million visitors (Evans, 2017, *para* 12).

Veterans also have another option to decrease wait times to schedule appointments: video telehealth. In 2017, the VA initiated patient scheduling as part of the Faster Care for Veterans Act of 2016 with the goal of expediting access to care for computer-literate Veterans who are able to use the internet or mobile application to schedule their own appointments (GAO, 2018b). Video Telehealth within the VHA has been described as an interactive and synchronous video conference between a healthcare provider and the Veteran (Abel et al., 2018). The most recent addition to the VA's expansion of video telehealth is the VA Video Connect platform which was launched in 2017. More than 20,000 Veterans used this service in the first year of operation (Mena Report, 2018). VA Video Connect, also established to increase access to VA healthcare and increase communication between Veterans and their healthcare providers, utilizes video technology for virtual healthcare appointments (US Department of Veterans Affairs, 2018b). In March of 2018, the VA released a consistently top 10 healthcare application version of VA Video Connect to expand access from laptops and desktops to Android and iOS users (Hale et al., 2018; Mena Report, 2018).

Limitations of digital health within this population

Nonadherence has been a longstanding issue in management of complex and chronic conditions. Almost 60% of recent studies on this topic have found that digital health results increased adherence to treatment plans in part due to increased self-efficacy in healthcare planning (Walsh & Groarke, 2019). Digital health can also increase patient engagement through closing gaps in patient/provider communication and consequently increasing patient self-care (Sharma et al., 2018). The use of telehealth has resulted in lower use of both inpatient and outpatient care and further evidence suggests that geriatric Veterans who participate in telehealth interventions are satisfied with this care (Whitten, 2006). Paradoxically, while digital health has been found to have an optimal impact on patient care, research has found it to have a potentially detrimental impact on provider productivity (Sharma et al., 2018; Whitten & Mackert, 2005). Currently, there is a paucity of research indicating whether this negative side effect will decrease with continued use of digital health once eHealth literacy increases.

Despite huge strides being made by the VHA to improve access to care, limitations still exist due to challenges inherent in the use of digital heath. The GAO notes that in certain areas, limitations of internet services, such as on the islands of Guam and American Samoa, lead to disruptions of telehealth appointments (GAO, 2018a). In efforts to address these concerns, the healthcare system is upgrading equipment and increasing bandwidths to support telehealth (GAO, 2018a). Another obstacle unique to the VHA relates to various state requirements establishing the license of a provider which may limit Veteran's access to care during the digital health implementation process (GAO, 2018a). While federal law requires VHA staff to hold a valid clinical license in any state (meaning that the provider is not required to be licensed in the state where the medical center is located), states can require providers to be licensed in the state where they are practicing. For instance, typical delivery of telehealth services by a licensed clinical healthcare provider in a Washington, DC medical center to a Veteran located in Virginia may not be permissible; this is also problematic in areas that are often more difficult to access, such as Guam and Samoa (GAO, 2018a). Thus, this has a detrimental impact on the expansion of telehealth by placing limits on those providers who may perform digital health in certain states.

Although there is a national push for implementation of digital health technologies among this population (Sharma et al., 2018), adaption rates continue to be low (Whitten, 2006), especially among the homebound population (Neter & Brainin, 2019; Turvey et al., 2014). Research indicates that there are several common themes/issues related to the use of digital health technologies in this population: eHealth literacy (Kahn, Aulaka, & Bosworth, 2009; Neter & Brainin, 2019), access to technology (Turvey et al., 2014), functional and cognitive status of the patient (Whitten, 2006; Woods et al., 2017), motivation for use of technology (Sharma et al., 2018), and the availability of caregiver/surrogate (Mishuris et al., 2014). Three of these can be viewed as limiting factors (functional and cognitive status, eHealth literacy, and access to technology) and two as factors that may facilitate digital use (motivation for use of technology and availability of a caregiver/surrogate) among homebound patients (Mishuris et al., 2014). In a qualitative study of HBPC Veterans with a small sample size, study respondents of Veterans and their surrogates have been found to be interested in and eager to participate in learning digital health technology, despite having limited eHealth literacy and access to technology (Mishuris et al., 2014).

Another element found to impact the use of digital health is that of socioeconomic status. Digital health has been found to be used less in minority populations and those of lower socioeconomic status (Chapman, 2010; Senecal, Widmer, Bailey, & Lerman, 2017). This is due, in part, to the cost associated with the use of these technologies. Additionally, surrogates for this population are vital in increasing the use of digital technology as often this population is much more likely to be dependent on their surrogate/caregiver for assistance. Of course, this situation can play a large role in the patient's ability to access technology.

Currently, several of the identified barriers to the use of this technology can readily and successfully be addressed through specific training and education to facilitate patient behavioral modification. This population is eager to learn and utilize digital health technologies so that additional efforts to provide Veterans with the skills to engage in digital services can easily be incorporated into existing services making this path both cost-effective and readily available.

Facilitating behavioral change for digital technology acuity

There are several factors that should be considered when the goal is to incorporate digital health technologies with traditional healthcare. While there are a plethora of theories to explain and predict behavioral change, one theory that has been found to be particularly applicable to facilitating behavioral change with the use of digital health is the multiphase optimization strategy (MOST) (Moller et al., 2017). This theoretical foundation postulates that thorough preparation, several factors of optimization, and continued evaluation result in increased efficiency, efficacy, and sustainability of the behavioral change and success of digital health (Collins, Murphy, & Strecher, 2007). Additionally, behavioral changes aimed at enhancing patient technology use should focus on both the patient and the healthcare provider in order to achieve optimal success.

Preparation

The task of preparation for the use of digital health technology among HBPC Veterans primarily falls to the team of healthcare providers. There are multiple barriers to the successful use of digital health technology and each must be adequately identified and addressed. Primary among these barriers is access to a computer, smartphone, tablet, or internet connectivity at the Veterans' place of residence and existing socioeconomic factors that may play a pivotal role (Abel et al., 2018; Chapman, 2010; Perzynski et al., 2017; Senecal et al., 2017; Woods et al., 2017). Failure to screen for any of these critical components impacts the Veterans' probability of successfully integrating the use of digital health technology. Additionally, the provider(s) should assess the Veterans' comfort with and motivation for the use of technology (Woods et al., 2017).

Healthcare providers should also assess their own technology literacy in preparation for the use of digital telehealth in the homecare setting. While technology is present in all aspects of everyday life, healthcare providers must first be comfortable using technology and familiar with the technology to educate, demonstrate, and problem solve digital health technology problems that they seek to incorporate into the Veteran's healthcare regime (Ahmadvand et al., 2018). In many circumstances, instruction or reinforcement of basic computer skills may be necessary for both the provider and the Veteran/caregiver prior to the introduction of a digital healthcare technology; however, there is no expectation that healthcare providers must be information technology (IT) experts. Similarly, patient and caregiver assessments should include their current level of technology literacy, willingness to accept new tasks, and ability to incorporate them. The introduction of new tasks and information can be a significant barrier and participants must be evaluated for their ability to accept change and utilize new innovations (Aref-Adib et al., 2019). Healthcare providers should assess (1) any preconceived ideas they may have regarding the patient or caregiver's ability to learn, and (2) their level of motivation or willingness, particularly as it relates to homebound patients and caregivers who may not have any or limited technology literacy. Acceptance of new tasks can be enhanced through providing education that explains the benefits, level of complexity, and time investment that is required to achieve success. Furthermore, using patient input to establish goals they view as both desirable and beneficial may enhance acceptance and consistency. In addition, emphasizing a desirable patient outcome may be more readily accepted, and therefore, an expectation of improvements in initial compliance.

Engaged caregivers, when available, are crucial to a successful digital health technology introduction even in cases where the Veteran is relatively independent at the onset. By utilizing this approach, caregivers can be educated along with the Veteran regarding the benefits and technology requirements. Additionally, caregivers should be educated on the use of this technology in the event that there is a decline in the Veterans' functional and cognitive function which requires the caregiver to play an increasing role or assume substantial responsibility for managing any equipment, reporting, or other requirements. Similarly, caregivers should be encouraged to provide feedback and share their feelings regarding the use of the technology. This includes specific questions that help to determine how adding additional responsibilities will impact them as these duties may result in the development of or increase in caregiver stress.

Optimization

Optimization is the next stage in this model whereby the goal is to make using digital health technology as user-friendly and effective as possible. Ease of digital health technology navigation is important for achieving this goal that is pivotal to sustaining ongoing Veteran utilization (Woods et al., 2017). Optimization of digital health technology for the patient, caregiver, and provider should focus on customizing education and training requiring an assessment of individual learning styles and how they incorporate new information. For example, individuals may be visual, auditory, or hands-on learners, while others may prefer independent self-study. Healthcare providers may prefer independent self-study or learning that is well-organized and delivered over a short span of time due to inherent time constraints and limited availability placed on their primary responsibilities. Short training bursts may also work for patients who become more easily distracted, experience pain, have other physical limitations, or need more frequent breaks, while their caregivers may require hands-on training.

Successful and effective learning must ultimately clearly delineate the goals and benefits that are tailored to meet the individual needs of all stakeholders involved in the educational learning process.

An additional element to consider when optimizing the use of digital health technologies in the HBPC population is attention to the numerous functional and cognitive factors that can impact the Veterans' use of digital health technology. Impaired health, declining physical and/or cognitive function, is prevalent among chronically homebound Veterans and can adversely impact their ability to integrate new skills or tasks. Veterans who currently use digital health technology with little difficulty may lose cognitive or physical function over time which may prevent them or impair their ability to input the necessary data or use any required equipment such as blood pressure or glucose monitoring equipment. Failing eyesight with resulting blindness or loss of hearing also impacts use of technology as equipment error messages or prompting cues may not be seen or heard. To optimize the use of digital technologies in these cases, the preparation stage of engaging caregivers who are educated regarding the positive aspects of the use of digital health technologies can become a positive asset in transitioning the user of the technology from the Veteran to the caregiver. This process ensures compliance with and longevity of the use of digital health technology even if the Veterans' functional or cognitive status declines.

Another factor to consider during this stage is the increased use of the technology beyond what the provider may have initially expected. Studies have found that Veterans who are engaged with their providers digitally have been found to demand greater independence in their healthcare choices and require greater shared decision-making than their counterparts (Ahmadvand et al., 2018). Consequently, providers must be prepared to address this shift from the traditional medical paradigm to that of a shared partner in the provision of healthcare services.

Ongoing evaluation

In populations with chronic health conditions that may decline over time, the patient may no longer be able to comply with the requirements of the digital health programs (Luptak et al., 2010; Mishuris et al., 2014; Neter & Brainin, 2019; Walsh & Groarke, 2019). Additionally, the caregiver who was once available to enter this data may now be overwhelmed with caregiver responsibilities and unable to perform these tasks (Senecal et al., 2017). Thus, optimal benefits from the use of digital health technologies require ongoing evaluation by all healthcare providers. Ultimately, there are numerous reasons why a digital health technology program may have poor compliance in addition to those noted above. Therefore, collaboration with the Veterans' caregivers, healthcare team members, and other programs is essential for the patient and program success.

As with most digital health technologies, the patient must be compliant to remain in the program (Bendixen et al., 2009; Blanford et al., 2018; Dang et al., 2019;

Sellers et al., 2018). As aforementioned, noncompliance with program requirements (i.e., failure to enter health data or use provided equipment to automatically update results) can result in the Veteran being discharged from digital health technology programs. Ensuring ongoing patient compliance to reach and achieve the optimal healthcare outcomes for patients within Home-Based Primary Care requires a flexible strategy with modifications resulting from ongoing evaluations. In turn, collaboration and coordination are essential to achieve established goals for the Veteran, healthcare provider, and the digital health technology program. Collaboration may include digital health technology program participants joining in the homebound patients' care planning meetings to (1) identify ongoing issues and common goals, and (2) develop a coordinate plan of care. HBPC teams that are visiting the Veteran in the home are able to accurately assess what barriers are preventing the Veterans' compliance and also serve as a tool for ongoing monitoring. Problem identification and collaboration serve as problem resolution tools by refocusing on shared goals, enhancing patient outcomes, and resolving compliance concerns.

Case studies: The health consequences of applying digital health

In this section, we introduce two case studies[1] demonstrating the application of digital health to homebound care. In this first scenario, we emphasize the phases of preparation and optimization (including engagement of the family and caregiver) to receive optimal benefits from digital telehealth. In the second, the concept of ongoing evaluation is important. The case studies below are just two examples of how digital healthcare technology is impacting healthcare delivery among homebound Veterans to achieve positive outcomes.

The case of Mr. Johnson

Mr. Johnson is a 76-year-old alert and oriented Navy Veteran who suffers from congestive heart failure (CHF). Historically, his blood pressures related to his CHF ranged around 160/64. His height is 68″ and his weight ranges from 192 to 199 lbs, indicating a body mass index (BMI) of 30.26. The Veteran resides independently in his own home with his 77-year-old wife who also has chronic health problems. Upon initiation of care with the VHA HBPC program, the Veteran required no assistance with his care needs and was totally independent in his activities of daily living. The couple had a very limited income and an excellent support system comprising family and friends. The dyad were visited at least weekly by a daughter who made prepared meals and

[1] The cases presented in this chapter are based upon actual patients enrolled in VHS HBPC programs. Participant names have been changed and elements of their health service experience modified to protect their personal identities in accordance with the Health Insurance Portability and Accountability Act of 1996 (HIPAA), Public Law 104-191, and subsequent revisions.

would freeze them as the couple were no longer able to prepare balanced meals. The Veteran and his wife are unable to drive, but can rely on their daughter or a neighbor to provide transportation to medical appointments.

In 2007, Mr. Johnson was admitted to HBPC after frequent emergency room visits and lengthy hospitalizations related to his CHF. After his HBPC admission, his ER visits became less frequent; however, over the next few years due to the management of his healthcare needs in the home, but he continued to require lengthy hospitalizations related to his heart failure. He also continued to experience shortness of breath (SOB) and general malaise. Several falls also occurred, but they did not result in serious injuries. In 2017, the HBPC team determined that the Veteran may benefit from the VHA Cardiology Telehealth digital health program for management of his acute CHF symptoms. The HBPC team was easily able to assess if the dyad would benefit the candidate and achieve optimal results for the program because of the long-term relationship established with this Veteran/caregiver. Analysis by the HBPC team determined that there was adequate support of the Veteran's family to assist the dyad in setting up internet connectivity and to assist with the financial burdens of obtaining a computer that would allow them to utilize the program. Through local engagement of the couple's children, the family was able to obtain a laptop and internet service that could be easily used by the Veteran. The dyad was not initially excited by the proposal to innovate to a digital health solution; however, their children were pivotal in developing acceptance and ease of use for this technology.

In 2018, the Veteran and his family became amenable to receipt of the digital health program with the goals of stabilization of symptoms, obtaining education and self-management tools for optimal home management, and prevention of hospital readmissions. Through dedication of his HBPC team and ongoing support from his children, Mr. Johnson now uses his internet access to report his health status. Based on this information, his PCP or another member of the HBPC team can review his symptoms and determine if an appointment should be made for a physical assessment. When significant changes are noted on Mr. Johnson's monitoring equipment, he also receives a call from the Cardiology Telehealth program to assess if an appointment is required for further evaluation. Additionally, the HBPC provider can notify the program via progress notes of any significant changes that are observed during home visits.

Over the past twelve months, both Mr. and Mrs. Johnson have frequently collaborated with the digital health program by self-reporting significant changes in his health condition including the onset of depression related to his functional decline. Consequently, he is now receiving home visits from his HBPC social worker every 3–4 months. Through increased monitoring and early intervention, Mr. Johnson has not experienced any unscheduled hospitalizations since his enrollment in the Cardiac Telehealth Program. His average weight is 185.5 lbs with stable blood pressure at 128/70 and BMI is 28.95. Additionally, he has experienced a slight increase in his energy level, an improvement in his mood, and has not experienced any falls within the past 8 months. He is able to verbalize a better understanding of his disease progress and is very engaged in his healthcare management. Furthermore, Mr. Johnson has become very proficient with technology to the point that both he and his wife are now able to email their grandchildren and order their groceries online for their daughter to pick up each week.

The case of Mr. Brown

Mr. Brown is an 86-year-old widowed Air Force Veteran. He carries the diagnoses of diabetes and severe osteoarthritis. The Veteran is also hard of hearing and has impaired vision due to cataracts. He is on a fixed income of social security and lived alone in his own home until recently. Due to his osteoarthritis, the Veteran was having increasing difficulty getting around in his home and there were concerns from the Veteran's family and healthcare provider about his ability to safely reside in his home alone. While the Veteran had been active in the local VFW, due to his functional impairments he was no longer able to participate in his prior level of involvement. Both of his children lived out of state and worked full time and so the Veteran's local support system was limited to his VHA-funded home health aide who assisted the Veteran with laundry and personal care for 2h per day, 5 days per week.

The Veteran began having frequent falls, several of which resulted in the need for transfer to the local emergency department and in-home physical therapy. As these falls became more frequent and the associated injuries more severe, the Veteran's primary care provider consulted the LAMP Telehealth program in 2010 regarding Mr. Brown. The Veteran was initially resistant to the use of new technology in his home, but later became agreeable to enrollment in the LAMP Telehealth program after he was prepared by his primary care team and received education and information provided by healthcare providers and the LAMP telehealth team. Once enrolled, the LAMP Telehealth program began to address concerns with his abnormality of gait, frequent falls, and need for routine monitoring of his functional status. Enrollment provided long-term support for durable medical equipment, frequent monitoring of the Veteran, and ongoing communication with his clinic-based healthcare team as he left the home less frequently. The Veteran was no longer able to ambulate when his osteoarthritis worsened to the extent that he required a wheelchair and assistance with transferring. Communication occurred with the primary care health team to trigger a consult to Home-Based Primary Care when his telehealth team became aware of these issues through their engagement with the LAMP program.

In 2014, the Veteran was admitted to HBPC and he began to receive in-home primary care services by his healthcare team every 6–8 weeks. The Veteran began to endorse feelings of depression as his functional status worsened and his vision became more impaired. During this time, the Veteran also developed increasing anxiety regarding the possibility of cataract surgery and a fear of a total loss of vision without the procedure. As part of HBPC services, the Veteran received an E-consult to the HBPC psychiatrist who was able to work with him to ensure adequate medication management and a resulting decrease in both reported mental health symptoms. However, "noncompliance" with the digital health program became a concern as his functional abilities continued to decline.

Six months ago, the Veteran's son retired from employment at his out of state location and moved in with the Veteran to act as Mr. Brown's caregiver who was facing continuing functional decline. As the Veteran's son was adjusting to his own relocation and learning the details of his father's health condition and needs, the HBPC team was able to ensure optimization of the available technology. The HBPC team focused on the education and training needs of the Veteran's son to enable him to utilize myhealthevet for his father's prescription refills and email communication with the HBPC provider. This, in turn, has enabled the Veteran's son to fully participate in the custodial care that his father now requires. Additionally, the HBPC social worker was able to work with the Veteran and his son to obtain a VA nonservice connected (NSC) pension to increase the Veteran's limited income. This action enhanced the Veteran's quality of life and virtually eliminated his financial concerns.

While the Veteran is now wheelchair-bound and dependent upon a motorized scooter, he is able to leave the home with the assistance of his son for specialty medical appointments scheduled by his son through the myhealthevet Blue Button portal. At this time, the Veteran no longer endorses symptoms of depression or anxiety nor has he been hospitalized since 2014 or sustained a significant fall with injury since 2011. Despite his functional impairments, the Veteran is aware of how and when to use his provided equipment and is able to safely transfer with his son's assistance. He continues to verbalize his goals of remaining in his home with the support of both his HBPC and digital health team. Even with these successes, the HBPC team will continue to engage in ongoing evaluation of the Veteran's healthcare needs, the ongoing use of digital health technology to maintain the Veteran's well-being, and the support provided to his caregiver.

Discussion

As one of the largest providers of healthcare in the country, the VHA makes an ideal venue for research, development, and implementation of new healthcare technologies. With the growing Veteran population and the increasing demand for care and services, the VHA has the ability to treat a variety of healthcare conditions and conduct longitudinal studies of chronic debilitating illnesses that are impacting the aging Veteran population. Furthermore, the introduction of digital technology to the Veteran population serves as an advanced trial for non-Veteran population utilization across the country. This is significant as many taxpayers who share the burden of the costs of Veteran care are also indirectly benefitting from ongoing VHA research.

The phenomenon of how VHA research impacts non-Veteran populations may have limited awareness, but the topic is not new. VHA innovations in solid organ transplants starting in 1962, bone marrow transplants in 1982, and the development of the Left Ventricular Assist Devices (LVAD) in 2012 for patients awaiting heart transplants are all examples of the VHA leading change. The use of digital healthcare technology is another innovation and is a long history of transferring knowledge from VHA care and research to the general population. Furthermore, digital health technologies surpass the typical limits of innovation resulting from research that produces a stand-alone product for a limited population having little impact. Instead, these innovations can be replicated with positive outcomes including increasing quality of life. Digital healthcare technologies exemplify all of these factors.

Ultimately, there are several factors that should be taken into consideration for digital healthcare technologies to be successful. Those who seek to be educators in the field of healthcare technology must be knowledgeable, keep their skills up to date, and be capable of demonstrating their proficiency in technology and problem resolution to avoid customer frustration and loss of interest. Additionally, healthcare providers must understand their customer needs and limitations which may entail cognitive and functional decline. Furthermore, there may be a need to teach basic technology with the objectives to lay a good foundation, boost patient confidence, enhance patient acceptance of technology, and increase the likelihood that new tasks can be learned and mastered. Singularly, a basic understanding of project management and implementation should be understood as coordination and collaboration across the spectrum must be accomplished to ensure optimal implementation.

In the field of healthcare, particularly in the VHA system, there are a multitude of disciplines involved in the delivery of care. The roles and perspectives of each discipline should be considered with the implementation and maintenance of digital health, as alienation and conflict can result in the patient receiving less than desirable results. For example, if the digital healthcare technology being utilized requires notification to the provider of a change in the patients' status, it would be helpful to review the chart. Chart review can determine whether or not the patient is currently experiencing any problems and whether or not the provider is aware of and already addressing these concerns. If this is the case, consideration should be given contextually in how the information is relayed to the provider. If these factors are not taken into account and frequent and urgent notifications are repeatedly forwarded to the provider, the provider may experience frustration and potentially become less supportive of the program or simply ignore the notifications, all of which place the patient at risk. Finally, ongoing quality reviews and evaluations should add an intrinsic factor in program evaluation to determine if program goals are being met and what if any changes are needed.

Recommendations

This chapter noted multiple constraints imposed on Veterans and caregivers as they seek to access healthcare, as well as constraints for healthcare providers who endeavor to deliver optimal healthcare. Whether healthcare is delivered in institutional venues, such as hospitals, clinics, or medical provider offices, or noninstitutional venues, such as in the home or other residential settings, the use of digital technology can enhance both the delivery of healthcare and healthcare outcomes. For the integration of digital healthcare to be successful, ongoing educational opportunities to increase access are essential; as such, the following recommendations are being presented and are applicable to both the Veteran and non-Veteran population. Additionally, these recommendations can be easily incorporated into already existing venues such as media and information campaigns, thereby only resulting in minor changes and minimal financial costs.

Healthcare providers should focus on identifying and reducing barriers to patient participation in preventative healthcare initiatives. For those who have limited transportation resources, the utilization of digital technology in the home to participate in preventive telehealth programs, such programs as instructor-led exercise sessions, chronic pain management classes, cardiac symptom management, diabetes management, weight management, and a whole hosts of other preventive programs, would result in fewer acute care hospital days and overall reduction in healthcare costs, not just in the VA system, but throughout the country. For Veterans who may not be computer-literate, success for increasing awareness may be found through the VHA incorporating information about digital health programs during new patient orientation programs and through placing pamphlets and brochures with digital health program information in VA clinic lobby areas, utilization of the MyhealtheVet website, Veteran's organizations/groups, Veterans Services Offices, Veterans Benefits Administration offices, and Veterans Health Administration offices. Furthermore, VHA enrollment events provide an excellent opportunity where information can be disseminated about VHA digital health programs. Moreover, when Veterans demonstrate interest in a program, they can be presented with a one source list, such as that found in Table 4.2 Digital Health Technologies Utilized by the Veterans Health Administration (VHA) to Improve Patient Health. The benefit of wide and varied distribution is the ability to increase awareness of these programs among all Veterans whether or not they are enrolled in the VHA system.

Currently, there is a clear push to incorporate digital healthcare messages into the already existing VA media campaigns and we recommended that these not only be continued, but increased as a means to ensure and maintain awareness among all Veterans. Private healthcare recommendations include focusing on provider education regarding costs benefit and increase in positive patient outcomes, such as reduced emergency room visits, less frequent hospitalizations, reduction in hospital stay days, and increased staff efficiency. Additionally, the public health sector would benefit from campaigns educating the population about the reduction in healthcare dollars spent on care for chronic illnesses and revenue achieved from preventative health care programs and services. This can be achieved in a number of ways including during community outreach efforts and through incorporation into local public health departments' already existing preventive healthcare programs.

Conclusions

In this chapter, we discussed the Veteran's Health Administration (VHA) integration of digital technology within Home-Based Primary Care (HBPC) services on behalf of US Veterans who have served in the Armed Forces and Coast Guard. Key aspects of this chapter focus on the ability of the VHA to facilitate behavioral changes to enable Veterans to engage in digital platforms to supplement care that is both cost-effective and able to produce positive health outcomes. We demonstrated the relevance of the process of preparation, optimization, and ongoing evaluation through case studies depicting the positive outcomes when digital health technologies are part of treatment programs. Additionally, these innovative applications are viable and easily transferrable to healthcare settings in the general population, which enhances opportunities to increase access to care and reduce health costs, while improving healthcare outcomes for everyone. In conclusion, digital health programs are helpful tools in managing multiple disease processes when used in conjunction with the delivery of primary healthcare; however, successful implementation is paramount. Education regarding provider and patient expectations should be a focal point, as the integration of digital health may change patient expectations and may increase the need for provider communication, resulting in provider burnout and patient frustration. Furthermore, as digital health interventions are likely the result of the consultative process, collaboration and coordination are vital to maintain effective communication and role clarification.

References

Abel, E. A., Shimada, S. L., Wang, K., Ramesey, C., Skanderson, M., et al. (2018). Dual use of a patient portal and clinical video telehealth by Veterans with mental health diagnoses: Retrospective, cross sectional analysis. *Journal of Medical Internet Research*, *20*(11) e11350. https://doi.org/10.2196/11350.

Ahmadvand, A., Gatchel, R., Brownstein, J., & Nissen, L. (2018). The biopsychosocial-digital approach to health and disease: Call for a paradigm expansion. *Journal of Medical and Internet Research*, *20*(5), 1–7. https://doi.org/10.2196/jmir.9732.

American Public Health Association (APHA). (2014). *Removing barriers to mental health services for Veterans. APHA Policy No. 201411.*

Aref-Adib, G., McCloud, T., Ross, J., O'Hanlon, P., Appleton, V., et al. (2019). Factors affecting implementation of digital health interventions for people with psychosis or bipolar disorder and their family and friends: A systematic review. *The Lancet Psychiatry*, *6*, 257–266. https://doi.org/10.1016/S2215-0366(18)30302-X.

Beales, J. L., & Edes, T. (2009). Veteran's affairs home-based primary care. *Clinics in Geriatric Medicine*, *25*(1), 149–154. https://doi.org/10.1016/j.cger.2008.11.002.

Bendixen, R., Levy, C., Olive, E., Kobb, R., & Mann, W. (2009). Cost effectiveness of a telerehabilitation program to support chronically ill and disabled elders in their homes. *Telemedicine and E-Health*, *15*(1), 31–38. https://doi.org/10.1089/tmj.2008.0046.

Berchick, E., Hood, E., & Barnett, J. C. (2018). *Health insurance coverage in the United States. United Census Bureau Report Number P60-264*. Retrieved from https://www.census.gov/library/publications/2018/demo/p60-264.html.

Blais, R. K., & Renshaw, K. D. (2013). Stigma and demographic correlates of help-seeking intentions in returning service members. *Journal of Traumatic Stress*, *26*, 77–85. https://doi.org/10.1002/jts.21772.

Blanford, A., Gibbs, J., Newhouse, N., Perski, O., Singh, A., & Murray, E. (2018). Seven lessons for interdisciplinary research on interactive digital health interventions. *Digital Health*, *4*, 1–13. https://doi.org/10.1177/2055207618770325.

Cahn, A., Akirov, A., & Raz, I. (2018). Digital health technology and diabetes management. *Journal of Diabetes*, *10*, 10–17. https://doi.org/10.1111/1753-0407.12606.

Campbell, C., Freytes, I., & Hoffman, N. (2015). A home-based intervention's impact on caregiver burden for Veterans with dependence performing activities of daily living:

An interdisciplinary approach. *Social Work in Health Care*, *54*(5), 461–473. https://doi.org/10.1080/00981389.2015.1030056.

Canfield, J., & Weiss, E. (2015). Integrating military and Veteran culture in social work education; implications for curriculum inclusion. *Journal of Social Work Education*, *51*(S), S128–S144. https://doi.org/10.1080/10437797.2015.1001295.

Chang, C., Jackson, S. S., Bullman, T. A., & Cobbs, E. L. (2009). Impact of a home-based primary care program in an urban Veterans Affairs Medical Center. *Journal of the American Medical Directors Association*, *10*(2), 133–137. https://doi.org/10.1016/j.jamda.2008.08.002.

Chapman, A. (2010). The social determinants of health, health equity and human rights. *Health and Human Rights*, *12*(2), 17–30.

Collins, L., Murphy, S., & Strecher, V. (2007). The MOST and the SMART: New methods for more potent eHealth interventions. *American Journal of Preventative Medicine*, *32*(5S), S112–S119. https://doi.org/10.1016/j.amepre.2007.01.022.

Cooper, D. F., Granadillo, O. R., & Stacey, C. M. (2017). Home-based primary care: The care of the Veteran at home. *Home Healthcare Nurse*, *25*, 315–322. https://doi.org/10.1097/01.NHH.0000269965.

Dang, S., Ruiz, D. I., Klepac, L., Morse, S., Becker, P., et al. (2019). Key characteristics for successful adoption and implementation for home telehealth technology in Veterans Affairs HBPC: An exploratory study. *Telemedicine Journal and E-Health*, *25*(4), 309–318. https://doi.org/10.1089/tmj.2018.0009.

Edes, T. (2010). Innovations in homecare: VA HBPC. *Generations*, *34*(2), 29–34.

Edes, T., Kinosian, B., Vuckovic, N. H., Nichols, L. O., Becker, M. M., & Hossain, M. (2014). Better access, quality and cost for clinically complex Veterans with Home Based Primary Care. *Journal of the American Geriatrics Society*, *62*(10), 1532–5415. https://doi.org/10.1111/jgs.13030.

Edes, T., & Tompkins, H. (2007). Quality measure of reduction of inpatient days during HBPC. *Journal of the American Geriatric Society*, *55*, 6–7.

Evans, N. (2017). Relationship building: Enhancing Veteran access and the digital healthcare experience. *Vantage Point*.

Fargo, J., Stephen, M., Byrne, T., Munley, E., Montgomery, A., et al. (2012). Prevalence and risk of homelessness among US Veterans. *Preventing Chronic Disease*, *9*, 1–9. https://doi.org/10.5888/pcd9.110112.

Fox, M. (2018). *Veterans more likely than civilians to die by suicide, VA study finds*. NBC Health News. Retrieved May 24, 2019 from https://www.nbcnews.com/health/health-news/Veterans-more-likely-civilians-die-suicide-va-study-finds-n884471.

Government Accountability Office (GAO). (2015, February). *High risk series, an update. Report to Congressional Committees. GAO-15-290*.

Government Accountability Office (GAO). (2016, March). *Actions needed to improve newly enrolled Veterans' access to primary care. Veterans administration healthcare report to the chairman, subcommittee on oversight and investigations, committee on Veterans affairs, house of representatives. GAO-16-328*.

Government Accountability Office (GAO). (2017, March). *Actions needed to better recruit and retain clinical and administrative staff. Veterans' health administration testimony before the subcommittee on health, committee on Veteran's affairs, house of representatives. GAO-17-475T*.

Government Accountability Office (GAO). (2018a, April). *Opportunities exist for improving Veteran's access to health care services in the Pacific Islands*. Veterans' Health Administration. Report to Congressional Addressees. GAO-18-288.

Government Accountability Office (GAO). (2018b, June). *Faster care for Veterans act.* Washington, DC: Government Accountability Office (GAO). GAO-18-442R.

Griffin, R. J. (2015). *Statement of the acting inspector general of the VA before the Committee on Veterans Affairs hearing on "The state of VA Health Care"*. Retrieved May 17, 2019 from http://www.va.gov/oig/pubs/statements/VAOIG-statement-20140515-griffin.pdf.

Haggstrom, D., Saleem, J., Russ, A., Jones, J., Russell, S., & Chumbler, N. (2011). Lessons learned from usability testing of the VA's PHR. *Journal of the American Medical Informatics Association, 18*, 113–117. https://doi.org/10.1136/amiajnl-2010-000082.

Hale, A., Haverhals, L., Manheim, C., & Levy, C. (2018). Vet Connect: A quality improvement program to provide telehealth subspecialty care for Veterans residing in VA-contracted community nursing homes. *Geriatrics, 9*(3), 57–69. https://doi.org/10.3390/geriatrics3030057.

Hicken, B. L., Daniel, C., Luptak, M., Grant, M., Kilian, S., & Rupper, R. W. (2017). Supporting caregivers of rural Veterans electronically (SCORE). *Journal of Rural Health, 33*, 305–313. https://doi.org/10.1111/jrh.12195.

Hicken, B. L., & Plowhead, A. (2010). A model for home-based psychology from the VHA. *Professional Psychology: Research and Practice, 41*(4), 340–346. https://doi.org/10.1037/a0020431.

Houston, T. K., Volkman, J. E., Feng, H., Nazi, K. M., Shimada, S. L., & Fox, S. (2013). Veteran internet use and engagement with health information online. *Military Medicine, 178*(4), 394–400. https://doi.org/10.7205/MILMED-D-12-00377.

Hughes, S., Weaver, F., Giobbie-Hurder, A., Manheim, L., Henderson, W., et al. (2000). Effectiveness of team-managed HBPC: A randomized multicenter trial. *Journal of the American Medical Association, 284*(22), 2877–2885. https://doi.org/10.1001/jama.284.22.2877.

Hynes, D. M., Whittier, E. R., & Owens, A. (2013). Health information technology and implementation science. *Medical Care, 51*(3), S6–12.

Kahn, J., Aulaka, V., & Bosworth, A. (2009). What it takes: Characteristics of the ideal PHR. *Health Affairs, 28*(2), 369–376. https://doi.org/10.1377/hlthaff.28.2.369.

Kramer, B. J., Creekmur, B., Cote, S., & Saliba, D. (2015). Improving access to noninstitutional long-term care for American Indian Veterans. *Journal of American Geriatric Society, 63*(4), 789–796. https://doi.org/10.1111/jgs.13344.

Lafontaine, M., Azzi, S., Paquette, D., Tasca, G. A., Greenmand, P. S., et al. (2018). Telehealth for patients with chronic pain: Exploring a successful and unsuccessful outcome. *Journal of Technology in Human Services*. https://doi.org/10.1080/15228835.2018.1491370.

Levine, D. M., Lipsitz, S. R., & Linder, J. A. (2018). Changes in everyday and digital health technology use among seniors in declining health. *Journal of Gerontology, 73*(4), 552–559. https://doi.org/10.1093/gerona/glx116.

Levitt, L., Neuman, T., & Brody, M. (2019, May 29). *Making sense of Medicare-for-All and other plans to expand public coverage. Kaiser Family Foundation*. Retrieved June 16, 2019 from https://www.kff.org/health-reform/event/may-21-web-briefing-making-sense-of-medicare-for-all-and-other-plans-to-expand-public-coverage/.

Lewis, C., Coleman, A., Abrams, M., & Doty, M. (2019). *The role of Medicaid expansion in care delivery at community health centers. In The commonwealth fund.* Retrieved from https://www.commonwealthfund.org/publications/issue-briefs/2019/apr/role-medicaid-expansion-care-delivery-FQHCs.

Liu, C. K., Hsu, C. Y., Yang, F. Y., Wu, J., Kou, K., & Lai, P. (2017). Population health management outcomes obtained through a hospital based and telehealth informatics enabled telecare service. In *IEEE biomedical circuits and systems conference*. ISBN: 9781-50058044.

Luptak, M., Dailey, N., Juretic, M., Rupper, R., Hill, R. D., et al. (2010). The care coordination home telehealth rural demonstration project: A symptom-based approach for serving older Veterans in remote geographical settings. *Rural and Remote Health*, *10*(2), 1375–1385.

Matthews, K. C., Vinson, L., DeWitt, M., McGuire, M., & Karel, M. (2019). Caring for older patients with complex problems: Challenge strategies and the VHA experience. *American Journal of Geriatric Psychiatry*, *27*(3), S45–S46. https://doi.org/10.1016/j.jagp.2019.01.198.

Miliard, M. (2019, June 19). *Telehealth set for big boost with $100M in new funding from FCC*. Healthcare IT News. Retrieved from https://www.healthcareitnews.com/news/telehealth-set-big-boost-100m-new-funding-fcc?mkt_tok=eyJpIjoiTWpFek16TmpNakZsT1RWbSIsInQiOiJSM3Qrb3RTeVlZUlJJV2czcFVpMGxraFpCZ0pwdnY0V3RTVlAzZUlJSW4xSHZNOVFvWTV0QUdIaEQrUm1VVU1BNFFlbHlGS21xK0NhWm1QMTBRWXgrXC9RT1JQMjl5eWN1eE56TDUzMjRheUtPQTRYUDRqOWFyeG4yRkxYR00ySDkifQ%3D%3D.

Miltner, R., Selleck, C., Moore, R., Patrician, P., Froelich, K., Eagerton, G., et al. (2013). Equipping the nursing workforce to care for the unique needs of Veterans and their families. *Nurse Leader*, 45–48. https://doi.org/10.1016/j.mnl.2013.05.013.

Mishuris, R., Stewart, M., Fix, G. M., Marcello, T., McInnes, D. K., et al. (2014). Barriers to patient portal access among Veterans receiving home-based primary care: A qualitative study. *Health Expectations*, *18*, 2296–2305. https://doi.org/10.1111/hex.12199.

Moller, A., Merchant, G., Conroy, D., West, R., Heckler, E., Kugler, K., et al. (2017). Applying and advancing behavior change theories and techniques in the context of a digital health revolution: Proposals for more effectively realizing untapped potential. *Journal of Behavioral Medicine*, *40*, 85–98. https://doi.org/10.1007/s10865-016-9818-7.

Moye, J., Harris, G., Kube, E., Hicken, B., Adjognon, O., Shay, K., et al. (2019). Mental health integration in geriatric patient-aligned care teams in the Department of Veterans Affairs. *American Journal of Geriatric Psychiatry*, *27*(2), 100–108. https://doi.org/10.1016/j.jagp.2018.09.001.

Mueller, S. (2018, September 2). *Delaware passes new law to help patient access to primary care*. Delaware Public Media. Retrieved June 4, 2019 from https://www.delawarepublic.org/post/delaware-passes-new-law-help-patient-access-primary-care.

Naditz, A. (2008). Telemedicine at the VA: VistA, MyHealtheVet, and other VA programs. *Telemedicine and e-Health*, *14*(4), 330–333. https://doi.org/10.1089/tmj.2008.9973.

National Coalition for Homeless Veterans (n.d.). FAQ about homeless Veterans. Retrieved May 24, 2019 from http://nchv.org/index.php/news/media/background_and_statistics/#demo.

Nelson, K. M., Helfrich, C., Sun, H., Herbert, P. L., Liu, C. F., et al. (2014). Implementation of the patient centered medical home in the VHA associations with patient satisfaction, quality of care, staff burnout, and hospital and emergency department use. *Journal of the American Medical Association Internal Medicine*, *174*(8), 1350–1358. https://doi.org/10.1001/jamainternmed.2014.2488.

Neter, E., & Brainin, E. (2019). Association between health literacy, eHealth literacy and health outcomes among patients with long-term conditions. *Adjustment to Chronic Illness*, *24*(1), 68–81. https://doi.org/10.1027/1016-9040/a000350.

North, L., Kehm, L., Bent, K., & Hartman, T. (2008). Can HBPC cut costs? *The Nurse Practitioner*, *33*(7), 39–44. https://doi.org/10.1097/01.NPR.0000325980.52580.f3.

Olenick, M., Flowers, M., & Diaz, V. J. (2015). US Veterans and their unique issues: Enhancing health care professional awareness. *Advances in Medical Education and Practice*, *6*, 635–639. https://doi.org/10.2147/AMEP.S89479.

Perzynski, A., Roach, M., Schick, S., Callahan, B., Gunzler, D., et al. (2017). Patient portals and broadband internet inequality. *Journal of the American Medical Informatics Association, 23*, 927–932. https://doi.org/10.1093/jamia/ocx020.

Petrovich, J., Pollio, D., & North, C. (2014). Characteristics and service use of homeless Veterans and nonveterans residing in a low-demand emergency shelter. *Psychiatric Services, 65*(6), 751–757. https://doi.org/10.1176/appi.ps.201300104.

Pew Research Center. (2017). *The changing face of America's Veteran population.* Retrieved May 21, 2019 from https://www.pewresearch.org/fact-tank/2017/11/10/the-changing-face-of-americas-Veteran-population/.

RAND Corporation (2015). *Assessment A (Demographic): A product of the CMS Alliance to Modernize Healthcare.* Prepared for US Department of Veteran's Affairs. Contract No HHS-M500-2012-00008I.

Razak, M. (2018). *VHA innovation experience delivers the future of Veteran healthcare.* Vantage Point.

Mena Report. (2018, May 2). *VA Video Connect expands Veterans Access to Healthcare.* Retrieved May 17, 2019 from http://go.galegroup.com.ezproxy.net.ucf.edu/ps/retrieve.do?tabID=T004&resultListType=RESULT_LIST&searchResultsType=SingleTab&searchType=AdvancedSearchForm¤tPosition=1&docId=GALE%7CA537133099&docType=Brief+article&sort=RELEVANCE&contentSegment=ZGRC-MOD1&prodId=GRGM&contentSet=GALE%7CA537133099&searchId=R3&userGroupName=orla57816&inPS=true.

Rickles, N., Dominguez, S., & Amaro, H. (2010). Perceptions of healthcare, health status and discrimination among African-American Veterans. *Journal of Health Disparities Research & Practice, 4*(2), 50–68.

Rosenheck, R., & Koegel, P. (1993). Characteristics of Veterans and nonveterans in three samples of homeless men. *Hospital Community Psychiatry, 44*(9), 858–863. https://doi.org/10.1176/ps.44.9.858.

Saha, S., Freeman, M., Toure, J., Tippens, K. M., & Weeks, C. (2007). *Racial and ethnic disparities in the VA healthcare system: A systematic review. Evidence synthesis pilot program.* Washington, DC: Department of Veterans Affairs Health Services Research and Development Service. Retrieved July 14, 2019 from https://permanent.access.gpo.gov/lps124979/RacialDisparities-2007.pdf.

Seervai, S. (2017). *How Hurricane Maria worsened Puerto Rico's healthcare crisis.* The Commonwealth Fund. Retrieved June 4, 2019 from https://www.commonwealthfund.org/publications/publication/2017/dec/how-hurricane-maria-worsened-puerto-ricos-health-care-crisis.

Sellers, E., Campbell, C., & Moore, K. (2018). Early referral and collaboration with low ADL monitoring program (LAMP) in Veterans with Amyotrophic Lateral Sclerosis. In *Poster presented at the Paralyzed Veterans of America 2018 healthcare summit & expo, Gainesville, Florida.*

Semeah, L. M., Campbell, C. L., Cowper, D. C., & Peet, A. C. (2017). Serving our homeless veterans: Patient perpetrated violence as a barrier to health care access. *Journal of Public and Nonprofit Affairs, 3*(2), 223–234. https://doi.org/10.20899/jpna.3.2.223-234.

Senecal, C., Widmer, R., Bailey, K., & Lerman, A. (2017). Socioeconomic environment and digital health usage and outcomes: Is there a digital divide? *Journal of the American Academy of Cardiology, 69*(11), S1754. https://doi.org/10.1016/S0735-1097(17)35143-4.

Sharma, A., Harrington, R. A., McClellan, M. B., Turakhia, M. P., Eapen, Z. J., et al. (2018). Using digital health technology to better generate evidence and deliver evidence-based care. *Journal of the American College of Cardiology, 71*(23), 2680–2690. https://doi.org/10.1016/j.jacc.2018.03.523.

Sharp, N. D., Pineros, S. L., Hsu, C., Starks, H., & Sales, A. E. (2004). A qualitative study to identify barriers and facilitators to implementation of pilot interventions in the VHA Northwest network. *Worldviews on Evidenced-Based Nursing*, *1*(2), 129–139. https://doi.org/10.1111/j.1741-6787.2004.04023.x.

Statista. (2019). *US national health expenditure as percent of GDP from 1960 to 2019*. Retrieved May 17, 2019 from https://www.statista.com/statistics/184968/us-health-expenditure-as-percent-of-gdp-since-1960/.

Sweet, C. M., Chiguluri, V., Gumpiina, R., Abbott, P., Madero, E. N., et al. (2018). Outcomes of a digital health program with human coaching for diabetes risk reduction in a Medicare population. *Journal of Aging and Health*, *30*(5), 692–710. https://doi.org/10.1177/0898264316688791.

Turvey, C., Klein, D., Fix, G., Hogan, T. P., Woods, S., et al. (2014). Blue button use by patients to access and share health record information using Department of Veterans Affairs' online patient portal. *Journal of the American Med Inform Assoc*, *21*, 657–663. https://doi.org/10.1136/amiajnl-2014-002723.

United States Bureau of Labor Statistics. (2014). *Current population survey [Database]*. Retrieved May 21, 2019 from https://www.bls.gov/cps/.

United States Bureau of Labor Statistics. (2017). *News release*. USDL-18-0453. Retrieved from https://www.bls.gov/news.release/archives/vet_03222018.pdf.

United States Census Bureau. (2015). *American community survey [Database]*. Retrieved May 17, 2019 from http://www.census.gove/programs-surveys/acs.

United States Census Bureau. (2017). *American community survey (ACS) 5-year estimates. Veterans: Definitions and concepts*. Retrieved May 23, 2019 from https://www.census.gov/topics/population/Veterans/about.html.

United States Department of Veterans Affairs. (2007). What is home based primary care? Retrieved February 13, 2020 from https://www.benefits.gov/benefit/302.

United States Department of Veterans Affairs. (2012). *Survey of Veteran enrollees' health reliance upon VA*. Veterans Health Administration. Retrieved May 17, 2019 from http://www.va.gov/healthpolicyplanning/soe2011/soe2011_report.pdf.

United States Department of Veterans Affairs. (2017, June). *Overview of VA research*. Retrieved May 17, 2019 from https://www.research.va.gov/pubs/docs/va_factsheets/varesearch_overview_factsheet.pdf.

United States Department of Veterans Affairs. (2018a). *Office of policy & planning. Veteran population—National Center for Veterans Analysis and Statistics*. Retrieved May 17, 2019 from www.va.gov/vetdata/Veteran_Population.asp.

United States Department of Veterans Affairs. (2018b). *VA video connect*. Washington, DC: US Department of Veterans Affairs. Retrieved from https://mobile.va.gov/app/va-video-connect.

United States Department of Veterans Affairs. (2019a). *Access and manage your VA benefits and health care*. Retrieved May 17, 2019 from www.va.gov.

United States Department of Veterans Affairs. (2019b). *Geriatrics and extended care. home based primary care*. Retrieved May 17, 2019 from https://www.va.gov/geriatrics/guide/longtermcare/Home_Based_Primary_Care.asp.

United States Department of Veterans Affairs. (2019c). *Rural Veterans*. Office of Rural Health. Retrieved May 17, 2019 from https://www.ruralhealth.va.gov/aboutus/ruralvets.asp.

United States Department of Veterans Affairs. (2019d). *Geriatrics and extended care. Patient Education Information Sheet*. Retrieved May 17, 2019 from https://www.northflorida.va.gov.

United States Department of Veterans Affairs. (2019e). *National center for Veterans analysis and statistics*. Retrieved May 21, 2019 from https://www.va.gov/vetdata/docs/SpecialReports/Profile_of_Veterans_2017.pdf.

Walsh, J. C., & Groarke, J. M. (2019). Integrating behavioral science with mHealth. *European Psychologist*, *24*(1), 38–48. https://doi.org/10.1027/1016-9040/a000351.

Ward, C., Cope, M., & Elmont, L. (2017). Native American Vietnam-era Veteran's access to VA healthcare: Vulnerability and resilience in two Montana Reservation Communities. *Journal of Community Health*, *42*, 887–893. https://doi.org/10.1007/s10900-017-0330-y.

Ward, C., Solomon, Y., & Stearmer, S. M. (2011). *Access to health care in rural communities and highly rural areas: Views of Veterans, community members and leaders and service providers. White Paper submitted to Department of Veterans Affairs.* Western Region, Salt Lake City, Utah: Office of Rural Health, Veterans Rural Health Resource Center.

Weaver, F., Hughes, S., Kubal, J., Ulasevich, A., Bonarigo, F., & Cummings, J. (1995). A profile of Department of Veterans Affairs hospital based home care programs. *Home Health Care Services Quarterly*, *15*(4), 83–97.

Webb, S. C., & Herrmann, W. J. (2002). *Historical overview of racism in the military*. Defense Equal Opportunity Management Institute Special Series Pamphlet 02-1.

Whitten, P. (2006). Telemedicine: Communication technologies that revolutionize healthcare services. *Generations*, *30*(2), 20–24. https://doi.org/10.1017/S0266462305050725.

Whitten, P., & Mackert, M. (2005). Addressing telehealth's foremost barrier: Provider as initial gatekeeper. *International Journal of Technology Assessment in Health Care*, *21*(4), 517–521.

Wolf, J. (2006). The benefits of diabetes self-management education of the elderly Veteran in the home care setting. *Home Healthcare Nurse*, *24*(10), 645–651.

Woods, S., Forsberg, C., Schwartz, E., Nazi, K., Hibbard, J., Houston, T., et al. (2017). The association of patient factors, digital access and online behavior on sustained patient portal use: A prospective cohort of enrolled users. *Journal of Medical Internet Research*, *19*(10), 1–14. https://doi.org/10.2196/jmir.7895.

Supplementary reading recommendations

Abroms, L., Allegrante, J., Aultd, M., Gold, R., Riley, W., & Smyser, J. (2019). Toward a common agenda for the public and private sectors to advance digital health communication. *American Journal of Public Health*, *109*(2), 221–222.

Campbell, C., Freytes, I., & Hoffman, N. (2015). A home-based intervention's impact on caregiver burden for Veterans with dependence performing activities of daily living: An interdisciplinary approach. *Social Work in Health Care*, *54*(5), 461–473.

Campbell, C., McCoy, S., Hoffman, N., & O'Neil, P. (2014). Decreasing role strain for caregivers of Veterans with dependence in performing activities of daily living. *Health & Social Work*, *39*(1), 1–64. https://doi.org/10.1093/hsw/hlu006.

Center for Disease Control and Prevention. (n.d.) Traumatic brain injury and concussion. https://www.cdc.gov/traumaticbraininjury/index.html.

Code of Federal Regulations. (2002). *Department of Veterans Affairs. Title 38. Part 17 medical benefits package.*

DeCherrie, L., & Soriano, 'Il, & Hayashi, J. (2012). Home based primary care: A needed primary care model for vulnerable populations. *Mount Sinai Journal of Medicine*, *79*(4), 425–432. https://doi.org/10.1002/msj.21321.

Veterans Crisis Line. https://www.Veteranscrisisline.net/.

Jha, A., Perlin, J., Kizer, K., & Dudley, R. Effect of the transformation of the Veterans Affairs Health Care System on the quality of care. New England Journal of Medicine, 348, 2218–2227. https://doi.org/10.1056/NEJMsa021899.

Karlin, B., Zeiss, A., & Burris, J. (2010). Providing care to older adults in the Department of Veterans Affairs: Lessons for us all. *Journal of the American Society on Aging, 34*(2), 6–12.

Mader, S. L., Medcraft, M. C., Joseph, C., Jenkins, K. L., Benton, N., et al. (2008). Program at home: A Veterans Affairs healthcare program to deliver hospital care in the home. *Journal of the American Geriatrics Society, 56*, 2317–2322. https://doi.org/10.1111/j.1532-5415.2008.02006.x.

Silver, G., Keefer, J., & Rosenfeld, P. (2011). Assisting patients to age in place: An innovative pilot program utilizing the patient centered care model (PCCM) in home care. *Home Health Care Management and Practice, 23*(6), 446–453. https://doi.org/10.1177/1084822311411657.

United States Department of Veterans Affairs. (2016). (i) Home based primary care, https://www.va.gov/geriatrics/Guide/LongTermCare/Home_Based_Primary_Care.asp;
(ii) Uniform geriatrics and extended care services in VA medical centers and Clinics. VHA Directive 1140.11. Veterans Health Administration, Washington, DC.

United States Department of Veterans Affairs. (2019). *Geriatrics and extended care programs.* www.va.gov/geriatrics.

United States Department of Veterans Affairs. (n.d.). (i) Connected care programs. https://connectedcare.va.gov/; (ii) Polytrauma/TBI System of care. https://www.polytrauma.va.gov/; (iii) VAntage Point. Official Blog of the U.S. Department of Veterans Affairs. https://www.blogs.va.gov/VAntage/; (iv) Veterans Benefits Administration https://benefits.va.gov/benefits/; (v) Veterans Health Administration https://www.va.gov/health/.

Definitions

Activities of daily living (ADL) these include key components of personal hygiene (specifically bathing), dressing, continence of bowel and bladder, feeding, transferring, and mobility. These are contrasted from the instrumental activities of daily living (IADLs) of the ability to communicate, transportation, meal preparation, shopping, management of medications, and management of finances.

Cognitive functioning the ability to cognitively process information received. This term encompasses elements of attention, language, perception, short- and long-term memory, and the ability to cognitively process information and decision-making.

Deployment occurs when active duty military members (or members of the US Reserves are activated) are assigned overseas to a current area of conflict or to a location requiring action of the US Armed Forces.

Dyad a group composed of two individuals.

Geriatric the aging population, typically those over the age of 65.

Geriatrics and extended care (GEC) the Veterans Health Administration includes GEC services as part of 38 CFR 17.38 (Medical Benefits Package) for all Veterans. These include the benefit of hospice and palliative care, in-home and inpatient respite services, geriatric evaluation, nursing home care, and adult daycare services.

Home-Based Primary Care primary care services (to include an ARNP/RN, licensed clinical social worker, psychiatrist, psychologist, dietician, occupational therapist, and pharmacist) delivered to the Veteran in their home rather than in the VA healthcare center. Population includes those who are frail or elderly with chronic medial illness and are primarily homebound as a result of these illnesses.

iOS the mobile operating system used for mobile devices manufactured by Apple Inc.

Palliative care in contrast to curative or aggressive care, palliative care focuses on comfort measures to relieve pain. The focus of this approach is on quality of life rather than extension of quantity of life.

Posttraumatic stress disorder (PTSD) as defined by the DSM-5, this is a trauma and stress-related disorder whereby an individual witnesses a traumatic event that was directly experience and the disturbance results in clinically significant distress impairing the individual's ability to function. Symptoms include reexperiencing the traumatic event, avoidance, negative cognitions and mood, and hyperarousal.

Traumatic brain injury often referred to as a TBI, the US Centers for Disease Control defines TBI as a disruption of normal functioning of the brain resulting from a traumatic injury to the head.

Vital signs clinical healthcare measurements to include temperature, pulse, blood pressure, and respiration rate.

The power of peer support in family/youth serving systems

Eve Bleyhl
Nebraska Family Support Network, Omaha, NE, United States

Abstract

The Family First Prevention Services Act of 2018 recognizes that children have a better chance of overcoming behavioral obstacles in the home with support from youth serving systems receiving limited allocated funding to teach interactive parenting skills and service delivery for mental health counseling and substance abuse treatment. But barriers to service delivery, professional relationships, and program needs persist in the midst of persistent labor shortages and rising numbers of families with service needs. Family peer support approach to Children's Behavioral Health, Child Welfare, and Juvenile Justice is a significant alternative that utilizes the lived experience of prior recipients of care to guide others entering the program. This chapter utilizes four case studies demonstrating the change from family members in need of services to those empowered to provide services. The Nebraska Family Support Network serves as a model of family peer support.

Keywords: Child trauma, Child Welfare Services, Peer support networks, Service delivery, Out-of-home care.

Chapter outline

Three Facets of Public Health and Paths to Improvements. https://doi.org/10.1016/B978-0-12-819008-1.00005-5

Do we want public systems raising America's children?

America's children are increasingly at risk due to a number of reasons including poor socioeconomic conditions and the rise of opioid and other addictions of parents or legal guardians. These determinants, such as poverty, trauma, domestic violence, mental illness, and addiction, are significant contributors to parents/youth becoming system involved (Donnelly, Baker, & Gargan, 2017; Farrell et al., 2017; Grohol, 2018; SAMHSA, 2017a). In 2017, more than 3.5 million or 4.7% of all American children aged 0 to 17 years were being raised in foster care or living apart from their families of origin in out-of-home care, while population projections and growing numbers of confined youth suggest that number could be close to 6% in 2025 (ChildStats.gov, 2017; U.S. Department of Health and Human Services, Administration for Children and Families, Administration on Children, Youth and Families, Children's Bureau, 2019). An estimated 437,465 children were in foster care in 2016 (Child Welfare Information Gateway, 2017b), while more than 20,000 youths annually face uncertain futures as they turn 18 years of age and age out of the system (Fowler, Marcal, Zhang, Day, & Landsverk, 2016).

The scope of the problem is daunting and falls across several systems of care. For example, the U.S. Department of Health and Human Services (USDHSS) Child Maltreatment Report for 2017 (2019) using rounded figures indicate the following:

- 10% rise in national number of children who received a child protective services investigation response or alternative response percent increase from 2013 (3,184,000) to 2017 (3,501,000) (USDHHS, 2019, p. 18).
- 74.9% of victims are neglected, 18.3% are physically abused, and 8.6% are sexually abused (USDHHS, 2019, p. 23). These victims may suffer a single maltreatment type or a combination of two or more maltreatment types that may also include medical neglect or psychological trauma (USDHHS, 2019, p. 46–47).
- Nearly 674,000 national victims of child abuse and neglect were reported for fiscal year 2017 establishing a national population rate of 9.1 victims per 1000 children that demonstrates a 2.7% increase from national data for 2013 (656,000) (USDHHS, 2019, p. 20).

Other youth are spending much of their childhood in residential treatment settings for mental health challenges. The National Mental Health Services Survey (N-MHSS), an organization under the Substance Abuse and Mental Health Services Administration (SAMHSA), reports 29,454 24-hour residential clients aged 12 and younger and 36,732 24-hour residential clients for clients 13–17 years old represent a daily snapshot of the count on April 29, 2016 (SAMHSA, 2017b). Finally, "On any given day, nearly 53,000 youth are held in facilities away from home as a result of juvenile or criminal justice involvement" (Sawyer, 2018, para 1). These additional sources, among many others, indicate the disproportionate number of families and youth struggling to remain a family unit in the United States.

Further, these neglected and abused youth that are raised in public systems (e.g., child protective services, residential treatment programs, group homes, or juvenile detention facilities) show a greater propensity for experiencing adult challenges in mental health, addictions, physical health, educational attainment, earning potential, homelessness, parenting, and criminal activity (Clemens, Helm, Myers, Thomas, & Tis, 2017;

Gypen et al., 2017). One study found that children raised in foster care "had twice the level of post-traumatic stress disorder as Gulf War veterans, nearly one-third were abused in foster care and, as young adults, only one in five was doing well," (Wexler, 2018, para 2). Yet, well-documented and persistent barriers remain in terms of access to required services and support systems that could positively impact a family's capacity to keep children in the home (Stroul, Blau, & Friedman, 2010).

America has an unfortunate history of pointing fingers and wanting to absolve public responsibility for circumstances that create the situations that these children are living by implying that our nation's challenges around supporting families and youth trace solely back to "bad" parents or "bad" kids. While this mindset explains America's historical punitive approach to addressing the needs of these families, the repositioning of children into the hands of "professionals" outside the home does not address the likely determinants of out-of-home placements nor can parents meet their child's needs when they are frequently shut out of their child's care.

An absence of a fundamental mindset that engages parents with children in their homes subjects children who are already traumatized in their family of origin to trauma in the systems whose intention is to intervene (Salazar, Keller, Gowen, & Courtney, 2012). The situation and the number beg the question, "Do we want public systems raising America's children?" The answer, of course, is a resounding, "No." But finding alternatives to care is a high hurdle in a series of barriers facing the existing system of care. These barriers span failed legislation (History Nebraska, 2017; O'Hanlon, 2013) to a persistent shortage of workers (Donnelly et al., 2017) that must be overcome to stop the cycle of childhood maltreatment and adverse childhood experiences often painting a child into a corner for their entire lifespan. However, the evolution of peer support offers a viable alternative to the status quo.

Donnelly et al. (2017, p. 8):

> *Family peer support is the instrumental, social and informational support provided from one parent to another in an effort to reduce isolation, shame and blame, to assist parents in navigating child serving systems and provide other relevant life experiences. Family peer support is the unrelenting focus on the parent/primary caregiver(s), while other team members focus on the identified child and family.*

The good news is that the tide is turning. Recent legislation such as the Family First Prevention Services Act, passed in February 2018, was founded on the belief that children ultimately do best with their biological parents following peer support methods used in Nebraska since 2003. Thus, the FFSA 2018 prioritizes keeping families together and puts more money toward services including parenting classes, mental health counseling, and substance abuse treatment (Torres & Rricha, 2018). The Act also puts limits on placing children in institutional settings such as group homes. This congressional affirmation, if appropriately enacted, is expected to change the face of support for struggling parents and youth in our society (Wiltz, 2018) and the methods utilized in the delivery of services. More importantly, the Act established peer support as a legislated service delivery component for children's behavioral health in the United States and will increase funding for this model in Nebraska.

For many families/youth, the foundation of family engagement may be the first time that families get the opportunity or impetus to work on the challenges they face. The objective to involve family members at the earliest possible point in the casework process promises to be instrumental in the eventual elimination of the need for a child to be placed outside of the home. Just as youth who have parents active in their educational setting do better in school, the expectation is that utilizing effective engagement strategies creates a trust in the process that increases the likelihood of family buy-in. If families and youth feel like they have a voice in the process and if they receive encouragement/validation vs shame/blame, they are more likely to resonate with identifying needs and setting goals to address the challenges they are facing.

We are addressing the significance of family peer support in this chapter beginning with a multilevel overview from the broad national perspective of the United States, the State of Nebraska, and at the organizational level focusing on the operations of the Nebraska Family Support Network. Then, four case studies demonstrate the application of lived experiences in the process of healing broken families. We move on to discuss some relevant challenges and demonstrate how new strategies through collaborative efforts are being formulated using the Family Run Organizations of Nebraska (FRON) model.

Multilevel view of peer support and childhood experiences

The progression of peer support is important to understanding the trickle down impact upon state and local organizations. In this section, we bring a national overview of the history of peer support. Then, we provide a state-level snapshot of Nebraska focusing on the influence of peer support participants on various policies. Finally, we view the perspective of peer support at the organizational-level, emphasizing the activities of the Nebraska Family Support Network (NFSN). This information is intended to lead to an expansive perspective on general premises that are carried throughout multiple levels of interaction in peer support.

National overview of the history of peer support

Peer support has national historical roots in 12-Step programs, such as Alcoholics Anonymous, beginning roughly 80 years ago. The transition of individual groups to the emergence of family engagement, lived experience, and peer support has been a relatively new option for struggling families. Family peer support, creating parent to parent relationships for those who share the common problem of placement of their child(ren) outside of the home (National Federation of Families for Children's Mental Health, 2017a), began to appear in the late 1970s. At the same time, organizations such as the National Alliance on Mental Illness (NAMI) were focusing on education and resources for mental health consumers and their families (Bodenheimer, 2016; Grohol, 2018).

The change in mindset from consumer peer support to family peer support in care interaction further developed through the 1980s bringing a novel focus on the family or Family Movement in landmark conferences sponsored by the Portland State University Research and Training Center on Family Support with a theme of

"Families and Professional Working Together," (National Federation of Families for Children's Mental Health, 2017a, para 4). By 1989, The Federation of Families for Children's Mental Health became the first family run organization in the United States according to the Family Run Executive Director Leadership Association (FREDLA) (2015). Now they are one of 118 today in which most include the criteria of lived experience in the systems in which we provide our services (FREDLA, 2019).

SAMHSA, a subgroup of the United States Department of Health and Human Services (USDHHS), in 1992, became a substantial federal agency and funding source enacted by Congress to advance attention to behavioral health in the United States (FREDLA, 2015; National Federation of Families for Children's Mental Health, 2017a). This family peer support milestone was the foundation of their establishment as a significant voice and resource for utilizing peer support in public systems (Donnelly et al., 2017). Today, peer support includes the attainment of physical health (Tang, 2013) built on a focus on child's behavioral health initiated when Congress funded the Child & Adolescent Service System Program in 1984 (FREDLA, 2015).

Additional milestones were important to the development of Family Run and Youth Organizations seen today. They include a certification commission formed for parent peer support providers in 2011 followed closely by the collaboration between SAMHSA and Centers for Medicare and Medicaid (CMS) in 2013. The collaboration was instrumental in "identifying parent peer support as a key service that can enable children with complex needs to live at home and participate fully in family and community," (FREDLA, 2015, p. 1). This established eligibility for Centers for Medicare and Medicaid Services (CMS) billing, providing urgent resources for parent peer support.

State of Nebraska snapshot of child maltreatment and adverse childhood experiences

On a state level, Nebraska families are facing their own challenges with youth that require sustained treatment for mental health and substance abuse. But children also face out-of-home placement for developmental problems such as autism and concurring developmental disabilities, and mental illness that may be exasperated due to the prevalence of single parenthood. More than 28.1% of Nebraska kids were living with a single parent in 2017, an increase from 12% in 1980 (Voices for Children in Nebraska, 2018, p. 33).

What is especially alarming is that there are many more children who require services but are not receiving them. ©Kids Count in Nebraska Report 2018 includes a summary of The National Survey of Children's Health 2016–2017, indicating that 57.6% of children needing mental health counseling actually received that service (Voices for Children in Nebraska, 2018, p. 40). While 3, 657 Nebraska children received mental health and substance abuse services through the CMS in 2017 (Voices for Children in Nebraska, 2018, p. 40), there were another 2692 needing treatment who did not receive services. However, the problem is larger than reflected in these numbers. The report also indicates another 44, 453 children diagnosed with mental/behavioral conditions who still require a treatment regimen (Voices for Children in Nebraska, 2018, p. 40).

Behavioral disorders also impact many Nebraska children from participating in normal childhood activities. The National Survey of Children's Health in the ©Kids Count in Nebraska Report 2018 indicates 65,752 children in Nebraska struggle with behavioral health problems including anxiety (16,462), Attention Deficit Disorder (ADD) and Attention Deficit Hyperactivity Disorder (ADHD) (25,323), depression (13,600), and are on the continuum of Autism Spectrum Disorder (10,367) (Voices for Children in Nebraska, 2018, p. 40). Empirical research has confirmed worse mental health status compared to the general population for youth in foster care (Greiner et al., 2017).

Nearly 10,000 (9878) Nebraska youths were arrested in 2017 (Voices for Children in Nebraska, 2018, p. 122). More than 10,000 (10,453) youth were involved in the child welfare system at some point in 2017 in which 7,157 (68.5%) from 3890 families were court-involved and another 464 noncourt-involved children became court-involved (Voices for Children in Nebraska, 2018, p. 22).

Various systems of care, such as Children's Behavioral Health, Child Welfare, and Juvenile Justice, are fighting to stay ahead of the needs. However, lack of adequate community-based resources and shortage of workers are falling short of the need.

Organizational history of parent to parent peer support at NFSN

The Nebraska Family Support Network (NFSN) was originally incorporated as a 501 (c) 3 in 1993 to support families with children struggling with emotional/mental/behavioral health issues whose parents were being excluded from participating in their child's treatment or plan of care. Nebraska was a forerunner incorporating a parent-to-parent peer support model as early as 2006 into family serving systems well ahead of other states. Essential funding from the Nebraska DHHS, Division of Behavioral Health, allowed NSFN to incorporate as independent 501(c) 3s to be a statewide education and resource agency for parents with children living with mental/behavioral health challenges. In 2006, other Family Organizations and family peer support were launched in Nebraska, resulting in a statewide presence serving six Behavioral Health Regions. The Family Organizations alignment with System of Care values and principles, a condition of funding, was important to service delivery and their goals to (1) ensure a parental voice, and (2) improve outcomes with parental participation.

Stroul et al. (2010, p. 1):

> *Systems of Care values are:*
>
> *Family driven and youth guided, with the strengths and needs of the child and family determining the types and mix of services and supports provided;*
>
> *Community based, with the locus of services, as well as system management, resting within a supportive, adaptive infrastructure of structures, processes, and relationships at the community level;*

Culturally and linguistically competent, with agencies, programs, and services that reflect the cultural, racial, ethnic, and linguistic differences of the populations they serve to facilitate access to and utilization of appropriate services and supports.

Critical to support services in fiscal year 2017/2018, in which NFSN provided direct peer support services to 431 families through Family Navigator, Family Peer Support, and Nebraska Child and Family Services (CFS) Peer Mentoring Programs, are the concepts of family engagement/feedback and lived experience. Together, these components define the role of an effective peer support provider through productive engagement strategies across systems of care.

Defining family engagement and assessing feedback

Family engagement is ultimately a family-centered and strengths-based approach to making decisions, setting goals, and achieving desired outcomes for children and families. In December 2017, NFSN hosted a Parent Focus Group at the request of Region Six System of Care to discuss the barriers to these generally recognized practices. NFSN hosts these Parent Focus Groups to help inform local systems of care on barriers that families experience in accessing services. Participants provided feedback categorized under service delivery, professional relationships, and program needs (Table 5.1).

Lived experience

Four Family Organizations in Nebraska distinguish themselves from other direct service providers because their primary qualification for service provision is lived experience. Lived experience normally encompasses a parent willing to face the

Table 5.1 Feedback from Parent Focus Groups at Nebraska Family Services Network, Region 6, Nebraska System of Care December 2017.

Service delivery	- inadequacy of available services - limited service in rural communities - length of time to get an evaluation and results - challenge of finding the right medication
Professional relationships	- establish relationships with community police - limited availability of cultural competent therapists - parents feeling discounted/threatened by system representatives - challenges associated with implementing an Individual Education Program (IEP) with school officials, particularly lack of training on traumatic-informed care
Program needs	- mental health-focused care options - drug treatment programs for juveniles - respite care for parent/caregivers (capacity to handle their own needs) - aftercare support for youth who were in residential treatment facilities

challenging circumstances with their child and other families/youth with such persistence as to become a strong advocate. In this case, all leadership and direct service staff demonstrate lived experience because they each have supported a child in one or more of Nebraska's child/family service systems. As a result, the NFSN primary mission focused on children's mental/behavioral health has evolved into a range of peer support services in child welfare and juvenile justice.

Therefore, there are several additional responsibilities associated with a Family Organization in addition to working directly with parents and their families. For example, ensuring that a "Family Voice," comprised of Family Organization representatives and parent leaders, is part of policy-making committees to shape new legislation; hire youth advocates with lived experience; provide family members growth and wellness opportunities in addition to support groups; educate family members of their system rights and provide training to develop leaderships and advocacy skills, recognition of mental health needs, and access to services; and to act as a community resource for other professionals and community members.

Role of an effective peer support provider

Despite the changing organizational status and regulatory policy, NFSN has held on to several criteria surrounding the role of an effective parent peer support provider. Lived experience remains a high priority. But there are other important considerations in the development of a peer support provider.

Business acumen, in terms of strategic thinking/problem resolution and an entrepreneurial spirit, also unleashes the necessary level of creativity, innovation, and service that is necessary to succeed. Certainly, a passion to help others, patience, resilience, flexibility, and the establishment of healthy boundaries enable the provider to serve and make sure they are at their best to do so.

Another important element is preparing staff to be successful by completing extensive training that relates to their role and family/youth well-being, following standards to earn national certification as parent support providers with the National Federation of Families for Children's Mental Health (2017b) and/or the Nebraska State Certification.

Specific tasks associated with the role of peer support provider include the following—

- Provides one-on-one parent support.
- Listens to the needs and concerns of families without blaming or judging.
- Helps families locate and best utilize available community resources.
- Assists parents in keeping their child in the family home with those who know them and their needs best.
- Supports reunification with parents whose children are in placement.
- Helps parents whose own mental health and addiction issues are putting their families at risk.
- Helps families keep their child from going further into the Juvenile Justice System.
- Attends meetings with families to assist them in communicating and advocating effectively for the needs of themselves and their children.
- Assists families in identifying and growing a lasting support system.
- Builds parent-to-parent networks to help strengthen a family's base of community support.

- Links to appropriate behavioral health services for children and to other services and supports to help meet basic needs of the family.
- Assists with developing a family Wellness Recovery Action Plan (WRAP), safety plan, or other plans the family may find helpful to identify and meet their needs.
- Assists parents who desire to increase their parenting skills and strategies to develop and grow in this area through education and role modeling.
- Encourages wellness and recovery principles.

Productive engagement strategies across systems of care

Historically, there have been 16 of 34 primary characteristics of family engagement that include more than one system such as Child Welfare (CW), Juvenile Justice (JJ), Behavioral Health (BH), Education (ED), and Early Childhood Education (ECE) with some noted exceptions (Child Welfare Information Gateway, 2017a, p. 1). One way to view these characteristics is to identify the dynamic within the group setting with key words or actions focusing on behavior, boundaries, communication, encouragement, and facilitation.

For example, peer support personnel learn to moderate their behavior, establish boundaries between and among the peer support personnel and family participants, and establish lines of communication between family members, peer support trainees, and to youth involved within a system. Peer support professionals also serve to encourage continued participation through techniques of validation and emphasizing family and individuals' strengths of participants. Finally, as a facilitator, the peer support leader must remain clear on the original reason for separation of the child from the home and provide resources to secure the best family outcomes.

Notable exceptions do not include the ED and ECE who would not involve youth in decision-making and service planning, nor would any focus be on the family strengths but instead on the student. In another exception, JJ would not be involved in assessing the family or providing resources. However, the NFSN experience differs in that JJ does collaborate and engage in helping families and youth identify resources and support to try and reduce further system engagement. Also, NFSN staff helps parents engage more productively with ED and ECE school representatives to develop IEPs. In this way, NFSN reinforces the family voice approach to facilitating interaction with all of the various systems of care.

Value of lived experiences to family and public systems in four case studies[a]

In practice, NFSN incorporates findings from evidenced-based resources, annual reporting, surveys, and other resources in order to facilitate daily operation needs and family peer support interactions. While the evidence, initiative, or method can

[a] Participant names have been changed and elements of their family services experience modified to protect their personal identities in accordance with the Health Insurance Portability and Accountability Act of 1996 (HIPPA), Public Law 104–191, and subsequent revisions. However, NFSN employees freely consent to the use of the case information in the spirit of lived experience.

Table 5.2 Evidence-based input for Peer Support Services decision-making and treatment practices.

40 Developmental Assets	Building blocks of healthy development—known as Developmental Assets—that help young children grow up healthy, caring, and responsible. Developmental Assets target age ranges from 3 to 18, but the example referred to is for adolescents (age 12–18).[a]
Active Parenting	Families in Action is a school- and community-based intervention for middle school-aged youth designed to increase protective factors that prevent and reduce alcohol, tobacco, and other drug use; irresponsible sexual behavior; and violence. Family, school, and peer bonding are important objectives. The program includes a parent element and a teen component that was added later to complement the program. The parent component uses the curriculum from Active Parenting of Teens that is based on Adlerian parenting theory advocating mutual respect among family members, parental guidance, and use of an authoritative (or democratic) style of parental leadership that facilitates behavioral correction.[b]
Motivational Interviewing	An approach that helps families find the motivation to make positive decisions for themselves and their children.[c]
Protective Factors Survey	Measures conditions in families and communities that, when present, increase the health and well-being of children and families.[d]
Trauma-Informed Care	Trauma-informed care is a strengths-based service delivery approach "that is grounded in an understanding of and responsiveness to the impact of trauma; that emphasizes physical, psychological, and emotional safety for both providers and survivors; that creates opportunities for survivors to rebuild a sense of control and empowerment."[e]
Wellness Recovery Action Plan (WRAP)	A self-designed prevention and wellness process that anyone can use to get well, stay well, and make their life the way they want it to be.[f]

[a] Search Institute (1997).
[b] California Evidence-Based Clearinghouse (CEBC) for Child Welfare (2017), NREPP (2010).
[c] Center for Evidence-Based Practices (CEBP) at Case Western Reserve University (2006, 2009, 2011), Miller and Rollnick (2012).
[d] CEBC (2015), Counts, Buffington, Chang-Rios, Rasmussen, and Preacher (2010).
[e] Hopper, Bassuk, and Olivet (2010, p. 82), CEBP (2006, 2009, 2011).
[f] Copeland Center for Recovery and Wellness (2018).

be drawn from a variety of sources (Table 5.2), peer support providers often rely on their own lived experience to persuade a parent to take appropriate action. In this section, we demonstrate the transition of lived experience from a peer support provider to help family/youth overcome obstacles and systemic issues that may be revealed in oversight and organizational reporting. The first case scenario, "A father engagement specialist is born," demonstrates a positive response to a program deficit in which the need to increase communication with father figures has been identified in the program. The case demonstrates the conversion from a father in need to a father providing family services. The second case scenario, "Wayne (and Rick's) first challenge," tests the skills of the Father Engagement specialist in a live situation. The third and fourth cases, "From ward of state to youth navigator national award" and "Family navigator establishes suicidal risk in parent to parent intervention," establish the value of lived experience to address process improvements and life-saving support services.

A father engagement specialist is born

Like many other states, Nebraska CFS received a poor rating for efforts to engage fathers into the process of taking custody of their children when the biological mother gets involved in the Child Welfare system and children are removed from her custody (Nebraska Department of Health and Human Services (DHHS), 2018). The deficit in father participation was apparent when the Nebraska Child and Family Services Review Round 3 Program Improvement Plan (2018) results were published. These results include a deficit in parent participation, in general, and specifically in father participation indicating that mothers were meeting 71% of the time, but fathers only 54% according to Family Team Meeting documentation (NDHHS, 2018, p. 14). Further, other data from the Performance Accountability and Quality Assurance Reviews indicates that case managers report 60% face-to-face contact with a child's mother in the last 4 months, while only 40% of the fathers were met in this manner (NDHHS, 2018, p. 14). These benchmarks demonstrate that parental contact, especially the father, was an area that required vast improvement.

Thus, NFSN reached out to a CFS partner to examine how the addition of peer support might help positively impact Father Engagement outcomes. NFSN engaged Wayne, a former client and a father who successfully pursued and gained full custody of his own four daughters, to take on this role. His story provides a glimpse of the transition from parent to a career as a peer support provider.

Wayne is a middle-aged, high-school educated, Black man who was born and raised in Omaha, Nebraska. His work experience primarily includes food service and retail settings throughout his adulthood. He was married and living a standard family life prior to fatherhood. Later, both he and his wife worked. She was employed in telemarketing during the daytime and he worked during the evenings. This arrangement was beneficial to avoid the costs of daycare, but contributed to the unraveling of the marriage. The marriage further dissolved when Wayne's wife became addicted to drugs and involved with a dangerous crowd. Wayne's children soon began to tell him about the weird people his wife brought home that was confirmed by neighbors questioning the activities of unsavory people at his home.

Wayne knew he had to act but a drug intervention for his wife backfired and led to his eldest daughter and himself being kicked out of their home. Wayne filed for divorce, went to court to fight for full custody of all four children, and sought support from Child and Family Services to be successful as a single parent. But when his daughter began to display significant mental health challenges, he knew that he was out of his depth and called the Nebraska Family Helpline (Nebraska Department of Health and Human Services (DHHS) Behavioral Health, 2018) who connected him with a Family Navigator at the Nebraska Family Support Network. His Navigator helped Wayne find mental health services for his daughter and provided support throughout the custody process, and ongoing encouragement. Persistence paid off when parenting classes and adherence to all Family Navigator recommendations led him to full custody of his other three daughters.

Nebraska DHSS (2018):

> *The Nebraska Family Helpline at (888) 866-8660 makes it easier for families to obtain assistance by providing a single contact point 24 hours a day, seven days a week. Trained Helpline operators screen calls to assess immediate safety needs, identify the potential level of a behavioral health crisis, make recommendations or referrals to appropriate resources, and help callers connect to emergency resources or providers. The Helpline is supervised by licensed mental health professionals.*

The positive influence of the Family Navigator on Wayne's family continued through in-home therapy to help them adapt to their new circumstances. She helped Wayne find community resources to stretch beyond limited means and to provide information on teenage employment programs to get his daughter engaged in positive activities. Wayne recalls that her belief in him and the encouragement that she offered when he became weary of the overwhelming process carried him to successful closure of the family case file in his favor.

Together with his case closed and willingness to participate in family events and children's mental health awareness, clearly Wayne's lived experience was an opportunity for NFSN to examine the potential for a Father Engagement Specialist to impact outcomes for families in the child welfare system. For a variety of reasons, Wayne was not convinced he was cut out for the role due to lack of formal experience helping others and his uncertainty about this role as his destiny. But apprehension quickly dissipated when he completed his initial interview and the implications of this intervention type sunk into his spirit. Wayne leapt in with both feet and began his training. Within 2 months, he was working independently with his own clients. Open and eager to continue learning, Wayne blossomed in his new role. His first challenge highlights the power of father-to-father engagement.

Wayne (and Rick's) first challenge

The referral from Child & Family Services of Rick, a white male in his 30s, was accompanied with several red flags. The father was separated from his family due to domestic violence. He did not have a history of stable employment, lacked a high-school diploma, and had a daily marijuana habit. Despite this, he did have some experience in a variety of labor-intensive, blue-collar jobs during his adult life. Rick was raised by his maternal grandmother and did not have extended family connections. Therefore, he had limited informal support.

A Child Protective Services (CPS) investigation was opened when Rick's wife called the police for domestic abuse. He was mandated to leave the family home and was not allowed to move back until he met the requirements of his CPS plan for reunification when the investigation revealed grounds for concern. Any violation of these terms would lead to the removal of his children from their mother's care resulting in the family moving deeper into the system. Rick perceived these actions as state interference and was furious about their intrusion in his family's business. Some of the red flags that accompanied Rick's profile came from an alarming description from his wife. She indicated that this dad was known to have problems with anger management, racism, and a demonstrated tendency for violence. The referral that came to NFSN recommended that Wayne should not meet with this client alone. CPS typically provides precautionary information, but all referral information must be read in entirety and taken into consideration by the peer support specialist.

While Wayne had been well-trained on client engagement and support, he was justifiably taken aback by this warning and assessed the real possibility that Rick might be dangerous. He was also aware that he must strive to listen to the parent's description of all incidents without bias. Wayne was counseled to set the meeting for a public place or at the very least meet outside the home, even if it was on a porch stoop, and to leave immediately should he feel in danger. With this guidance, he sent an introductory letter with his picture, a standard procedure for NFSN peer support providers, and had a coworker attend the first meeting, not as protection per se, but as a witness who could call for help if the encounter escalated to violence. Peer Support specialists are trained to err on the side of caution, but thankfully there was no cause for concern.

At the first encounter, Wayne and his coworker met Rick at his grandmother's house. They report a successful meeting and observations that Rick was deeply humbled even though he remained extremely angry. Nevertheless, Rick demonstrated that he loved his family and was willing to do what was necessary to reunite his family. Wayne's relatable personality was important to the success of this meeting as well as his lived experiences. He understood and clearly projected that he was not there to judge Rick, but to listen to him via the utilization of motivational interviewing.

Wayne also gleaned other important information from the meeting. What quickly became apparent was that Rick did not feel heard at all and consequently became frustrated and bewildered. Part of the frustration was that Rick loved his family and for the most part believed he was a good husband and father. He also felt that his wife and kids understood his love for them. But mixed with this love was a great remorse for what happened and a great desire to return to the life he had before the domestic incident. Wayne listened with empathy and told him he understood. However, he also explained to Rick that recapturing the time before the event was not going to be easy. The family was under investigation per child safety and Rick had to do what the Family Preservation Team ordered him to do if he wanted to keep his family intact. Meanwhile, Wayne also assured Rick that he would walk him through this process step by step and so, by the end of the meeting, Rick received the message and they got to work.

Wayne helped Rick identify Domestic Violence (DV) and Anger Management programs. During the 4-week wait on both classes, Wayne attended Family Team

Meetings (FTM) with Rick. Wayne skillfully and privately pointed out that Rick often tried to dominate and talk over his wife. This often took the form of Rick finishing his wife's sentences before she got a chance to share what was on her mind. The direct approach that Wayne used with Rick was a continuous learning process that sometimes took one step forward followed by two steps back. At one point, Rick's frustration with the process boiled over at a FTM and his projected reunification date was pushed back a month. Rick was devastated. Wayne just kept encouraging him to stay the course until completing his requirements. Wayne proved that he could honestly relate to what Rick was feeling by reaching out father to father. This outreach allowed Rick to trust Wayne and embrace the path that they were walking together.

Wayne used the WRAP (Wellness Recovery Action Plan) model to help Rick identify his triggers and think proactively about how he could react in a more appropriate way instead of becoming upset. Role playing was a big part of working through this process. Wayne earned Rick's trust by believing in him, viscerally relating to him per Wayne's own lived experience, and investing in his outcomes. Rick began to see Wayne as his partner toward rejoining his wife Marla who had expressed her desire for Rick to return home. But Marla, who was also attending DV classes and therapy, made it clear that this was contingent on being able to trust that he could control his anger and stop reacting with physical or verbal abuse.

Other interactions were changing throughout the process. For example, Rick's CFS case manager changed early in the process at his request. This development turned out to be a real blessing because the new case manager was more seasoned, no-nonsense, action-oriented, amenable, and supportive to Wayne's input on behalf of Rick's commitment to accomplishing the requirements for returning to his family. She appreciated the partnership between Wayne and Rick, was not threatened by Wayne's input, and maintained communication with Wayne almost daily. They both believed that this case could be resolved without the family having to go deeper into the system. Wayne's ability to drive the process of monitoring parental adherence coupled with the commitment of the CFS worker to remove mountains of red tape improved the pace to family reunification. Thus, Rick completed all requirements within months of getting a new case worker. Rick's first intervention allowed him to gain insight into how his anger affected his family so that he began to own his behaviors, move beyond his conditioning, and make the necessary behavioral modifications to engage with his family. Wayne was there to support and encourage him during every step of the way including encouraging further participation by suggesting that he attend a NFSN focus group for parents. The goal was to discuss challenges and successes in accessing services for children who were struggling with mental/behavioral health challenges. Wayne told Rick that participating in something like this would be beneficial for his character because participation would demonstrate to the team his willingness to learn and grow.

Rick, previously convinced that family business was not for public consumption, stepped out of his comfort zone to attend a public format for discussing family challenges. Observers noted that Rick was quiet for most of the meeting, but participated in the introductions and listened attentively. He saw parents, not unlike himself and his wife, showing up to share what they had been through in trying to get appropriate services for their child. He heard about their challenges with the schools and the toll on their time, energy, and resources in their efforts to obtain services for their child. He saw parents who loved their children just like he loved his children and he recognized that these parents were not perfect, either. Many of them had challenging lives, but they were willing to complete the appointed tasks before them to help their child and their family to improved outcomes. Finally, Rick heard how involvement with the system had helped some of these parents get their home situation stabilized.

While Rick observed many similarities in his reunification process, he also noticed some real differences. The most glaring difference in his case was that *he* was considered "the problem" not his children or the system. As the discussion began to wrap up and the facilitator asked if anyone had anything else that they wanted to share, Rick felt a need to be heard. He didn't tell his whole story. But he did mention that the State was keeping him out of his home and how much he just wanted to be back with his family. With genuine eagerness, a desire to change, and puzzlement he asked, "How could I work on challenges with my family if I am not allowed in the home? How could I make things better if I am separated from them?" Two key items that serve as evidence of the power of parent-to-parent interaction to engage and encourage growth were shared with these statements: (1) there was no judgment around that table, and (2) his sharing was met with compassion and encouragement. These were important elements for his family, but also for those who deliver support services.

When Wayne discharged this family, Rick had completed DV and Anger Management courses, was employed, and continuing therapy. He was also home. Marla had also completed DV and Women's Empowerment classes through the Women's Center for Advancement. Growth was evident in both of them, but clearly Rick's whole mindset was changed as he had a greater respect for his wife and family. Wayne was amazed and blessed to see the progress this man had made with his support and assistance. With in-home family therapy in place, the case closed with child welfare.

In a follow-up survey with our CFS partner, the caseworker who worked with Rick and Wayne gave the following comment: "The Father Engagement Specialist I worked with was able to effectively keep a difficult father on track to successfully complete the services adding the skills he required to return home to his family."

From ward of state to youth navigator national award

The power of lived experience can have a profound impact on systems of care beyond providing direct support to help parents find their voice and make progress in healing their family situation. For this reason, NFSN hires parents and youth who have lived experience in a range of public systems translating into a literal claim that we have walked in our client's shoes. Therefore, we encourage staff to take advantage of every opportunity to use their lived experience to inform policy development and influence practices in public systems.

Marie, a Youth Navigator within the system of Juvenile Justice, came with a range of lived experience in foster care and juvenile justice. At the age of 10, she became a ward of the state due to parental charges of child abuse. Statistics confirm that children in foster care are much more likely to enter the juvenile justice system than children living with their families of origin (Anspatch, 2018; Juvenile Law Center, 2018).

The progression of Marie's story is not unusual or surprising, but rather a heartbreaking journey that could have been avoided if counselors had deployed simple exploratory questions about her condition when unusual circumstances became apparent outside of the home.

Marie entered the juvenile justice system at the age of 11 just 1 year after entering the child welfare system. She was arrested at her elementary school for threatening an assistant principal. The circumstances of the incident were not taken into consideration by any adults in the education system that led to her arrest at the Youth Center and release back to her foster parent. Marie's court appearance directed her to court-ordered meetings with her probation officer twice a month as a condition of her probation. Her foster mother, consumed with the general care of up to eight other children and not leaving much time or energy for individual care, placed the responsibility to keep those appointments on 11-year-old Marie. Marie was expected to walk more than a mile from the foster home to the probation center regardless of weather conditions.

The adult responsibilities and negative experiences were teaching Marie to develop a tough exterior even as she lived in a constant state of fear and hypervigilance. Part of those experiences includes memories of waiting to meet with her probation officer in a big room with few children and a large proportion of adults. She also recalls feeling overwhelmed and alone in her required sessions with the probation officer because she was unable to comprehend most of the discussion, expectations, or how she was supposed to achieve them. "Looking back I have such compassion for that young child! I was a troubled kid and I was causing trouble, but I was still just a child," said Marie.

Marie quit attending the sessions after several months because she felt that she couldn't meet the probation mandates. Violation of probation brought her before the Judge a second time. This time she was ordered back to the youth center. Marie underwent several evaluations there and one assessment revealed a diagnosis of Hypomania Bipolar. Consequently, Marie was placed in a new foster home and began treatment for this condition. At the same time, she was also ordered to take an anger management class.

The weekly anger management class which she was required to attend was a resource with unexpected outcomes. Marie was dropped off to attend the meeting alone and was the only youth in the class. She believes this experience stripped her of any childhood innocence that remained as she was forced to sit in a class full of adults, mostly male, and listen to how they were there for beating their women or choking their mothers. This child was terrified which led her to a state of mind that blocked the fear and substituted a *fake it to make it* frame of mind. Faking became a second nature defense mechanism including acting hard to protect herself from bullies, fit in, or to relate based on faked emotions. Eventually, the thug persona transitioned from a survival strategy to another doorway to trouble in her life.

Consequently, Marie experienced 88 placements in Nebraska and Iowa as she spiraled deeper into the juvenile justice system until she aged out of foster care at 19 years old. She experienced trauma throughout her placement in the various systems of care, including events in her foster care setting, which left her "torn and damaged" and without normal teenage experiences such as first dates or attending social events (e.g., homecoming, prom). She was a victim of rape resulting in impregnation at age 13. Yet, she gives credit for care in the residential treatment programs because they empowered her with tools that support her recovery to this day. Marie is a survivor but the end of foster care brought on new challenges.

Aging out of foster care meant figuring out how to live without any tangible resources, services, skills, or positive support. Given these limitations, there was no surprise when she entered the adult justice system. But one pass through prison was enough for Marie. She knew that she had to break this cycle and find a different way to live. So, Maria became committed to her personal growth and mental health recovery. Her first steps were seeking help to deal with the traumatic things that happened to her during childhood and adolescence. She entered therapy. A prior caseworker kept in touch with her throughout her journey and she was encouraged to seek employment in the system where she could apply her lived experiences.

Today, Marie is a young woman employed at NFSN who loves reaching out to youth who face the same challenges she has faced—loss, confusion, and no sense of personal power. Marie offers youth a difference because they have someone who can relate to them, give them support, and demonstrate that you can come out the end of the foster care system with hope. Honest, tough, and compassionate advocacy are a part of her character that she uses as a foundation for care.

Ironically, Marie participates in a cross-over diversion team cochaired by the same juvenile judge who presided over her case when she was a child. This team represents people who work in various areas of the juvenile justice system. Marie has come to trust that these team members truly want to help kids have better outcomes and she loves having the opportunity to positively influence youth as part of the team.

Her lived experience is also instrumental in changing some of the condition of standard services. Marie was able to relay her bad experiences with adult/youth interaction in the anger management classes and team members were shocked to learn of the practice of mixed youth and adults in classes. The practice was quickly verified and corrected. Recently, Marie was asked to be the keynote speaker for an adolescent mental health conference targeted at School Resource Officers. Marie had so much lived experience to draw on that she wasn't sure what to focus on. She was advised to consider what she most wanted these officers to understand about youth with mental/behavioral health challenges when they are acting out in a school setting. The conference attracted many law enforcement officers in addition to the school resources officers.

Marie's story held her audience spellbound. Marie began by sharing that, at age 11, she got kicked out of her foster care placement for leaving the lights on, the final straw that broke her foster parent's patience. Marie acknowledged that she was very combative, defensive, and did not take redirection well at that stage. It was near bedtime when her interactions with her foster parent escalated to a point of no return. The foster parent told Marie that she was no longer welcome at her home, and then, she was told to wait on the front porch steps until her caseworker arrived. She was not allowed to take anything with her and was told that her caseworker would have to get her belongings. For Marie, the wait seemed like an eternity, but eventually her caseworker showed up, took her to a holding location, and worked on a new placement. At about 4 a.m., Marie arrived at the new placement and was ushered into a room to lay down to get some rest before she would leave for school. Her history of child sexual abuse made her apprehensive and she was restless because she didn't know who was in the bed next to her. She didn't have any pajamas, so she slept in the clothes she'd worn that day. The next morning the new foster parent told her to get ready for school. Marie had no clean clothes, toothbrush, or hairbrush and so her ability to "get ready" was going to be extremely limited.

Marie goes on to share how she fell asleep in class. She fondly remembered how she often slept at school because the classroom was the only place where she felt safe and without the need to be hyper-vigilant of her surroundings. But when her teacher tried to rouse her she could not stay awake. So, the teacher sent her to the Assistant Principal's office. On this day, Marie was marched down to see the assistant principal. There was a row of chairs where kids were supposed to sit. The Assistant Principal walked by and told Marie to sit on the floor because she was dirty. Marie had her coat on and she knew that the coat was quite dirty. The principal started questioning her as to why her coat was so dirty and why her hair wasn't presentable. This questioning didn't come across in a way that conveyed concern, to Marie it was shaming and blaming—she felt this adult's disgust. There were other kids and adults in the vicinity who could hear everything the assistant principal said. At that point, Marie leaned over to a kid sitting next to her and spoke loud enough to be heard by everyone, "I will be the first 11 year-old to be on Ripley's Believe It or Not for killing her school principal." The principal called the police saying that Marie had threatened her. This led to Marie getting arrested and put in the Youth Center, which launched her journey with juvenile justice.

The room was silent as Marie implored the School Resource Officers to ask the right questions and to not let appearances and defiance distort the reality that these are kids whose lives are in turmoil. Marie said that looking back; no one ever asked, "What happened to you?" with any real compassion or interest. No one helped by sharing her burden of pain, fear, and frustration. No one acknowledged that anger was a normal response considering the history of abuse and assault. Instead, she was treated like a nuisance, a problem to be solved vs a kid who had been through hell. Marie asked these officers to err on the side of compassion. If any adult had said to her, "Wow, you had a rough night. No wonder you are tired. Why don't you lie down in the nurses' office?," how different her journey might have gone.

Many officers lined up to thank her for her courage and to express admiration for her resilience after her presentation was complete. Marie was recently awarded a significant national award for her work in helping to change outcomes for children and youth in the juvenile justice and foster care systems. The lived experience, from ward of the state to Youth Navigator national award winner, is an ongoing reminder that we can use the past to heal the future of others and ourselves.

Family navigator establishes suicidal risk in parent-to-parent intervention

Many of the Helpline referrals received by agencies such as NFSN come when parents are warned or ticketed for educational neglect when their child has excessive unexcused absences from school. A child in Nebraska with 20 unexcused absences becomes a matter for the court. Children are required to attend school if they are 6 years old prior to January 1st of the current school year and until they are 18 years of age (Neb. Rev. Stat. 79-201(1)). As a result, students cannot legally "drop out" at age 16 or otherwise as may be legal in other state regulations.

In this case, the school in which Juanita's 15-year-old was attending reported her daughter's excessive absences. This event triggered the parent to reach out for help in managing her child's behaviors including her unwillingness to attend school. The youth was not defiant and she did not find herself in trouble outside the home, but she would ditch school every day. She claimed that she did not fit in at school and just wanted to be at home. Employment for both mom and her live-in boyfriend during school hours presented a major challenge for monitoring the youth's attendance.

NFSN peer support providers responded to her request for help by asking Juanita where and when she was comfortable meeting. This was especially important since she was a migrant worker. While other families may prefer alternative locations, migrant workers have a particular aversion to home meetings even if they have appropriate documentation because they prefer to keep anything that potentially relates to public systems away from their personal lives. But, Juanita was out of her comfort zone and desperate to help her daughter. Despite a lack of flexibility in job hours at the meat packing plant where she was employed, she agreed to take time off work to take her daughter to meet with a Family Navigator at NFSN offices.

The staff conducted a Suicide Behaviors Questionnaire—Revised (SBQ-r) assessment on youth with parental permissions as part of the intake process. Any score above 7 is a cause for concern and this young woman scored a 10. Juanita's Family Navigator, April, stressed to this mom that her daughter was at a significant risk for suicide and that she needed to take her for an emergency assessment immediately. Juanita was resistant for two reasons. First, she thought her daughter was just being dramatic. Second, she had to return to her job. Juanita reasoned that she did not have time to take her for an assessment on this day. April, a parent who had a great deal of lived experience with a child who self-harmed, was adamant that this situation called for immediate intervention. She told Juanita that she would go with her to have her daughter assessed. They went to the ER at a local hospital that has an adolescent behavioral health unit. They had to wait 4 hours before the youth was taken back for assessment. The youth was admitted for observation after the assessment was completed.

The next day, April called Juanita to see how things were going. Juanita was very grateful that she had listened and taken action. She shared that when she spoke with her daughter's doctor, he stated that this was an extremely serious situation because he believed that her daughter had every intention of acting on her suicidal ideation and, without intervention, could well have succeeded. April applauded Juanita's commitment to her child's well-being and gave her a range of information on where she could go to learn more about adolescent depression and suicide. She also provided a list of therapists and invited Juanita to come to a parent support group hosted by NFSN. April's knowledge was an example of Parent-to-Parent interventions which are an important part of the application of lived experiences that saves lives.

Case summary

The case studies illuminate how peer support, especially those with lived experiences, can add value across the system of care. Clearly, the element of personal experience can guide peers to assessments, services, and other resources to change negative family behavior patterns to positive family support. Standardizing engagement techniques from these learned experiences provides a sound basis from which to overcome inconsistencies and prior reports of "lack of efforts to build relationships between bio [logical] parents and foster parents," (NDHHS, 2018, p. 14), the number one root cause of systemic cultural failure in the system of care.

Challenges in family peer support collaborations

There are both internal and external challenges in peer support collaborations. The challenges that NFSN continues to face with internal support staff and external collaborations are not unique to our organization, but demonstrate the systemic challenges facing all Family Run Organizations utilizing personnel with lived experience. Here are the most common challenges in a long list of experiences.

Internal challenges with peer support staff:

- In Nebraska, peer support providers are frequently single parents who must address challenges with their own children which can impact attendance and their ability to serve others.
- Parents who participate in peer support may be triggered when facing direct client activity in familiar systems of care that they navigated with their own children.
- Peer support staff may fall into the "fixer" mode and take on their clients' activities or outcomes despite staff training on boundaries.
- Parents and youth obtain characteristics of strength and resilience resulting from surviving navigating systems of care. However, the clash of strong personalities can cause drama in the workplace.
- The influence of a single toxic employee can have tremendous negative fallout in family organizations that have a smaller number of employees.
- Entrepreneurial and creative characteristics that define peer support staff may come with a lack of orientation for detail required for comprehensive and complex documentation requirements for parent/family peer support personnel to complete as part of their case management. Documentation maintains organizational compliance and personnel must be closely supervised in this area.

One way that NFSN staff has addressed these problems is to participate in the Adverse Childhood Experiences (ACES) Provider Survey (National Center for Injury Prevention and Control, Division of Violence Prevention, n.d.; SAMHSA, 2018; U. S. Centers for Disease Control and Prevention, Kaiser Permanente, 2016). At NFSN, the average ACES (Adverse Childhood Experiences Study) staff score is 6, which is higher than your average workplace and indicates serious implication for trauma impact. A Family Run Organization work environment has to be sensitive and committed to being a trauma-informed workplace. Therefore, understanding the organizational baseline score is important to managing an organization with individuals who utilize lived experiences. Recognizing the strength and weaknesses of NFSN personnel aids in the development of additional interaction skills to support, encourage, and address observed behavior that may be indicative of a coworker's lack of attention to their own wellness or losing sight of their own behavioral triggers. These help in maintaining healthy boundaries both in and out of work situations.

External partnership challenges:

- Other providers and behavioral health partners don't understand the peer support role.
  - Staff is asked to perform tasks that are out of our model's purview that is designed to engage, support, and empower parents and youth.
  - Peer support is wrongly identified as "mini-clinicians."
  - Peer support is underutilized in support of the client's fullest benefit.
  - Peer Support is considered an informal support permitting some professionals to resist including them in every aspect of the case. This action generates mutual mistrust.
- System and behavioral health partners.
  - Do not recognize that they and the parent(s) they are working with can benefit from the presence of peer support staff during meetings.
  - System professionals can have a hard time acknowledging that an informal support significantly contributed to desired outcomes.
- Potential deficiency in case manager self-awareness and experience.
  - Case managers may have unexamined biases that impact their ability to stay objective and support the reunification process.
  - Newly graduated case managers have valuable education and training, but may not relate to the practical reality of this vulnerable population because they have very young or no children in their own life experience. This position may escalate a parent's anxiety or trigger their defensiveness that requires a peer support intervention. Unfortunately, peer support interventions with case managers may cause them to become defensive or to become resistant to attempts to deescalate the situation.
  - While many case managers are very appreciative of how well a peer support provider can engage a challenging parent, others can feel threatened and disengage from full partnership.
- Families, professionals, and peer support
  - Depending on their level of trauma, mental illness, and/or substance use, some parents can manipulate providers against each other. This can include sabotaging peer support with the professional and vice versa.
  - The NSFN process of educating and empowering parents and youth to speak up for themselves can cause friction for themselves in the systems they are navigating. Some system professionals prefer an obedient/submissive client vs an educated, empowered client.
- Family-run organizations are a family advocacy organization and overseer.
  - Reporting incidents and practices that violate the rights of the parents we serve is an important mission of family advocacy. In the same way, we are accountable as an organization for our actions. Discretion and care are utilized in addressing incidents directly. However, NSFN may take necessary steps to inform the appropriate levels of administration in our partnering organizations about any problems that cannot be directly resolved. This role may, at times, inhibit relationships at the direct service level.

Many of the external challenges experienced by NFSN are remedied through staff training (1) focusing on diplomacy while advocating for and supporting the parents they are serving, (2) discouraging a combatant "us against them" mindset, and (3) understanding the behavioral health and therapeutic processes that impact the families that we serve. A fourth method encompasses generating awareness of the role of peer support and the likely possibility of parent(s) being caught in the line of fire of interdisciplinary collaboration. This is especially true when implementing the Wraparound Model (Debicki, 2008; National Wraparound Initiative, Regional Research Institute, School of Social Work, Portland State University, 2019) bringing multiple participants to the table to collaboratively address the most productive approach for serving families and youth. The model can suffer from the old adage of "too many cooks can spoil the broth." Therefore, professionals and service providers must take extra precautions in their effort to overcome these potential opportunities for egos to clash, and instead, choose to take actions to facilitate progress for broken families.

Discussion and recommendations

We have demonstrated the benefits and drawbacks of applying lived experience in family peer support services. Alternative solutions to out-of-home placement are important to enabling parents to gain knowledge and experience and apply those skills toward promoting the health, wellness, and safety of their families. We provide a brief summary of Family Run Organizations and the Family Support Coaching model that targets effective supervision for Family Peer Support providers. The model, funded by a SAMHSA Grant number 1HR1SM063017-01 Statewide Family Network Program, was developed as a product of collaboration between Nebraska's Family Run Organizations and Dan Embree, a consultant with EnRoute. (For access to the full report, please contact Karla Bennetts at Families Care—Fiscal Agent for the grant.) The Family Run Organizations of Nebraska (FRON) consider the model under collective ownership.

Family run organizations in a nutshell

- Peer support is a cost-effective means of engaging the family to work on tangible goals that address family/youth challenges.
- The "lived experience" component of family peer support helps establish trust and offers hope that motivates the families we serve and motivated parents are much more likely to achieve productive outcomes for their child.
- Peer support services give other options to overwhelmed families who believe their options are limited to out-of-home placement. Instead, they identify community-based resources and provide the opportunity to learn strategies for helping the child be successful in the home.
- Well-trained peer support providers can help ensure that the Family Voice remains the focus of attention. They become a channel for key information on events, protected rights, and education resources to teach parents how to interact productively with the professional realm so that the family voice is not lost in professional settings and processes.

The family peer support coaching model

Peer support personnel take a team approach to modeling collaborative teamwork to facilitate communication and ensure that the family voice is present in all activities. Supervisors must instill adherence to common themes to support this activity with the ultimate goal to develop family and professional partnerships. The list below demonstrates the foundation of the active duties of a family support provider in this role.

(1) Respect and value the role and expertise of each team member.
(2) Develop understanding of the responsibilities and expectations of each team member.
(3) Recognize the limits of one's knowledge and skills and seek assistance from others.
(4) Exhibit leadership and model collaboration.
(5) Assertively and collaboratively represent opinions and ideas and actively listen to the opinions of others.
(6) Empower family voice and advocate for family choice in team processes.
(7) Foster shared-decision-making with family members and other team members.
(8) Demonstrate practicality, flexibility, and adaptability in the process of working with others, emphasizing strengths, and celebrate successes.
(9) Respond immediately, if at all possible, to requests for consultation or intervention from other providers.
(10) Operationalize Systems of Care values and principles.
(11) Advocate for and foster the use of peer support approaches and peer support providers in the healthcare setting as a component of healthcare delivery.

Conclusion

In this chapter, we recognize the importance of family engagement to the future of child welfare and the knowledge that a single entity cannot address the challenges facing American families. But, identifying the span of youth living in out-of-home placements in various systems of care draws attention to the crucial element of family/youth engagement in supporting families to remain intact. Years of trauma in the home compiled by trauma in the system of care present significant mental/behavioral health challenges requiring support beyond the boundaries of adolescence. We suggest the productive potential of collaborations between families, peer support providers, and the systems that serve families/youth is a cost-effective alternative, addresses current limitations (e.g., available budgets, shortage of skilled workers), and supports better outcomes. The Nebraska Family Support Network utilizes a Family Peer Support Coaching Model as an example of this direction.

References

Anspatch, R. (2018, May 25). The foster care to prison pipeline: What it is and how it works. *Teen Vogue*. Fostered or Forgotten Series. Retrieved from https://www.teenvogue.com/story/the-foster-care-to-prison-pipeline-what-it-is-and-how-it-works.

Bodenheimer, G. A. (2016). The forgotten illnesses: The mental health movements in modern America. *Inquiries Journal*, *8*(7). Retrieved from http://www.inquiriesjournal.com/a?id=1428.

California Evidence-Based Clearinghouse (CEBC) for Child Welfare. (2015). *Protective factors survey*. Retrieved from https://www.cebc4cw.org/assessment-tool/protective-factors-survey/.

California Evidence-Based Clearinghouse (CEBC) for Child Welfare. (2017). *Active parenting of teens: Families in action*. Retrieved April 10, 2019 from https://www.cebc4cw.org/program/active-parenting-of-teens-families-in-action/detailed.

Center for Evidence-Based Practices (CEBP) at Case Western Reserve University. (2006, 2009, 2011). Motivational interviewing. Retrieved April 10, 2019 from https://www.centerforebp.case.edu/practices/mi—Trauma-informed care. Retrieved April 10, 2019 from https://www.centerforebp.case.edu/practices/trauma.

Child Welfare Information Gateway. (2017a). *Family engagement inventory: Commonalities across the practice level strategies domain*. Retrieved April 10, 2019 from https://www.childwelfare.gov/pubPDFs/Common-practice.pdf.

Child Welfare Information Gateway. (2017b). *Foster care statistics 2016*. Washington, DC: U. S. Department of Health and Human Services, Children's Bureau. Retrieved April 10, 2019 from https://www.childwelfare.gov/pubPDFs/foster.pdf.

ChildStats.gov. (2017). *POP1 child population: Number of children (in millions) ages 0–17 in the United States by age, 1950–2017 and projected 2018–2050*. Retrieved April 10, 2019 from https://www.childstats.gov/americaschildren/tables/pop1.asp.

Clemens, E. V., Helm, H. M., Myers, K., Thomas, C., & Tis, M. (2017). The voices of youth formerly in foster care: Perspectives on educational attainment gaps. *Children and Youth Services Review, 79*, 65–77. https://doi.org/10.1016/j.childyouth.2017.06.003.

Copeland Center for Recovery and Wellness. (2018). *WRAP® Wellness Recovery Action Plan*. Retrieved from https://copelandcenter.com/wellness-recovery-action-plan-wrap.

Counts, J. M., Buffington, E. S., Chang-Rios, K., Rasmussen, H. N., & Preacher, K. J. (2010). The development and validation of the protective factors survey: A self-report measure of protective factors against child maltreatment. *Child Abuse & Neglect, 34*(10), 762–772.

Debicki, A. (2008). A best practice model for a community mobilization team. In E. J. Bruns & J. S. Walker (Eds.), *The resource guide to wraparound*. Portland, OR: National Wraparound Initiative, Research and Training Center for Family Support and Children's Mental Health.

Donnelly, T., Baker, D., & Gargan, L. (2017). *The benefits of family peer support services: Let's examine the evidence*. Developed [in part] under contract number HHSS283201200021I/HHS28342003T from the Substance Abuse and Mental Health Services Administration, U.S. Department of Health and Human Services (HHS). Retrieved April 10, 2019 from https://www.nasmhpd.org/sites/default/files/Benefits%20of%20Family%20Peer%20Support%20FIC%20SAMSHA%20Updated.pdf.

Family Run Executive Director Leadership Association (FREDLA). (2015). *Timeline of the family movement in children's behavioral health*. Retrieved April 10, 2019 from http://www.fredla.org/wp-content/uploads/2015/09/Timeline-infographic.pdf.

Family Run Executive Director Leadership Association (FREDLA). (2019). *Family run organizations*. Retrieved April 10, 2019 from https://www.fredla.org/wp-content/uploads/2019/02/FRO-List-At-a-glance-2_2019.pdf.

Farrell, C. A., Fleegler, E. W., Monuteaux, M. C., Wilson, C. R., Christian, C. W., et al. (2017). Community poverty and child abuse fatalities in the United States. *Pediatrics, 139*(5), e20161616.

Fowler, P. J., Marcal, K. E., Zhang, J., Day, O., & Landsverk, J. (2016). Homelessness and aging out of foster care: A national comparison of child welfare-involved adolescents. *Children and Youth Services Review*, *77*, 27–33. Retrieved April 25, 2019 from https://doi.org/10.1016/j.childyouth.2017.03.017.

Greiner, M. V., Beal, S. J., et al. (2017). Foster care is associated with poor mental health in children. *The Journal of Pediatrics*, *182*, 401–404. https://doi.org/10.1016/j.jpeds.2016.12.069.

Grohol, J. M. (2018). *Harriet Shetler, co-founder of NAMI. In Pysch Central*. Retrieved March 25, 2019 from https://psychcentral.com/blog/harriet-shetler-co-founder-of-nami/.

Gypen, L., et al. (2017). Outcomes of children who grew up in foster care: Systematic-review. *Children and Youth Services Review*, *76*, 74–83. https://doi.org/10.1016/j.childyouth.2017.02.035.

History Nebraska. (2017, September 21). *Safe Haven Act sign for original and amended law, 2008. [Blog post]*. Retrieved April 26, 2019 from https://history.nebraska.gov/blog/safe-haven-law-2008.

Hopper, E. K., Bassuk, E. L., & Olivet, J. (2010). Shelter from the storm: Trauma-informed care in homeless service settings. *The Open Health Services and Policy Journal*, *3*, 80–100.

Juvenile Law Center. (2018, May 26). *What is the foster care to prison pipeline?* [Blog Post]. Retrieved from https://jlc.org/news/what-foster-care-prison-pipeline.

Miller, W. R., & Rollnick, S. (2012). *Motivational interviewing: Preparing people for change* (3rd ed.). New York, NY: The Guilford Press.

National Center for Injury Prevention and Control, Division of Violence Prevention. (n.d.). About the CDC-Kaiser ACE study. [Last reviewed April 2, 2019]. Retrieved April 26, 2019 from https://www.cdc.gov/violenceprevention/childabuseandneglect/acestudy/about.html.

National Federation of Families for Children's Mental Health. (2017a). *History*. Retrieved March 25, 2019 from https://www.ffcmh.org/history.

National Federation of Families for Children's Mental Health. (2017b). *National certification for parent family peers*. Retrieved March 25, 2019 from https://www.ffcmh.org/certification.

National Registry of Evidenced-based Programs and Practices (NREPP). (2010). *Active parenting of teens: Families in action*. Substance Abuse and Mental Health Services Administration (SAMHSA).

National Wraparound Initiative, Regional Research Institute, School of Social Work, Portland State University. (2019). *Wraparound basics or what is wraparound: An introduction*. Retrieved from https://nwi.pdx.edu/wraparound-basics/.

Nebraska Department of Health and Human Services (DHHS). (2018, May 4). *Nebraska child and family services review. Revised Program Improvement Plan* http://dhhs.ne.gov/children_family_services/Documents/NE%20PIP%20Re-Submission%205-14-18.pdf.

Nebraska Department of Health and Human Services (DHHS) Behavioral Health. (2018). *Nebraska family helpline*. Retrieved March 25, 2019 from http://dhhs.ne.gov/behavioral_health/Pages/nebraskafamilyhelpline_about.aspx.

O'Hanlon, K. (2013, Jan 21). 5 years later, Nebraska patching cracks exposed by safe-haven debacle. *Lincoln Journal Star*, Retrieved from https://journalstar.com/special-section/epilogue/years-later-nebraska-patching-cracks-exposed-by-safe-haven-debacle/article_d80d1454-1456-593b-9838-97d99314554f.html.

Salazar, A. M., Keller, T. E., Gowen, L. K., & Courtney, M. E. (2012). Trauma exposure and PTSD among older adolescents in foster care. *Social Psychiatry and Psychiatric Epidemiology*, *48*(4), 545–551.

Sawyer, W. (2018, February 27). *Youth confinement: The whole pie*. [Blog Post-Prison Policy Initiative]. Retrieved from https://www.prisonpolicy.org/reports/youth2018.html.

Search Institute. (1997). *40 Developmental Assets® for Adolescents (ages 12–18)*. Retrieved from https://www.search-institute.org/our-research/development-assets/developmental-assets-framework/.

Stroul, B., Blau, G., & Friedman, R. (2010). *Updating the system of care concept and philosophy*. Washington, DC: Georgetown University Center for Child and Human Development, National Technical Assistance Center for Children's Mental Health. Retrieved April 10, 2019 from https://gucchd.georgetown.edu/products/Toolkit_SOC_Resource1.pdf.

Substance Abuse and Mental Health Services Administration (SAMHSA). (2017a). *Family and caregiver support in behavioral health*. Retrieved March 25, 2019 from https://www.samhsa.gov/sites/default/files/programs_campaigns/brss_tacs/family-parent-caregiver-support-behavioral-health-2017.pdf.

Substance Abuse and Mental Health Services Administration (SAMHSA). (2017b). *National Mental Health Services Survey (N-MHSS): 2016*. Data on Mental Health Treatment Facilities. BHSIS Series S-98, HHS Publication No. (SMA) 17-5049 Rockville, MD: Substance Abuse and Mental Health Services Administration.

Substance Abuse and Mental Health Services Administration (SAMHSA). (2018). *Adverse childhood experiences*. [Last updated July 9, 2018]. Retrieved March 25, 2019 from https://www.samhsa.gov/capt/practicing-effective-prevention/prevention-behavioral-health/adverse-childhood-experiences.

Tang, P. (2013, June 7). *A brief history of peer support: Origins*. Retrieved March 25, 2019 from http://peersforprogress.org/pfp_blog/a-brief-history-of-peer-support-origins/. Peers for Progress, Peer Support Around the World, Chapel Hill, NC: University of North Carolina.

Torres, K., & Rricha, M. (2018, March 9). *Fact sheet: Family first prevention services act*. [Blog post First Focus Campaign for Children]. Retrieved April 6, 2019 from https://campaignforchildren.org/resources/fact-sheet/fact-sheet-family-first-prevention-services-act/.

U.S. Centers for Disease Control and Prevention, Kaiser Permanente. (2016). *The ACE study survey data* [Unpublished Data]. Atlanta, GA: U.S. Department of Health and Human Services, Centers for Disease Control and Prevention.

U.S. Department of Health and Human Services, Administration for Children and Families, Administration on Children, Youth and Families, Children's Bureau. (2019). *Child maltreatment 2017*. Retrieved March 25, 2019 from https://www.acf.hhs.gov/cb/research-data-technology/statistics-research/child-maltreatment.

Voices for Children in Nebraska. (2018). *©Kids count in Nebraska 2018 report*. Retrieved April 10, 2019 from https://voicesforchildren.com/wp-content/uploads/2019/01/2018-Kids-Count-in-Nebraska-Report.pdf.

Wexler, R. (2018, February 24). *Studies show foster care a toxic environment*. [Times Union]. Retrieved April 10, 2019 from https://www.timesunion.com/opinion/article/Studies-show-foster-care-a-toxic-environment-12706517.php.

Wiltz, T. (2018, May 2). *This new federal law will change foster care as we know it*. [Blog post PEW Charitable Trusts]. Retrieved from https://www.pewtrusts.org/en/research-and-analysis/blogs/stateline/2018/05/02/this-new-federal-law-will-change-foster-care-as-we-know-it.

Supplementary reading recommendations

Adverse Childhood Experiences (ACES) Survey for Providers. (2015). The National Crittenton Foundation. Retrieved April 26, 2019 from http://www.nationalcrittenton.org/wp-content/uploads/2015/10/ACEs_Toolkit.pdf.

FRIENDS National Resource Center. (2019). Retrieved from https://friendsnrc.org/protective-factors-survey. (i) Protective factors survey (PFS); (ii) Protective factors survey, Second Edition (PFS-2).

Hambrick, E. P., Oppenheim-Weller, S., N'zi, A. M., & Taussig, H. N. (2016). Mental health interventions for children in foster care: A systematic review. *Children and Youth Services Review*, *70*, 65–77. https://doi.org/10.1016/j.childyouth.2016.09.002.

Osman, A., Bagge, C. L., Guitierrez, M., Konick, L. C., Kooper, B. A., et al. (2001). The suicidal behaviors questionnaire-Revised (SBQ-R): Validation with clinical and nonclinical samples. *Assessment*, *5*, 443–454. https://doi.org/10.1177/107319110100800409.

Shah, M. F., Liu, Q., Mark Eddy, J., Barkan, S., Marshall, D., et al. (2017). Predicting homelessness among emerging adults aging out of foster care. *American Journal of Community Psychology*, *60*, 33–43. https://doi.org/10.1002/ajcp.12098.

Substance Abuse and Mental Health Services Administration (SAMHSA). (2019). *Evidence-based resource center [Last Updated: April 14, 2019]*. Retrieved May 21, 2019 from https://www.samhsa.gov/ebp-resource-center.

Substance Abuse and Mental Health Services Administration (SAMHSA). (1999). Stable resource toolkit. *In The suicide behaviors questionnaire revised SBQR—Overview*. Retrieved April 26, 2019 from https://www.integration.samhsa.gov/images/res/SBQ.pdf.

U.S. Department of Health and Human Services. (2017). The AFCARS report: Preliminary FY 2016 estimates as of October 2017 (No. 24). Retrieved April 10, 2019 from https://www.acf.hhs.gov/cb/resource/afcars-report-24.

University of Nebraska Medical Center. (2019). *Nebraska system of care (NeSOC)*. Retrieved from https://www.unmc.edu/bhecn/education/nebraska-system-of-care/index.html.

Villegas, S., & Pecora, P. J. (2012). Mental health outcomes for adults in family foster care as children: An analysis by ethnicity. *Children and Youth Services Review*, *34*(8), 1448–1458. https://doi.org/10.1016/j.childyouth.2012.03.023.

Definitions

Adverse childhood experiences encompasses a variety of emotional and physical trauma including emotional neglect, physical or sexual abuse. Conditions of trauma include parental indications of domestic violence, substance abuse, single parent household, or incarceration of a parent or other family member. These environments contribute to long-term negative health implications.

Child protective services a branch of a state's Health and Human Services, responsible for the assessment, investigation, and intervention on cases of child abuse and neglect, including sexual abuse.

Foster care a state-approved temporary placement option for children who, for a variety of reasons, have been removed from the home and care of their bio parents or primary care givers. Foster care can also refer to group homes and residential care centers.

Maltreatment types neglect or other types such as physical or sexual abuse; see examples of adverse childhood experiences

Victim a child for whom the state determined at least one or more maltreatment(s) is substantiated or indicated.

Sustainability education, not just STEM but the root of business

Naz Onel[a], *Amit Mukherjee*[a], *Beth Ann Fiedler*[b]
[a]School of Business, Stockton University, Galloway, NJ, United States,
[b]Independent Research Analyst, Jacksonville, FL, United States

Abstract

The acceleration of environmental, economic, and social issues across the global marketplace has stimulated a variety of responses from many entities, including global companies and business schools, as they race to adapt to increasingly significant economic, demographic, and technological changes. Following these dynamic trends, several business schools have come to recognize the importance of implementing sustainability in different areas of business education. Sustainability has been taught in university education programs in Science, Technology, Engineering, and Mathematics, but the application to business schools is a recent expansion. Case study methodology investigates the significance of incorporating sustainability concepts into business education in terms of (1) meeting overall educational goals, and (2) illustrating how implementation is an opportunity to reinforce adopted learning goals of business schools. Findings illuminate best practices and sustainable strategy initiatives in 20 business schools. We recommend an enhanced four-point strategy to integrate Education for Sustainable Development into business schools.

Keywords: Sustainability education, Business students, Triple bottom line, Business sustainability.

Chapter outline

Three Facets of Public Health and Paths to Improvements. https://doi.org/10.1016/B978-0-12-819008-1.00006-7

New marketplace objectives require new business education

Today, the longstanding notion that corporations function first and foremost to maximize their profits and keep their shareholders content is changing rapidly. As members of the Business Roundtable organization, nearly 200 chief executive officers (CEOs) of major U.S. corporations recently issued a "purpose of a corporation" statement that outlines a new set of objectives for the modern-day corporation. The re-imagined ideas by America's top business leaders now at the forefront of their business objectives replace the age-old notion of maximizing shareholder value with employee investment, ethical conduct of business with suppliers, delivering the highest value to customers, supporting surrounding communities, and improving their existence in a positive way (Fitzgerald, 2019). This important shift in business mindsets signals changes in leadership styles, repurposing business existence, and a growing need for future leaders who are prepared to operate in the new sustainable marketplace that is conscientious, responsible, and compassionate.

The term "sustainability" has long been associated with the Brundtland Commission's definition of sustainable development from *Our Common Future*, also known as the Brundtland Report, which defines sustainable development as "development that meets the needs of the present without compromising the ability of future generations to meet their own needs," (United Nations General Assembly, 1987, p. 43). "Sustainability" is now an umbrella term that encompasses all the dimensions that contribute to sustainable business practices that are referred to as the triple bottom line (Ten Bos & Bevan, 2011). Thus, the emphasis on sustainability requires an overall shift in an institutional mindset to the three pillars of sustainability—an organization's social, environmental, and economic performance (Kiron, Kruschwitz, Haanaes, & von Streng Velken, 2012).

The shift in business purpose and consequent need to prepare leaders to embrace the concept of sustainable development into business education and practice did not go unnoticed by sensible leaders, business managers, C-suite executives, and business school deans (Hommel, Painter-Morland, & Wang, 2012). Their efforts were instrumental in evolving academic culture to respond in real time to the complex and large-scale sustainability challenges of contemporary times. Today, we see evidence of incorporating the issue of sustainability and growing momentum in the mainstream business curriculum in many Universities' schools of business (hereinafter referred to as business schools) (Rasche, Gilbert, & Schedel, 2013), despite a number of obstacles to the integration of these issues, such as continued economic crisis and public criticism of business schools (Muff et al., 2013). This chapter focuses on the gradual incorporation of sustainability and triple bottom line principles into the business school curriculum of leading institutions of higher education. Triple bottom line includes social well-being, environmental impact, and financial results to measure an organization's performance. This idea is also referred to as 3P's: people, planet, and profits (Lawrence & Weber, 2014). Moving in the direction of Education for Sustainable Development (ESD) in business schools can have a positive impact on the world which is already demonstrated by the persistent growth both of sustainable business and in the number in academic programs, teaching, and learning of the subject matter.

However, pockets of resistance to implementing ESD at any level remain, despite many positive outcomes demonstrated in this study. Attempts to integrate sustainability into the curriculum suffer from some of the persistent core problems of business education faced by many schools. These problems are summarized as business schools' myopic orientation, worry of accreditation, publishing criteria, and rankings (Painter-Morland, 2015). However, we suggest that the objectives of sustainability education could be merged with the current educational strategy of business schools in order to reinforce overlapping learning goals. Eventually, the process can help with accreditation efforts as well as publishing and rankings of schools. For example, one of the international accreditation bodies (Association to Advance Collegiate Schools of Business—AACSB) General Skill Area standards requires "ethical understanding and reasoning" competency of the undergraduate students, which is also one of the main aims of ESD.

The goals of this chapter are threefold. First, the chapter demonstrates the significance of incorporating sustainability concepts into business education in terms of meeting the educational goals outlined by AACSB. Concurrently, this illustrates how implementation could be an opportunity to reinforce adopted learning goals of programs and schools. Second, this chapter contributes to the literature by reflecting on the progression from enshrinement of "sustainability" in mission or vision statements of schools of business to integrating sustainability into their business curricula. The analysis is conducted through a case study approach in which we develop a matrix of business schools that have integrated sustainability into their curricula. Finally, the chapter elaborates on the best practices and strategies of successful business schools with sustainability initiatives. Major findings spotlight the best practices of the Copenhagen Business School in Denmark and aid in the development of a four-point strategy to integrate Education for Sustainable Development into business schools.

Planting sustainability seeds in business education

Higher education has successfully developed and implemented sustainability programs in Science, Technology, Engineering, and Mathematics (STEM) fields of study, but the same implementation in business administration, management, entrepreneurship, and related programs has only recently been the focus of attention. Increasing methods to incorporate sustainability concepts into business education presents an opportunity to limit harmful environmental disruption, increase economic prospects to overcome health problems related to poverty, and reduce social degradation. Further, instilling the foundation of sustainability in current and future business leaders can enhance long-term community economic stability with the embedded notion of environmental protection in a balanced approach to business development.

The United Nations Educational, Scientific and Cultural Organization (UNESCO) define Sustainability Education, also referred to as Education for Sustainable Development (ESD), as "including key sustainable development issues into teaching and learning; for example, climate change, disaster risk reduction, biodiversity, poverty reduction, and sustainable consumption. It also requires participatory teaching and learning methods that motivate and empower learners to change their behavior and take action for sustainable development," (2014, *para* 2–3). The organization also suggests that the ESD "allows every human being to acquire the knowledge, skills, attitudes and values necessary to shape a sustainable future," (United Nations Educational, Scientific and Cultural Organization (UNESCO), 2014, *para* 1). UNESCO (2014) indicates that collaboration is important to student achievement in some key ESD competencies, such as critical thinking, imagining future scenarios, and making decisions. These skills are foundational for future business leaders.

Watering the sustainability education seed in business should be relatively easy since colleges and universities have the tools and reasons to integrate sustainability into their curricula. In fact, over the past decade, several studies have been published on how to integrate sustainability in higher education, including studies with the focus of business education (Porter & Córdoba, 2009; Rusinko, 2010; Setó-Pamies & Papaoikonomou, 2016; Slager, Pouryousefi, Moon, & Schoolman, 2018; Swaim, Maloni, Napshin, & Henley, 2014). These studies were conducted in response to business school deans' and corporate CEOs' acknowledgement of the importance of sustainability as a strategic concern that should be considered as a part of business education (Hommel et al., 2012). Leadership in education concurs that through research and teaching associated with sustainability, higher education can be redesigned with deliberate institutional evolution and large-scale academic reorganization. There is general agreement that business education must undergo an evolution, but there is an absence of teaching philosophy and education strategies to allow the sustainability concept to grow as methodically as in the STEM fields.

Other important shortcoming of today's business education is the lack of understanding of the important facets of ESD. These items include (1) implementing participatory teaching and learning methods, (2) teaching critical thinking, (3) informing about current sustainability issues (e.g., sustainable consumption, water conservation), (4) motivating and empowering learners to change their behavior to achieve a more sustainable environment, and (5) teaching knowledge skills, attitudes, and values necessary to shape a sustainable future. The latter involves teaching other imperative competencies such as imagining future scenarios and making decisions in a collaborative way. All these important aspects of ESD highlight the significance of incorporating sustainability concepts into business education. We argue that these objectives of ESD could be merged into the current educational strategy in business schools in a manner that will not be perceived as an additional burden. In fact, they appear as an opportunity to reinforce long-lived Learning Goals of business schools, outlined by the AACSB International discussed in the next section.

Matching international business school standards with Education for Sustainable Development (ESD): Is there common ground?

AACSB (2018, p.2):

> *The business environment is undergoing profound changes, spurred by powerful demographic shifts, global economic forces, and emerging technologies. At the same time, society is increasingly demanding that companies become more accountable for their actions, exhibit a greater sense of social responsibility, and embrace more sustainable practices. These trends send a strong signal that what business needs today is much different from what it needed yesterday or will need tomorrow.*

The AACSB International is an independent accrediting agency that oversees the standardization of collegiate schools of business (AACSB, 2018). The program has been in operation since 1916 and many of the top business schools in the United States are accredited by the organization. As of August 2019, there are 856 business institutions in 56 countries and territories that have earned AACSB accreditation (additionally, 188 institutions hold a specialized AACSB accreditation for their accounting programs) (AACSB, 2019, *para* 1–2). The accreditation is designed to ensure that all business students are learning material that is most relevant to their field of study and preparing them to be effective leaders upon graduation. Earning accreditation is the first step in a long-term commitment to continually innovate and have community/global impact in order to maintain accreditation during periodic reviews every 5 years (AACSB, 2018). The AACSB's *Preamble: Engagement, Innovation, and Impact* section of the Accreditation Standards Report acknowledges the changes that are taking place in the business environment and their necessary reflection in business education.

AACSB International's overall vision displays profound similarities with the ESD approach. The organization's main purpose is to transform business education for global prosperity (AACSB, 2018). Additionally, the organization views business and business schools as forces "for good, contributing to the world's economy and to society," and AACSB exists to play "a significant role in making that benefit better known to all stakeholders – serving business schools, students, business and society," (AACSB, 2018, p. 1). With these aims, the AACSB lists various Learning Goals for business schools to emphasize.

High-quality business schools have processes in place for determining relevant learning goals for specific degree programs (Gerstein & Friedman, 2016) that lead to curricula development in line with AACSB (2018) definition of learning goals as an educational expectation for each degree program that maximizes the opportunity to reach outcomes linked to achieving goals. "Subsequently, these schools have systems in place to assess whether learning goals have been met," (AACSB, 2018, p. 32). If these schools or the programs cannot meet the predetermined learning goals, they must attempt to improve by putting alternative processes in place. Otherwise, they sustain the risk of losing the accreditation for schools in case of poor performance on these predetermined goals and objectives or other required initiatives and implementations set by the AACSB International.

Under the Standard 9 entitled, "Curriculum content is appropriate to general expectations for the degree program type and learning goals," AACSB highlights the schools' responsibility to translate the general areas into expected competencies consistent with the degree program learning goals. For Bachelor's Degree Programs and Higher, this section of the Standards categorizes general skill areas and general business and management skill areas under three different areas of focus: (1) general skills, (2) general business knowledge, and (3) technology agility. These three focus areas list 19 specific learning expectations at the bachelor's level and an additional 5 learning expectations at the master's level (AACSB, 2018).

Table 6.1 shows a number of learning expectations listed under the Standard 9 of the AACSB International and their matching ESD goals with corresponding Business Program or School-Specific Learning Goals for both undergraduate and graduate business schools. For the purpose of this study, we used corresponding Business Program or School-Specific Learning Goals from Stockton University, School of Business, Business Studies and MBA Program Learning Goals (2019a, 2019b), and Rutgers University Business School Goals and Objectives for Undergraduate Programs (New Brunswick-Newark) (Rutgers University, 2019). For example, one of the learning expectations listed under the Standard 9 is: *Analytical thinking: Students will be able to analyze and frame problems*, which corresponds with *Critical thinking and Imagining future scenarios* of the ESD goals. Both goals could be matched with Business Program (or school-specific) Learning Goals of *Critical Thinking and Problem-Solving* and *Solving Business Problems*. Overall, the comparison across three guiding principles demonstrates the significance of incorporating sustainability concepts into business education in terms of meeting the educational goals outlined by the international accreditation body. In fact, the ESD appears as an opportunity to directly reinforce and meet many different learning goals of business schools.

Table 6.1 Learning expectations listed under the Standard 9 of the Association to Advance Collegiate Schools of Business (AACSB) International and their matching Education for Sustainable Development (ESD) goals with corresponding Business Program or School-Specific Learning Goals for both undergraduate and graduate business schools.

AACSB list of learning expectations for bachelor's degree programs and higher[a]	**Education for Sustainable Development (ESD) goals**	**Business Program or School-Specific Learning Goal examples**[b]
AACSB general skill areas		
Oral communication Students will be able to communicate effectively orally	Participatory teaching and learning	*Communication skills (oral communication)* Graduates will be able to deliver information in a persuasive, logical, and organized manner with a professional demeanor using appropriate supportive visual aids
Written communication Students will be able to communicate effectively in writing	Participatory teaching and learning	*Communication skills (written communication)* Graduates will be able to create informational, analytical, and technical documents which are organized, precise, and relevant
Ethical understanding and reasoning Students will be able to identify ethical issues and address the issues in a socially responsible manner	To acquire the knowledge, skills, attitudes, and values necessary to shape a sustainable future	*Ethics* Graduates will be able to demonstrate ethical reasoning when faced with moral dilemmas in business situations *Recognizing moral dilemmas* Graduates will be able to recognize ethical issues in business situations *Identify appropriate stakeholders* Graduates will be able to identify the parties affected by the moral dilemma

Continued

Table 6.1 Continued

AACSB list of learning expectations for bachelor's degree programs and higher	Education for Sustainable Development (ESD) goals	Business Program or School-Specific Learning Goal examples
		Identify alternative responses Graduates will be able to identify and briefly describe at least two recognized theories of ethical decision making *Apply ethics theory in a business setting* Graduates will be able to apply, with analysis, a selected recognized ethics theory to a life-like ethics dilemma found in a business setting
Analytical thinking Students will be able to analyze and frame problems	Critical thinking and imagining future scenarios	*Critical thinking and problem-solving* Graduates will be able to logically interpret, analyze, and summarize the results of information examined and will be able to apply appropriate analytic, problem-solving, and decision-making skills in business situations *Solving business problems* Graduates will be able to analyze real-world business situations and make informed decisions
Interpersonal relations and teamwork Students will be able to work effectively with others and in team environments	Making decisions in a collaborative way	*Management-specific learning goal* Teamwork—Graduates will be able to facilitate interaction with team members and contribute their expertise toward the creation and development of group projects

Reflective thinking Students will be able to understand oneself in the context of society	Participatory teaching and learning (motivating and empowering learners to change their behavior and take action for sustainable development)	*Critical thinking learning goal* Graduates will be able to logically interpret, analyze, and summarize the results of information examined and will be able to apply appropriate analytic, problem-solving, and decision-making skills in business situations
Application of knowledge and integration of real-world business experiences Students will be able to translate knowledge of business into practice	Including current key sustainable development issues into teaching and learning (e.g., climate change, biodiversity, sustainable consumption)	*Critical thinking and problem-solving learning goal* Graduates will be able to logically interpret, analyze, and summarize the results of information examined and will be able to apply appropriate analytic, problem-solving, and decision-making skills in business situations. *Solving business problems* Graduates will be able to analyze real-world business situations and make informed decisions
AACSB general business knowledge areas		
Economic, political, regulatory, legal, technological, and social contexts of organizations in a global society	Including key sustainable development issues into teaching	*Global perspective learning goal* Knowledge of the diversity of past and current economic, legal, political, and social structures Understanding of the impact of cultural and demographic diversity on business interactions
Social responsibility, including sustainability, diversity, and ethical behavior and approaches to management	Including key sustainable development issues into teaching and learning; such as, climate change, disaster risk reduction, biodiversity, poverty reduction, and sustainable consumption	*Ethics* Graduates will be able to demonstrate ethical reasoning when faced with moral dilemmas in business situations *Recognizing moral dilemmas* Graduates will be able to recognize ethical issues in business situations

Continued

Table 6.1 Continued

AACSB list of learning expectations for bachelor's degree programs and higher	Education for Sustainable Development (ESD) goals	Business Program or School-Specific Learning Goal examples
		Identify appropriate stakeholders Graduates will be able to identify the parties affected by the moral dilemma *Identify alternative responses* Graduates will be able to identify and briefly describe at least two recognized theories of ethical decision making *Apply ethics theory in a business setting* Graduates will be able to apply, with analysis, a selected recognized ethics theory to a life-like ethics dilemma found in a business setting
Additional AACSB list of learning expectations for general business master's degree programs	ESD goals	**MBA learning goal examples**
Leading in organizational situations	Acquiring the knowledge, skills, attitudes, and values necessary to shape a sustainable future	*Management-specific knowledge and its application* Graduates will understand, be able to apply, and will work toward integrating and evaluating perspectives and techniques of strategic thinking for managing organizations
Managing in a diverse global context	Critical thinking and imagining future scenarios	*Management-specific knowledge and its application* Graduates will understand, be able to apply, and will work toward integrating and evaluating the unique contributions made by diversity in organizations

Thinking creatively	Critical thinking	*Critical thinking* Graduates will evaluate, integrate, and synthesize management information
Making sound decisions and exercising good judgment under uncertainty	Critical thinking and imagining future scenarios	*Critical thinking* Graduates will think strategically about the organization in a global, economic, environmental, political, ethical, legal, and/or regulatory context *Ethical judgment* Student will demonstrate to critically evaluate business decision-making scenarios and develop innovative and ethical solutions
Integrating knowledge across fields	Sustainability covers many different disciplines. Hence, teaching with an interdisciplinary approach is necessary	*Critical thinking* Graduates will frame organizational problems from a variety of functional and stakeholder perspectives in an integrative and interdisciplinary manner

[a] This is not the exhaustive list of learning expectations covered under the Standard 9 of AACSB report. For the full list see AACSB (2018, p. 35).

[b] Examples are from Stockton University, School of Business, Business Studies and MBA Program Learning Goals (2019a, 2019b), and Rutgers University Business School Goals and Objectives for Undergraduate Programs (New Brunswick-Newark) (Rutgers University, 2019).

In the next section, we will explain the importance of voluntary actions of institutions and their collaborative acts with the Principles for Responsible Management Education (PRME) of the United Nations.

Moving beyond curriculum: Voluntary actions can change the world

Moving beyond curriculum for a systemic institutional integration is crucial (Painter-Morland, Sabet, Molthan-Hill, Goworek, & de Leeuw, 2016). Successfully embedding sustainability into different business programs, as they suggest, would necessitate (1) systemic thinking and systemic leadership; (2) connectedness to stakeholders, mainly, business, the society, and environment; and (3) building an institutional capacity. Although we agree with their approach on moving beyond curricula and the need for policy change, we have a different perspective on the importance of voluntary actions by institutions. "Policy-change requires more than just signing up to, for example, the Principles for Responsible Management Education (PRME) initiative from the United Nations Global Compact, and implementing a sustainability policy," (Painter-Morland et al., 2016, p. 743). Similarly, Burchell, Kennedy, and Murray (2015) argue that the soft governance processes of the PRME cannot produce the change and development for which it is aimed if agents within institutions are not empowered to effect change.

We believe that true advancement towards more sustainable business schools and business education is possible by implementing institutional efforts that are beyond curriculum requiring voluntary actions by the institutions. This is similar to Corporate Social Responsibility (CSR) activities undertaken by many organizations today as a soft governance initiative. CSR is a self-regulating business model that helps a company be socially and environmentally accountable to all its stakeholders.

During the 1970s, corporations were moved by social pressures to think beyond improving monetary value for the corporate stockholders or improving upon statutory obligations to minimally comply with legislation. As a result, organizations increasingly began to take voluntary actions to improve the quality of life for employees, their families, the local community, and society at large. Today, this concept is known as CSR that is mainly defined as a concept whereby organizations consider the interests of society by taking responsibility for the impact of their activities on customers, employees, shareholders, communities, and the environment in all aspects of their operations (Onel & Fiedler, 2018).

Since its factual appearance in the U.S. in the 1970s, CSR has come a long way from being "a nice thing to do" to a key to business success (ACCP, 2019). Changing the mindsets of corporations, small or large, could be possible through individual voluntary actions, leading to an overall systematic change in the marketplace.

This progression of CSR demonstrates the importance of voluntary actions in order to bring changes to educational institutions as well. Similarly, contrary to Burchell et al. (2015) and Painter-Morland et al.'s (2016) arguments, we observe that change is possible through broad and voluntary approaches of schools that add sustainability initiatives. For example, working together with UN PRME moves institutions to think broader, help them gain trust, and build a strong community and global reputation. This initiative also helps institutions to move ahead by applying a strategy of continuous improvements in sustainability activities which influences all aspects of the institution. The process of increasing sustainability, as a whole, creates a sense of being a part of something bigger and more valuable that becomes evident in environmental and social well-being.

Founded in 2007, PRME is a UN-supported platform to raise global higher education focus on sustainability business or management schools' focus of sustainability globally and "to equip today's business students with the understanding and ability to deliver change tomorrow," (UNPRME, 2019f, *para* 1). More than 650 signatories have become a voluntary participant of PRME establishing the initiative as "the largest organized relationship between the United Nations and management-related higher education institutions," (UNPRME, 2019f, *para* 2).

Case studies, challenges, and opportunities for growing future business leaders

Businesses are becoming increasingly concerned with implementing sustainable objectives in their practices. As society becomes more interested in responsible sourcing and ethical reasoning, companies are beginning to change their behaviors, leading universities to change business schools' curriculum. This section focuses on different approaches adopted by institutions of higher education in the United States, United Kingdom, Australia, Canada, and Denmark. Of the 20 business schools under examination, 18 hold AACSB International accreditation (exceptions are Bard College and Western Colorado University) and eight are listed as PRME participants at different levels (i.e., noncommunicating signatory, basic signatory, advanced signatory, and PRME Champion). The schools are presented in alphabetical order.

American University-Washington, DC, USA; private research, 12,000 combined enrollment

American University's Kogod School of Business offers a specialized Master of Science (MS) degree in sustainability management (MSSM). The MSSM curriculum is business-based integrated with social science, public policy, and international issues. Students are taught to analyze organizational problems in their environmental milieu and come up with economic, ecological, and socially responsible solutions. Together with an in-depth interdisciplinary exploration of sustainability theories and practices, students are introduced to modern managerial methods that are applicable to sustainable business practices. MSSM students also participate in an international residency each year to learn about best practices in sustainability management across multiple industry and governmental organizations (American University, 2019a).

The MBA and BBA programs of the Kogod School offer few sustainability-related courses. Examples are undergraduate electives "Marketing for Social Change" and

"Nonprofit and Social Entrepreneurship" and the MBA course "Business at the Private-Public Intersection" (American University, 2019b). The Kogod School of Business is a basic signatory of the UN's PRME initiative.

Bard College-Annandale-on-Hudson, NY, USA; private liberal arts, 2500 combined enrollment

Bard College offers its MBA program held in New York City and is unique in its exclusive focus on sustainability. Sustainability is not a curricular track or concentration available to interested students, but the MBA program is designed for students focused on sustainability with integration of sustainability into every class. The Bard MBA curriculum provides a foundation in management fundamentals with a focus on the triple bottom line: economic success, environmental integrity, and social equity. The program teaches students to scrutinize not just the economic impacts of the decisions they make, but the social and environmental impacts as well, while also providing them the expertise to make the business case for sustainability. In addition to the specialization in sustainable business earned by all MBA graduates, the Bard MBA offers two Focus Areas in Impact Finance and Circular Value Chain Management (Bard College, 2019a). The school's MBA program is counted as one of the few schools globally that fully integrates sustainability into its core business curriculum (UNPRME, 2019b).

One may also pursue a dual MS/MBA degree offered through the Bard Center for Environmental Policy and Bard MBA. This is a three-year program designed for students keen on careers at the intersection of environmental policy and business. Students graduate with master's-level training in environmental or climate policy and a business degree focused on sustainability and social justice (Bard College, 2019b). The Bard MBA in Sustainability is a participant of PRME initiative as a basic signatory.

Brandeis University-Waltham, MA, USA; private research, 6000 combined enrollment

The Heller School for Social Policy and Management is a graduate school at Brandeis committed to progressive social policy and development. Heller's Social Impact MBA is designed for students who want to lead organizations that create positive, sustainable social change. Sustainable Development—a concentration available to Social Impact MBA students—provides them with a solid foundation in sustainable development concepts and methods and gives them a broad understanding of the political, social, and policy factors that affect the sustainability of projects. A key feature of the program is a Consulting Project in which teams of students consult on crucial management problems like developing water management strategies or drawing up marketing strategies and distribution plans for green products at a mission-driven organization or government agency. Some of the classes offered in the program include "Managing the Triple Bottom Line", "Business and the Environment," and "Investing in Energy" (Brandeis University, 2019a).

In addition to the Social Impact MBA, the Heller School offers an MA in Sustainable International Development (MA-SID) preparing students for professional-level positions in international development. An intensive, two-year, 82 credits Dual-Degree Social Impact MBA/MA in Sustainable International Development is also offered. (Brandeis University, 2019b).

In addition to the Heller School for Social Policy and Management, Brandeis has an International Business School that offers traditional business-focused MBA, MA, and undergraduate degrees. The offerings do not include sustainability-related programs or courses. There are no Joint or Dual-degree offerings between programs offered by the Heller School and the International Business School (Brandeis University, 2019c). Brandeis does not participate in the PRME initiative.

Copenhagen Business School-Copenhagen, Denmark; public, 22,000 combined enrollment

The Copenhagen Business School (CBS) set up a research center, CBS Sustainability, to study sustainable practices and developments in organizations, markets, and the society. The program specializes in public policy, corporate communication, consumer behavior, social innovation and entrepreneurship, business and human rights, and corporate social responsibility (Copenhagen Business School, 2019a).

Responsible Management and Sustainability are the dominant themes of the Copenhagen MBA. Students begin the program by developing their skillset through a rigorous 50-h course entitled "Managing Sustainable Corporations" that is divided into four broad topics: Corporate Responsibility and Sustainability; The New Global Rules of the Game; Business Imperative; and Earning Trust with Stakeholders. In addition, CSR and sustainability are incorporated into most of the core courses, training students with a 360-degree vision, and the ability to evaluate and implement CSR across business units. (Copenhagen Business School, 2019b).

CBS offers a variety of courses for the Bachelor of Science, Master of Science, and MBA degrees related to sustainability which are either contributed to or coordinated by CBS Sustainability's faculty. At the undergraduate level, examples of such courses are "Introduction to Sustainable Business," "Scandinavian Sustainability and Corporate Social Responsibility," and "The Corporation in Society-Managing Beyond Markets." At the Master level, over 24 courses related to sustainability are offered. Some examples are "Achieving the SDGs: Feeding the Future of Sustainability," "Circular Economies and Sustainable Development Goals," "Consulting for Sustainability-Harnessing Business Models and Innovation," and "Critical Perspectives on Sustainability" (Copenhagen Business School, 2019c). The CBS is a participant of PRME initiative as a "PRME Champion." This highest participatory status is a result of its continuous efforts and communicating these improvements via the progress report on the United Nations Global Compact's PRME, since joining the initiative in 2008 (UNPRME, 2019c).

Cornell University-Ithaca, NY, USA; private research, 23,000 combined enrollment

Cornell's SC Johnson College of Business offers a variety of sustainability initiatives, courses, and minors. The School houses the Center for Sustainable Global Enterprise that frames global sustainability challenges as business opportunities and works with firms to specify innovative, entrepreneurial, and new business alternatives they can implement in the marketplace.

Cornell's Two-Year MBA program includes Immersion Learning which is a rigorous semester of integrated course and field work in a specific industry or career interest. One of the Immersion programs available is Sustainable Global Enterprise (SGE). Through real-world, company sponsored applied projects, SGE students gain conceptual and practical knowledge of a complex set of interconnected economic, social, and environmental issues. Students enrolled in the SGE immersion work in multidisciplinary teams on apprentice consulting assignments with business enterprises focused on exploring new business opportunities related to sustainability. Students take the SGE Immersion Practicum and a case-based course on strategies for sustainability and choose five credits of electives (within Johnson or across Cornell) relevant to their specific interests and career goals in sustainability. An outstanding aspect of the program is the high potential of placement after graduation due to the partnership with several companies such as Citi, Danone, Deloitte Consulting, GE Energy Finance, General Mills Liberty Mutual, and Unilever (Cornell University, 2019). Although the Johnson School of Business is listed as noncommunicating signatory of the PRME since its last report in 2015, the other management-related schools from Cornell University, Charles H. Dyson School of Applied Economics and Management and ILR School of Applied Social Sciences, have basic and advanced signatory status, respectively. Overall, three of their schools participate in the program at different levels (UNPRME, 2019a).

Duke University-Durham, NC, USA; private research, 16,000 combined enrollment

Fuqua School of Business of Duke University offers a master's dual-degree program with a Master of Business Administration combined with a Master of Environmental Management (MEM) from the Nicholas School of the Environment. The dual-degree program allows for 1 year of classes under the MEM degree, 1 year under the MBA degree, and 1 year of blended courses. Under the MEM degree, some classes a student may take are "Business and the Environment," "Natural Resources Economics," "Business Strategy for Sustainability," and "Corporate Environment Strategy" (Duke University, 2019a). In addition to this dual-degree program, they have a concentration within their MBA program on Energy and Environment. This concentration sets out to help students learn and "understand the dynamics of the new energy economy, to balance corporate sustainability considerations with business objectives, to manage risk and innovation, and to act on energy and environmental challenges in an innovate way" (Duke University, 2019b). According to PRME, Fuqua School of

Business is not a member of the UN's initiative; however, the school's Nicholas School of the Environment is. Currently, the Nicholas School of Duke is listed as a noncommunicating participant (UNPRME, 2019d).

Griffith University-Southport, Australia; private research, 45,000 combined enrollment

The Griffith Business School of the university offers sustainability programs at the undergraduate and postgraduate levels, priding themselves for having a core curriculum focused on sustainability, social responsibility, and ethics (Erskine-Shaw, 2019). Griffith's sustainable enterprise major teaches students the triple bottom line approach to business, balancing profits, people, and the planet. Students gain the skills and knowledge necessary to manage the sustainability needs of organizations and develop the analytical, communication, and decision-making skills to implement sustainable business solutions.

The school teaches the 17 United Nations Sustainable Development Goals (SDG) guided by the SDG Compass. The first step of the SDG Compass is to understand SDGs. The second is to define priorities. The third is to set goals. The fourth is to integrate those goals. The final step is to report and communicate (Erskine-Shaw, 2019). The Griffith Business School's main priority of teaching SDGs in the classrooms makes the program unique. Regardless of the course variations that individuals take, they will still be taught about SDGs. The Griffith Business School is a participant of PRME initiative as a basic signatory.

Harvard University-Cambridge, MA, USA; private research, 22,000 combined enrollment

Harvard Business School (HBS) focuses their research and funds towards corporate social responsibility and sustainability. Required MBA program curriculum includes "Business, Government and the International Economy" and "Leadership and Corporate Accountability" that cover sustainability issues. In addition, students have several choices of elective courses that focus on sustainability and corporate social responsibility (Harvard University, 2019a).

The school currently offers the Sustainable Business Strategy online certificate. The certificate program consists of approximately 20–25 h of material delivered over a three-week period fully online (Harvard University, 2019b). Harvard does not participate in the PRME initiative.

INSEAD—Fontainebleau, France; graduate business school, 13,000 enrollment

INSEAD (Institut Européen d'Administration des Affaires) is an international school with campuses in Europe (Fontainebleau), Asia (Singapore), and the Middle East (Abu Dhabi). The INSEAD Hoffmann Global Institute for Business and Society explores the intersection of business and society, bringing together a diverse body of leading INSEAD scholars convened under "Corporate Responsibility and Ethics", "Sustainability," "Humanitarian Research Group," and "Social Impact Initiative." The Institute attempts to educate business leaders of tomorrow who create value both for their organizations and for society—from INSEAD MBA, other masters and PhD programs to Executive Education (INSEAD, 2019a). Required MBA courses include "Business and Society" mainly devoted to sustainability issues, and four others with some sustainability content. Students also have a choice of many sustainability-related elective courses including "Business Sustainability," "Impact Investing," "Ethical Decision-Making in Business," "Social Entrepreneurship and Innovation," and "SDG Bootcamp-Building Impact Business" (INSEAD, 2019b). Similarly, the Global Executive MBA curriculum has a required ethics course, tailored to the needs of a more senior audience. The Executive Education portfolio to corporate clients includes a dedicated Social Entrepreneurship Program about growing business or other kinds of ventures to solve social problems and effect societal change. Another highly specialist program for healthcare executives addresses issues of ethics and compliance, while the Advanced Management Program for very senior business leaders offers an elective on business sustainability (INSEAD, 2019c). INSEAD is a participant of PRME initiative as a basic signatory.

New York University-New York, NY, USA; private research, 51,000 combined enrollment

Stern School of Business of NYU houses the Center for Sustainable Business aiming "to help current and future business leaders embrace proactive and innovative mainstreaming of sustainability, resulting in competitive advantage and resiliency for their companies as well as a positive impact for society," (New York University, 2019a). The Center for Sustainable Business embeds sustainability into the MBA and undergraduate curricula as well as provides learning opportunities outside of the classroom. The center also provides executive education courses in sustainability.

A Sustainable Business Co-Concentration is offered for undergraduate business students. Students have over 20 sustainability-related courses to choose from. Among them are "Social Entrepreneurship," "Sustainable Business in the New Economy," "Law and Business of Social Enterprise," and "Sustainability for Competitive Advantage." MBA students may choose to specialize in Sustainable Business and Innovation. Students have a choice of over 24 sustainability-related courses, such as "Investing for Environmental and Social Impact," "Social Venture Capital," "Strategy: A Social Purpose," and "Social Enterprise in Sustainable Food Business." Executive Education offerings include on-campus short courses such as "Sustainability Training for Business Leaders," "Sustainable Finance," and "ESG Investing"; custom-tailored programs for organizations on sustainability and a comprehensive 14-week primer "Corporate Sustainability" delivered online (New York University, 2019b). New York University is not a part of the PRME initiative.

Stockton University-Galloway, NJ, USA; public, 9000 combined enrollment

Stockton's School of Business offers an undergraduate Business elective course in 'Sustainability Marketing' and an MBA elective course in "Corporate Sustainability Strategy." In addition to contributing to the Sustainability degree program discussed in the next paragraph, Business School faculty with interest in sustainability issues contributes to the Liberal Arts General Studies requirements by offering related courses such as "Environment, Society and Business" and "Business, Government and Society." (Stockton University, 2019a).

Stockton's School of Natural Sciences and Mathematics offers B.S. and B. A degrees in Sustainability. With a curriculum that integrates the natural sciences, technology, economics, policy, ethics, and management, students can pursue areas of study such as public policy and law as well as sustainability management in business and industry. Professors from every School in the University, from Business, to Public Health, to Social Sciences teach courses in the degree. Students specializing in Sustainability Management gather the experience and knowledge to develop an economic framework that integrates environmental and social justice into sustainable business practices and policies. The coursework includes training in management, green finance and accounting, sustainable supply chain management, sustainability marketing, and green entrepreneurship (Stockton University, 2019b). Stockton University does not participate in the PRME initiative.

University of Michigan-Ann Arbor, Michigan, USA; public research, 46,000 combined enrollment

The University of Michigan's commitment to sustainability research and practice is embodied in its Erb Institute for Global Sustainable Enterprise. ERB is a partnership between the Ross School of Business and the School for Environment and Sustainability (SEAS). Drawing on faculty expertise from both the schools, the institute offers a dual-degree program in which students receive their MBA from Ross and their MS from SEAS. The MS from SEAS provides students an in-depth understanding of science, technology, and policy issues related to sustainability, while the MBA from Ross provides an understanding of core management principles such as finance, operations, strategy, and marketing. There are over 30 courses between the two schools focused on social sustainability and entrepreneurship, corporate responsibility and change management, and environmental energy sustainability (University of Michigan, 2019a).

The Ross School hosts the Business + Impact initiative, which is an Information Clearing house connecting business students to the varied social impact resources, researchers, internship/volunteer opportunities, and relevant courses throughout the university. In the Ross School MBA program, relevant courses include "Sustainable Operations and Supply Chain Management," "Strategies for Sustainable Development I-Competitive Environmental Strategy," and "Strategies for Sustainable Development II-Managing Social Issues." Relevant BBA Ross School courses include "Finance for Societal

Good," "Base of the Pyramid: Business Innovation for Solving Society's Problems," and "Business Innovation for Social Impact" (University of Michigan, 2019b). The University of Michigan does not participate in the PRME initiative.

University of Pennsylvania-Philadelphia, Pennsylvania, USA; private, 21,000 combined enrollment

The University of Pennsylvania's Wharton School of Management of the university hosts a Social Impact Initiative "inspired by the vision of business and capital markets working together to create sustainable solutions to the world's greatest social and environmental challenges" and whose mission is to "build the evidence, talent and community to advance business solutions for a better world." (University of Pennsylvania, 2019a).

One of the majors available to MBA students at Wharton is Business, Energy, Environment and Sustainability (BEES). The BEES major is designed to provide solid foundations for those interested in the evolving relationships between business and the natural environment, management of environmental risks, and the business and economics of energy. Relevant courses such as "Environmental Sustainability and Value Creation," "Marketing for Social Impact," and "Environmental Management, Law and Policy" are available (University of Pennsylvania, 2019b).

The Wharton School undergraduate degree in business offers an interdisciplinary Concentration in Environmental Policy and Management, especially relevant to those interested in pursuing careers in the environmental sector of the economy—in business, government, or environmental consulting. Wharton also offers a Minor in Sustainability and Environmental Management, providing students opportunities to learn about the nature of environmental constraints facing organizations and how to effectively consider these limitations as part of the decision-making process for for-profit as well as nonprofit organizations (University of Pennsylvania, 2019c). The school does not participate in the PRME initiative.

University of Toronto-Toronto, Ontario, Canada; public research, 61,000 combined enrollment

The University of Toronto's Rotman School of Management houses several research institutes, speaker series, clubs, and organizations that promote sustainability in the business world mainly driven by the Michael Lee-Chin Family Institute for Corporate Citizenship providing guidance on sustainability strategy, social enterprise, and responsible investment/impact investing (University of Toronto, 2019a).

MBA students at the Rotman School may major in Social Impact and Sustainability, exploring environmental, social, and governance issues in business; business strategies that pursue sustained, long-term resource efficiency; innovation strategies that employ sustainability as a source of new, profitable growth; corporate social responsibility or corporate citizenship strategy and programs; social entrepreneurship or social innovation and responsible investment or social finance. MBA courses available for this major (or as Electives to students who do not major in Social Impact and Sustainability) include "Sustainability Strategy," "Sustainable Finance," "Leading Social Innovation," "Social Entrepreneurship," "Clean Energy: Policy Context and Business Opportunities," "Social Value and Impact Investing, Designing for Equality," and "Environmental Finance and Sustainable Investing" (University of Toronto, 2019b).

Undergraduate business students may pursue a four-year Bachelor of Commerce (B.Com.) degree offered jointly by the Rotman School of Management and the Faculty of Arts and Science. There is no dedicated track or major in sustainability. Students do have the choice of several sustainability-related courses including "Investing for Impact," "Social Entrepreneurship, The Socially Intelligent Manager," and "Environmental and Social Responsibility" (University of Toronto, 2019c). The University of Toronto is not a part of the PRME initiative.

University of California, Berkeley-Berkeley, California, USA; public research, 42,000 combined enrollment

The University of California, Berkeley's Institute for Business & Social Impact at the Haas School of Business at Berkeley is comprised of six centers and programs with a shared goal to achieve social impact. The Center for Responsible Business of the Institute connects students, businesses, and faculty to focus on sustainable finance and investment, human rights and business, sustainable supply chains, and sustainable food (University of California, 2019a).

Berkeley Haas MBA students can choose to have "Social Impact" as their area of specialization either because they seek to bring a social impact perspective to their future work, or to pursue careers in corporate social responsibility, nonprofit and public leadership, or social entrepreneurship. They may access related courses that include "Social Investing," "Human Rights and Business," and "Social Impact Marketing." Social impact experiential learning opportunities include Impact Investing Practicum where students help impact investing firms develop solutions; Social Sector Solutions where student teams partner with McKinsey & Company coaches to provide strategic consultation to a nonprofit, social enterprise, or public organization; Social Lean Launchpad that applies the Lean Launchpad and Social Blueprint Business Design methodologies to frame the insights, strategies, and practices that distinguish social ventures; and Strategic & Sustainable Business Solutions in which teams work on live consulting projects, delving into sustainability and corporate responsibility challenges, while honing their project and client management skills (University of California, 2019b).

The undergraduate Business Administration program at Haas does not offer a sustainability-related major, minor, or concentration. However, the Haas School offers more than eleven courses directly related to sustainability. Examples include "Social Entrepreneurship," "Strategic Approaches to Global Social Compact," "Applied Impact Evaluation," "Topics in Social Sector Leadership," and

"Sustainable Business Consulting Projects" (University of California, 2019c). The Haas School of Business is a participant of PRME initiative as a basic signatory.

University of North Carolina-Chapel Hill, North Carolina, USA; public research, 30,000 combined enrollment

The University of North Carolina, Chapel Hill's Kenan-Flagler Business School offers a Full-Time MBA program in Sustainable Enterprise drawing on the resources of the Kenan-Flagler Center for Sustainable Enterprise (CSE). The Sustainable Enterprise MBA is designed to complement the traditional disciplines of marketing, operations, finance, entrepreneurship, and consulting. Students are encouraged to take advantage of extracurricular opportunities like attending conferences, guest speakers and networking events, student competitions, and experiential learning programs to further engage in Sustainable Enterprise. Some of the classes offered in the Program are: "Strategy & Sustainability," "Systems Thinking for Sustainable Enterprise," "Sustainable Operations," and "Sustainability Consulting." These courses are also available as Electives to MBA students not specializing in sustainability (University of North Carolina, 2019). The school is not a part of the PRME initiative.

University of Warwick-Coventry, UK; public research, 26,000 combined enrollment

The Warwick Business School's Master of Business Administration program has been ranked as one of the best in the world for sustainability and advancing environmental, social, and economic goals. The Business School has gained its reputation in sustainability through core sustainability classes and onsite learning opportunities. Undergraduate Business students may pursue a BA/BSc (Honors) degree in Global Sustainability Development and Business. The MBA Business and Sustainability module teaches the United Nations' Sustainable Development Goals (SDGs). The program also has a course on corporate responsibility. However, all of the Warwick Business School's teachings are not always inside the classroom. Through the MBA program, they allow students to visit the greenest city in the world, Vancouver. This learning experience gives students a first-hand view on business sustainability. Overall, the Warwick Business School ensures their students are well-versed in different aspects of sustainability (University of Warwick, 2018). The school is listed as a noncommunicating signatory of the PRME since its last report in 2012.

Western Colorado University-Gunnison, Colorado, USA; public, 3000 combined enrollment

Western Colorado University's School of Business and the School of Environment and Sustainability jointly offer a double major in Business Administration and Environment and Sustainability. They provide a foundation in both fields so that they can help businesses pursue their profit-making objective with an environmental conscience. Students may take several sustainability-related courses for this dual major including "Business and the Environment," "Environmental Law," "Science of Sustainability and Resilience," and "Applied Sustainability" (Western Colorado University, 2019a).

The Graduate School of the university offers an MBA specializing in the "Outdoor Industry." There is no sustainability specialization. However, students may take a few sustainability-related MBA courses including "Sustainable Finance," "Sustainable Accounting," and "Natural Resource Regulation and Economics" (Western Colorado University, 2019b). The Western Colorado University is not a part of the PRME initiative.

Western Washington University-Bellingham, Washington, USA; public, 16,000 combined enrollment

Western's Business and Sustainability program is housed in the Department of Management of the College of Business and Economics and offered jointly with the Huxley College of the Environment. In this undergraduate program, students gain foundational knowledge of economics, giving them the skills to apply economic analysis to problems in sustainable business; knowledge of environmental science and policy, giving them the ability to comprehend the social and political milieu as well as the scientific issues impacting sustainable business; and thorough knowledge of business and management in the context of business sustainability (Western Washington University, 2019a).

Western's MBA program does not have a sustainability focus. Students do have the choice of a related elective course entitled "Business and its Environment" (Western Washington University, 2019b). The Western Washington University is not a part of the PRME initiative.

Yale University-New Haven, CT, USA; private research, 13,000 combined enrollment

Yale School of Management's MBA for Executives (EMBA) offers sustainability as an area of focus. During the first year in the program, students are immersed in an Integrated Core Curriculum that is built around a series of stakeholder perspectives, traversing traditional business functions to explore the distinct points of view of multiple stakeholders across private and public realms. The first-year experience also includes a series of speakers through participation in the Colloquium on Sustainability Leadership. This allows students to hear from top CEOs, policymakers, and other individuals in the field on the sustainability issues businesses encounter. In the second year, students take courses to deepen their knowledge about sustainability that they can then bring into the business world. The courses include "Climate Change: Law, Policy, & Opportunity," "Corporate Environmental Management & Strategy," "Managing Sustainable Operations," "Corporate Finance and Risk Management in Sustainability," "Sustainability Systems," and "The Theory & Practice of Sustainable Investing" (Yale University, 2019). Yale School of Management also offers two sustainability-focused joint-degree programs—an MBA in combination with either a Master of Environmental Management (MEM) or a Master of Forestry (MF). Yale is not a signatory of the PRME initiative.

Table 6.2 provides an overview of the 20 case studies of business schools we examined for the purpose of this study. We highlight their AACSB accreditation status and PRME participation and demonstrate best practices in business sustainability education. Overall, we find that five case study universities were categorized as global sustainability beginners with their PRME basic signatory status and sustainability-focused curriculum, while the remainder fell into various statuses. These include business schools with some sustainability focus, but without a global approach. In total, there were 12 schools categorized into this group with the need of global sustainability focus improvement. Two schools in the study were identified as non-communicating members of the global initiative. Although these schools started as a member and initially reported their actions a few times to PRME, they have ceased further communication. Consequently, these business schools require improvements in their relationship with global sustainability initiatives. Out of 20 business schools under examination, only one school, Copenhagen Business School from Denmark, was recognized as a "sustainability leader" because of their continuous efforts in integration. This was evidenced by their internal emphasis on responsible management and sustainability and extensive selection of sustainability-related courses at the MBA, MS, and BS levels. Finally, Copenhagen's strong ties to global initiatives with worldwide impacts make this business school a model for sustainable business education.

Discussion

The Statement on the Purpose of a Corporation from the association of CEOs of leading U.S. companies—Business Roundtable—released in August 2019 is a driver that can invigorate reluctant business schools to modify their curricula to embrace ESD. Given the practical and prescriptive nature of this momentous statement to lead their companies for the benefit of all stakeholders—customers, employees, suppliers, communities, and shareholders (Business Roundtable, 2019), there is an increasing need to match that vigor in the core business school programs as companies embrace environmental, social, and governance (ESG) strategies and programs. Indeed, these strategies cannot remain the exclusive purview of sustainability professionals. Rather, the need is for professionals to be trained in sustainability principles and policies as applied to their roles in various enterprises. Operationalized, this transition applies equally to engineers, hedge fund managers with command over impact investing, or procurement managers knowledgeable about greening the supply chain. Business school students are increasingly aware of this orientation of business towards embracing the triple bottom line requiring concomitant classes and programs teaching the additional job skills associated with enveloping green methods into traditional professions. The need is evident as many business students are seeking careers in the

Table 6.2 Higher education institutions and their AACSB accreditation denoted by bold-face in the higher education institution column, PRME participation, and sustainable education focus.

Higher education institution[a]	PRME membership[b]	Notes[c]	Highlights[d]
American University DC, USA	Basic signatory	Global sustainability beginner	Specialized Master of Science (MS) in sustainability management (MSSM). Few undergrad and grad electives offered to mainstream students
Bard College NY, USA	Basic signatory	Global sustainability beginner and Needs business school standardization improvement	MBA program fully integrates sustainability into its core business curriculum. Dual MS/MBA degree offered in conjunction with the Bard Center for Environmental Policy. Few finance/economic courses through Division of Social Studies;-no undergrad program in Business
Brandeis University MA, USA	No	Needs global sustainability focus improvement	Social Impact MBA and MA in Sustainable International Development offered by the Heller School for Social Policy and Management. International Business School's business-focused MBA and undergraduate degrees do not currently offer sustainability-focused courses
Copenhagen Business School[e] Copenhagen, Denmark	PRME Champion	Sustainability leader with its continuous efforts	Responsible Management and Sustainability are the dominant themes of the MBA program. Incorporation of sustainability into most of the core courses, including "Managing Sustainable Corporations." Extensive choice of sustainability-related courses at the MBA, MS, and BS levels
Cornell University NY, USA	Noncommunicating (Johnson School of Business) Basic (Charles H. Dyson School of Applied Economics)	The business school needs to improve its relationships with global sustainability initiatives	MBA students have the choice of a semester-long rigorous specialization in Sustainable Global Enterprise. Other MBA and undergraduate students have choice of sustainability management courses

Continued

Table 6.2 Continued

Higher education institution	PRME membership	Notes	Highlights
Duke University NC, USA	No	Needs global sustainability focus improvement	Offers an MBA dual-degree with Master of Environmental Management from the Nicholas School of the Environment. Also, a concentration on Energy and Environment with emphasis on sustainable management available to MBA students
Griffith University Southport, Australia	Basic signatory	Global sustainability beginner	Griffith Business School offers sustainability programs at the undergraduate and postgraduate levels, with a core curriculum focused on sustainability, social responsibility, and ethics. Offers a sustainable enterprise major to undergraduates
Harvard University MA, USA	No	Needs global sustainability focus improvement	Required curriculum in the MBA program includes two sustainability-related courses. Extensive choice of sustainability-related electives. Offers a Sustainable Business Strategy online certificate
INSEAD Fontainebleau, France (also campuses in Asia & the Middle East)	Basic signatory	Global sustainability beginner	Required MBA courses include "Business and Society" mainly devoted to sustainability issues, and four others with some sustainability content. Extensive choice of sustainability-related elective courses
New York University NY, USA	No	Needs global sustainability focus improvement	The Stern School embeds sustainability into the MBA and undergraduate curricula. MBA students may choose to specialize in "Sustainable Business and Innovation." A Sustainable Business Co-Concentration is offered to undergraduate business students

Stockton University NJ, USA	No	Global sustainability beginner	School of Business offers undergraduate and graduate elective courses focused on sustainability issues. Interdisciplinary undergraduate degree in “Sustainability Studies” offers an area of study in sustainability management in business and industry
University of Michigan-Ann Arbor MI, USA	No	Needs global sustainability focus improvement	The Ross School of Business and the School for Environment and Sustainability jointly offer an MBA/MS dual-degree focused on sustainable management. Extensive elective courses are offered to MBA and undergraduate students
University of Pennsylvania PA, USA	No	Needs global sustainability focus improvement	Wharton MBA students may major in business, energy, environment, and sustainability. Undergraduate program in business offers an interdisciplinary concentration in environmental policy and management as well as a minor in sustainability and environmental management
University of Toronto Ontario, Canada	No	Needs global sustainability focus improvement	MBA students at the Rotman School may major in Social Impact and Sustainability. Extensive selection of sustainability-related electives for both MBA and undergraduate business students
University of California, Berkeley CA, USA	Basic signatory	Global sustainability beginner	Haas School of Business MBA students can choose to have “Social Impact” as their area of specialization. Extensive selection of sustainability-related electives for both MBA and undergraduate business students
University of North Carolina-Chapel Hill NC, USA	No	Needs global sustainability focus improvement	Kenan-Flagler Business School offers a Full-Time MBA program in Sustainable Enterprise. Extensive selection of sustainability-related electives for both MBA and undergraduate business students

Continued

Table 6.2 Continued

Higher education institution	PRME membership	Notes	Highlights
University of Warwick Coventry, UK	Noncommunicating signatory	The business school needs to improve its relationships with global sustainability initiatives	The Warwick Business School has gained its reputation in sustainability through core sustainability classes, the MBA Business and Sustainability module, and the BA/BSc (Honors) degree in Global Sustainability Development and Business
Western Colorado University CO, USA	No	Needs global sustainability focus and standardization improvements	The School of Business and the School of Environment and Sustainability jointly offer a double major in Business Administration and Environment and Sustainability. The MBA electives include few sustainability-related courses
Western Washington University WA, USA	No	Needs global sustainability focus improvement	Business and Sustainability undergraduate degree is jointly offered by the College of Business and Economics and the Huxley College of the Environment. The MBA electives include a sustainability-related course
Yale University CT, USA	No	Needs global sustainability focus improvement	Yale School of Management's MBA for Executives (EMBA) offers sustainability as an area of focus. The school also offers two sustainability-focused joint-degree programs-an MBA in combination with either a Master of Environmental Management (MEM) or a Master of Forestry (MF)

[a] AACSB.net (2019).
[b] UNPRME (2019e).
[c] UNPRME (2019e); results of case analysis.
[d] Results of case analysis.
[e] Denotes best practice.

traditional functions of finance, marketing, operations, or accounting, but in business organizations where they can make a social impact, do something for environmental sustainability, or be involved in corporate social responsibility (Kline, 2019).

However, along with the obvious need and positive impacts on a variety of stakeholders and the environment, there are known obstacles to institutional change that must be overcome to avoid stunting the future growth of sustainability and sustainability education. Sterling's (2004) three levels of institutional responses to the challenges of sustainability teaching could be helpful in overcoming some of these challenges. They are:

1. *Educating about sustainability—an accommodative response*: At this basic level, sustainability modules are being added to the educational offer.
2. *Education for sustainability—a reformative response*: At the second level which takes education further, the institution is being transformed by more sustainability methods of adoption.
3. *Capacity building—a transformative response*: At the highest level, the educational institution becomes a place where students are transformed by the adoption of skills as sustainability becomes the main goal.

We find that the first level is apparent in some of our case studies, such as Stockton University, who are designated as a "global sustainability beginner." The university is accommodating the new demands of teaching sustainability in an incremental, "add on" approach, but the undergraduate program offers only one sustainability-related course that is discipline-based "Sustainability Marketing." MBA students have the choice of only one related elective "Corporate Sustainability Strategy" and the university is not a member of PRME.

The second level, "Education for Sustainability," is exemplified by the Stern School of Business of New York University. The hub of sustainability education at Stern is the Center for Sustainable Business with a mandate to transform and mainstream sustainability education by embedding sustainability into the MBA and undergraduate curricula. In addition, for those who wish to specialize in Sustainability, a Sustainable Business co-concentration is available to undergraduates, and a Sustainable Business and Innovation concentration is available to MBA students. The university, however, is not a member of PRME.

The last institutional response, capacity building through transformative sustainability approach as rapid changes occur in the environment reflected in the appearance of social issues, is found in the best practices of the Copenhagen Business School. A comprehensive understanding of the centrality of sustainability in all business decisions is distilled to all MBA students who are immersed in a rigorous 50-h core course entitled "Managing Sustainable Corporations" and find that CSR and sustainability are incorporated into most core courses. The entire business school builds and continually enhances their capacity to transform business school education by focusing on sustainability as the guiding force, while sustainability instruction and research is coordinated through the knowledge hub-CBS Sustainability Center. Consequently, CBS has achieved the coveted participatory status of "PRME Champion."

Notable is that the institutions cannot teach their students to do what they themselves cannot muster within their own institutional settings. Successful institutions will engage in visibly sustainable activities, such as protecting natural habitats on school property, engaging in the practice of recycling, or active participation in community service. If the school is treating their current surroundings in a way that harms the natural environment, then it is difficult to convince students to respect and treat the earth systems and natural environment in a sustainable and conscientious way.

Recommendations: Case study findings and enhanced model underpinned by Rusinko's matrix

In response to variety of challenges, different business schools from all around the world have tried to find effective ways to progress in different areas of sustainability at varying levels. In this study, we employ Rusinko's (2010) matrix showing the different levels of adoption of sustainability in education. She suggests four ways of implementing sustainability into the curriculum:

- The first method is integration of sustainability within existing structures and through a narrower, more discipline-specific focus. Integrating sustainability into a currently existing business course could be a good example of this method. The instructor can add the subject as a new topic, assignment, case study, module, or service-learning project.
- The second method represents integration of sustainability in a narrower term, with the focus of discipline-specific approach, but at the same time, through a new structure development. A good example of this could be a stand-alone marketing discipline course development on sustainability, such as Sustainability Marketing.
- The third approach represents integration of sustainability into the business curriculum through a broader, cross-disciplinary focus within the existing structures. For example, sustainability could be taught as a topic, with the help of cases, modules, or service-learning methods in different courses.
- Lastly, sustainability education can have a cross-disciplinary approach that could be implemented through new structures. A cross-disciplinary required course in sustainability for all business students could be a good example to this approach. Also, developing cross-disciplinary sustainability programs, majors, or minors within the business school could be categorized as the same.

In addition to these four dimensions, we suggest three further aspects. First, the integration from first method to second one, meaning that the schools are able to use their existing courses (or minors, majors, programs) to build new structures, such as new sustainability minors, majors, or programs. Second, integration from third method to fourth one, meaning that they can integrate existing core requirements into new cross-disciplinary sustainability courses, minors, majors, or programs. Third, the institutional efforts can move beyond curriculum to support national and/or global initiatives, such as Environmental Protection Agency (EPA) and United Nation (UN) initiatives. Table 6.3 illustrates an enhanced four-point strategy to integrate Education for Sustainable Development into business schools advancing Rusinko's categorical foundation by enveloping strategies discussed herein.

Table 6.3 Integration of Education for Sustainable Development (ESD) in business schools matrix (adapted from Rusinko, 2010).

	ESD delivery	
	Existing structures	**New structures**
	I. Integrate into existing course(s), minor(s), major(s), or programs(s)	II. Create new, discipline-specific sustainability course(s), minor(s), major(s), or programs(s)
Narrow (discipline specific)	I & II. Using existing courses in a new sustainability minor, major, or program	
	III. Integrate into common core requirements	IV. Create new, cross-disciplinary Sustainability course(s), minor(s), major(s), or programs(s)
ESD focus broad (cross-disciplinary)	III & IV. Integrating existing core requirements into new sustainability courses, minors, majors, or programs	
ESD focus beyond curriculum (global voluntary initiatives)	Integration of sustainability in existing and new structures, integration into common core requirements, creating new, cross-disciplinary curricula, supported by institutional voluntary actions	

Conclusion

In this chapter, we discussed the potential of business students to address global problems in social, environmental, and economic development through adaptations in sustainability education. In case studies of 20 business schools, we determined that the best practice for sustainability education was the Copenhagen Business School (CBS) in Copenhagen, Denmark. The case for an international business guideline following a common curriculum with embedded concepts of sustainability was demonstrated in the comparison of various guidelines. The best practices gleaned from the case studies of business schools which have integrated sustainability into their business programs confirmed the need to centralize global responsible management standards, outlined by PRME. We recommend a four-point strategy to integrate Education for Sustainable Development into business schools. These include optimizing existing structures to narrow curriculum options that are discipline-specific and engaging in cross-disciplinary course structures with modifications for ESD delivery for existing and new structures.

References

AACSB. (2018). *2013 eligibility procedures and accreditation standards for business accreditation. AACSB international—The association to advance collegiate schools of business*. Retrieved August 19, 2019 from https://www.aacsb.edu/-/media/aacsb/docs/accreditation/business/standards-and-tables/2018-business-standards.ashx?la=en&hash=B9AF18F3FA0DF19B352B605CBCE17959E32445D9.

AACSB. (2019). *AACSB—Accredited Universities and Business Schools*. Retrieved August 19, 2019 from https://www.aacsb.edu/accreditation/accredited-schools.

AACSB.net (2019). Schools Accredited in Business. Retrieved August 20, 2019 from https://www.aacsb.net/eweb/DynamicPage.aspx?Site=AACSB&WebKey=ED088FF2-979E-48C6-B104-33768F1DE01D.

ACCP. (2019). *Association of Corporate Citizenship Professionals. Corporate social responsibility: A brief history*. Retrieved September 7, 2019 from https://www.accprof.org/ACCP/ACCP/About_the_Field/Blogs/Blog_Pages/Corporate-Social-Responsibility-Brief-History.aspx.

American University. (2019a). *American University, MS sustainability management*. Retrieved August 26, 2019 from https://www.american.edu/kogod/graduate/sustainability-management/.

American University. (2019b). *Business administration (MBA): Full-time*. Retrieved August 26, 2019, from https://www.american.edu/kogod/graduate/mba/curriculum.cfm.

Bard College. (2019a). *Bard MBA in sustainability*. Retrieved August 24, 2019 from https://www.bard.edu/mba/.

Bard College. (2019b). *M.S./M.B.A. program in environmental policy/climate science and sustainability*. Retrieved August 24, 2019 from https://www.bard.edu/cep/program/ms-mba/.

Brandeis University. (2019a). *Social impact MBA*. Retrieved August 8, 2019 from https://heller.brandeis.edu/mba/.

Brandeis University. (2019b). *Dual MBA and MA in sustainable international development*. Retrieved August 8, 2019 from https://heller.brandeis.edu/academics/dual-joint-degrees/sid-mba.html.

Brandeis University. (2019c). *Master of Business Administration*. Retrieved August 8, 2019 from https://www.brandeis.edu/global/academics/mba/index.html.

Burchell, J., Kennedy, S., & Murray, A. (2015). Responsible management education in UK business schools: Critically examining the role of the United Nations principles for responsible management education as a driver for change. *Management Learning*, *46*(4), 479–497.

Business Roundtable. (2019). *Business roundtable redefines the purpose of a corporation to promote 'An Economy That Serves All Americans'*. Retrieved August 29, 2019 from https://www.businessroundtable.org/business-roundtable-redefines-the-purpose-of-a-corporation-to-promote-an-economy-that-serves-all-americans.

Copenhagen Business School. (2019a). *CBS sustainability*. Retrieved July 27, 2019 from https://www.cbs.dk/en/research/departments-and-centres/department-of-management-society-and-communication/cbs-sustainability.

Copenhagen Business School. (2019b). The Copenhagen full-time MBA sustainability in business. Retrieved July 27, 2019 from https://www.cbs.dk/en/executive-degrees/mba/the-copenhagen-full-time-mba/your-mba-experience/responsible-management.

Copenhagen Business School. (2019c). *CBS courses*. Retrieved July 28, 2019 from https://www.cbs.dk/en/research/departments-and-centres/department-of-management-society-and-communication/cbs-sustainability/courses.

Cornell University. (2019). SC Johnson College of Business, Sustainable Global Enterprise, Cornell University, SC Johnson College of Business. Retrieved August 2, 2019 from https://www.johnson.cornell.edu/programs/full-time-mba/two-year-mba/curriculum/immersion-learning/sustainable-global-enterprise/.

Duke University. (2019a). *MEM/MBA and MF/MBA*. Retrieved August 3, 2019 from https://centers.fuqua.duke.edu/edge/education/memmba/.

Duke University. (2019b). *Concentrations + certificates*. Retrieved August 3, 2019 from https://www.fuqua.duke.edu/programs/daytime-mba/concentrations-certificates.

Erskine-Shaw, G. (2019). *Learning and teaching*. Retrieved from: https://www.griffith.edu.au/griffith-business-school/griffith-centre-for-sustainable-enterprise/learning-and-teaching.

Fitzgerald, M. (2019, August 19). *The CEOs of nearly 200 companies just said shareholder value is no longer their main objective*. CNBC News. Retrieved August 19, 2019 from https://www.cnbc.com/2019/08/19/the-ceos-of-nearly-two-hundred-companies-say-shareholder-value-is-no-longer-their-main-objective.html?__source=newsletter%7Cmorningsquawk.

Gerstein, M., & Friedman, H. H. (2016). Rethinking higher education: Focusing on skills and competencies. *Psychosociological Issues in Human Resource Management*, *4*(2), 104–121 ISSN 2332-399X, eISSN 2377-0716.

Harvard University. (2019a). *MBA Experience. Required curriculum (1st year) cases*. Retrieved August 5, 2019 from https://www.hbs.edu/environment/mba-experience/Pages/curriculum.aspx.

Harvard University. (2019b). *Become a purpose-driven leader*. Retrieved August 5, 2019 from https://online.hbs.edu/courses/sustainable-business-strategy/.

Hommel, U., Painter-Morland, M., & Wang, J. (2012). Gradualism prevails and perception outbids substance. *Global Focus*, *6*(20), 30–33. https://doi.org/10.5848/EFMD.978-1-909201-16-3_31.

INSEAD. (2019a). *The Hoffmann Global Institute for Business and Society*. Retrieved August 7, 2019 from https://www.insead.edu/centres/the-hoffmann-global-institute-for-business-and-society#impact.

INSEAD. (2019b). Academics—MBA Programme. Retrieved August 6, 2019 from https://www.insead.edu/master-programmes/mba/academics#curriculum-overview.

INSEAD. (2019c). *INSEAD Master Programmes*. Retrieved August 6, 2019 from https://www.insead.edu/master-programmes.

Kiron, D., Kruschwitz, N., Haanaes, K., & von Streng Velken, I. (2012). Sustainability nears a tipping point. *MIT Sloan Management Review*, *53*(2), 68–75.

Kline, M. (2019). *Why business schools need to teach sustainability*. Retrieved August 29, 2019 from https://www.inc.com/maureen-kline/why-business-schools-need-to-teach-sustainability.html.

Lawrence, P. R., & Weber, J. F. (2014). *Business and society stakeholders, ethics, public policy*. New York, NY: McGraw-Hill Irwin.

Muff, K., Dyllick, T., Drewell, M., North, J., Shrivastava, P., & Haertle, J. (2013). *Management education for the world: A vision for business schools serving people and planet*. Northampton, MA: Edward Elgar Publishing.

New York University. (2019a). Center for Sustainable Business. About the NYU Stern Center for sustainable business. Retrieved August 29, 2019 from https://www.stern.nyu.edu/experience-stern/about/departments-centers-initiatives/centers-of-research/center-sustainable-business/about.

New York University. (2019b). *Sustainable business educational offerings at Stern*. Retrieved August 29, 2019 from https://www.stern.nyu.edu/experience-stern/about/departments-centers-initiatives/centers-of-research/center-sustainable-business/educational-offerings.

Onel, N., & Fiedler, B. (2018). Green business—Not just the color of money. In B. A. Fiedler (Ed.), *Translating National Policy to improve environmental conditions impacting public health through community planning* (pp. 171–202). Cham, Switzerland: Springer Publishing International. https://doi.org/10.1007/978-3-319-75361-4_10.

Painter-Morland, M. J. (2015). Philosophical assumptions undermining responsible management education. *Journal of Management Development, 34*(1), 61–75. https://doi.org/10.1108/JMD-06-2014-0060.

Painter-Morland, M., Sabet, E., Molthan-Hill, P., Goworek, H., & de Leeuw, S. (2016). Beyond the curriculum: Integrating sustainability into business schools. *Journal of Business Ethics, 139*(4), 737–754.

Porter, T., & Córdoba, J. (2009). Three views of systems theories and their implications for sustainability education. *Journal of Management Education, 33*(3), 323–347. https://doi.org/10.1177/1052562908323192.

Rasche, A., Gilbert, D. U., & Schedel, I. (2013). Cross-disciplinary ethics education in MBA programs, rhetoric or reality? *Academy of Management Learning & Education, 12*(1), 71–85. https://doi.org/10.5465/amle.2011.0016A.

Rusinko, C. A. (2010). Integrating sustainability in management and business education: A matrix approach. *Academy of Management Learning & Education, 9*(3), 507–519. https://doi.org/10.5465/AMLE.2010.53791831.

Rutgers University. (2019). *Program learning goals and objectives—Newark and New Brunswick.* Retrieved September 21, 2019 from https://www.business.rutgers.edu/sites/default/files/documents/program-learning-goals-objectives.pdf.

Stockton University. (2019a). *Business studies program.* Retrieved July 27, 2019 from https://www.stockton.edu/business/business-studies-program.html.

Stockton University. (2019b). *Policy concentration and sustainability management concentration.* Retrieved July 27, 2019 from https://stockton.edu/sciences-math/sustainability/sust-policy.html.

Setó-Pamies, D., & Papaoikonomou, E. (2016). A multi-level perspective for the integration of ethics, corporate social responsibility and sustainability (ECSRS) in management education. *Journal of Business Ethics, 136*(3), 523–538. https://doi.org/10.1007/s10551-014-2535-7.

Slager, R., Pouryousefi, S., Moon, J., & Schoolman, E. D. (2018). Sustainability centres and fit: How centres work to integrate sustainability within business schools. *Journal of Business Ethics*, 1–17. https://doi.org/10.1007/s10551-018-3965-4.

Sterling, S. (2004). Higher education, sustainability, and the role of systemic learning. In P. B. Corcoran & A. E. J. Wals (Eds.), *Higher education and the challenge of sustainability* (pp. 49–70). Dordrecht: Springer.

Swaim, J. A., Maloni, M. J., Napshin, S. A., & Henley, A. B. (2014). Influences on student intention and behavior toward environmental sustainability. *Journal of Business Ethics, 124*(3), 465–484. https://doi.org/10.1007/s10551-013-1883-z.

Ten Bos, R., & Bevan, D. (2011). Sustainability. In M. J. Painter-Morland & R. ten Bos (Eds.), *Business ethics and continental philosophy* (pp. 285–305). Cambridge: Cambridge University Press.

United Nations Educational, Scientific and Cultural Organization (UNESCO). (2014). *What is ESD? [Online].* Retrieved February 16, 2020 from http://www.unesco.org/new/en/unesco-world-conference-on-esd-2014/resources/what-is-esd/.

United Nations General Assembly. (1987). *Report of the world commission on environment and development: Our common future.* Oslo, Norway: United Nations General Assembly, Development and International Co-operation: Environment.

University of California. (2019a). *Challenging the status quo to build a more equitable, sustainable, & inclusive society*. Retrieved August 21, 2019 from https://haas.berkeley.edu/ibsi/.

University of California. (2019b). *Full-time MBA program. Social impact*. Retrieved August 21, 2019 from https://mba.haas.berkeley.edu/careers/social-impact.

University of California. (2019c). *Undergraduate program. curriculum*. Retrieved August 21, 2019 from https://haas.berkeley.edu/undergrad/academics/curriculum/.

University of Michigan. (2019a). *The ERB MBA/MS dual-degree experience*. Retrieved August 4, 2019 from https://erb.umich.edu/programs/erb-mbams-dual-degree-experience/.

University of Michigan. (2019b). *Full-time MBA curriculum*. Retrieved August 5, 2019 from https://michiganross.umich.edu/graduate/full-time-mba/curriculum.

University of North Carolina. (2019). *MBA sustainable enterprise: At a glance*. Retrieved August 27, 2019 from https://www.kenan-flagler.unc.edu/wp-content/uploads/2019/04/FTMBA19-016-SustainabilityAAGDigitalAccess.pdf.

University of Pennsylvania. (2019a). *Wharton social impact initiative*. Retrieved August 15, 2019 from https://socialimpact.wharton.upenn.edu/about-wsii/.

University of Pennsylvania. (2019b). *MBA major. Business, energy, environment and sustainability MBA major*. Retrieved August 15, 2019 from https://igel.wharton.upenn.edu/education/mba-major/?_ga=2.142064882.1130719762.1566908511-860439187.1565369083.

University of Pennsylvania. (2019c). *Environmental policy & management*. Retrieved August 16, 2019 from https://undergrad-inside.wharton.upenn.edu/envp/.

University of Toronto. (2019a). *Michael Lee-Chin family institute for corporate citizenship*. Retrieved August 3, 2019 from http://www.rotman.utoronto.ca/FacultyAndResearch/ResearchCentres/LeeChinInstitute.aspx.

University of Toronto. (2019b). *Social impact & sustainability*. Retrieved August 3, 2019 from http://www.rotman.utoronto.ca/Degrees/MastersPrograms/MBAPrograms/Majors/Sustainability.

University of Toronto. (2019c). *2018–19 course outlines*. Retrieved August 3, 2019 from https://rotmancommerce.utoronto.ca/resource/2018-19-course-outlines/.

University of Warwick. (2018). *Warwick MBA tops global ranking for sustainability*. Retrieved August 19, 2019 from https://www.wbs.ac.uk/news/warwick-mba-tops-global-ranking-for-sustainability/.

UNPRME. (2019a). *Cornell University*. Retrieved August 28, 2019 from https://www.unprme.org/participation/search-participants.php?nameparent=cornell&from=&to=&utype=&search=Search.

UNPRME. (2019b). *Details for Bard MBA in sustainability*. Retrieved August 28, 2019 from https://www.unprme.org/participation/view-participants.php?partid=2904.

UNPRME. (2019c). *Details for Copenhagen Business School*. Retrieved August 28, 2019 from https://www.unprme.org/participation/view-participants.php?partid=279.

UNPRME. (2019d). *Details for Nicholas School of the Environment—Duke University*. Retrieved August 28, 2019 from https://www.unprme.org/participation/view-participants.php?partid=4548.

UNPRME. (2019e). *Participant reports*. Retrieved September 21, 2019 from https://www.unprme.org/reporting/participant-reports.php.

UNPRME. (2019f). *What is PRME?* Retrieved September 7, 2019 from https://www.unprme.org/about-prme/.

Western Colorado University. (2019a). *Bachelor of arts business administration and environment & sustainability*. Retrieved August 12, 2019 from https://www.western.edu/academics/undergraduate/school-business/business-administration-and-environment-sustainability.

Western Colorado University. (2019b). Master of Business Administration (MBA) outdoor industry MBA. Retrieved August 12, 2019 from https://www.western.edu/academics/school-graduate-studies/outdoor-industry-mba.

Western Washington University. (2019a). *College of Business and Economics. Business & sustainability*. Retrieved August 14, 2019 from https://cbe.wwu.edu/mgmt/business-sustainability.

Western Washington University. (2019b). *Daytime MBA program*. Retrieved August 14, 2019 from https://cbe.wwu.edu/mba/daytime-mba.

Yale University. (2019). *Yale School of Management. Sustainability*. Retrieved August 21, 2019 from https://som.yale.edu/programs/emba/curriculum/areas-focus/sustainability.

Supplementary reading recommendations

College Choice. (2018, June 21). Best masters in sustainability degrees, Retrieved September 15, 2019 from https://www.collegechoice.net/rankings/best-masters-in-sustainability-degrees/.

Czykiel, R., Figueiró, P. S., & Nascimento, L. F. (2015). Incorporating education for sustainability into management education: How can we do this? *International Journal of Innovation and Sustainable Development*, *9*(3–4), 343–364.

Delong, D. F., & McDermott, M. (2013). Current perceptions, prominence and prevalence of sustainability in the marketing curriculum. *Marketing Management Journal*, *23*(2), 101–116.

Doh, J. P., & Tashman, P. (2014). Half a world away: The integration and assimilation of corporate social responsibility, sustainability, and sustainable development in business school curricula. *Corporate Social Responsibility and Environmental Management*, *21*(3), 131–142.

Exeter University. BSc Business and Management. (n.d.). Retrieved August 3, 2019 from http://www.exeter.ac.uk/undergraduate/degrees/business/businessman/.

Fukukawa, K., Spicer, D., Burrows, S. A., & Fairbrass, J. (2013). Sustainable change: Education for sustainable development in the business school. *Journal of Corporate Citizenship*, *49*, 71–99.

Fuqua Duke (2019). Duke Fuqua School of Business, Dual Degrees. Retrieved August 25, 2019 from https://www.fuqua.duke.edu/programs/daytime-mba/dual-degrees.

Godemann, J., Herzig, C., Moon, J., & Powell, A. (2011). *Integrating sustainability into business schools–analysis of 100 UN PRME sharing information on progress (SIP) reports*. Nottingham: International Centre for Corporate Social Responsibility (58-2011) .

Nonet, G., Kassel, K., & Meijs, L. (2016). Understanding responsible management: Emerging themes and variations from European business school programs. *Journal of Business Ethics*, *139*(4), 717–736.

Online.hbs, *Harvard Business School online. Become a purpose-driven leader,* Retrieved August 21, 2019 from https://online.hbs.edu/courses/sustainable-business-strategy/.

Princeton Review, 2019 *Best green MBA,* Retrieved July 27, 2019 from https://www.princetonreview.com/business-school-rankings?rankings=best-green-mba.

Wersun, A. (2017). Context and the institutionalisation of PRME: The case of the University for the Common Good. *The International Journal of Management Education*, *15*(2), 249–262.

Definitions

Association to Advance Collegiate Schools of Business (AACSB) International a global independent nonprofit association that oversees the standardization of collegiate schools of business whose accrediting mission is to ensure that all business students are learning material that is most relevant to their field of study, preparing them to be effective leaders upon graduation, fostering engagement, improving innovation, and increasing impact in business education.

C-suite senior chief executive officers.

Combined enrollment reflects the number of undergraduate and graduate students.

Education for Sustainable Development (ESD) a United Nations initiative embracing sustainable development education that respects cultural diversity and secures intergenerational resources; empowers students by providing a foundation of environmental integrity, economic viability, and just society to make responsible and informed decisions.

Principles for Responsible Management Education (PRME) a United Nations-supported initiative founded as a platform based on raising the profile of sustainability in management schools around the world; the largest organized association between UN and management higher education, program objectives draw attention to the incorporation of UN Sustainable Development Goals (SDGs) in economic development decisions.

PRME basic signatory publicly recognized degree-granting academic institutions, identified by legal/government recognition such as accreditation, become Basic Signatories to PRME by signing a Letter of Commitment; Advanced Signatory status can be achieved by contributing an Annual Service Fee that allows access to additional benefits and opportunities for global participation.

PRME Champion PRME Champions are experienced and engaged PRME signatories committed to working collaboratively to develop and promote activities that address shared barriers to making broad scale implementation of sustainability principles a reality. Leaders in the space of responsible management education, PRME Champions undertake advanced tasks and game changing projects that respond to systemic challenges faced by the PRME community, as well as to key issues identified by the United Nations and the UN Global Compact.

PRME Noncommunicating Signatory a central commitment of any institution participating in the PRME initiative is to share, at least once in 24 months, information with its stakeholders on the progress made in implementing the Six Principles of PRME through the Sharing Information on Progress (SIP) reporting protocol. Signatories that fail to comply with the SIP policy by missing their reporting deadline are designated as "noncommunicating."

Purpose of a corporation employee investment, ethical conduct of business with suppliers, delivering the highest value to customers, and community engagement that demonstrates a focus on the positive aspects of life; embraces organizational social, environmental, and economic performance.

The Triple Bottom Line (TBL) postulates that companies should emphasize on more than one bottom line; in addition to profit, there should be people and the planet focus.

The United Nation's Sustainable Development Goals (SDG) the SDGs are considered to be the blueprint for a better and more sustainable future for all. Current 17 SDGs address some important global challenges, such as poverty, inequality, climate, environmental degradation, prosperity, and peace and justice and call for global actions by developed and developing countries.

Section B

Cultural impact on public health

Culture, cultural diversity, and enhanced community health

Beth Ann Fiedler
Independent Research Analyst, Jacksonville, FL, United States

Abstract

Culture, and thus cultural diversity, can identify social determinants of health (SDOH), but they are unlikely candidates as the basis of reconciliation and well-being. We examine Jacksonville Community Council, Inc.'s community-driven framework, community quality of life indicators, and the new role of data to shape community partnerships addressing nonclinical challenges such as homelessness. The case study demonstrates how JCCI led the City of Jacksonville on a 40+ year journey of community data and research that influenced policy leaving a legacy of international influence built on civic engagement. JCCI's pioneering spirit is evident in the shift towards effective data collection on culture and social needs permitting data to be positioned to address problems identified in SDOH to shape partnerships, link resources to patient care, and address social issues that impact community health. Weaving social care into SDOH gives individuals a place at the table and opportunity for enhanced community health and stability.

Keywords: Cultural diversity, Civic engagement, Community-driven, Public health, Data.

Chapter outline

Three Facets of Public Health and Paths to Improvements. https://doi.org/10.1016/B978-0-12-819008-1.00007-9

Seeking a seat at the table

Gore (1984): *People are people so why should it be, you and I should get along so awfully?*

Culture unites. There are both social and cultural benefits of diversity, such as innovation, that are linked to experiential interaction. Culture separates. People or people groups can be oppressed, pigeon-holed into categories and occupations, or even lose their lives because of religious, ethnic, or gender differences. A culture clash can cause miscommunication that, in turn, leads to escalating prejudice for any preexisting stereotypes that could undermine organizational social cohesion with far-reaching population effects. Given this reality, cultural diversity must be more than a common theme of coexistence sometimes expressed as 'tolerance' or government support of cultural preservation. These concepts, though a step forward, cannot elicit the positive aspects of culture and could, in fact, work against the potential for positive outcomes. Instead, bringing forth the best aspects of culture presupposes an element requiring diverse cultures to come to the same table, as it were, even if they first choose only to satisfy their appetites with their own familiar foods. "Together at the same table" is just the beginning. The idea being that, eventually, diners may consider a taste of what else may be on the table, "exchanging ideas as easily as exchanging recipes," and through cooperation prepare something new that both can accept as palatable, perhaps even enjoyable. Thus, the value of culture comes at the intersection of cultural diversity and even external institutions where people with unique perspectives choose to unite in ways that could be mutually beneficial. We focus on the health benefits of cultural diversity, but develop a parallel understanding of how too much emphasis on any one aspect of cultural diversity can represent a negative impact on health. Discussing problems to seek solutions is healthy, while focusing on categories is not.

This chapter first unfolds a sampling across various disciplines demonstrating the dichotomy of diverging literature and other media on culture, cultural diversity, and public health. We revisit a community-driven framework, value of community data indicating quality of life, and fostering communication from the case study of the groundbreaking work of the Jacksonville Community Council, Inc. The methodology includes problem identification and measuring progress to address health problems at the community level and the introduction of data innovation to create an avenue to link patients to resources. Finally, we can expect improved health outcomes, stronger families, communities, and work environments when we focus on the common denominator of public health through cultural interaction.

Addressing the tug of war

Culture and cultural diversity might be explained as a game played by youth: tug of war. On one side, like-minded individuals grab hold of a rope with a center line normally identified with some kind of handkerchief, while the opposing team grabs hold of the other side. Each side tugs the rope to pull the other team to their side and thus win the game. While life situations often present more than two opposing views, in

Fig. 7.1 Culture and cultural diversity often play a role in an adult version of tug of war where many perceive to be starting on unequal ground in policy decision-making.
Reproduced by permission from Shutterstock.com.

many ways the game resembles policy making, but the stakes are higher and some would argue that the game is not played on level ground (Fig. 7.1). Another important aspect of this approach is that there are always winners and thus, losers. Therein rests the problem with most policy development. In this section, we address this problem by defining key terminology and demonstrating the tug and pull of literature that often accompanies policy development and decision-making.

Defining culture, cultural diversity, and a culture of health

Cultural diversity is first explained by culture which, not surprisingly, has multiple dimensions and global connotations. For example, Sociology generally defines culture as encompassing common characteristics and ideas of self (e.g., belief systems, behaviors, and customs associated with artifacts, symbols, or social norms) within a wider spectrum of a group, community, or nation. Each level of association (e.g., ethnicity, community, national) may represent vastly different and complex cultural mores.

Anthropology and Sociology study the role of how these sociocultural constructs and ideas are created as well as exchanged within specific group and historical contexts that serve as a means to maintain social order through various mediums (e.g., philosophy, art, beliefs). In the workplace, culture can represent the personality of the business entrepreneur, accepted methods of communication, and employee response to that atmosphere, while a business development culture focuses on common goals such as earnings or meeting market demand. There are many different aspects or dimensions to ascribe, characterize, and define specific sociocultural constructs in virtually innumerable ways of human life. However, the concept of a Culture of Health is relatively new and spurred on by the Robert Wood Foundation. "Culture of Health is broadly defined as one in which good health and well-being flourish across geographic, demographic, and social sectors; fostering healthy equitable communities guides public and private decision making; and everyone has the opportunity to make choices that lead to healthy lifestyles," (Evidence for Action.org, n.d., *para* 2–3). Thus, cultural diversity can be defined as recognizing the multiple ways in which culture impacts our daily lives through social exchange and relationships, respectfully acknowledging differences and intentionally seeking ways to value and celebrate these cultural differences with the end goal of creating avenues of communication and enhanced community health.

The tug and pull of literature and various media

Numerous benefits and drawbacks of cultural diversity have been proclaimed in academic literature, across nontraditional works and media, and through many spectrums spanning migration population shifts impacting economic development (Kwakwa & Peña-Vasquez, 2019; Lian & O'Neal, 1997; Syrett and Sepulveda, 2011; Tubadji, Osoba, & Nijkamp, 2015), healthcare (Hall et al., 2015; Preda & Voigt, 2015; Shepherd, 2019), and in the work place (Bove & Elia, 2017; DiTomaso, 2015; Forbes Insights, 2011; HULT, 2019). The benefits of multiculturalism include a positive community influence, but also a long history of increased tension related to migratory diaspora and new local, geographic integrations, forced or otherwise, across the globe by economic, cultural, and social conditions exasperated by blight, famine, genocide, war, or disease. Some of these points and counterpoints to the benefits of cultural diversity are expressed in the citations herein. They are not inherently value or judgment-laden as either right or wrong; however, they do express unique contextual perspectives and sociohistorical interactions that are important towards gathering a better understanding of human development influencing research and policy development.

While cultures interweave, subsequent impacts on society can elicit mixed responses, depending on either a personal bias or perhaps simply as the problem of inherent difference or otherness; when cultures clash, what holds true is that communication, interaction, and engagement are necessary to move forward. Without interaction or engagement, we have singular-focused distraction and likely inaction towards problem resolution. What we bring attention to is that a topical focus on cultural differences can be harmful in terms of long-term policy development and social well-being. For example, some argue that the perception of racial bias may instead be a product of favoritism through the concept of networking with like associates so that "racial inequalities can exist without racism," (DiTomaso, 2015, p. 60), offering additional perspectives beyond the idea of institutionalized, systemic racism. There is also critical analysis of the normative assumptions in SDOH research conducted in the United Kingdom which presupposes an inherent social injustice in health inequalities. Preda and Voigt (2015, p. 25) investigate the UK research and suggest that the "normative underpinnings of the approach are insufficiently supported and that the policy recommendations do not necessarily follow the arguments provided and may be inconsistent." While the authors do not dispute the social justice problem stemming from the influence of high socioeconomic status and power associated with wealth, they do find fault with policy derivations from the skewed framework that links health to health equality in the health equity social exchange model (Preda & Voigt, 2015).

On the other hand, persistent problems with school integration are likely the residual effect of school district maps impacted by discriminatory practices such as redlining (Superville, 2019) and racial hierarchy that has historically prevented home ownership for minorities (Wells, 2019). Others identify that advances in recognizing gender may be at the expense of attention to disability, age, or variations on gender in the lesbian, gay, bisexual, and transgender (LGBT) population (Butler et al., 2016; Forbes Insights, 2011; North & Fiske, 2015). These various and conflicting perspectives tug at one another in the process of creating awareness to demonstrate that (1) the exchange of ideas is an important process, (2) overemphasizing any one factor from any perspective loses the potential of a positive effect of cultural diversity overall and is likely to contribute to further bias, and (3) a transition from focusing on the imbalance must eventually shift to problem resolution in order to achieve value.

"The complexity of interaction and behaviour cannot be reduced to facile insider-outsider, majority-minority, privileged-under privileged narratives," (Shepherd, 2019, p. 8). Achieving cultural awareness is only the first step and solutions cannot be achieved with endless meetings with colleagues (Shepherd, 2019). Instead, problematic health communication associated with cultural differences suggests a prescriptive remedy of increased interaction among people, such as between the patient/provider, which can help to define cross-cultural communication in a positive light. The implications for local implementation focused on listening skills and compassion portray how minor changes in the clinical environment can reduce the perception of cultural mistreatment (Shepherd, 2019). The same can hold true in other community relationships, such as with political leadership/constituents, business owners/employees, and in other cultural contexts.

By the same token, we cannot ignore notable differences in cultural upbringing that can influence public health. The foremost of which is suggested by PEN-3 advocates which place culture as central to the health model and health intervention evaluation (Airhihenbuwa, Ford, & Iwelunmor, 2014; Iwelunmor, Newsome, & Airhihenbuwa, 2014). "The PEN-3 cultural model consists of three primary domains: (1) Cultural Identity, (2) Relationships and Expectations, and (3) Cultural Empowerment", (Iwelunmor et al., 2014, p. 21). The PEN-3 model focuses on Person, Extended Family, and Neighborhood which are components of the Cultural Identity domain. We find this example of particular interest due to the inherent multilevel aspects of culture, impact of relationships, and how culture can be a positive or negative driver of empowerment.

Further, the emphasis on utilizing social determinants is undeniable in the application of healthcare even if the methods of incorporating social information remain problematic for many care systems (Maliard, 2019a, 2019b). "Momentum for more forward-looking SDOH initiatives is building across healthcare, as stakeholders recognize the key role of socioeconomic factors, environment and community in better health and wellness," (Maliard, 2019a, *para* subheading). Major healthcare organizations and agencies, such as the Massachusetts General Hospital and the United States

Center for Digital Health Information, indicate the need to incorporate nonclinical data for treatment decision-making, linking community advocacy and support groups and aligning organizational mission objectives with other local care organizations (Maliard, 2019a). These and other factors synchronize with the concept of a culture of health (Weil, 2019) emphasizing health problem resolution, while understanding contributing factors to poor health without placing patients into categorical silos.

Finally, Gerring, Thacker, Lu, and Huang (2015) introduce the negative impact of diversity on national public health and discuss various ways by which ethnicity can be politicized that contributes to negative outcomes in several areas including social trust, macroeconomic stability, and participation. Their study utilized the 2012 Demographic and Health Survey data containing input from 90 countries to study human development factors such as wealth, child mortality, and education; sample size varies with country, region, or district-level analysis. However, their findings indicate that the negative level of diversity diminished with subnational scales. In other words, community level diversity may be an avenue for unity and not a point of contention in recognizing and addressing public health problems close to home. This leads to the next section where we examine the City of Jacksonville in a case study on community level solutions to community level problems.

Case study: Jacksonville Community Council, Inc.

In 1985, something remarkable happened in the often quiet and sometimes stormy Northeast corner of the State of Florida in the City of Jacksonville. The Jacksonville Quality of Life Progress Report was birthed by the Jacksonville Community Council, Inc. (JCCI) as the world's first community indicator project. JCCI, a nonpartisan organization, felt strongly that a citizen-driven didactic supported by quality of life measurements could serve as a planning and evaluation tool for public policy and budget decision-making (Swain, 2002; Warner, 2014). During their tenure, the nonprofit produced more than 30 quality of life assessments since 1985, laying claim to the oldest indicators project, and 80 community investigations since 1975 have addressed a wide range of topics such as aging, mental health, and prostitution. Issues were open to public scrutiny, an important aspect of the JCCI process, allowing for contrasting perspectives to be included in solutions proposed by civic leadership in response to debates.

JCCI was influential in establishing national and international reporting of quality of life indicators as part of their legacy. They were among the founders of the Community Indicators Consortium, Inc. (CIC, 2017) that continues the tradition of projects focused on well-being across the US and internationally in Australia, Canada, and Germany (CIC, 2019a, 2019b). They currently have more than 300 projects in progress, encourage membership, and offer annual conferences to establish meaningful didactic and timely and credible metrics to address community problems.

During their tenure, JCCI brought forth methods of community-driven action and metrics to address and measure success. The legacy of JCCI continues in the promise of community activism embedded in multiple organizations in the City of Jacksonville through the Citizen Engagement Pact of Jacksonville in 2017. Unfortunately, the nature of the PACT does not permit any one organization to assume the role of measuring quality of life in Jacksonville. However, local organizational outreach to data-rich agencies and organizations, such as Florida Department of Health (www.floridahealth.gov), US Center for Disease Control (www.cdc.gov), and the National Association of County and City Health Officials (NACCHO, 2019) (www.nccd.org), provide opportunities for consistent analysis to identify trends. To get a better understanding of JCCI and their legacy, this section will discuss the framework, community quality of life indicators, and the power of communication that may aid in addressing local issues experienced in your community.

Community-driven framework

From humble beginnings sprouting from a Jacksonville Community Planning Conference in 1974 when the city government underwent major restructuring, JCCI's inception in 1975 immediately applied the backbone of community members to their mission "to bring people together to learn about our community, engage in problem solving and act to drive positive change," (Citizen Engagement PACT of Jacksonville, 2017, *para* 3). However, changes in JCCI leadership, financial support, and a self-assessment analyzing the feasibility of a "stand-alone organization leading the civic engagement process" (Citizen Engagement PACT of Jacksonville, 2017, *para* 8) were instrumental in bringing JCCI to a close. The self-assessment found that, "the community needs can be met if the JCCI mission is served through multiple organizations versus one organization taking on the entire body of work," (Citizen Engagement PACT of Jacksonville, 2017, *para* 7). This led to the formation of the Citizen Engagement Pact of Jacksonville in 2017.

Brune Matthis (2017, *para* 3):

> *The Citizens Engagement Pact of Jacksonville was described...as an effort "to ensure the values that foster a culture of civic engagement to improve the quality of life for all citizens will continue to be ingrained in the Jacksonville and Northeast Florida community."*

Notable is that the community benefits when individuals engage in political and community activities. Just as important are the benefits of individuals from participating in civic engagement that is recognized by the United Nations as a self-created condition towards human development that enhances human abilities to achieve well-being (UN Human Development Program, n.d.) (Fig. 7.2).

The PACT carries important tenets of the JCCI community-driven framework embodied in their commitment to community solutions through community engagement. Swain (2002, p. 11–16) provides a general overview of the framework that aided in the establishment of measureable community quality of life indicators and successful community research. An overarching theme in JCCI was flexibility to learn from the progress of other implementations, limit comparisons to City of Jacksonville

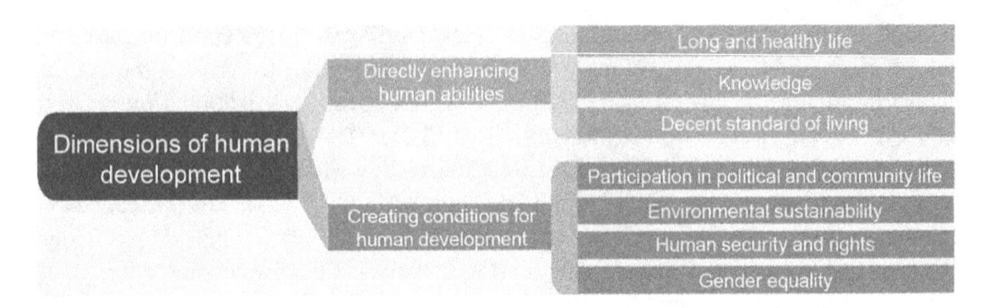

Fig. 7.2 The United Nations Envelops Civic Engagement as an important dimension of human development where political and community participation can allow individuals and communities to achieve well-being.
Used by Permission under Creative Commons Attribution 3.0 IGO: UN Human Development Program (n.d.).

trends, and focus on incorporating best practices and metrics to aid in policy direction and decision-making such as inclusion of regional health and human services indicators in 1995.

Basic tenets of the community-driven framework revolved around quality of life project indicator selection representing constituents in the City of Jacksonville who have specific expertise in leadership, problem resolution, or first-hand knowledge of the problem. The last report (JCCI, 2015) focuses attention on rates spanning the environment, economics, education, and health along with other trends. An important aspect of any longstanding measure is the incorporation of project planning, goal setting, and dedicated involvement and commitment expressed in inclusive civic debate. Additional components of the framework include fundamental research, an emphasis on measuring with a positive outcome orientation, and consistent data collection, reconciliation, and geographic scale. A final component is the focus on the future vs. lagging trends so that policy can be influenced to offset negative trends. Ultimately, "the framework used will continue a culture of inquiry, convening and implementation," (Citizen Engagement PACT of Jacksonville, 2017, *para* 7).

Community quality of life indicators

The City of Jacksonville quality of life indicators were comprised of nine top tier elements that include fundamental concerns about education, economy, and government. However, the environment (i.e., natural, social) became important to other elements such as health, mobility, public safety, and culture/recreation (Swain, 2002). "These criteria reflect the Jacksonville project's commitment to a broadly inclusive, balanced definition of the quality of life and to a citizen-based, consensus-based process of defining the quality of life consistent with the community's vision and understanding of its own values. One of the most important criteria is policy relevance," (Swain, 2002, p. 14).

Local business developers, Jacksonville Chamber of Commerce, and other JCCI collaborators placed value on the environment and the economy with an important

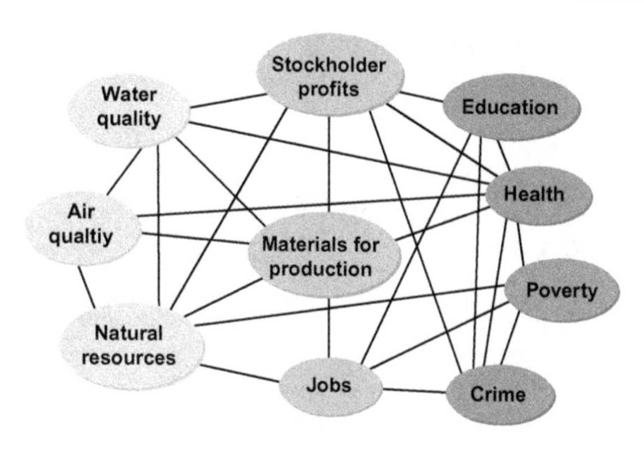

Fig. 7.3 Communities Are a Web of Interactions among the Environment, the Economy, and Society.
Reproduced with permission from the Guide to Sustainable Community Indicators by Maureen Hart, (c) 1995, Sustainable Measures (2010).

understanding that these indicators were linked in a unique way that reflected current community health and future potential. Fig. 7.3 demonstrates the important interaction of society with economic and environmental considerations. They were also adamant in creating quality of life indicators that could be measured accurately to reflect poor development choices that cost taxpayers' money in environmental cleanup. Instead of focusing on the global definition of Gross Domestic Product (GDP) (Chappelow, 2019) that would reward cleanup costs as revenue, business-minded constituents believed that utilizing the Genuine Progress Indicator (GPI) (Hayes, 2019) would more accurately reflect the negative impact on the community by deducting these costs against positive economic activity. This concern is reflected in the US agencies and the World Health Organization definition of quality of life. "Quality of life (QOL) is a broad multidimensional concept that usually includes subjective evaluations of both positive and negative aspects of life," (CDC, 2018, *para* 3; WHOQOL, 1998).

While JCCI has ceased tracking quality of life indicators for the City of Jacksonville, health-related quality of life indicators (HRQOL) are an important part of federal agency oversight and national funding. "On the community level, HRQOL includes community-level resources, conditions, policies, and practices that influence a population's health perceptions and functional status," (CDC, 2018).

The type of indicators selected play a pivotal role in the level of accuracy achieved in policy development. Warner (2014) indicates that both good and great indicators represent a trending directional difference between target objectives. If you expect a 10% increase in income to keep pace with inflation and the cost of living but only achieve a 5% increase, the distance between the actual and desired goal is 5% less representing half of target goal and thus, trending low. "Great indicators add context and allow for projection of future outcomes; by examining anticipated trend lines, policy development and action can be implemented to bend the trend lines," (Warner, 2014, p. 44). Some of the original indicators include the percentage of vacant housing units, healthy tributary water resources, and residents who believe that they can influence local government policy and decision-making. The predictive nature of trends should form policies that reduce the difference between desired and actual objectives before the problem spins out of control. Though we may be playing catch up in this arena, there are some positive changes in artificial intelligence, information technology, and geography which can stop the spin in general quality of life and health indicators.

Community data led partnerships: Nonclinical challenges of value-based care

"Investing in the social determinants of health is the only way to address the true causes of illness, behavioral health, readmissions and emergency room care," (S. Morse, senior editor, Healthcare Finance News, personal communication, September 16, 2019). Just as JCCI worked in conjunction with the Chamber of Commerce and business developers to exact change in Jacksonville using quality of life indicators, business and technology innovation in Artificial Intelligence (AI) and Geography can elevate cultural and social data to actionable policy and actual healthcare. Organization and company investments in community and health support innovative technology that transforms SDOH patient information to usable data points address nonclinical challenges.

Indirect and direct community investments have brought forth needed solutions to pressing social needs. For example, preserving affordable housing is important to community and individual/family stability. In the San Francisco housing market where the valuation is about $1.3 million, community investment has been important to improving community health. Lack of housing or any type of housing concern or instability has been linked to costly delays in preventive care and treatment (Enterprise Community Partners, Inc., 2019; Peters, 2019).

"Homelessness is just one of the social issues driving up healthcare utilization and costs. Safe housing, local food sources, transportation, public safety and many more factors all contribute to a person's health – or lack of it," (Padarthy, Knudsen, & Vatikuti, 2019, p. 4). Addressing the behavior and social factors, including community cohesion and employment, that contribute to as much as 60% of individual health status (Padarthy et al., 2019, p. 3) requires caregiver/patient rapport. Until recent developments in Artificial Intelligence and Machine Learning, incorporating SDOH into patient care workflows and referrals to community agencies was impossible. Today, new alliances optimizing the link between SDOH and community services are working with existing institutions demonstrating an improved impact on community health across the US (HappiDoc, 2019).

Where we live and how we interact with our community is important to health. Current data analysis trends inform us that the cultural and environmental conditions impact our lifestyle and behavior choices that can affect our well-being. Responding to this knowledge, companies like Esri, a leader in geographic information system (GIS) software headquartered in Redlands, California, founded originally in 1969 as the Environmental Systems Research Institute, understands the value of location. The company formulates health solutions by coordinating data from the ArcGIS Living Atlas of the World (https:/livingatlas.arcgis.com/en/Geography) to spot trends in human, environmental, and animal health. International data from more than 130 countries have contributed valuable data, such as the CDC Social Vulnerability Index, to help others utilize local geographical data to build unique maps that can be used to guide business development decisions or to help stop local disease outbreaks. (HIMSS TV, 2019b).

Summarily, data identifying and segmenting individuals are transitioning to data linking individuals to community assets and information for a variety of decision-making. Curiously, the next iteration of indicators is not necessarily new factors, but how that data is being positioned to address nonclinical challenges expressed in SDOH such as low income, homelessness, and high mortality rates in certain populations. The shift from the sometimes politicized categorical cultural silos to the actionable use of data to solve community problems in community partnership pioneered by JCCI is reflected in the innovation of data to shape partnerships and link resources to patient care by addressing social issues. The new direction of data involves the process of weaving social care into the community health imperative to enhance community health.

Discussion

Given the twist and turns of cultural diversity interspersed with an emphasis on SDOH, ending the policy tug of war is not an easy proposition. But, we can opt not to choose sides or pick up the rope to drag our opponents stumbling and exhausted over some imaginary barrier. This is not winning. Instead, we can choose to expend energy on solving the problem that the rope represents. Otherwise, the incongruence between both ends of the rope will persist if any one factor continues to have disproportionate attention skewing policy that may not be optimal or advantageous to most stakeholders. An alternative to single focus demographic identifiers is the utilization of data towards diagnosis and viable treatment options to improve health. We must understand that the key to improving public health is relationships built in close proximity such as within a local community. To recognize this path, we must reexamine the multifaceted problem of culture within our daily routines, such as work and meal preparation; foundational beliefs and health systems that support our physical, mental, and spiritual health; how we view gender roles; and the opportunity to develop solutions that meet the basic essentials of divergent populations.

But there are known obstacles down this road. "A consensus regarding the need to orient health systems to address inequities is emerging, with much of this discussion targeting population health interventions and indicators. We know less about applying these approaches to primary health care," (Ford-Gilboe et al., 2018, p. 635). Yet, we see innovative technology that dares to use data to solve hard and sometimes painful problems experienced in society.

Understanding that someone is culturally different does not require remarkable skills. After all, just about everyone can pick up one end of a rope. Choosing to treat people remarkably does! Focusing data on effective interventions and treatment solutions, not differentiating or categorizing patient populations, allows communication to reemerge in the quest for alleviating the negative impact of culture on health.

Communicating SDOH to healthcare workers who appropriately act on personal data which connect the patient to community services is one step closer to dropping the rope.

Recommendations: Weaving social care into the community health imperative

Weaving social care metrics into community health solutions is the direction for quality care facilities. If your data is a barrier to improved patient health outcomes, resolve to get better data as a collaborative project with your IT department (HIMSS TV, 2019a). Sometimes better data means attending community meetings to understand what problems local constituents are facing. Alternatively, enhanced data can be achieved by developing local partnerships that include data specialists (e.g., statisticians, mathematicians), business leaders, and local representatives to respond to problems with innovative technology and policy interventions, respectively.

In Cuba, social care is integrated into medical education and healthcare professionals actively engage in community meetings to hear and act on problems. For example, a lapse in garbage collection discussed in civic meetings can turn a simple problem into a local health crisis. However, healthcare professionals engaged in civic participation can use that information to intervene on behalf of citizens to stem the problem. Further, healthcare professionals enlist the aid of statisticians as a valuable contribution to health services on equal footing with clinical practitioners and other health professionals (Seervai, 2019).

Though we don't like to say the words out loud, healthcare and social service programs are a business that requires sustainable return on investments (ROIs). General consensus from business analysts suggests that investments with a social focus contained in SDOH data can produce value-based care (e.g., positive ROI, better patient outcomes, healthier communities) (HIMSS TV, 2019c).

Engaging culturally diverse professionals and subject matter experts to solve SDOH social needs brings people to the table in a very special way that effectively ends the tug of war. Sustainable business, participation, and data analysis remain fundamental to human development, health treatment, and community health. But organizations, political leadership, and patients require certain criteria to reach the path to building sustainable and effective health partnerships. "They will need three things: (1) information on who needs which specific supports; (2) data on the costs and resources required to provide these social service interventions; and (3) tools to estimate the financial risks and benefits from investing in health-related social needs," (Tsega, Shah, & Lewis, 2019, *para* 5). The Commonwealth Fund provides an ROI Calculator (https:/www.commonwealthfund.org/roi-calculator) as a tool to support and enable partnerships to make sound business decisions and inform policy by understanding how investing in social needs can save money and improve patient outcomes.

Finally, sometimes the simplest solutions are the ones we already know. For example, the Three Phases of Disparities Research model remains a single-minded approach that focuses on disparities: detect, understand, and address disparities by applying information to create interventions that inform and direct (Kilbourne, Switzer, Hyman, Crowley-Matoka, & Fine, 2006). Some have already set down the rope and set the table. We can bring our different foods to chew on. Now, we just have to come to the table and eat.

Conclusion

Gathering information from business, literature, and organizations including higher education, we discuss the notion of cultural diversity in community from the common problem of health. We underline the overarching theme of the value of cultural diversity in community public health by fostering participation and inclusive debate in a case study of the now defunct Jacksonville Community Council, Inc. However, the JCCI legacy remains in spirit as a social needs approach to healthcare has emerged in line with the pioneer nonprofit organizations' history of tracking trends using quality of life indicators. We do not reject that bias, racism, and other obstacles exist. Instead, we suggest that weaving social care into the community health imperative by applying social determinants of health to provide health services is a way to achieve improved patient outcomes and better community health. Recognizing the negative health impact of culture and using information to resolve disparities is the key to achieving results. Civic engagement is a major theme to understand perspective, generate partnerships, address social needs, and direct policy with valid data and analysis.

Acknowledgment

Special thanks to Matt Thomas Bagwell of Tarleton State University for his insightful comments on the final first draft.

References

Airhihenbuwa, C. O., Ford, C. L., & Iwelunmor, J. I. (2014). Why culture matters in health interventions: Lessons from HIV/AIDS stigma and NCDs. *Health Education & Behavior, 41*(1), 78–84. https://doi.org/10.1177/1090198113487199.

Bove, V., & Elia, L. (2017). Migration, diversity, and economic growth. *World Development, 89*, 227–239. https://doi.org/10.1016/j.worlddev.2016.08.012.

Brune Mathis, K. (2017, February 17). *JCCI restructuring after over 40 years; citizen engagement pact being announced Wednesday*. JaxDaily Record. [online]. Retrieved September 6, 2019 from https://www.jaxdailyrecord.com/article/jcci-restructuring-after-over-40-years-citizen-engagement-pact-being-announced-wednesday.

Butler, M., McCreedy, E., Schwer, N., Burgess, D., Call, K., et al. (2016). *Improving cultural competence to reduce health disparities. Comparative effectiveness review no. 170 (prepared by the Minnesota)*. .

Chappelow, J. (2019). *Gross domestic product—GDP. [Investopedia] [online]*. Retrieved September 14, 2019 from https://www.investopedia.com/terms/g/gdp.asp.

Citizen Engagement PACT of Jacksonville. (2017). *JCCI history*. Retrieved September 7, 2019 from https://jaxpact.org/history/.

Community Indicators Consortium (CIC). (2017). *Citizen engagement PACT of Jacksonville (formerly Jacksonville Quality of Life Indicators—JCCI)*. https://communityindicators.net/indicator-projects/citizen-engagement-pact-of-jacksonville-formerly-jacksonville-quality-of-life-indicators/.

Community Indicators Consortium (CIC). (2019a). *Citizens engagement PACT of Jacksonville, FORMERLY JCCI, Inc*. Retrieved September 10, 2019 from www.communityindicators.net/community-health.

Community Indicators Consortium (CIC). (2019b). *Indicator projects*. Retrieved September 11, 2019 from https://communityindicators.net/indicator-projects/.

DiTomaso, N. (2015). Racism and discrimination versus advantage and favoritism: Bias for versus bias against. *Research in Organizational Behavior*, *35*, 57–77. https://doi.org/10.1016/j.riob.2015.10.001.

Enterprise Community Partners, Inc. (2019, April 3). *Survey: When housing costs undermine health and peace of mind*. Retrieved September 17, 2019 from https://www.enterprisecommunity.org/blog/health-survey-renters.

Evidence for Action.org (n.d.). What is a culture for health? Retrieved August 28, 2019 from http://www.evidenceforaction.org/what-culture-health.

Forbes Insights. (2011). *Global diversity and inclusion fostering innovation through a diverse workforce*. Retrieved August 25, 2015 from http://images.forbes.com/forbesinsights/StudyPDFs/Innovation_Through_Diversity.pdf.

Ford-Gilboe, M., Wathen, C. N., Varcoe, C., Herbert, C., Jackson, B. E., et al. (2018). How equity-oriented health care affects health: Key mechanisms and implications for primary health care practice and policy. *The Milbank Quarterly*, *96*(4), 635–671.

Gerring, J., Thacker, S. C., Lu, Y., & Huang, W. (2015). Does diversity impair human development? A multi-level test of the diversity debit hypothesis. *World Development*, *66*, 166–188.

Gore, M. (1984). People are people [Recorded by Depeche Mode]. On *Some great reward* [Produced by Daniel Miller, Depeche Mode, and Gareth Jones]. Music Work at Highbury, London and Hansa Mischraum, Berlin, Germany. Mute.

Hall, W. J., Chapman, M. V., Lee, K. M., Merino, Y. M., Thomas, T. W., et al. (2015). Implicit racial/ethnic bias among health care professionals and its influence on health care outcomes: A systematic review. *American Journal of Public Health*, *105*(12), e60–e76.

HappiDoc. (2019, September 5). *New alliance negotiates with health plans to address the social determinants of health*. Retrieved September 17, 2019 from https://www.happidoc.com/news-news/new-alliance-negotiates-health-plans-address-social-determinants-health.

Hayes, A. (2019, July 18). *Genuine progress indicator (GPI) [Investopedia] [online]*. Retrieved September 12, 2019 from https://www.investopedia.com/terms/g/gpi.asp.

HIMSS TV. (2019, August 28a). *Health equity: IT's role in abolishing unequal care*. Retrieved September 16, 2019 from https://www.healthcareitnews.com/video/health-equity-its-role-abolishing-unequal-care?mkt_tok=eyJpIjoiTm1VM05ESXhNamczT0dKaiIsInQiOiJWbWo5ZWRzb3JwNnJhUzFMRlBXM3FUc0FnMTM0TVA3YnVqZ1FkQWppVlBcL2hXQmVwOVwvU0I0TWF3S1R5a05NUHpuQWdLZmxIMmNLdnM2SXlubFgxNWxPQlZRMk9aTm9qbFZ5dDVhNm9RbXpEXC90eGpNaWx1TDh3RDA5bGlrQW9nZiJ9.

HIMSS TV. (2019, August 22b). *How Esri is leveraging geography to make better decisions for health*. Retrieved September 16, 2019 from https://www.healthcareitnews.com/video/europe/how-esri-leveraging-geography-make-better-decisions-health?mkt_tok=eyJpIjoiTm1VM05ESXhNamczT0dKaiIsInQiOiJWbWo5ZWRzb3JwNnJhUzFMRlBXM3FUc0FnMTM0TVA3YnVqZ1FkQWppVlBcL2hXQmVwOVwvU0I0TWF3S1R5a05NUHpuQWdLZmxIMmNLdnM2SXlubFgxNWxPQlZRMk9aTm9qbFZ5dDVhNm9RbXpEXC90eGpNaWx1TDh3RDA5bGlrQW9nZiJ9.

HIMSS TV. (2019, August 14c). *Identifying ROI for social determinants of health*. Retrieved September 17, 2019 https://www.healthcareitnews.com/video/identifying-roi-social-determinants-health?mkt_tok=eyJpIjoiTm1VM05ESXhNamczT0dKailsInQiOiJWbWo5ZWRzb3JwNnJhUzFMRlBXM3FUc0FnMTM0TVA3YnVqZ1FkQWppVlBcL2hXQmVwOVwvU0I0TWF3S1R5a05NUHpuQWdLZmxIMmNLdnM2SXlubFgxNWxPQlZRMk9aTm9qbFZ5dDVhNm9RbXpEXC90eGpNaWx1TDh3RDA5bGlrQW9nZiJ9.

Hult International Business Schools (HULT). (2019, January). *13 benefits and challenges of cultural diversity in the workplace*. Retrieved August 25, 2019 from https://www.hult.edu/blog/benefits-challenges-cultural-diversity-workplace/.

Iwelunmor, J., Newsome, V., & Airhihenbuwa, C. O. (2014). Framing the impact of culture on health: A systematic review of the PEN-3 cultural model and its application in public health research and interventions. *Ethnicity & Health*, *19*(1), 20–46.

Jacksonville Community Council, Inc. (JCCI). (2015). *Quality of life progress report* (31st ed.). Retrieved September 11, 2019 from https://communityindicators.net/wp-content/uploads/2018/11/31st-Annual-Quality-of-Life-Report.pdf.

Kilbourne, A. M., Switzer, G., Hyman, K., Crowley-Matoka, M., & Fine, M. J. (2006). Advancing health disparities research within the health care system: A conceptual framework. *American Journal of Public Health*, *96*(12), 2113–2121. https://doi.org/10.2105/AJPH.2005.077628.

Kwakwa, M., & Peña-Vasquez, A. C. (2019). A neighbor like me: The effect of fractionalization on subjective well-being. *Politics, Groups, and Identities*. https://doi.org/10.1080/21565503.2018.1564054.

Lian, B., & O'Neal, J. R. (1997). Cultural diversity and economic development: A cross-national study of 98 countries, 1960–1985. *Economic Development and Cultural Change*, *46*(1), 61–77. https://doi.org/10.1086/452321.

Maliard, M. (2019, September 1a). *Investments in social determinants will pay off, with better outcomes and value [HealthcareITNews] [online]*. Retrieved September 11, 2019 from https://www.healthcareitnews.com/news/investments-social-determinants-will-pay-better-outcomes-and-value.

Maliard, M. (2019, March 20b). *Time to tackle IT and culture barriers to social data integration, says Jacob Reider [HealthcareIT News] [online]*. Retrieved September 9, 2019 from https://www.healthcareitnews.com/news/time-tackle-it-and-culture-barriers-social-data-integration-says-jacob-reider.

National Association of County and City Health Officials (NACCHO). (2019). *The essential elements of local public health*. Retrieved September 17, 2019 from http://essentialelements.naccho.org/.

North, M. S., & Fiske, S. T. (2015). Modern attitudes toward older adults in the aging world: A cross-cultural meta-analysis. *Psychological Bulletin*, *141*, 993–1021. https://doi.org/10.1037/a0039469.

Padarthy, S., Knudsen, K., & Vatikuti, S. (2019). *The social determinants of health: Apply AI and Machine Learning to Achieve Whole Person Care*. Retrieved September 17, 2019 from https:/www.cognizant.com/whitepapers/the-social-determinants-of-health-applying-ai-and-machine-learning-to-achieve-whole-person-care-codex4379.pdf.

Peters, A. (2019, January 19). *This healthcare giant invests millions in affordable housing to keep people happy*. Fast Company. Retrieved September 17, 2019 from https://www.fastcompany.com/90291860/this-healthcare-giant-invests-millions-in-affordable-housing-to-keep-people-healthy.

Preda, A., & Voigt, K. (2015). The social determinants of health: Why should we care? *The American Journal of Bioethics*, *15*(3), 25–36. https://doi.org/10.1080/15265161.2014.998374.

Seervai, S. (2019, June 28). Cuba: Where primary care is all about community. In J. Tallman & S. Seervai (Eds.), Vol. 25 *The dose* (p. 49). https://doi.org/10.26099/z6r9-6020 podcast, MP3 audio.

Shepherd, S. M. (2019). Cultural awareness workshops: Limitations and practical consequences. *BMC Medical Education, 19*(1), 14. https://doi.org/10.1186/s12909-018-1450-5.

Superville, D. R. (2019, January 8). Is there a path to desegregated schools? [online]. *Education Week*. Retrieved August 26, 2019 from https://www.edweek.org/ew/articles/2019/01/09/is-there-a-path-to-desegregated-schools.html.

Sustainable Measures (2010). Sustainability indicators 101. Retrieved September 15, 2019 from http://www.sustainablemeasures.com/node/89.

Swain, D. (2002). *Measuring progress: Community indicators and the quality of life*. Retrieved September 1, 2019 from http://www.managingforimpact.org/sites/default/files/resource/measuring_progress__community_indicators_and_the_quality_of_life.pdf.

Syrett, S., & Sepulveda, L. (2011). Realising the diversity dividend: Population diversity and urban economic development. *Environment and Planning A: Economy and Space*, *43*(2), 487–504. https://doi.org/10.1068/a43185.

The WHOQOL Group. (1998). The World Health Organization quality of life assessment (WHOQOL). Development and psychometric properties. *Social Science & Medicine*, *46*, 1569–1585.

Tsega, M., Shah, T., & Lewis, C. (2019, July 22). The importance of sustainable partnerships for meeting the needs of complex patients: Introducing the return-on-investment calculator. *To the Point (blog), Commonwealth Fund*. https://doi.org/10.26099/9c0c-nj93.

Tubadji, A., Osoba, B. J., & Nijkamp, P. (2015). Culture-based development in the USA: Culture as a factor for economic welfare and social well-being at a county level. *Journal of Cultural Economics*, *39*(3), 277–303. https://doi.org/10.1007/s10824-014-9232-3.

UN Human Development Program, Human Development Reports (n.d.). What is human development? Retrieved September 14, 2019 from http://hdr.undp.org/en/content/what-human-development. Creative Commons Attribution 3.0 IGO License

US Centers for Disease Control & Prevention (CDC), National Center for Chronic Disease Prevention and Health Promotion, Division of Population Health. (2018). *Health-related quality of life (HRQOL)*. Retrieved September 1, 2019 from https://www.cdc.gov/hrqol/concept.htm.

Warner, B. J. (2014). The future of community indicator systems. In N. Cytron, et al. (Ed.), *What counts: Harnessing data for America's communities* (pp. 42–57). USA: Federal Reserve Bank of San Francisco.

Wells, A. S. (2019). How to redefine 'good' in education. *Education Week*, *38*(17). 21, 23.

Supplementary reading recommendations

Carey, G., & Friel, S. (2015). Understanding the role of public administration in implementing action on the social determinants of health and health inequities. *International Journal of Health Policy and Management*, *4*(12), 795–798. https://doi.org/10.15171/ijhpm.015.185.

Civic Way (2019). News summary: Achieving American one community at a time. Retrieved September 5, 2019 from https://www.civicway.org/track-it/civic-engagement-summary/.

County Health Rankings. (2019). *Key findings report*. Retrieved September 17, 2019 from https://www.countyhealthrankings.org/.

Dluhy, M., & Swartz, N. (2006). Connecting knowledge and policy: The promise of community indicators in the United States. *Social Indicators Research, 79*(1), 1–23.

East-West Gateway Council of Governments. (2019). *Public involvement plan*. Retrieved September 15, 2019 from https://www.ewgateway.org/wp-content/uploads/2019/07/Public-Involvement-Plan-May-2019.pdf.

Eckersley, R. (2015). Beyond inequality: Acknowledging the complexity of social determinants of health. *Social Science & Medicine, 147*, 121–125. https://doi.org/10.1016/j.socscimed.2015.10.052.

Evidence-based Practice Center under Contract No. 290-2012-00016-I. AHRQ Publication No. 16-EHC006-EF. Rockville, MD: Agency for Healthcare Research and Quality.

Federal Reserve Bank of San Francisco, & Urban Institute (2014). *What counts: Harnessing data for America's communities*. In N. Cytron, K. L. S. Pettit, G. T. Kingsley, D. Erickson, & E. S. Seidman (Eds.), Federal Reserve Bank of San Francisco, USA. Retrieved September 1, 2019 from http://www.whatcountsforamerica.org/wp-content/uploads/2014/12/What-Counts-11.25.14.pdf.

ICF, International (2012). Demographic and health survey sampling and household listing. Manual, ICF International, Measure DHS, Calverton, MD.

Markus, A. (2011). Attitudes to multiculturalism and cultural diversity. In M. Clyne & J. Jupp (Eds.), *Multiculturalism and integration: A harmonious relationship* (pp. 89–100). ANU Press.

National Coalition for Dialogue & Deliberation (NCDD) (2008). Community indicator. Retrieved September 1, 2019 from http://ncdd.org/rc/item/1489/.

Seervai, S. (2019, June 28). Cuba: Where primary care is all about community. In J. Tallman & S. Seervai (Eds.), *Vol. 25 The dose* (p. 49). https://doi.org/10.26099/z6r9-6020 podcast, MP3 audio.

Sustainable.Org (2019). Jacksonville Community Council, Inc. Retrieved September 5, 2019 from https://www.sustainable.org/creating-community/inventories-and-indicators/150-jacksonville-community-council-inc.

Swain, D., & Hollar, D. (2003). Measuring progress: Community indicators and the quality of life. *International Journal of Public Administration, 26*(7), 789–814.

Weil, A. A. (2019). Neighborhoods and health, Medicaid, and more, *Health Affairs. 38*(9), 1419. https://doi.org/10.1377/hlthaff.2019.01015.

Definitions

Artificial intelligence (AI) computer science software programming that enables machines to perform tasks associated with human beings and to learn from experience.

Community health a clinical field of study focusing on how community structures impact health; study focuses on improving the health of a community without focusing on shared characteristics (e.g., age, diagnosis) in a given area.

Cultural diversity how different people groups choose to acknowledge differences and agree to find ways to unite.

Genuine Progress Indicator (GPI) "an alternative metric to…gross domestic product (GDP) economic indicator. The GPI indicator takes everything the GDP uses into account, but adds other figures that represent the cost of the negative effects related to economic activity (such as the cost of crime, cost of ozone depletion and cost of resource depletion, among others).

The GPI nets the positive and negative results of economic growth to examine whether or not it has benefited people overall." (Hull, YYYY, *para* 2–3).

Gross Domestic Product (GDP) "total monetary or market value of all the finished goods and services produced within a country's borders in a specific time period. As a broad measure of overall domestic production, it functions as a comprehensive scorecard of the country's economic health." (Chappelow, 2019, *para* 1).

Human development creating conditions (e.g., environmental sustainability, civic engagement, gender equality, and human security) that enhance human abilities (e.g., earnings, longevity, and knowledge (UN Human Development Program, n.d., Fig. 7.1)).

Machine learning (ML) branch of AI that presupposes machines can learn from data analysis and statistical models instead of specific computer science software programming.

Racial hierarchy privilege of power by racial association obtained by one race over another based on the belief that some racial groups are superior in behavior, intelligence, or other factors.

Redlining a discriminatory practice by which banks and/or insurance companies refuse or limit loans, mortgages, insurance, and other products within defined geographic areas such as inner-city neighborhoods.

Culture, language, and health care professionals

Dawood Ahmed Mahdi[a], *Beth Ann Fiedler*[b]
[a]Faculty of Languages and Translation, King Khalid University, Abha, Saudi Arabia,
[b]Independent Research Analyst, Jacksonville, FL, United States

Abstract

Cultural competence is an acknowledged skill set for health care professionals (HCP) because differences in cultural values and practices are a recognized barrier to communication between the HCP and patient. These differences are apparent in Asian Arab states comprised of diverse populations and dialects where foreign health professionals hold little or no knowledge of the local Arabic language or its dialects, even though Arabic is the primary language. Bilingual Arab HCPs also face communication barriers relating to their foreign counterparts. Overcoming compounding factors by developing a culturally competent workforce can be accomplished if HCPs adopt a cost-effective strategy of personal assessment, communication, and self-management in relation to caregiver attitudes and systemic practices. The exceeding number of expatriates and diverse demographic trends in Arab Asian countries emphasize the need for HCPs to focus on patient needs and respect their cultural values.

Keywords: Cultural competence, Transcultural health care, Health communication, Immigration, Diversity

Chapter outline

Three Facets of Public Health and Paths to Improvements. https://doi.org/10.1016/B978-0-12-819008-1.00008-0

Cultural competence enabling a quality workforce

A culturally competent workforce acts as a prerequisite for meeting the diverse needs of the people who belong to a different culture, particularly in the context of the Arab Asian Countries, where there exists an exceeding number of expatriates. The diverse demographic trend necessitates that professionals focus on the ethnic and racial health disparities to ensure provision of quality and equitable care as per the preference of the patients. For instance, countries like the United Arab Emirates (87%), Kuwait (73%), and Qatar (68%) lead in the proportion of immigrant population, while Saudi Arabia, Oman, and Bahrain have immigrant population ranging from 34% to 51% in 2015 (Pison, 2019b, *para* 4). By comparison, the "United States has the highest number of immigrants (foreign born individuals), with 48 million in 2015, five times more than in Saudi Arabia (11 million) and six times more than in Canada (7.6 million)" but "in proportion to their population size, these two countries have significantly more immigrants: 34% and 21%, respectively, versus 15% in the United States," (Pison, 2019a, *para* 4).

However, the significant problem facing some individual Arabs seeking health care is the presence of foreign health care professionals (HCP) who have little or no knowledge about the primary Arabic national language or distinguishing cultural factors (Alananzeh, Ramjan, Kwok, Levesque, & Everett, 2018). Similarly, lack of linguistic and cultural competence is difficult for the Arab health practitioners who are well-versed in both Arabic and English. However, communicating with foreign employees whose native languages are neither English nor Arabic is problematic.

Unfortunately, there is limited information on the scope of this problem in Arab countries since the majority of health professional migration has been to nations such as the United States that are well-documented. Even well-established databases, such as the International Labor Organization (www.ilo.org), do not track the required level of details for medical professionals (i.e., specific profession, place of birth, nation of current employment), while other database resources, such as the Migration Data Portal (https://www.migrationdataportal.org/overviews), focus on refugee data or limited information related to migrant workers. The World Health Organization (WHO) has commendable studies on the migration of health workers to European countries (Siyam & Poz, 2014), but not migration to Arab countries. Nevertheless, some anecdotal and qualitative resources confirm the concern to overcome language and other cultural barriers as the following example from the United Arab Emirates illustrates. "Emiratis reflect a general lack of confidence in public medical facilities due to lack of local expertise and perceived high costs of care" (Allianz Care, 2019, para 3).

Linguistic competence promotes patient involvement in care, feedback to the HCP, and valuable information sharing in the didactic between HCP and patient. However, absence of such competence can serve as a challenge due to a predictable decline in the frequency of general patient visits leading to preventable medical complications (Bowen, 2015). Further, literature has linked the significance of communication in relation to patient satisfaction and effective health outcomes (Betancourt, Green, Carrillo, & Owusu Ananeh-Firempong, 2016), highlighting the need to disseminate the relevance of HCP cultural competence in health care. Thus, fostering of effective communication is essential for achieving positive health outcomes (Fig. 8.1).

Facilitation of care infused with an understanding of the culture and language in medical interactions reduces behaviors that are misunderstood or perceived as inappropriate, while corrective action taken in accordance with cultural heritage can produce optimum quality of care (Attum & Shamoon, 2018; McBain-Rigg & Veitch, 2011; Meuter, Gallois, Segalowitz, Ryder, & Hocking, 2015). Consequently, cultural and linguistic competence must be a derivative of understanding the HCP's role in comprehending communication barriers in the establishment of care quality based on gaps in four primary areas: (1) cultural orientation, (2) health problem stigmatization, (3) gender roles, and (4) transfer of medical knowledge.

The linguistic competence of the health care worker is essential for the communication of the diagnosis or treatment details based on the fact that improper communication can lead to the occurrence of the associated risk factors (Shishehgar, Gholizadeh, DiGiacomo, & Davidson, 2015) and negative health outcomes (Raddawi, 2015). The language barrier also increases the risk of errors in terms of relaying medication instructions on life-saving treatments resulting in adverse complications (van Rosse, de Bruijne, Suurmond, Essink-Bot, & Wagner, 2016), reduces patient's satisfaction of the treatment provided (Albahri, Abushibs, & Abushibs, 2018), and the effectiveness of assistance provisioned by the staff (Eskes, Salisbury, Johannsson, & Chene, 2013).

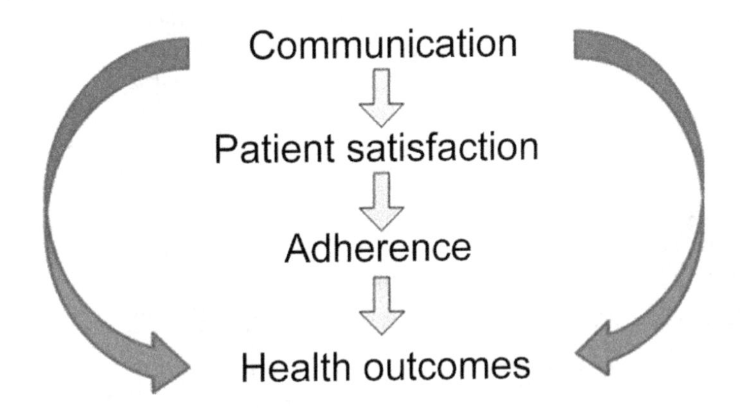

Fig. 8.1 Link between communication and health.
Reproduced with permission from Betancourt et al. (2016).

Foreign HCPs whose native language is English may find cultivating cultural competence less problematic, but with unique challenges in countries where English is a leading secondary language such as Kuwait, Jordan, Qatar, or Yemen as opposed to countries with multiple secondary languages. These nations include Syria, Iraq, and Oman (Jenkins, 2000). But, the widespread use of Arabic dialects in addition to the secondary languages can also add to the complexity of HCP/patient communication (Young, 2013). For example, Saudi Arabia has three Arabic dialects (e.g., Najdi found in the central region with four location-specific dialects, Hejazi in cities such as Mecca, Jeddah, and Medina, and Gulf in the coastal region) in addition to secondary languages. The Arabic dialect of Maghrebi is found in Jordan and Lebanon, while both Bahrain and the United Arab Emirates have a Gulf Arabic dialect, but the United Arab Emirates is known as Emirati Arabic and may contain variations. Finally, Palestine has a unique division in secondary languages with additional cultural implications (Table 8.1).

Developing cultural competence in a variety of circumstances is important to formulate a cost-effective and efficient training solution that includes recognizing how languages may be local, regional, national, or apply to several countries who share borders or waterways. Thus, characterizing modest methods to introduce cultural

Table 8.1 Asian Arab countries and population impact where the primary language is Arabic.

Country	Population[a,b]	Arabic dialects[c]	Secondary language[d]
Bahrain	1,641,000	Gulf	English, Farsi, Urdu
Iraq	39,310,000	Mesopotamian/Iraqi	Kurdish, Turkmen, Syriac (Neo-Aramaic), Armenian
Jordan	10,102,000	Levantine	English
Kuwait	4,207,000	Gulf	English
Lebanon	6,856,000	Levantine	French, English, Armenian
Oman	4,975,000	Gulf	English, Baluchi, Urdu, Swahili, Indian dialects
Palestine	4,981,000	Levantine	Hebrew, English[c]
Qatar	2,832,000	Gulf	English
Saudi Arabia	34,629,000	Gulf	English, Hindi, Urdu[c]
Syria	17,070,000	Levantine, Eastern Syria-Mesopotamian/Iraqi	Kurdish, Armenian, Aramaic, Circassian, French, English
United Arab Emirates	9,771,000	Gulf	English, Hindi, Malayalam, Urdu, Pashto, Tagalog, Persian
Yemen	29,162,000	Yemeni	English, Socroti, Mahri

[a] United Nations, Department of Economic and Social Affairs, Population Division (2019).
[b] Note: Population information as of July 1, 2019.
[c] Jenkins (2000).
[d] Central Intelligence Agency (2019).

and linguistic competency may provide the best platform to train HCP professionals in the Asian Arabic regions with greatest impact.

Given the scope of the problem and potential patient risk, this chapter focuses attention on the impact of cultural diversity and language barriers to health care by identifying common communication barriers, developing basic knowledge of cultural competence, and defining the concept in terms of health quality of care. Then we examine the role of clinicians in achieving effective HCP-patient communication through cultural and linguistic competence by generating awareness of special health conditions and overcoming key cultural factors (e.g., religion, family, and context). We move on to focus on the HCP provider process to address language and cultural barriers followed by a discussion and summary highlighting important concepts of this chapter.

Cultural diversity, language barriers, and pathways to cultural competence in health care

Communication is fundamental to patient care. But the diversity in languages, even in cultures in which the primary language and dominant religion is the same, is a great challenge to health care providers who have limited access to adequate resources to overcome the treatment barrier to multilingual patients. This service problem is prominent in at least 12 Asian Arabic countries (of 22 Arabic-speaking nations) for 3 main reasons: (1) the secondary language in the nation greatly differs from the primary Arabic language, (2) the variance in Arabic dialects, and (3) cities or regions within a nation where different languages dominate. The problem becomes more complex when the health care professional is foreign to the area. Generating awareness of the dominant patient clinical communication behaviors resulting from HCP/patent cultural conflict and the impact of miscommunication is important for defining and developing cultural competence.

Dominant patient clinical communication behaviors

Two types of clinical communication behaviors dominate the need for HCP training. They are persistent and demanding behavior and interpretation barriers. The first type occurs when domineering family members inappropriately confront the HCP in cases where the wife is adamant about staying with the husband in adherence to Arab culture. Demanding behavior in clinical care is an observed deficiency in lack of trust in the interaction between the HCP and patient/family members which often leads to problems with providing care. Perceived differences in sociocultural areas of health beliefs, medical practice, or levels of faith often contribute to this barrier. The HCP can utilize skills in moderating the situation once there is recognition of this common communication problem in which the wife may feel some noticeable irritation or discomfort. The HCP can request the presence of a third party (i.e., husband), reduce eye contact, and limit contact during medical examination by receiving consent for any individual removal of clothing. Clothing is immediately returned upon completion of examination to reduce discomfort. HCPs may also consider actions that do not require complete removal of an article of clothing, such as shifting a shirt sleeve, which may be more appropriate. The best approach is to continue rearranging the sheets as well as a gown to minimize skin exposure.

Further, the HCP can respectfully inquire of the wife if she would be more comfortable if her husband addressed any health questions. In this manner, respect for both the patient and quality of care remains intact. Modesty and privacy are highly respected in the Arab society and this is of highest concern when the HCP is same gender as the patient undergoing treatment (Epner & Baile, 2012; Padela & Del Pozo, 2011).

Interpretation barriers lead to misunderstanding that will impact the quality of service delivery attributed to several factors. First, the intensity of the problems amplifies when the interpreter lacks sufficient training or competence (Eklof & Ahlborg Jr., 2016); second, limited understanding of medical terminology (Hadziabdic, Lundin, & Hjelm, 2015). Another important aspect of interpretation error is the misunderstanding of cross-cultural usage of seemingly familiar words. Misunderstanding key cultural variants can produce mortality and morbidity.

A recent tragic example of an English/Spanish interpretation barrier was when William Ramirez, 18, was admitted to a Florida hospital after commenting to his friend, "Me Siento intoxicado," (Sulaiman, 2019). Tragically, the young man underwent treatment for a drug overdose "intoxicated" instead of for the nausea "sick in the stomach" which was an underlying symptom for a brain aneurysm. The ineffective communication and interpretation led to a malpractice lawsuit costing up to $71 million and permanent quadriplegic disability for Ramierez (Sulaiman, 2019). The HCP must take measures to avoid these situations since the responsibility for care falls on them. As the HCP is building their own cultural competency, one method to avoid common misunderstandings is to utilize cultural negotiators, known as culture brokers, and other knowledgeable local staff to overcome care obstacles through bicultural awareness, language, and medical knowledge.

Other clinical behaviors are worth mentioning in which both the patient and the HCP may miss important opportunities to communicate. The intercultural communication barrier in a health environment of care takes into consideration that patients may be too frightened or too ill to initiate communication. In these cases, however, the practitioner may not take notice because the HCP is pressed for time or is extremely reliant on and preoccupied with technology (Ulrey & Amason, 2001).

HCP/patient communication may also be impeded due to exaggerated cultural differences when doctors perceive they are in a position of power when their status is significantly higher than that of the patient (Yeandle, Rieckmann, Giovannoni, Alexandri, & Langdon, 2018). The HCP may also struggle with a myopic view called medical ethnocentrism that inclines the HCP to address the care needs of the dominant culture (Meehan, Menzies, & Michaelides, 2017). In this case, the individuals belonging to a minority group may perceive these as a factor that thwarts their access to care. These examples demonstrate how cultural perceptions of personal modesty and comfort may impede the process of sharing information that is essential for provisioning of proper care and the reason why incorporating cultural competence, which overlooks cultural differences, is essential to providing discrimination-free treatment.

Developing cultural competence

In general, development of cultural competence for the health care professional begins with self-awareness in the context of the patient environment. Growth in cultural competence comes with the expansion of knowledge of the prevailing social and cultural problems, communication competencies, and health convictions in the region. This knowledge accumulates through a process of identifying styles of communication, decision-making, assessing the family, gender issues, and other prominent concerns relating to race, potential prejudice, and mistrust which can impact care. Together, these items can be considered an assessment of basic cultural values.

A solid understanding of cultural values provides the caregiver with the necessary sensitivity to ensure good patient care. Moreover, this also influences the perspective of the patients towards health care sector (Attum & Shamoon, 2018). But, the learning curve may be vast before the health professional and patient can achieve this goal. A common definition of cultural competence in health care may be helpful.

Defining cultural health care competence and quality of care

Cultural competence in health care represents the convergence of policy based on a set of congruent behaviors, attitude, and knowledge to form a system which prompts professionals to effectively provide quality patient care (Chu & Goode, 2009; Cross, Brazron, Dennis, & Issacs, 1989; Truong, Paradies, & Priest, 2014). Another component requiring further consideration impacting cultural competency from the patient perspective is a general lack of awareness of their rights, while HCP core considerations include (1) variation in the concept of disability, (2) amount of training associated with cultural competence, (3) development of a self-reflective and adaptable attitude toward diverse patients, and (4) communication skills that best fits according to the given culture (Olaussen & Renzaho, 2016). Thus, the individual words in this term provide another level of understanding.

For example, culture is generally regarded as an integrated set of characteristics defined by prevailing language, thoughts, actions, and beliefs. However, there is an expanding awareness of cultural heterogeneity within populations, especially in the mobility of health care professionals to overcome global shortages, redefining culture in terms of public health services. Therefore, culture is further defined as the institution that caters to diverse social, ethnic, and religious groups (Desmet, Ortuño-Ortín, & Wacziarg, 2017). Thus, cultural competence within the institution of a hospital must embrace the cultural environment surrounding that location including language, beliefs, and customs.

The term competence indicates the capacity of an individual or organization to possess and effectively apply these characteristics in the conduct of care. In this way, effectiveness is a function within the dynamics of set patient cultural beliefs, practices, and needs which extends from the individual to their community. Consequently, cultural competence in health care stems from assimilating the founding tenets of patient/family-centered care which, in turn, have their basis in the social and cultural understanding that impacts the quality of the medical and treatment services (Butler et al., 2016). This definition manifests in the context of the Middle East where health care professionals must nurture their patients back to health by valuing the dominant cultural perspective of the Muslim religion.

However, any deficiency in cultural understanding can raise various challenges based on different systems of beliefs or religion and their normal practices. The HCP path of cultural competence (Fig. 8.2) has been described as a continuum beginning with an initial level of cultural destructiveness, variable with individual personnel, and ending with cultural proficiency (Cross et al., 1989; Lim & Mortensen, 2015; Waitemata District Health Board, 2019) (Fig. 8.2). "Research tells us that most service providers fall between cultural incapacity and cultural blindness on the following cultural competence continuum," (Waitemata District Health Board, 2019, para 2).

Cultural incapacity in an institution or public service is identified as an inequitable distribution of resources which favors specific groups, while cultural blindness is a notable discomfort unless all health care professionals apply universal care to all patients (Cross et al., 1989; Waitemata District Health Board, 2019) (Table 8.2). Unfortunately, there are very few organizations that have formally and successfully incorporated culture and language proficiency into various organizational policy, practices, service deliverance, administration, or patient engagement. Recognizing the need for change has led to the establishment of short-term programs (i.e., cultural precompetence), but without concurrent changes in organizational policy and patient engagement strategies. However, adaption of the cultural competence continuum is preferred because utilization results in enhancements in the service quality of the health care provider service in a culturally diverse environment (Matthews & Van Wyk, 2018).

Facilitating the incorporation of cultural health care competence and quality of care can be achieved by measuring several prominent concepts to demonstrate they can

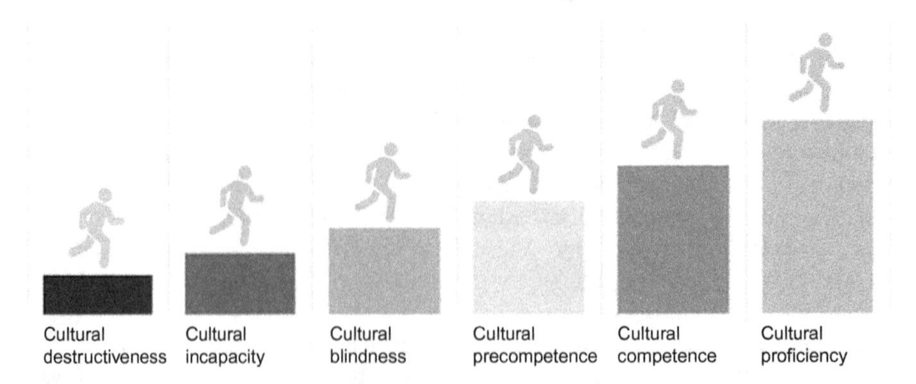

Fig. 8.2 Path of cultural competence continuum.
Reproduced with permission from Waitemata District Health Board (2019).

Table 8.2 Cultural competence continuum (Chu & Goode, 2009, p. 6; Cross et al., 1989; Waitemata District Health Board, 2019).

Cultural destructiveness	The initial stage of the continuum that is categorized by culturally oppressive attitudes, structures, policies, as well as practices within an institute or a system causing a destructive effect on a cultural group(s).
Cultural incapacity	Represents the lack of systems' capacity to effectively respond to the interests, needs, and preferences of groups which differ in terms of culture and linguistics.
Cultural blindness	Achieving blindness allows individual health care professionals to reach a philosophy that objectively views and treats everyone in the same way.
Cultural precompetence	Characterizes the level of individual or system awareness of cultural and linguistic variances to recognize strengths and weaknesses in order to effectively respond to different populations; creates short-term solutions vs long-lasting policy.
Cultural competence	Symbolizes the level of acceptance and respect toward a culturally diverse environment by formulating policy, mission statements, structures, procedures, and systems which assist in meeting the diverse patients' needs in accordance with their preferences; advocating quality of equitable care for every patient.
Cultural proficiency	A level of achievement demonstrating that the institute/system is enhancing practices pertaining to cultural competence by introducing novel therapeutic approaches for caring and sharing methods that are beneficial to positive health outcomes.

become formal policy. These foundational concepts are safety, effectiveness, patient centeredness, equity, timeliness, and efficiency (Santé en Français (French Health), 2012, p. 18) that may already serve as metrics for other health care quality indicators.

First, patient safety is a major concern when a language or cultural barrier arises because of the likelihood of an adverse event that could cause patient death, disability, or other harm. Therefore, the concept of patient safety is not confined to the error-free delivery of care but inclusive of the desire to reduce misdiagnosis of the patient, shielding the patient from unnecessary risk and assuring that the patient consents to diagnosis and care (Santé en Français (French Health), 2012). Thus, the establishment of an HCP/patient relationship in which proper communication is a priority is an important step in promoting patient participation in treatment and diagnosis decisions. This exemplifies the HCP patient-centered approach in which their compassion, responsiveness, values, and empathy are focused on the patient's preferences and individual needs.

Next, Betancourt (2006) highlights that cultural competence posits effect on the care quality in two ways: (1) health care systems must develop strategies to access patient linguistic and cultural competence to facilitate the detection of care and health disparities, and (2) an effective health care system enables the care provider to ensure treatment which aligns with the patients' preference as well as values

(Santé en Français (French Health), 2012). In this way, effectiveness is a pathway to equity so that care does not vary based on personal patient attributes (e.g., language, ethnicity, culture, geographic literacy, gender, or socioeconomic status).

Finally, the patient's hospital length of stay may be adversely affected by lack of cultural and linguistic competence. On the other hand, timeliness and efficiency can increase when tangential procedures, such as scheduling timely doctor's appointments for relevant follow-ups, can improve recovery and reduce costly time in the hospital (Santé en Français (French Health), 2012).

Clinicians' role in medical care in view of cultural and individual patient factors

Clinical effectiveness is significantly affected by the accreditation of organizations by setting standards for their performance (Baum, 2016). HCP duties may encompass direct patient support, guidance, instruction regarding medication, and selecting appropriate health promotion materials and audiovisual aids for the betterment of the patients. But, the capacity to effectively make informed decisions requires cultural and linguistic competence to bridge the gap in communication, establish trust, and make life-saving decisions. These are evident in the following overview of several clinical situations (e.g., special health condition, multiple cultural considerations, and the role of the clinician in addressing patient competence) in which decision-making in relation to specific patient needs is critical to quality care.

Special health conditions

The National Student Nurses' Association reported a case in which the clinical specialist disregarded the diabetic patients' spiritually informed dietary restriction (Hughes, 2017). If a patient is not provided with a suitable food alternative, the lost caloric intake together with an insulin medication injection can result in dangerous blood sugar levels and serious, potentially deadly, patient reaction. In this case, religious food restriction was a special health consideration for a patient with the condition of diabetes and serves as an example of overcoming cultural disparities which can impede the provisioning of quality care leading to adverse outcomes (Hughes, 2017).

The Association of American Medical Colleges (AAMC) has recognized certain cross-cultural skills which may be present among the clinicians practicing in the Arab regions (Fig. 8.3). This concept has evolved from making assumptions about a patient based on their background toward the implementation of patient-centered care. AAMC highlights competency in communication skills inclusive of the exploration, empathy, and other techniques that are helpful in understanding the patients' care needs, values, and preferences. The effective HCP assimilates the scientific knowledge with equivalent communication skills for understanding the patients' individual need with their concern for compassion and mutual respect to make the best-informed care selection. Similarly, in many Arab medical communities, certain skills have been recognized that include the important mix of clinical skills along with intercultural

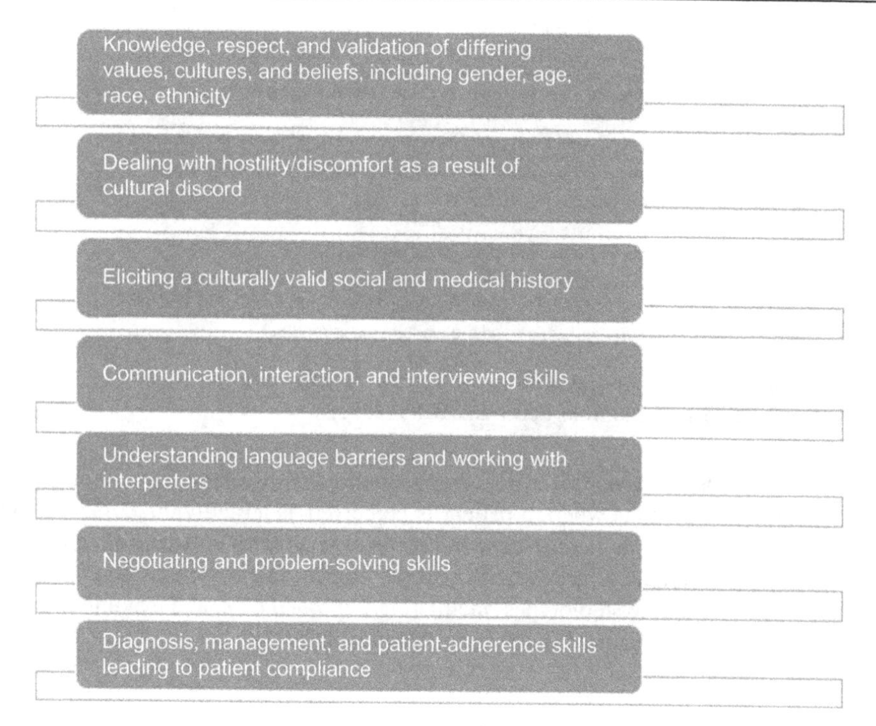

Fig. 8.3 Cross-cultural clinical skills.
Reproduced with permission and modified from the Association of American Medical Colleges (AAMC) (2005).

skills that were present among medical experts. These skills were used to facilitate patient improvement by providing effective health care treatments. However, the study notes that there remains a need to provide proper training regarding the understanding and acquisition of common skills that are required to obtain cross-cultural competency (Atwa & Nasser, 2016).

The foundational skills of the clinician are also categorized and portrayed in the form of an acronym "PREP" (Epner & Baile, 2011). The acronym represents noncognitive, fundamental blocks such as preparation, respect, environment, and presentation (Fig. 8.4). The four elements personify the basis of individual skills, overall presence, and attitude of the care provider by setting the tone of clinical encounters that are carried out independently by the HCP.

The preparation block is quite apparent and intuitive, whereas the other three are multidimensional and require further interpretation. For example, what are some actions that show respect? How can the clinical environment be optimized during suboptimal circumstances (i.e., intensive care unit, emergency department)? Finally, the concept of presentation may allow the HCP to determine how to present themselves to the patient by considering the symbolism of their white coat, personal grooming, and how these might help to build rapport. All these concepts can further be reflected upon to portray an effective and culturally competent image.

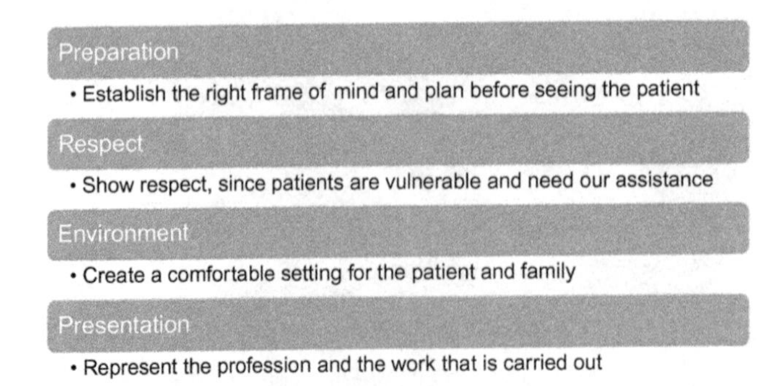

Fig. 8.4 Clinicians' foundational skills.
Reproduced with permission from Epner and Baile (2011).

In a broader perspective, the HCP can apply cultural competence by adopting two simple rules of thumb (1) where a patient is perceived as an individual, and (2) the HCP focuses on providing patient-centered services regardless of background or culture. Placing the individual needs first in all the clinical interactions is considered a critical step for avoiding stereotypes, facilitating improved care, and evading the pitfalls of deficiency in cultural competence (Hughes, 2017).

Overcoming key cultural factors

There are several cultural factors that impact the capacity of the HCP to conduct quality care. The foremost of which may surprisingly be religious diversity in Middle Eastern cultures. However, several other factors have significant roles in understanding and refining the clinical role of cultural competence. Others discussed in this section include affiliation and family style and contextual factors in patient engagement.

Religious and ethnic diversity

Religious perspectives are an important cultural consideration depending on the physician location, religious beliefs, and national origin or location characteristics, such as urban or rural designation, or length of time the physician must stay there. In addition to Islam, Arab-speaking nations include various religions such as Christian denominations, Melkites, Maronites, Coptics, Chaldean Catholics, and Syrians, whereas Zoroastrians, Moslems, Baha'is, Chaldeans, and Jews may be found in Iran.

The capacity to recognize variations in ethnicity also plays an important role in the health care sector. Ethnic identification, as well as acculturation awareness in a patient, helps the HCP to assess the overall picture by including a spectrum of components such as socioeconomic culture and education. The clinician skill in discerning these variations can impact the HCP/patient relationship.

However, the foremost religious consideration occurs in the treatment of Arabic women which requires great alertness and cultural sensitivity. Eye contact, though not physical, can be considered immodest and must be severely limited even when the HCP is communicating health information. Further, the HCP must take clues from the female patient that is calibrated on the eye contact which the patient establishes herself. Physical contact must be reduced to bare minimum and be accompanied by clear communication to the patient of what she should expect. The HCP can ask the following questions to a female patient to ensure a culturally comfortable environment:

HCP to patient: "For ensuring best health plan, I need to examine you. Is that OK?"

HCP to patient: "In order to maintain modesty, I will ensure that a sheet covers you the entire time and a nurse will also be present throughout the examination procedure."

HCP to patient: "If you would like your sister/another family member to be with you, we can ask them to come into the examination room with you."

HCP to patient: "Your private parts may need to be examined. This will be done with sensitivity and the most efficient way for sustaining your modesty. You will be treated the same way as I would treat my own family."

Preparing the female for contact and respecting her desire to maintain modesty are important concepts for any HCP, but especially in the case of an Arabic female. This requires the health care provider to develop skill and sensitivity for sustaining trust and rapport among the Arab patients, particularly women, who are covered from head to toe.

Affiliation and family style

A dominant factor for the Middle Eastern populations is affiliation with another person or family. This is recognized as a universal need that also considers the level of intensity of this aspect which may vary for the individual and cultural group. Clinical practitioners in the Asian Arab Region must understand that the patient seeks fulfillment of their need for affiliation. Consequently, a patient suffering from any illness or disease heavily relies upon the caregiver as a coping mechanism and so clinicians must recognize the patriarchal family structure and the crucial role that family plays in the sociological, economic, and psychological well-being of the patient (Halligan, 2006). In the health care setting, the clinician may expect to include the patient's family in decisions pertaining to health who may dictate the degree and extent of care. This stems from the Middle Eastern tradition of passing down perceptions of illness, disease, and death through one generation to the next. So, the family must be involved in patient decisions and information since the majority of the population relies on the family for care.

There are certain situations where families may insist on receiving details of the diagnosis even before the patient in order to advise the HCP how much information should be relayed to them. The use of euphemisms is suggested for answering the questions which are asked by the family (Bushnaq, 2008). For example, when a family member asks the hospital clinician about the prognosis of the patient, he may respond that the patient, "...is really in a critical condition, and it is the right time for him to meet his family and to prepare for the hereafter in case he/she deteriorates," (Bushnaq, 2008, p. 1292).

Clinicians may carry this awareness of the importance of family to the patient and inquire about his life and family when greeting the patient in a clinical setting. This practice exhibits the notion that the HCP genuinely cares about him and is not just concerned about his symptoms as a patient (Martin, 2009). Notable is that this type of family inquiry should only occur when the HCP moves from the position of stranger to an insider. Otherwise, these types of personal questions may be perceived by people in the Middle East as dubious behavior. However, the courteous respect for the patient's family can positively alter the behavior of the patient that is apparent in reduced demands.

Time, space, and context

The HCP/patient relationship in the Arabic culture has a few unique characteristics. Punctuality, for example, is not measured as a form of commitment, but rather in comparison to the perception of the relative importance of other tasks. Thus, punctuality is perceived less among the Arab patients as compared to those in Western cultures. However, the patient must embrace a formal approach to medical appointments in order to contribute to the establishment of the HCP/patient relationship just as the caregiver must adopt new interaction strategies.

When the patient arrives for a scheduled appointment, another unique characteristic becomes apparent as the HCP must sustain an appropriate distance when communicating with the patient. This is because the cultural norms and values differ among the Arabs. The Arab conversational space is more than twice as close (2 ft) than an American counterpart who is more comfortable at a distance of 5 ft. The close proximity allows each person to more easily read the reaction of the other during a conversation. The tendency to use touch while communicating is apparent among the Arabs in family settings (Lipson & Dibble, 2005) and acceptable when HCP and patient are of the same gender and have established trust.

Gender differences require a different approach because this type of close proximity is only allowed with parents and the spouse in the Islamic religion. The inability of the HCP to maintain an appropriate distance may hinder the female patient from seeking health care services (Salman, 2012). Here the most preferred method is to seek verbal information from the patient for examination or assign the case to the similar gender HCP. The latter is typical example of cultural precompetence in medical facilities engaging in short-term solutions. Moreover, common methods of greeting in Western culture, such as handshaking, may be perceived as inappropriate, and therefore, the HCP should be on the side of caution and not use this method of initial introduction. An effective approach which can be adopted includes greeting her with a smile and respectful nod. However, a handshake initiated by the female is acceptable. This principle applies generally to both men and women.

HCPs must also be sensitive to the cultural and religious beliefs regarding the physical or mental health diagnosis of the patient. Arab cultures view illness as a test from God (Wallin & Ahlström, 2010) and so the HCP must avoid discussing terminal disease with the patient. This practice by the HCP is based on Muslim belief that only God possesses the power to create life and cause death (Attum & Shamoon, 2018).

This requires an extension of the delicate understanding of the importance of family in communicating prognosis that may be unfamiliar to foreign HCP.

A caregiver unfamiliar with these practices must adopt these behaviors or mold the practices in a certain way which helps them to nurture a better HCP/patient relationship. Clearly, the Arab HCP medical practice and knowledge must be in context with cultural sensitivity given the frequency and intensity of medical visits. Still, gathering medical information is important, but understanding the web of circumstances while respecting patient values (e.g., individual experiences, attitudes, personality) provides an optimum HCP/patient relationship that enhances quality care (Wahass, 2005).

Recommendations

Tangential clinical competencies (e.g., language, culture) can aid in improved clinical outcomes through effective HCP/patient interactions based on clinician improvements in communication and self-management support built upon prevailing Middle Eastern culture. The three primary components defining the role of clinicians, senior health executives, and patients include assessment, communication, and self-management (Fig. 8.5), while performance standards, use of technology, and facility navigation may play a role. HCPs, clinical and nonclinical support, and the health care organization play an important role in addressing health care communication challenges (National Research Council, 2010). This section recommends that many aspects of existing assessment, communication, and self-management models can apply to HCP cultural and linguistic development.

Assessment

Just as HCPs may overestimate the patient's ability to read which leads to an argument for the use of tools to assess literacy in various languages (Batterham, Hawkins, Collins, Buchbinder, & Osborne, 2016; Bishop et al., 2016; Bloom-Feshbach et al., 2016;

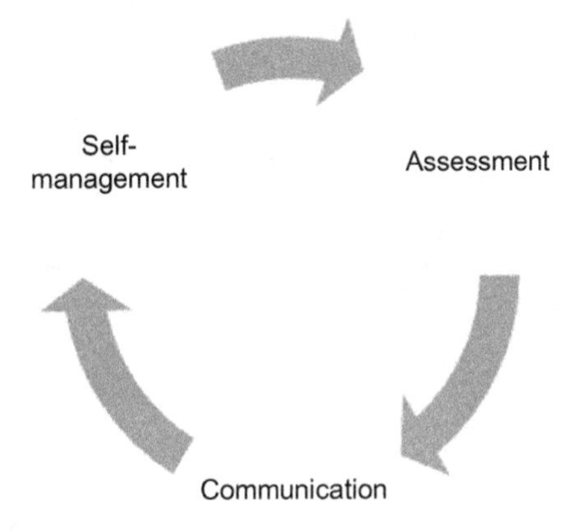

Fig. 8.5 Supporting components of clinician role of culturally responsible care.

Busch, Martin, DeWalt, & Sandler, 2015; Guzys, Kenny, Dickson-Swift, & Threlkeld, 2015; Health Research and Educational Trust, 2012; Nair, Satish, Sreedharan, & Ibrahim, 2016), health care administrators focusing on medical skills in the hiring process may not consider evaluation of cultural competence as an important workforce prerequisite. Given the vast number of Arabic dialects, languages, and cultural factors, a rational approach to a solution may likely begin with administrators and human resources enabling the process of cultural and linguistic competence beginning with the HCP assessment.

Taking the lead from researchers who recommend collecting detailed patient information to identify cultural group membership (Hasnain-Wynia & Baker, 2006), health care administrators and human resources can improve clinical insight on language and cultural diversity beyond the medical conditions that characterize patient health. Given this insight, the HCP can identify cultural deficiencies and assess any personal or professional bias, if any, that may disrupt quality of care. In order to ensure quality care, the health care professionals are required to unpack and understand ones' own belief, values, and any cultural bias to formulate answers to important questions such as why they seek employment in a culturally divergent environment. Uncovering an answer to this and other questions in this personal process is one method to equip individuals to practice patient care in a particular culture (Johnson, MacDonald, & Oliver, 2016). For HCPs to be competent, they must make a commitment to a continual, honest, and self-reflective journey to widen their cultural spectrum, conscious of their behavior and practices, and willing to alter them according to the cultural perspective of the patient undergoing treatment.

The development of cultural and linguistic competence serves as a great challenge to health care providers who have limited access to adequate resources. Utilizing the local and regional knowledge of executive level health care professionals and HCP dedication of the HCP may be an important inroad into systemic solutions to offset quality problems known to manifest in environments of care that are culturally diverse.

Another advantage of reviewing patient cultural data from a personal HCP perspective is the benefit of developing sensitivity to both patient and HCP deficiencies. Deficits in language deficiency (Greenhalgh, 2015), health literacy (Paasche-Orlow, Schillinger, Greene, & Wagner, 2006), or cultural deficiency may be a source of embarrassment. Therefore, this method can avoid miscommunication by the HCP taking universal precautions rather than patient screening which can be deemed inappropriate (Hasnain-Wynia & Baker, 2006). HCP assessment with the aid of senior health care executives providing key cultural information can remedy the patient experience which, in turn, can influence comprehension and willingness to adhere to medical advice.

Communication

Improvement in clear HCP/patient communication is encouraged in the environment of care because less communication can bring about the unnecessary outcomes in treating patients or result in more painful events (Cherednychenko, Povoroznyuk,

Povoroznyuk, & Dzerovych, 2016). However, there is a notable inconsistency in the approach to better communication from other perspectives not discussed in this paper, such as health literacy stressing the tenets of clear communication, versus the approach from the perspective of cultural competence.

Clear communication defined in health literacy literature limits the number of messages delivered, reduces jargon, and has the goal of patient understanding. However, this method focuses on the instructional skills of the clinician. Though this has some advantages to some culturally diverse patients, culturally based communication emphasizes the individual and their health perspective. Consequently, this impacts the way in which a patient may be open to receiving, processing, or rejecting medical information from an HCP. Thus, general questions that demonstrate concern for the patient and their health benefits (Kleinman, Eisenberg, & Good, 1978) are still practical and reinforced in current literature (Betancourt et al., 2016). Several HCP to patient questions follow:

- What is the reason (cause) of your illness?
- What are the effects of illness?
- How does your illness work?
- Which treatment do you think would be better for you?

Cross-cultural exploration between the HCP and patient is important for reaching consensus for treatment. However, low language proficiency in patients remains an obstacle as a greater number of reported problems have been associated with this deficiency (Malloy-Weir, Charles, Gafni, & Entwistle, 2015). Of course, this problem extends to HCP, administrative, and other caregiver support personnel including interpreters. Clearly, even speaking the same language does not presume the same culture, and therefore, misunderstandings can occur. For example, a Farsi speaking health and financial coordinator may be unable to help a Farsi-speaking patient because they are from different tribes. The extension of communication for support services beyond the HCP/patient encounter is another important aspect of health care to reduce preventable errors and increase compliance with medical directives. A comprehensive cultural competence and linguistic educational program still remains an important consideration for senior health care executives (Andrulis & Brach, 2007).

Self-management

"Self-management has evolved beyond the practice of merely providing information and increasing patient knowledge," (Grady & Gough, 2014, e26) in chronic diseases such as asthma, diabetes, and others. HCP, senior executive health care management, and patients all have a role in supporting self-management from their unique perspectives. Today, clinicians increasingly identify and engage patients as partners in their own health care. The HCP can facilitate cooperation by managing their own attitudes and response in relation to their patients by applying cultural and linguistic knowledge to the practice of medicine. Consequently, cooperation elicits adherence to medical recommendations which, in turn, sustains a trustworthy relationship from which to continue quality medical care. The cycle continues as quality care reduces risk and increases patient satisfaction that is a win-win for all!

Inclusion of support personnel in this patient-centered focus of health care is also important to this cultural and linguistic competency perspective. Senior health care administration working with human resources can act on behalf of both HCPs and their patients by vetting and indoctrinating interpreters and other support personnel to culturally relevant material. Tracking foreign health care professionals to better understand the scope of the problem at the national level and creating an accessible knowledge database of case studies demonstrating practical HCP/patient encounters on the job can also facilitate the important concept of communication through information exchange. This dissemination of information to HCP and support personnel should logically lead to the generation of specific patient tools associated with the documented encounters to limit future recurrence. Each self-management encounter becomes an opportunity for growth which contributes to the overall quality of the spectrum of care. Ironically, self-management is not the patient, the caregiver, or administration acting alone. Self-management encompasses engaging in resources, asking questions, and doing so in a way that is socially acceptable.

"Self-management…requires a partnership between patients and providers that fosters mutual understanding and shared decision-making," (Fritz et al., 2016, para 1). Arabic HCPs should easily recognize the inherent value of assessment, communication, and self-management because of the prominent similarity between the embedded characteristics of trust and community to the cultural values of family and belonging.

Discussion

The difference in language, cultural values, and practices serves as a barrier to HCP/patient communication because these differences act as a catalyst for conflict. These events may leave the patient with the impression of deficiency based on misinterpretation that escalates to missed opportunities for care and health deterioration. Moreover, the beliefs and values, social and kinship networks, and taboo and obligations can further amplify this communication disparity. As previously noted, the communication of disease information may be restricted due to the patient perception of illness as a test from God. Thus, the HCP in a clinical setting must understand and respect the differences which can drive poor interaction between the patients and the provider. However, there are other factors that were not the focus of this paper such as social determinants in Arabic-speaking nations or the detailed contribution of health literacy or informed consent that may color the perception of care.

Another item that requires consideration is that the Arab health care system focuses on providing treatment to the symptoms rather than preventing the disease occurrence. This perspective limits the role of self-management in the Western sense of the individual taking measures to prevent disease through screening or other testing. A prime example is that Arab Muslim women are less likely to use preventive services, such as screening for breast cancer, because of the belief that these procedures conflict with their cultural or religious beliefs, despite understanding and awareness of the significance of the testing.

Catering to the needs of patients following different culture and language requires one to improve use of certain strategies which help the practitioners to overlook the different values and beliefs. This competence allows the HCP to develop sensitivity to their patients' beliefs and values obliging them to willingly develop cultural and linguistic competence in nonverbal and verbal communication behaviors (Khan et al., 2017). Effective communication by the health care provider helps the patients in their recovery while also enhancing their satisfaction. Effective intercultural competence results in positive outcomes including delivery of effective patient care and relief from stress (Meuter et al., 2015). More importantly, HCP cultural diversity training to improve patient communication discussed herein is cost-effective in general terms compared to the potential savings on litigation related to medical errors and the reduction of patent morbidity and mortality.

Just as the HCP/patient relation does not improve in isolation, developing cultural and linguistic competence among the health care workforce must engage to bring forth and develop an alliance with medical educators, researchers, and social scientist. Their common goal would be to initiate educational programs that improve this competence as a whole with the primary responsibility of ensuring quality within the health care system. Expected output would include establishing measures related to culture and linguistic competence within the framework of delivering a high-quality medical care which would establish a regulatory institution. In addition, efforts must be made by the patients to make the most of each opportunity and supply in the feedback (such as participate or communicate with the health care providers) to enhance the health care system design, in the pursuit of an integrated approach for health care that reflects the patient's diverse need and preferences. Achieving cultural competence will not rest on memorizing information, but built on a professional platform that encompasses sensitivity, respect, curiosity, astuteness, partnership, and tolerance. One thing that must be remembered by the personnel is that all people really care about is that someone cares about them.

Of course, the development of a health care institution embodying the vast subtleties in Arabic culture and language does not occur overnight or without funding. During the interim period when funding and other dogma are under debate, hospitals and service providers can collect data as described in "Recommendations" section. Senior health administrators and human resources can work directly with HCPs, nonclinical, and clinical support in local development of cultural competencies to discuss events and develop methods to enhance the HCP/patient relationship in a culturally sensitive manner. In the short-term, each institution can develop a regionally specific matrix of methods and resources that are cost-effective and relevant to patient care. In the long-term, the output of local collection of case studies and methods to overcome imbalances in care can feed into the higher concept of institutional development.

Conclusion

This chapter discusses the important but sometimes overlooked aspect of cultural competence in a clinical setting to bring forth the concept of quality of care by promoting equitable care that mitigates disparities in culturally diverse populations. We suggest that the health care provider can accomplish better outcomes by a patient-centered approach to health care through a process of assessment, communication, and self-management with the cooperation of senior health executives and human resource personnel. The chapter explains that foreign and all health care professionals, particularly in the Asian Arab countries, must deliver patient care that is sensitive to their system of values, culture, and language. The health care system as a whole should include measures of cultural competence and utilization of health care provider training strategies drawn from multicultural experiences to avoid preventable morbidity and mortality. Building local data to educate foreign HCPs is an important step for human resources, risk management, and patient quality of care, leading to an overriding organization to monitor, address, and equip these professionals to better serve their patient populations.

Acknowledgment

The authors would like to thank the United Nations for providing relevant information to supplement this text. We take full responsibility for interpretation of same.

References

Alananzeh, I., Ramjan, L., Kwok, C., Levesque, J. V., & Everett, B. (2018). Arab-migrant cancer survivors' experiences of using health-care interpreters: A qualitative study. *Asia-Pacific Journal of Oncology Nursing*, *5*(4), 399.

Albahri, A. H., Abushibs, A. S., & Abushibs, N. S. (2018). Barriers to effective communication between family physicians and patients in walk-in centre setting in Dubai: A cross-sectional survey. *BMC Health Services Research*, *18*(1), 637.

Allianz Care. (2019). *Healthcare in the United Arab Emirates (UAE)*. Retrieved June 1, 2019 from https://www.allianzworldwidecare.com/en/support/view/national-healthcare-systems/health care-in-uae/.

Andrulis, D. P., & Brach, C. (2007). Integrating literacy, culture, and language to improve health care quality for diverse populations. *American Journal of Health Behavior*, *31* (1), S122–S133.

Association of American Medical Colleges (AAMC) (2005). *Cultural competence education*. Retrieved April 22, 2019 from https://www.aamc.org/download/54338/data/.

Attum, B., & Shamoon, Z. (2018). Cultural competence in the care of Muslim patients and their families. In *StatPearls*: StatPearls Publishing.

Atwa, H., & Nasser, A. A. (2016). Physicians' self-assessment in intercultural clinical communication in Jeddah, Saudi Arabia: A pilot study. *Education in Medicine Journal*, *8*(2), 15–25. https://doi.org/10.5959/eimj.v8i2.420.

Batterham, R. W., Hawkins, M., Collins, P. A., Buchbinder, R., & Osborne, R. H. (2016). Health literacy: Applying current concepts to improve health services and reduce health inequalities. *Public Health*, *132*, 3–12.

Baum, F. (2016). *The new public health* (4th ed.). Oxford University Press.

Betancourt, J. R. (2006). *Improving quality and achieving equity: The role of cultural competence in reducing racial and ethnic disparities in health care*. The Commonwealth Fund.

Betancourt, J. R., Green, A. R., Carrillo, J. E., & Owusu Ananeh-Firempong, I. I. (2016). Defining cultural competence: A practical framework for addressing racial/ethnic disparities in health and health care. *Public Health Reports*, *118*, 292–302.

Bishop, W. P., Craddock Lee, S. J., Skinner, C. S., Jones, T. M., McCallister, K., & Tiro, J. A. (2016). Validity of single-item screening for limited health literacy in English and Spanish speakers. *American Journal of Public Health*, *106*(5), 889–892.

Bloom-Feshbach, K., Casey, D., Schulson, L., Gliatto, P., Giftos, J., & Karani, R. (2016). Health literacy in transitions of care: An innovative objective structured clinical examination for fourth-year medical students in an internship preparation course. *Journal of General Internal Medicine*, *31*(2), 242–246.

Bowen, S. (2015). *The impact of language barriers on patient safety and quality of care*. Société Santé en Français.

Busch, E. L., Martin, C., DeWalt, D. A., & Sandler, R. S. (2015). Functional health literacy, chemotherapy decisions, and outcomes among a colorectal cancer cohort. *Cancer Control*, *22*(1), 95–101.

Bushnaq, M. (2008). Palliative care in Jordan: Culturally sensitive practice. *Journal of Palliative Medicine*, *11*(10), 1292–1293.

Butler, M., McCreedy, E., Schwer, N., Burgess, D., Call, et al. (2016). *Improving cultural competence to reduce health disparities*. Comparative Effectiveness Review No. 170. (Prepared by the Minnesota Evidence-based Practice Center under Contract No. 290–2012-00016-I.) AHRQ Publication No. 16EHC006-EF Rockville, MD: Agency for Healthcare Research and Quality. March 2016 www.effectivehealthcare.ahrq.gov/reports/final.cfm.

Central Intelligence Agency. (2019). *The world fact book*. Retrieved July 20, 2019 from https://www.cia.gov/index.html.

Cherednychenko, O., Povoroznyuk, V., Povoroznyuk, R., & Dzerovych, N. (2016). *Patient empowerment: Translating PILS, pre/post-op instructions, and ICS*. ATA57TH, 37.

Chu, G. Y. K., & Goode, T. (2009). Cultural and linguistic competence. In S. Hatch, et al. (Eds.) *Optometric care within the public health community*. Cadyville, NY: Old Post Publishing.

Cross, T. L., Brazron, B. J., Dennis, K. W., & Issacs, M. R. (1989). *Towards a culturally competent system of care: A monograph on effective services for minority children who are severely emotionally disturbed*. Washington DC: Georgetown University Child Development Center.

Desmet, K., Ortuño-Ortín, I., & Wacziarg, R. (2017). Culture, ethnicity, and diversity. *American Economic Review*, *107*(9), 2479–2513.

Eklof, M., & Ahlborg, G., Jr. (2016). Improving communication among healthcare workers: A controlled study. *Journal of Workplace Learning*, *28*(2), 81–96.

Epner, D. E., & Baile, W. F. (2011). Wooden's pyramid: Building a hierarchy of skills for successful communication. *Medical Teacher*, *33*(1), 39–43.

Epner, D. E., & Baile, W. F. (2012). Patient-centered care: The key to cultural competence. *Annals of Oncology*, *23*(Suppl. 3), 33–42.

Eskes, C., Salisbury, H., Johannsson, M., & Chene, Y. (2013). Patient satisfaction with language-concordant care. *Journal of Physician Assistant Education (Physician Assistant Education Association)*, *24*(3), 14–22.

Fritz, H., et al. (2016). Diabetes self-management among Arab Americans: Patient and provider perspectives. *BMC International Health and Human Rights*, *16*, 22. https://doi.org/10.1186/s12914-016-0097-8.

Grady, P. A., & Gough, L. L. (2014). Self-management: A comprehensive approach to management of chronic conditions. *American Journal of Public Health, 104*(8), e25–e31.

Greenhalgh, T. (2015). Health literacy: Towards system level solutions. *British Medical Journal*, 350.

Guzys, D., Kenny, A., Dickson-Swift, V., & Threlkeld, G. (2015). A critical review of population health literacy assessment. *BMC Public Health, 15*(1), 215.

Hadziabdic, E., Lundin, C., & Hjelm, K. (2015). Boundaries and conditions of interpretation in multilingual and multicultural elderly healthcare. *BMC Health Services Research, 15*(1), 458.

Halligan, P. (2006). Caring for patients of Islamic denomination: Critical care nurses' experiences in Saudi Arabia. *Journal of Clinical Nursing, 15*(12), 1565–1573.

Hasnain-Wynia, R., & Baker, D. W. (2006). Obtaining data on patient race, ethnicity, and primary language in health care organizations: Current challenges and proposed solutions. *Health Services Research, 41*(4p1), 1501–1518.

Health Research and Educational Trust. (2012). *A toolkit for collecting race, ethnicity, and primary language information from patients.* Retrieved April 24, 2019 from http://www.hretdisparities.org/hretdisparities-app/index.jsp.

Hughes, J. (2017). *Why is cultural competence in healthcare so important?* Retrieved April 23, 2019 from https://www.healthcarestudies.com/article/Why-Is-Cultural-Competence-in-Healthcare-So-Important/.

Jenkins, O. B. (2000). *Population analysis of the Arabic languages.* [Last reviewed July 16, 2011]. Retrieved July 11, 2019 from http://strategyleader.org/articles/arabicpercent.html.

Johnson, J. M., MacDonald, C. D., & Oliver, L. (2016). Recommendations for healthcare providers preparing to work in the Middle East: A Campinha-Bacote cultural competence model approach. *Journal of Nursing Education and Practice, 7*(2), 25.

Khan, S. M., Suendermann-Oeft, D., Evanini, K., Williamson, D. M., Paris, S., Qian, Y., et al. (2017). MAP: Multimodal assessment platform for interactive communication competency. In *Practitioner track proceedings.*

Kleinman, A., Eisenberg, L., & Good, B. (1978). Culture, illness, and care: Clinical lessons from anthropologic and cross-cultural research. *Annals of Internal Medicine, 88*(2), 251–258.

Lim, S., & Mortensen, A. (2015). Preparing a Culturally Competent Health Workforce for working with clients and families from refugee backgrounds. In *Auckland University of Technology (AUT)—Refugee research symposium, December 3.*

Lipson, J. G., & Dibble, S. L. (Eds.), (2005). *Culture & clinical care.* San Francisco: UCSF Nursing Press.

Malloy-Weir, L. J., Charles, C., Gafni, A., & Entwistle, V. A. (2015). Empirical relationships between health literacy and treatment decision making: A scoping review of the literature. *Patient Education and Counseling, 98*(3), 296–309.

Martin, S. S. (2009). Healthcare-seeking behaviors of older Iranian immigrants: Health perceptions and definitions. *Journal of Evidence-Based Social Work, 6*(1), 58–78.

Matthews, M., & Van Wyk, J. (2018). Towards a culturally competent health professional: A South African case study. *BMC Medical Education, 18*(1), 112.

McBain-Rigg, K. E., & Veitch, C. (2011). Cultural barriers to health care for Aboriginal and Torres Strait Islanders in Mount Isa. *Australian Journal of Rural Health, 19*(2), 70–74. https://doi.org/10.1111/j.1440-1584.2011.01186.x.

Meehan, J., Menzies, L., & Michaelides, R. (2017). The long shadow of public policy; Barriers to a value-based approach in healthcare procurement. *Journal of Purchasing and Supply Management, 23*(4), 229–241.

Meuter, R. F., Gallois, C., Segalowitz, N. S., Ryder, A. G., & Hocking, J. (2015). Overcoming language barriers in healthcare: A protocol for investigating safe and effective communication when patients or clinicians use a second language. *BMC Health Services Research, 15*(1), 371.

Nair, S. C., Satish, K. P., Sreedharan, J., & Ibrahim, H. (2016). Assessing health literacy in the eastern and middle-eastern cultures. *BMC Public Health, 16*(1), 831.

National Research Council. (2010). *The role of human factors in home health care: Workshop summary*. Washington, DC: The National Academies Press. https://doi.org/10.17226/12927.

Olaussen, S. J., & Renzaho, A. M. (2016). Establishing components of cultural competence healthcare models to better cater for the needs of migrants with disability: A systematic review. *Australian Journal of Primary Health, 22*(2), 100–112.

Paasche-Orlow, M. K., Schillinger, D., Greene, S. M., & Wagner, E. H. (2006). How health care systems can begin to address the challenge of limited literacy. *Journal of General Internal Medicine, 21*(8), 884–887.

Padela, A. I., & Del Pozo, P. R. (2011). Muslim patients and cross-gender interactions in medicine: An Islamic bioethical perspective. *Journal of Medical Ethics, 37*(1), 40–44.

Pison, G. (2019a). The number and proportion of immigrants in the population: International comparisons. *Population & Societies, 563*, 1–4.

Pison, G. (2019b). *Which countries have the most immigrants*. Retrieved May 30, 2019 from http://theconversation.com/which-countries-have-the-most-immigrants-113074.

Raddawi, R. (2015). Intercultural (mis-) communication in medical settings: Cultural difference or cultural incompetence? In *Intercultural communication with Arabs* (pp. 179–195). Singapore: Springer.

Salman, K. F. (2012). Health beliefs and practices related to cancer screening among Arab Muslim women in an urban community. *Health Care for Women International, 33*(1), 45–74. https://doi.org/10.1080/07399332.2011.610536.

Santé en Français (French Health). (2012). *Linguistic competence and quality of services business case for quality French-language health care services*. Retrieved April 22, 2019 from https://santeenfrancais.com/sites/ccsmanitoba.ca/files/attachments/cahier_eng_2014.pdf.

Shishehgar, S., Gholizadeh, L., DiGiacomo, M., & Davidson, P. M. (2015). The impact of migration on the health status of Iranians: An integrative literature review. *BMC International Health and Human Rights, 15*(1), 20.

Siyam, A., & Poz, M. R. D. (Eds.), (2014). *Migration of health workers WHO code of practices and the global economic crisis*. Retrieved June 1, 2019 from https://www.who.int/hrh/migration/migration_book/en/.

Sulaiman, A. (2019). *The impact of language & cultural barriers on patient safety & health equity*. Retrieved April 23, 2019 from http://www.wapatientsafety.org/impact-of-language-cultural-barriers-on-patient-safety-health-equity.

Truong, M., Paradies, Y., & Priest, N. (2014). Interventions to improve cultural competency in healthcare: A systematic review of reviews. *BMC Health Services Research, 14*(1), 99.

Ulrey, K. L., & Amason, P. (2001). Intercultural communication between patients and health care providers: An exploration of intercultural communication effectiveness, cultural sensitivity, stress, and anxiety. *Journal of Health Communication, 13*(4), 449–463.

United Nations, Department of Economic and Social Affairs, Population Division. (2019). *World population prospects 2019*. Online Edition.

van Rosse, F., de Bruijne, M., Suurmond, J., Essink-Bot, M. L., & Wagner, C. (2016). Language barriers and patient safety risks in hospital care. A mixed methods study. *International Journal of Nursing Studies, 54*, 45–53.

Wahass, S. H. (2005). The role of psychologists in health care delivery. *Journal of Family & Community Medicine*, *12*(2), 63.

Waitemata District Health Board. (2019). *Cultural competence continuum (Online image)*. Retrieved May 30, 2019 from https://www.ecald.com/resources/cultural-competence-assessment-tools/.

Wallin, A. M., & Ahlström, G. (2010). From diagnosis to health: A cross-cultural interview study with immigrants from Somalia. *Scandinavian Journal of Caring Sciences*, *24*(2), 357–365. https://doi.org/10.1111/j.1471-6712.2009.00729.x.

World Health Organization (WHO). (2019). *What is quality of care and why is it important?* Retrieved June 2, 2019 from https://www.who.int/maternal_child_adolescent/topics/quality-of-care/definition/en/.

Yeandle, D., Rieckmann, P., Giovannoni, G., Alexandri, N., & Langdon, D. (2018). Patient power revolution in multiple sclerosis: Navigating the new frontier. *Neurology and Therapy*, *7*(2), 179–187.

Young, J. S. (2013). Resources for Middle Eastern patients: Online resources for culturally and linguistically appropriate services in home healthcare and hospice, Part 3. *Home Healthcare Now*, *31*(1), 18–26.

Supplementary reading recommendations

Alhamami, M. (2019). Switching of language varieties in Saudi multilingual hospitals: Insiders' experiences. *Journal of Multilingual and Multicultural Development*, *1*(15). https://doi.org/10.1080/01434632.2019.1606227.

Buchan, J., Wismar, M., Glinos, I. A., & Bremmer, J. (Eds.), (2014). *Health professional mobility in a changing Europe: New dynamics, mobile individuals and diverse responses. Vol. 2*. United Kingdom: World Health Organization on behalf of The European Observatory on Health Systems and Policies. Retrieved June 1, 2019 from https://www.who.int/workforcealliance/03.pdf.

Lipson, J., & Meleis, A. I. (1983). Issues in health care of Middle Eastern patients. *The Western Journal of Medicine*, *139*(6), 854–861. Retrieved May 30, 2019 from http://www.pubmedcentral.nih.gov/articlerender.fcgi?artid=1011016.

Loney, T., Aw, T. C., Handysides, D. G., Ali, R., Blair, I., et al. (2013). An analysis of the health status of the United Arab Emirates: The 'Big 4' public health issues. *Global Health Action*, *6*, 20100. https://doi.org/10.3402/gha.v6i0.20100.

Mujallad, A., & Taylor, E. J. (2016). Modesty among Muslim women: Implications for nursing care. *Medsurg Nursing: Official Journal of the Academy of Medical-Surgical Nurses*, *25*(3), 169–172.

Nair, S. C., & Ibrahim, H. (2015). Assessing subject privacy and data confidentiality in an emerging region for clinical trials: United Arab Emirates. *Accountability in Research*, *22*(4), 205–221.

Nair, M., & Webster, P. (2012). Health professionals' migration in emerging market economies: Patterns, causes and possible solutions. *Journal of Public Health*, *35*(1), 157–163. https://doi.org/10.1093/pubmed/fds087.

Tackett, S., et al. (2018). Barriers to healthcare among Muslim women: A narrative review of the literature. *Women's Studies International Forum*, *69*. https://doi.org/10.1016/j.wsif.2018.02.009.

Definitions

Consent (informed consent) a person/patient who is able to make reasonable decisions based on mental stability and adult status to voluntarily agree to undertake health procedures proposed by a competent physician; Informed consent is a growing global patient right to receive all information from a health care professional before allowing a procedure, test, or other health service.

Culture brokers Emerging paraprofessional who facilitates an increase in the quality of medical professional and patient relationship through social services, information, and clarification; generally bilingual, culturally competent, and able to communicate across cultural barriers to facilitate patient care.

Cultural competence Professional self-management approach to equitable, inclusive, and quality care without prejudice toward culture or language differences.

Cultural heterogeneity the presence of multiple cultural influences (e.g., languages, traditions, beliefs) in a given population.

Demanding behavior A clinical communication behavior observed in a dominant patient when trust is not developed between a health care professional and the patient.

Interpretation barriers A clinical communication behavior observed in a dominant patient when the health care professional and the patient have different language skills; an interpreter lacks sufficient language or medical terminology training; or when similar words have distinct and separate meanings in different cultures.

Morbidity Reference to a portion of a specific population that suffers with mental health or physical communicable or noncommunicable disease (i.e., cancer) or unhealthy state (i.e., morbidly obese); complication resulting from miscommunication involving medical treatment can also be a morbidity factor

Mortality Rate of death measured by certain personal characteristics (e.g., race, ethnicity, religion, age, gender) or geographic location (e.g., city, state, region, country, urban, rural) of a specific population diagnosed with a specific disease (i.e., infant mortality); complication resulting from miscommunication involving medical treatment can also be a mortality factor.

Quality of care "The extent to which health care services provided to individuals and patient populations improve desired health outcomes. In order to achieve this, health care must be safe, effective, timely, efficient, equitable and people-centred" (World Health Organization, 2019, para 4).

Acts and actions: The church as an institution of cultural change

Beth Ann Fiedler
Independent Research Analyst, Jacksonville, FL, United States

Abstract

Significant strides in addressing cultural public health obstacles such as gender inequality, unequal distribution of wealth, and racism have advanced through research and development of government policy; but achieving harmony across compounding diversity factors remains a 21st Century challenge. This chapter explores the Book of Acts written by Luke, a Jewish physician, as an account of urban institutional development in the 1st Century formation of the Christian church. First, in the context of existing religious and political influence; and second, in relation to the development of modern governing structures in the United States which recognize equality and civil rights. Finally, the urban case study of the Book of Acts serves as the main focus of discussion and the example of long-term cultural influence in modern community outreach and concern for public health. The case study informs a framework for institutional development based on the congruence of common beliefs and equality.

Keywords: Christianity, Institution, Community activism, Sovereignty, Urban.

Chapter outline

Three Facets of Public Health and Paths to Improvements. https://doi.org/10.1016/B978-0-12-819008-1.00009-2

(New International Version, 2011):

> *He said to them, "Then give back to Caesar what is Caesar's, and to God what is God's."*[a]
>
> *Luke 20:25*

Ethnic diversity and cultural changes in the formation of the Christian church

The infusion of new ideas into steadfast cultural values and traditions is, at best, difficult. Obstacles to information dissemination exist across multiple races, ethnicities, nations, and many other demographic factors that divide populations. In turn, they compound the plausibility of successful change and can defeat perseverance in the best circumstances even with the most steadfast representatives. Nevertheless, the capacity to overcome these obstacles is a significant part of the history of mankind and a stepping stone for the advancement of civilization.

One historical account of the capacity to prompt change in an unreceptive environment appears in the Book of Acts in the Holy Bible. Acts depicts the account of the commission of the Apostles (Acts 1:8) and the beginning of the formation of the early Christian church (Fig. 9.1). This set the stage for the action and reactions of the blossoming church to an oppressed populace which were placed in written form beginning about 10 years after the crucifixion of Jesus (ca. 30–33) until about 70 (Barnett, 2005). The historical accuracy of biblical accounts has been repeatedly corroborated by learned scholars with a plethora of scholarly documents numbering more than 24,000 in various languages from Armenian to Slavic (Strobel, 2016, p. 66), archeological artifacts (Biblical Archeology Society, 2019; Strobel, 2016, pp. 99–117), and other evidence (Barnett, 1986; Barnett, 2005; Strobel, 2016).

According to biblical scholars such as Craig L. Blomberg, many ancient writings contained "outlandish flourishes and blatant mythologizing" (Strobel, 2016, p. 43), but biblical books were written and amply noted with intention and rigor to depict an accurate account. The purity of text is "unrivaled among ancient writings" according to Dr. Benjamin Warfield (Warfield, 1907 as cited in Strobel, 2016, p. 74) because of persistent use and accuracy in accordance with the original translation. In fact, the balanced accounts, even with the inclusion of discernible personal viewpoints, reflect modern research methodologies, while the majority of antiquity historical writing demonstrates two major problems. First, a bias that is evident in the dedication to a specific location or providence without other context, and second, exclusion of general history because of the intentional focus on important people and their events (Barnett, 2005). Blomberg also stresses that while each inspired author of the Books in the Bible had unique motivation, specifically in the New Testament

[a] Scripture quotations marked (NIV) are taken from the Holy Bible, New International Version, NIV. www.zondervan.com. The "NIV" and "New International Version" are trademarks registered in the United States Patent and Trademark Office by Biblica, Inc.

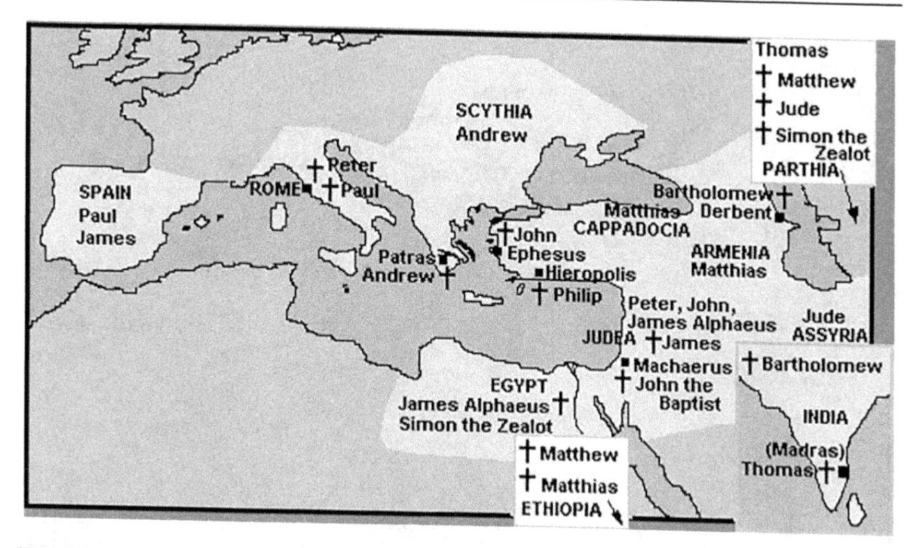

Fig. 9.1 Map depicts the period of the apostles and their travels during the early church approximately AD30-100.
Reproduced by permission from Gordon Smith, available at https://ccel.org/bible/phillips/CN610CHRONO.htm.

gospels, they endeavored to give an accurate account. For example, the perspective of Luke, "the theologian of the poor and of social concern" (Strobel, 2016, p. 32) prominently reflects his desire to provide accurate information to everyone without discrimination in the first chapters of the Book of Luke and Acts.

Still, biblical scholars recognize that there are some deficits in the narrative of Acts such as the focus on the ministry of Paul resulting from Luke's participation in his ministry, especially in Troas (ca. 49) and Philippi (ca. 57–62) (Barnett, 2005, pp. 191–191) (Fig. 9.2). Nevertheless, his own letters to the various churches that appear in the New Testament address the "absence of chronological markers for Paul's early years" (Barnett, 2005, p. 1) in the Book of Acts. The fact that the New Testament books do not follow a chronological order represents another hurdle given that Paul's letters to various churches are believed to predate the Gospels, but appear after them (Strobel, 2016). Nevertheless, the undisputed letters attributed to Paul (e.g., I Thessalonians, I and II Corinthians, Galatians, and Romans) illustrate the important details in other New Testament books that fill the timeline gaps. The addition of historical and general information surrounding the geographical territory and coinciding history during that period of time, also known as a modern research method of including context, also confirms the relevance (Barnett, 2005; Bock, 2002; Van Voost, 2000) and reliability (Blomberg, 2007, 2016; Habermas & Flew, 2005; Williams, 2010) of the Book of Acts, Luke's earlier self-titled work, and other New Testament letters.

On the other hand, some exert great effort to dismiss the historical and academic legitimacy of scripture as mere "fairy tale" (American Atheists, n.d.; DeMar, 2014), claiming that there is insufficient evidence to support the legitimacy of the document.

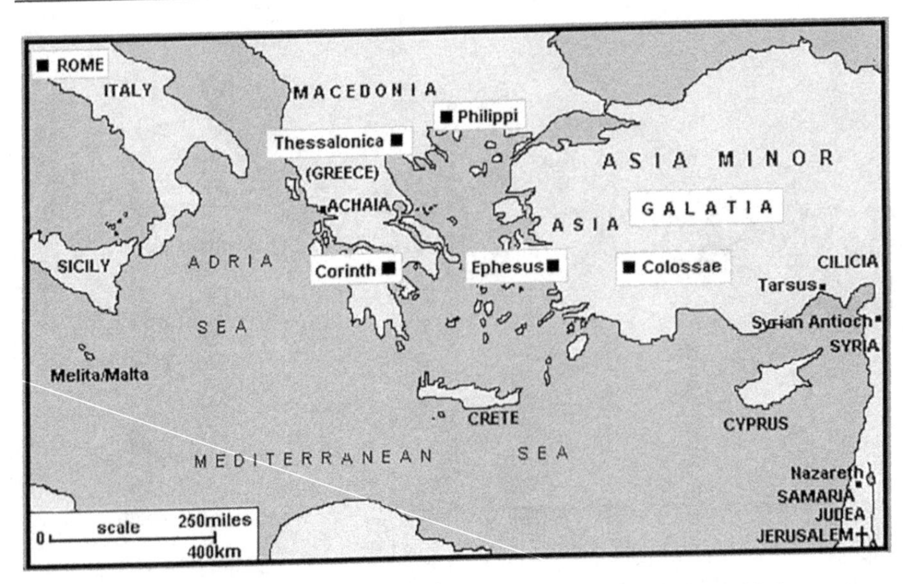

Fig. 9.2 Map references the primary locations where Paul traveled and sent his letters, important to the development of the early church, as depicted in the Book of Acts. Key: Boxes represent the six cities and province of Galatia where the letters from Paul were received. Reproduced by permission from Gordon Smith, available at https://ccel.org/bible/phillips/CN203ACTSJewishPeriod.htm.

However, they encapsulate their negative perception of religion, especially Christianity, while portraying their own doctrine and the promotion of their own agenda (Society, 2017). Perhaps for this reason, some scholars have deliberated various perspectives on the discipline of religious study to encourage new methods for confirmation of existing paradigms or to increase the proliferation of religious studies with renewed innovation and rigor (Führding, 2017). Still, others claim that religion serves as a catapult for political interests in the formation of modern government in nations such as the United States, Israel, India, Japan, France, and others (Stack, Goldenberg, & Fitzgerald, 2015).

Notable is that within the Christian apologist texts are a variety of opposing positions in which the reality of Jesus Christ is initially dismissed as mythology, but who later find Jesus Christ as their personal Savior. Ironically, what remains consistent in the general theme of naysayers is the persistent influence of religious beliefs on culture, social order, and ethics. Further, modern research methodologies, such as scholarship, research, and first-hand eyewitness accounts in context, are prevalent in biblical texts and unsuccessfully disputed. Together, this gives rise to reading the Book of Acts and supporting New Testament scripture for important information on the development of the Christian church.

The birth of Christianity in the midst of Jerusalem, the center of Judaism, is important to understand the longevity of the church as an institution of cultural change. The process offers clues to overcome persistent problems in modern society, such as racism and class prejudice, by establishing enduring elements of a stable society (e.g., community, equality, value of all human life, social welfare). This chapter highlights the relevance of the Book of Acts as a case study in the expression of individual freedoms, or civil rights, which form the urban and then global expansion of the institution of the early Christian church. Using the Book of Acts and other informative scripture, we focus on elements specific to Christianity that define the church as an institution and then discuss the cultural and legal context of Jerusalem, the Roman Empire, Christians, and modern civil rights in the United States. Then, we isolate specific content of the Book of Acts, identify the actions of the Apostles in the case study to formulate institutional and individual characteristics that unify the people of Christ, and establish how that foundation of unity brings forth the 1st Century church through contemporary cultural impact. The case study informs a framework for institutional development based on the congruence of common beliefs. The foremost of this is the development of positive cultural environments unencumbered by prejudice which are reflected in daily concern for the social unit and their community. These guidelines (e.g., service, fellowship, resource sharing) enable the early 1st Century Christian church to overcome cultural barriers and other destructive elements (e.g., idol worship, promiscuity, social unrest) by embracing the disabled, foreigners, poor, powerful, military, commoners, and wealthy. The cultural impact on public health is demonstrated tangentially by access to common resources, spirit of community, and general expression of joy. However, these concepts have universal applications to health and well-being common in modern literature, particularly in sociology and psychology (Hordern, 2016; Oman, 2018).

We begin this look into the past by defining the church in terms of classic Institutional Theory and in relation to the context of this paper.

Defining the church as an institution

Internal conflict and change, as the people of Israel were among the first converts transitioning to Christianity, are two important predictive elements in Institutional Theory (Scott, 1987; Selznick, 1948). Isomorphism, the underlying structure and associated processes within an organization undergoing change in relation to the emerging institution, is another recognized element of the theoretical premise (Scott, 1987; Selznick, 1948). These predictive constructs are causal in nature to the process of change and help to define an institution (Udo-Akang et al., 2012); in this case, the common understanding in Judaism and the relationship to the emerging institution of the Christian church. Finally, ongoing conflict leads to conformity that becomes recognizable outcomes in terms of social norms, regulation, and a cohesive mindset or schema of the emerging institution or organization. In other words, the origination of common rules together with a series of causes or conditions (e.g., conflict, change, conformity, isomorphism, and institutional emergence) leads to outcomes of social norms and stability which are the effects leading to the development of a new institution. These new social norms separate and distinguish Judaism and Christianity over a period of time (Scott, 1987; Selznick, 1948; Udo-Akang et al., 2012) which is prominent in the emergence of the Christian church.

There are similarities in the underlying foundation of the early church and modern institutions in the same way that the Christian church emerged from Judaism. But, there are also significant differences. We create a basic definition to underscore the meaning of the word institution in the context of this paper because many scoff at the idea as a result of negative association of the word institution with politics or political influence.

Modern society generally considers an institution taking the form of an organization (e.g., hospital, bank, prison, or university) and associates that organization with a physical structure, a specific purpose, and guiding principles. In this case, the structure can house individuals who are subject to various rules and regulations. Thus, institutions create law or policy to rule those individuals who are part of that institution. These laws and policies represent acceptable behavior or the consequences of unacceptable behavior within a cultural environment which becomes common practice over a period of time. For example, banking customers, hospital patients, and prisoners are subject to general guidelines to protect themselves and others within the institution. Of course, a prisoner who is incarcerated has less freedom due to the inherent nature of the institution of a prison. Nevertheless, all individuals within an institution are subject to certain protocols and cultural behaviors. Thus, ruling institutions establish order and behavioral conventions through the implementation of policy and law that benefits society in terms of social order.

In the same way, the church has self-regulating beliefs and patterns of cultural behavior which are evident in the institutions of family, marriage, and community. However, a distinguishing factor of the church purpose is the organizational power to serve and not to rule. Neither does the church require a specific building or place of worship, but a submission of service based on the foundation of belief in Jesus Christ. For example, loving thy neighbor represents an application of service that can occur anywhere without physical, cultural, or other limitations. Addressing the needs of all without concern for cultural barriers has also proven to be a method of establishing equality among the people resulting in social order. The notion of equality has evolved over millennium in the common law of governments and this important factor may be the most significant contribution of the institution of the church that is evident today.

The context of prevailing 1st Century religion, culture, and law

Cultural diversity, a consequence of Roman conquests and increased mobility, is the backdrop of the emergence of the Christian church in Jerusalem at the beginning of the 1st Century. The complex social environment surrounding the city, in the midst of scholarly pursuit, the laws that govern Jewish religious belief, and Roman rule provide the necessary context to differentiate Christian culture. Probably, the easiest explanation of the differences between and among leading cultural influences can be found in

the letter to the church at Corinth by Paul, a converted Jew who had spent years persecuting the church.
(NIV, 2011):

> [22] *Jews demand signs and Greeks look for wisdom,* [23] *but we preach Christ crucified: a stumbling block to Jews and foolishness to Gentiles,* [24] *but to those whom God has called, both Jews and Greeks, Christ the power of God and the wisdom of God.* [25] *For the foolishness of God is wiser than human wisdom, and the weakness of God is stronger than human strength.*
>
> *(I Corinthians 1:22–25):*

These verses depict how the crucifixion of Christ neither fit into the Mosaic Law (Law of God) known by the people of Israel, nor the law of people under Roman rule. The new Law of Christ pierced the established institutions much like the nails pierced the hands and feet of Jesus on the cross. We introduce the three prominent cultural perspectives of the occupants of the ancient city of Jerusalem: (1) Judaism, (2) Roman, and (3) budding Christianity, in association with the legal principles that motivate them in the next sections. Then, we illuminate some similarities in the struggle with diversity and cultural clashes in modern civil rights in the United States, reminiscent of the cultural environment in which Christianity arose.

Jews and the Law of God

The Old Testament teaches us that the Jewish people of the nation of Israel were called by God to separate themselves from other nations, but were often at the mercy of foreign leadership due to spiritual failings. The predicament was clear in their Egyptian enslavement (Book of Exodus). Of course, Jewish captivity from Babylonian exile (586–331 BC) to Roman rule (63 BC–AD 35) in the preceding 600 hundred years prior to the arrival of Jesus may have clouded their expectations (Bock, 2002). "In this period, when Israel was occupied, Jews faced the central question of how they could best live in the midst of Gentiles and still be faithful to God," (Bock, 2002, p. 81). The problem remained a dominant issue when the Law of Christ was added into the mix that differed from the Mosaic Law of Judaism.

Mosaic law

The Old Testament Book of Leviticus embodies the Mosaic Law that focuses on maintaining a pure body by adhering to strict regulations about acceptable food and certain ceremonial practices such as the washing of hands, observance of the holy day of Sabbath, and abstaining from unclean items including those who did not follow these rules. But, the observance of dietary requirements was knocked off balance when Jesus proclaimed, "What goes into someone's mouth does not defile them, but what comes out of their mouth, that is what defiles them" (Matt 15–11) because what you speak comes from the heart. Mosaic Law also required males to be circumcised (Acts 7:8) and forbid mingling with uncircumcised men (Acts 11:3) or those from another nation (Acts 10:28).

Other spiritual conflicts with the Mosaic Law include the notion of serving all men under the Law of Christ (I Cor 9: 19–23) because Christ's church was built on service and how to treat others well. Instead, the objective of Mosaic Law was to maintain purity, while the Law of Christ commands participation and compassionate response to the needs of others. Love was not an emotion, but a physical response in the form of actions on behalf of a person in need (I John 3:4). Of course, that is not to say that the Jewish community does not love and show compassion. There remains a significant thread of charitable giving throughout the Jewish community with the highest level being assistance to other Jews followed by anonymous gifts to the poor (Chabad-Lubavitch Media Center, 2019; Mishneh Torah 10:7–14). These "Laws of Charity" are referred to as the Maimonides' Eight Levels of Charity.

The most distinguishing factor in the perspective of the Jews versus the Christian faith was that the Jews held fast to the embedded cultural law, including reliance on ritual temple worship and performance of works, while Christians held fast to faith. Despite the oral history of the Jews foreshadowing a Christ presented by Stephen (Acts 7) and Paul (Acts 13:13–41), the Jewish people could not relate to a sacrificial ruler when they desired a political messiah. The stoning of Stephen was not for his profession of Christ, but for blasphemy in the eyes of the Jews for claiming that there was no longer a need for temple worship or other rituals because of the sacrificial offering of Christ.

Roman oppression

Judaism was recognized as a legal religion in the Roman Empire indicating that Jews could follow their religious practices. Many Jews prospered causing economic tension with other citizens of Rome. Notable is the Jewish practice of freeing indebted slaves every 7th year in Jubilee (Leviticus 25: 1–55). But, the absence of modern bankruptcy laws in Rome meant that anyone, regardless of race, who could not honor their debts, was subject to slave labor ranging from physical toil to tutoring children in reading, science, and mathematics. Nevertheless, anyone in Jerusalem could be indebted due to war or inability to pay. The Romans exacted a Jewish tax that was so oppressive that the likelihood of indebtedness could occur. Further, the Romans forced the Jews to worship the Roman Emperor as a God and this was appalling to them. These diversions from the Mosaic Law certainly contribute to their desire to be free from Roman rule in Jerusalem in the early part of the 1st Century (Book of Acts, Romans).

Just as the Jews asked for a king/political leader in Saul in the early history of the people of Israel (Acts 13:20–21), the Jews following the Mosaic Law were still looking for an earthly, conquering king and counting on their own righteousness through good deeds and adherence to the law to make themselves acceptable to God (Rom 10: 3). The Jewish allegiance to the Mosaic Law was hampered by the national allegiance that was tied to citizenship in Rome. While many purchased their citizenship to garner economic, social, and political opportunities that noncitizens could not access, the gap in ideology and daily living was enormous.

Israel's rejection of the lawgiver

Paul explains to the Jews and to newly converted Christians in his various letters to the churches that they have missed the point by focusing on the law instead of the lawgiver. The purpose of the law is "through the law we become conscious of our sin" (Rom 3:20), while laws that were part of rituals, such as circumcision or ritual sacrifices, act only as a sign that may "profit" the people of Israel because of God's blessing upon them (Rom 3:1–2). But following the law does not save them from their sin. Only faith in Jesus, not any right of cultural passage or personal activities associated with worship, can lead them to righteousness (Rom 4:9–12).

Israel's desire to hold onto the methods fixed in Mosaic Law that should have permitted them to see their Messiah as the living temple and lasting sacrifice instead made them stumble and blind to the "lawmaker" (Isaiah 8:14, 28:16; Romans 9:32–33). Instead, they are unable to recognize Him. The rejection opens the door for Gentile believers and salvation to the ends of the earth (Acts 13:47; Isaiah 49:6).

Christians and the Law of Christ

Christian's relationships to the Old Testament law are complex and subject to ongoing debate since some parts (e.g., most of the 10 Commandments) are still applicable, while other laws have been abrogated in the New Testament (e.g., food laws). Still, Christians were freed from the Mosaic Law of Israel because they believed in Jesus Christ (Rom 7; Gal 5:1). Jesus was not a prophet, but one who obeyed and fulfilled the Old Testament laws, prophecy, and promises.

Nevertheless, their allegiance to Christ instead of a nation or tribe was of the utmost importance giving them citizenship in heaven (Phil 3:20) and a common theme of unity in which existing cultural divisions embedded in the law of the people, such as nationality or slavery (Rom 3:28; Col 3:11), did not impede their core faith founded on Christ. Instead, the law of faith in Christ (Rom 3:27–31; Rom 4:1–4) instills the desire to "Carry each other's burdens, and in this way you will fulfill the law of Christ" (Gal 6:2) as "faith expressing itself through love" (Gal 5:6). This adherence to Christ and freedom from the law through grace (Rom 6:14) was a clear path for all to receive the free gift of salvation and the Holy Spirit (Acts 10:44–46; Acts 11:16–18) unencumbered by cultural restrictions or taboos. For Christians, the fulfillment of the law is faith working through love (Gal 5:6). The law of love, faith, and liberty fuels the Law of Christ. The Apostles also extend that love by recognizing that rulers were given by God and sought to give both earthly rulers and God their just respect (Luke 20:25; Acts 23:4–5; Romans 13:1–7).

Law of love

Instead of emphasizing separation, unity was a central theme repeated in the Book of Acts and confirmation of this theme is found in other parts of the bible. The Book of Philippians, for example, demonstrates the theme of unity, joy, and prayer in the church at Philippi as well as the extension of gentleness and peace to one another

(Phil 4). Unity plays a role in fulfilling the law of Christ and that role was to recognize that they were not under the law but under grace (Rom 6:14); that all were able to receive the free gift of salvation and the Holy Spirit (Acts 10:44–46, Acts 11:16–18). Peter's revelation was clear. He said to them: "You are well aware that it is against our law for a Jew to associate with or visit a Gentile. But God has shown me that I should not call anyone impure or unclean," (Acts 10:28). The directive to love your neighbor is given such high esteem as to be alternatively named the royal law (James 2:8). The closing lines of Philippians tell us that Paul's companions and fellow believers were "of Caesar's household" (Phil 4:22), indicating the reach extends beyond existing cultural boundaries.

The theme of unity is coupled with lifestyle evangelism in which good works glorify the Father in heaven (Matt 5:16). These actions include choosing God over law (Acts 4:19; Acts 5:29; Acts 10:15; Acts 10:28; Acts 10:34–35), telling of Christ (Acts 1:8), and loving Him on His terms through Jesus Christ by loving others (I Cor 14). Loving also includes forgiveness as Egyptians who had enslaved the Jews were in the crowd who received salvation and the gift of the Holy Spirit (Acts 2:10, Acts 10:44–46, Acts 11:16–18). Salvation and individual autonomy were extended to women such as Lydia in Macedonia (Acts 16:11–15); Greeks and women in Thessalonica (Acts 17:4); and prominent women and men in Berea (Acts 17:10–12). Yet, local customs permitted that young women could be sold as slaves or forced to marry by their family at the age of 12.

Law of faith

The transition from the prevailing theme of good deeds and adherence to many other rules and regulation in Mosaic Law to faith in a Savior was difficult for many to comprehend. Paul often spoke that the purpose of the law was to show transgression not to save (Romans 2:12–16). However, the theme of faith as a gift to enable service is prevalent in New Testament scripture.

Ephesians 2:8–10:

> [8]*For it is by grace you have been saved, through faith—and this is not from yourselves, it is the gift of God—*[9]*not by works, so that no one can boast.* [10]*For we are God's handiwork, created in Christ Jesus to do good works, which God prepared in advance for us to do.*

Consequently, Gentiles were embraced into the fold of Christianity (Ephesians 2:19–20) along with the converted people of Israel under the law of faith (Romans 3:27–31). Just as Abraham was justified by faith before circumcision (Romans 4:1–4) and David was blessed by faith (Romans 4:5–8), faith will justify the Gentile and others who choose to believe. That faith is placed in Jesus Christ.

Romans 4: 13–15:

> [13]*It was not through the law that Abraham and his offspring received the promise that he would be heir of the world, but through the righteousness that comes by faith.* [14]*For if those who depend on the law are heirs, faith means nothing and the promise is worthless,* [15]*because the law brings wrath. And where there is no law there is no transgression.*

Law of liberty

Ironically, the law of liberty is linked to the law of service creating a juxtaposition of inherent freedom from the law (Romans 7) and the desire to serve others because of that freedom (I Corinthians 9:19–24). Paul uses his personal autonomy to offer himself as a slave in service to the believers who were comprised of converted Jews and Gentiles in the developing Christian church and to those who struggled to convert. In these noted passages, he also explains his own struggle with sin, but later demonstrates that building faith by loving others in their weakness will strengthen faith that will help them move away from judging others on the previous tenets of Judaism, such as food regulations, and moving toward peace and reconciliation which is the better hope (Romans 14).

Overall, Christians faced violence from the people of Israel who remained faithful to the Mosaic Law and were also a target for Roman leaders. Formerly considered just a faction within Judaism, Christians faced growing opposition from Roman leaders as they split from the religion of Judaism that was recognized by Rome. Major spiritual beliefs also continued to divide the different systems of belief. One of the most pervasive items that Jews lobbed at the Christians was their belief that, like Jesus, the body will be resurrected (I Corinthians 15), while the Jews remained steadfast in their belief that the soul rises. Another cultural trap for converted Jews and Gentiles was that philosophy from the Greeks, legalism from the Jews, and carnality (Colossians 2:8) entered into the Christian church. Doing good deeds without Christ also persisted in Colossae where Paul praised their actions, but explained that without Christ their deeds were empty.

Other internal and external cultural elements were evident in the church which is addressed with great urgency. For example, there was zero tolerance for deceit in the handling of tithe offerings. The case of Ananias and Sapphira who lied about the amount of their contribution to the church in order to boast led to their unseemly and supernatural death (Acts 5:1–11). Also, when Simon (a sorcerer who professes faith in Jesus) attempts to buy the gift of the Holy Spirit in Samaria, the Apostles Peter and John cursed him and told him to repent this great wickedness (Acts 8:9–25). There was much to learn while the oral tradition of spreading the Word of Christ gave way to written documents. Many will argue that we still struggle with these basic lessons.

US citizen modern civil rights

The notion of personal liberties and freedom is a derivative of national allegiance and embodied in the definition of civil rights. These concepts are entrenched in sovereign national law in the United States. While state laws have an element of sovereignty or authority, they cannot overturn US Constitutional Law, Amendments, and Acts of Congress which represent the legal provision for individual personal liberties and minority groups, such as the disabled, who may otherwise be subject to discrimination. Yet, gender inequality, skin color, taxation without representation, unequal distribution of wealth, legalism, and religious persecution persist today as a source of division despite legislative directives.

The foremost freedoms include speech and religion (1st Amendment) and equal protection of laws to citizens and residents under the law afforded by the provision of "due process" in the 14th Amendment protecting life, personal liberty, and property. Though the Emancipation Proclamation declared those of color to be free in 1862 and the 13th Amendment eliminated slavery and involuntary servitude in 1865, another century would pass before black people in America would see real changes in terms of legal, social, and economic opportunities. Notable is that American and other global slavery was not dependent on unpaid debt as in the 1st Century, but more likely the result of an overt, oppressive opinion of those of color that pushed them into a lesser status. The landmark case of *Brown v. Board of education, 347 U.S. 483, 74 S. Ct. 686, 98 L. Ed. 873* ruled in favor of Brown in 1954. This Supreme Court ruling marked a significant advancement in aligning the civil liberties of Black Americans. The ruling began by rectifying the violation of the 14th Amendment by ordering consolidation of separate but equal facilities that had segregated this population in public facilities, education, and other areas. This change was substantial in their establishment of self-worth and opportunity to improve their quality of life. Of course, the ruling was not without controversy or opposition as court-ordered desegregation in public school districts also led to busing students out of their district in order to achieve mandated racial proportions. Seventy years later, the policy remains an area of debate to reach racial balance and is still an open area of debate.

Nevertheless, Table 9.1 demonstrates the long and dusty road to the development of US sovereign legislation toward the elimination of bias based on demographic characteristics or beliefs. We surmise that liberty, equality, and unity are just as important today and represent a comparable struggle revealed in the early development of the Christian church and one that continues today. The US Bill of Rights is the most prominent example of individual rights and freedoms declared to citizens (Bill of Rights, n.d.). Meanwhile legislation continues to undergo scrutiny and debate. A prime example is whether or not the 14th Amendment infringes upon other classes of people while favoring another. This debate is one of many in a nation of diverse individuals striving to achieve the American dream under various social, economic, and individual classifications.

In the case study to follow, we view some of the difficulties encountered by the developing 1st Century Christian church and the methods they used to overcome prejudice and discrimination in an era fraught with religious, political, and social upheaval.

The case study of the Book of Acts and the formation of the early church

The Book of Acts is the account of responsive action of the apostles to Jesus' commission to spread the word of God. They communicate this new information across many races and ethnicities including women, foreigners, military, wealthy, powerful, poor, and commoners to form the foundation of the 1st Century church. The work of the apostles extends upon the notion of Jesus as a bridge between man and God. In a similar

Table 9.1 Important US sovereign legislation emphasizing the path to civil rights and other obstacles of discrimination.

Year	Federal action/acts	Protection
1866	Civil Rights Act (42 United States Code Annotated (U.S. C.A.) § 1982)	Pre-14th Amendment to secure federal rights for all who legally live within US jurisdiction; property rights; provision to cease bias in racial hiring but not for racial harassment
1871	1871 Ku Klux Klan Act (17 Stat. 13)	Violation of civil rights became a criminal offense; citizens could take legal action
1885	Jim Crow Laws (late 19th Century)	Symbolically retracts rights of black people in the South by segregating public facilities, schools, voting, and travel under "separate but equal" clause
1920	Women's Suffrage	Voting rights in the 19th Amendment to the US Constitution
1932	Prisoner Rights	Grants prisoner's civil rights protection while incarcerated
1935	National Labor Relations Act (29 U.S.C.A. §151 et seq.)	Provides protection for labor union members; amendment extends to other forms of discrimination (e.g., race, sex)
1957	Civil Rights Act (42 U.S.C.A. § 1975)	Establishes Commission on Civil Rights oversight on race relations for Congressional and Presidential consideration; paves way for formal civil rights
1960	Civil Rights Act (42 U.S.C.A. § 1971)	Removes obstacles to "separate but equal" voter registration; denials subject to litigation
1963	Equal Pay Act (29 U.S.C.A. § 206)	Wage parity; Men/Women wages same for similar work
1964	Civil Rights Act (42 U.S.C.A. §§ 2000a et seq.)	Comprehensive civil rights parity without limitations to access public facilities, education; introduces Title VII prohibiting federal employment discrimination; denials subject to litigation; implementation of federal Equal Employment Opportunity Commission (EEOC)
1965	Affirmative Action (AA) Equal Opportunity 11246 (*several prior AA plans*)	Corrective action for discriminatory hiring patterns (e.g., race, color, sex, creed, and age); enforcement of equal opportunity employment hiring classifications through federal programs
1968	Civil Rights Act (25 U.S.C.A. § 1301 et seq.)	Prohibits discrimination on lending, leasing/purchase of property or housing; denials subject to litigation
1991	Civil Rights Act (Pub. L. No. 102–166, 105 Stat. 1071 [portions U.S.C.A. 42, 29, 2])	Extends Title VII to protect congressional employees; Permitted women and disabled to recover damages under Title VII per US Supreme Court reversals of decisions
1992	Americans with Disabilities Act (ADA) (Pub. L. No. 101–336, 104 Stat. 327) [portions U.S.C.A. 42, 29, 47]	Civil rights for citizens with physical, mental, and other impairments that limit quality of life; provides equal opportunities for employment and wheelchair access to public facilities, transportation, and telecommunications

Modified from Collins Dictionary of Law (2006); West's Encyclopedia of American Law (2008).

way, the apostles must bridge the gap in ethnic, racial, and cultural prejudices in order to be effective in their ministry. The success of the Apostles is evident throughout the New Testament scripture including the account of Philip teaching the representative of Ethiopian royalty on the road to Gaza (Act 8:26–40), Peter healing Aeneas in Lydda and the restoration of Dorcas, a charitable woman of great reputation for service in Joppa (Acts 9:36–43), Peter teaching Cornelius, a centurion in the Italian regiment, and his family in Caesarea (Acts 10:17–43), and the shock and intent to harm Paul preaching to the people of Damascus (Acts 9:20–29) after his reported miraculous conversion from persecutor of the members of the Christian church (Acts 8:1–3) to believer and Apostle. Curiously, Paul's early persecution of the Christian church was instrumental in scattering Christians across many nations (Fig. 9.3) which enabled the spread of the very Word of God he had attempted to thwart.

The Book of Acts reads like an institutional mission statement to educate the uninformed, value humanity without discrimination, and then to meet them at their point of need by introducing a new system of beliefs into the established culture. Of course, the social normalization emerging from this institution is often rejected because this new paradigm foils the status quo of both Roman political leadership and Judaism. Thus, the formation of the Christian church by the Apostles set off a series of events triggering misunderstandings, prejudice, envy, and escalated violence as they led the disruption of political, religious, and cultural order of the time. Concurrently, this span of time for the Christian church (6 BC–AD 70) also represents a period of sacrifice, joy, gratitude, and collaboration. Within this mindset is a dedication to help others. The overarching result of their movement would both influence and outlast empires, business, social, and spiritual conventions. This information is contained in the writings of Luke in what has become known as his two-part series encompassing the Book of Luke and then Acts.

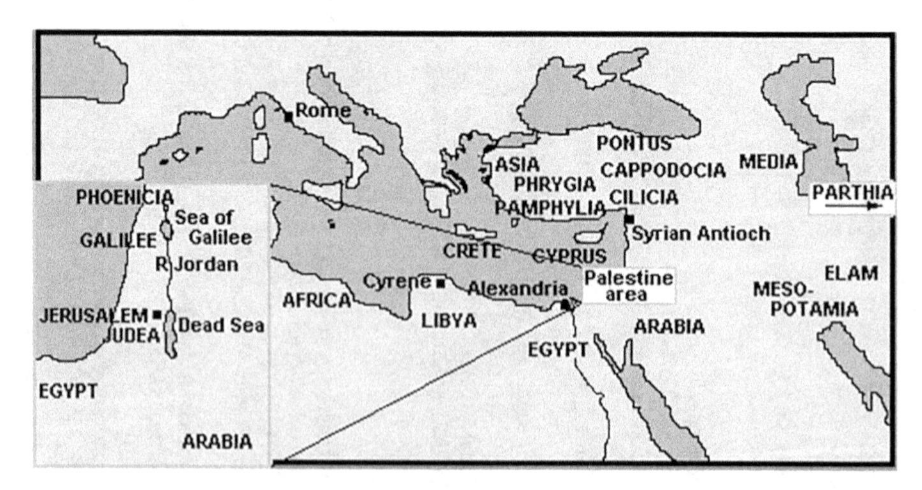

Fig. 9.3 The diaspora of the Jewish population contributed to the spread of the Gospel and created Jewish centers outside of Israel.
Reproduced by permission from Gordon Smith, available at https://ccel.org/bible/phillips/CN203ACTSJewishPeriod.htm.

Luke's contribution to future churches

Luke, a prominent physician and historian, set out to separate fact from fiction regarding the Apostle's journeys utilizing a qualitative method of research through interviewing that establishes multiple eyewitness accounts (Acts 1:3) accurately transcribed through the centuries. Luke's investigation of existing written documents could be easily placed into the category of a modern literature review. His short-term objective provides unbiased educational material regarding the birth, life, death, and resurrection of Jesus Christ for his friend Theophilus. But Luke's long-term objectives serve as a reference to the development of similar outcomes, such as developing inclusive institutions, in other cities much like the application of urban case studies today.

Theophilus, who is identified as a likely Roman official in the Gospel of Luke, is a Christian convert who later loses his position (Acts 1:1–4). Scholars generally concur that the greeting in Luke to Theophilus as "most excellent" that is later reduced to "O Theophilus" (Luke 1: 1–4) is indicative of this demotion from political service. Loss of position was one example of religious persecution for Christian converts when transitioning from their connections to Judaism and/or the Roman Empire. Imprisonment, brutal beatings, and stoning to death are also part of the experience of Christian believers. The most prominent is the stoning of Stephen (Acts 7:54–60) who was accused of blasphemy when he indicated that Christ was the end of the Jewish priesthood and rituals performed at the Jewish temple because of His perfect sacrifice and resurrection.

Still, Luke reveals how the Apostles recognize common history to introduce new knowledge and to bridge a gap between the old and the new. This methodology is utilized in classic Institutional Theory such as Paul (Acts 13:13–41) and Stephen (Acts 7) linking the history of the people of Israel to bridge the seemingly completely different perspectives.

The unlikely formation of the church

The formation of any institution that relies on communication between and among races, ethnicities, and nations seems unlikely to succeed, especially in an era in which Jews and others did not mingle among other nations (Acts 10:27–28). Further, the first of the people of Israel to be converted to Christianity faced charges of heresy. As others outside of Judaism who had previously been held at a distance by Jewish customs were welcomed into Christianity, this led to some internal strife, especially in terms of traditional food regulations and other aspects of the Mosaic Law. Nevertheless, the actions of those who profess their belief in Jesus result in a profound outcome—the blessings from Jesus' Apostles and disciples to men from many nations including Arabs, Asians, Cretans, Egyptians, Elamites, Parthians, Medes, and others (Acts 2:7–11). Overcoming national differences and opening lines of communication were just the beginning of actions that define the lack of prejudice that the Apostles and disciples demonstrate throughout the document.

Church growth was eminent as the news of Jesus spread and believers began to conduct themselves in a certain manner (Acts 2:38–45). Table 9.2 lists some behaviors that stand out against ancient conventions of separation between different people and

Table 9.2 Normalization of internal (Act 2:38–45) and external behavior in the formation of common beliefs in the 1st Century Christian church.

Internal	Prayer Fellowship Everyone has access to God Belief in a resurrected Christ Communion/Gathering for Meals
External (both)	Forgiveness (Act 7:54) Shared wealth (Act 4:32–37) Community service (Acts 6:1–7; Acts 11:27–30)
	Establishing common doctrine through reading the Word of God (Acts 5:12)
	Ethics, intolerance for deceit about giving (Acts 5:1–11) or sorcery (Acts 8:9–25)

nations compelling others to join the church. The Apostle Peter was privy to a great revelation and important aspect of the ministry—"I now realize how true it is that God does not show favoritism but accepts from every nation the one who fears him and does what is right," (Acts 10:34–35). Joy was the overwhelming response of those who became a part of the early church. We see a consistency in the actions of members with the church and the extension of welcome outside the church.

On the other hand, Acts demonstrates that there was also a negative reaction to the changes in cultural traditions and social order. In fact, so great was the opposition of members of the Sanhedrin (rulers, elders, scribes of Israel in Jerusalem) in all but a few cases, Peter and John were arrested (Acts 4:1–4) and issued a series of warnings not to speak of Jesus (Acts 4:5–12; Acts 4:13–22; Acts 5:17–42). Eventually, others conspired to beat the Apostles (Acts 5:33–42), while actual violence escalated to the extent that Stephen was stoned to death on the false accusation of blasphemy (Acts 6:11). There are several reports, including Acts 5:17, revealing that the Apostles' ability to freely speak of Jesus was being challenged by the elders of the Jerusalem church. Other actions of opposing leaders, such as King Herod's violence against the people of the church, led to the death of James, the brother of John (Acts 12:2), and imprisonment of Peter (Acts 12:3). But the Apostles and their disciples carry on even as Herod and the Jews stood against the church for political and religious reasons, respectively. Perhaps their persistence has helped to establish critical freedoms of religion and speech in modern history.

The Jerusalem decree

Amidst the backdrop of clashing political and religious cultural perspectives is a glimmer of light to come. In Acts 15: 22–29, the Jerusalem Council of AD 50 was facing a dilemma when Gentile Christians were being held accountable for upholding Jewish religious customs. These traditions include submitting to circumcision, ceremonial killing of animals, and others. Sproul (2019) and other biblical scholars believe this event to be a lasting example of the appropriate decision-making process in a church based on the ethical consideration of others and not necessarily a formal guideline. Moreover, this council meeting represents a significant demarcation between the customary practices of Judaism and the formation of lasting tenants of the Christian church.

Sproul (2019):

> *What we learn here in this historic circumstance is a principle of consideration that we have to apply everywhere the Gospels goes, to the mission field, and so on, where the church meets up against cultures that are radically different from our own.*

While the church recognized the freedom of the Gentile believers to partake of items that would offend Jewish converts, their solution was to take into consideration the impact of your actions on others in determining your freedom in these matters. Thus, they recommended the following:

(NIV, 2011):

> *You are to abstain from food sacrificed to idols, from blood, from the meat of strangled animals and from sexual immorality. You will do well to avoid these things.*
> *(Acts 15: 29)*

The Law of Moses was fading, but the liberty of Christians on behalf of those who were making the delicate transition of Jews from centuries of dutiful ritual to a new way of service to others was slowly emerging. Making ethical choices for the good of the collective body of Christ and for those who were entering that body was an important element of the founding principles. Freedom to choose to do the right thing to help others through their journey replaced legalistic practices upon which the Christian church of the 1st Century was born.

Discussion and recommendations

Opposing positions to God or law still require an individual choice, but today that right to religious freedom without persecution, regardless of choice, is reflected in the US Constitutional Law and the Bill of Rights. However, these rights are not internationally guaranteed—a problem with too great a scope for this paper.

Nevertheless, the Book of Acts provides a purpose, mission goal, tools of policy (e.g., equity, education, public service announcements), and collaboration (e.g., meetings with public officials, Council officials) to unify the community and introduce helpful guidelines. Thus, the Book of Acts serves to demonstrate the necessary response through public actions that purposefully object to and overcome religious persecution, gender inequality, oppressive business influence, unequal distribution of wealth, legalism, and false teachers and practices such as idol worship. This contribution from Luke also demonstrates how (1) failing to embrace cultural change or at least, introduce tolerance, can impede progress, and (2) embracing individuals without regard to differences can result in positive institutional change and individual expression of a joyous and purposeful life. A cultural revolution and the foundation of the

Christian church emerge from the common beliefs found in community actions (e.g., fellowship, service, and shared resources). One man can change everything (as long as we are confident of the credibility of that individual).

Previously, eyewitness accounts and urban case studies that form the book of the Holy Bible in the Abrahamic religious scripture have not often been viewed as literature, despite corroboration of the historical accuracy of other scholarly documents, artifacts, and scientific discovery. But, the Book of Acts draws attention in an historical case in urban institutional development as an example of multicultural exchange of ideas without regard to race, ethnicity, gender, wealth, power, national affiliation, or disability. Further, corroborating evidence emerges of a peaceful framework for community activism that even Freedom of Speech and Religious activists citing the US Constitution and the Bill of Rights could not find offense.

Paul explains in Romans 4:13–15 that the purpose of the Old Testament law was to show transgression, not to save. In the same way, legislated law informs of an infraction but does not necessarily prevent the violation. The Book of Acts teaches us that there are other methods to problem resolution apart from escalated violence including self-accountability and a conscience for the well-being of others. The next time someone is drinking coffee at your local shop, ask if you can join them. If they are afraid, that is ok. Be patient. A lot of people are afraid. But the next time you see them they may say hello. Mow your neighbor's overgrown lawn—maybe you did not know that their son was sick. Volunteer at the animal shelter or hold babies at your local hospital who were born addicted to crack. Overcome the public cultural taboo of intermingling and live an orderly life. Be a part of something bigger than yourself—maybe even the church. **(NIV, 2011):**

> [9] *Now about your love for one another we do not need to write to you, for you yourselves have been taught by God to love each other.* [10] *And in fact, you do love all of God's family throughout Macedonia. Yet we urge you, brothers and sisters, to do so more and more,* [11] *and to make it your ambition to lead a quiet life: You should mind your own business and work with your hands, just as we told you,* [12] *so that your daily life may win the respect of outsiders and so that you will not be dependent on anybody.*
>
> *(1 Thessalonians 4:9–12)*

There is truth in the old adage that actions speak louder than words. It is a simple lesson in life but also the most difficult.

Conclusion

Luke, a Jewish physician from Samaria, relays his intense examination and investigation of Jesus Christ in the Book of Acts that is the general topic of this chapter. His methodologies include various historical resources of the time period and eyewitness accounts comparable to the acceptable methods of modern qualitative research giving plausibility to his account of the emergence of the 1st Century Christian church. Deciphering the prevailing legal and religious perspectives during this time (e.g., Jews and the Mosaic Law, Christians and the Law of Christ, and Roman rule) and comparing them to sovereign law in the modern United States set the stage for the urban case study of the Book of Acts and the early church. We find consistent actions, such as community service and inclusion without discrimination in small group meetings comprised of people from many cultures, and examples of methods that permit this institution to stand when greater political institutions, such as the Roman Empire, have long ceased to exist. The quest for equality is certainly an undeniable Christian motif and one in which survives a part of the modern quest for civil rights, liberty, and freedom. We recognize the modern application of community service present in civic activism. We recommend the application of the theme of unity coupled with several actions attributed to the Apostles in the 1st Century to support a simple, modern framework to achieve personal peace and the potential for significant impact in modern institutions. Finally, we find that researchers cannot eliminate the Holy Bible as a reliable source of cultural and contextual information simply because you do not profess that faith in the same way you cannot eliminate medicine if you are not a doctor. They have inherent value because of the underlying foundation.

References

American Atheists (n.d.). Retrieved April 6, 2019 from https://www.atheists.org/.

Barnett, P. (1986). *Is the New Testament reliable? A look at the historical evidence.* Downers Grove, IL: InterVarsity Press.

Barnett, P. (2005). *The birth of Christianity: The first twenty years.* Grand Rapids, MI: W. B. Eerdmans Publishing Company.

Biblical Archeology Society. (2019). *Bible history daily.* Retrieved March 23, 2019 from https://www.biblicalarchaeology.org/category/daily/biblical-topics/new-testament/.

Bill of Rights. (n.d.). *Collins dictionary of law.* (2006). Retrieved April 8, 2019 from https://legal-dictionary.thefreedictionary.com/Bill+of+Rights.

Blomberg, C. L. (2007). *The historical reliability of the gospels* (2nd ed.). Downers Grove, IL: InterVarsity Press.

Blomberg, C. L. (2016). In R. B. Stewart (Ed.), *The historical reliability of the New Testament: Countering the challenges to Evangelical Christian beliefs.* Nashville, TN: B&H Academic.

Bock, D. L. (2002). *Studying the historical Jesus: A guide to sources and methods.* Grand Rapids, MI: Baker Academic.

Chabad-Lubavitch Media Center. (2019). *Maimonides' eight levels of charity.* Retrieved May 21, 2019 from https://www.chabad.org/library/article_cdo/aid/45907/jewish/Eight-Levels-of-Charity.htm.

Collins Dictionary of Law. (2006). Retrieved April 8, 2019 from https://legal-dictionary.thefreedictionary.com/Civil+Rights; Civil Rights.

DeMar, G. (2014, December 7). *American atheists hide behind the fairy tale of evolution.* Retrieved April 6, 2019 from https://americanvision.org/11504/american-atheists-hide-behind-fairy-tale-evolution/.

Führding, S. (Ed.), (2017). *Method and theory in the study of religion: Working papers from Hannover.* https://doi.org/10.1163/9789004347878.

Habermas, G. R., & Flew, A. G. N. (2005). In J. Ankerberg (Ed.), *Resurrected? An atheist and theist dialogue.* Lanham, MD: Rowman & Littlefield Publishers, Inc.

Hordern, J. (2016). Religion and culture. *Medicine, 44*(10), 589–592.

Oman, D. (Ed.), (2018). *Why religion and spirituality matter for public health: Evidence, implications, and resources*. New York, NY: Springer-Verlag.
Scott, W. R. (1987). The adolescence of institutional theory. *Administrative Science Quarterly*, *32*(4), 493–511. https://doi.org/10.2307/2392880.
Selznick, P. (1948). Foundations of the theory of organizations. *American Sociological Review*, *13*(1), 25–35. https://doi.org/10.2307/2086752.
Society (2017, March 5). *The new atheist delusion: Dawkins and movement 'Lost the Heart of the Cause'*. Retrieved April 6, 2016 from https://sputniknews.com/society/20170503105324 7150-new-atheism-movement-alienation/.
Sproul, R.C. (Host). (2019). Renewing Your Mind [Radio program]. In Ligonier Ministries (Producer) The Jerusalem Decree. Sanford, FL.
Stack, T., Goldenberg, N., & Fitzgerald, T. (2015). *Religion as a category of governance and sovereignty. Series—Supplements to methods & theory in the study of religion: (Vol. 8)*. Leiden: Brill.
Strobel, L. (2016). *The case for Christ: Solving the biggest mystery of all time*. Grand Rapids, MI: Zondervan.
Udo-Akang, D., et al. (2012). Theoretical constructs, concepts, and applications. *American International Journal of Contemporary Research*, 2(9), 89–97.
Van Voost, R. E. (2000). *Jesus outside the New Testament: An introduction to the ancient evidence*. Grand Rapids, MI: Wm. B. Eerdmans Publishing Co.
Warfield, B. B. (1907). *Introduction to textual criticism of the New Testament*: (pp. 12–13). London: Hodder & Stoughton.
West's Encyclopedia of American Law (2nd ed.). (2008). Retrieved April 8, 2019 from https://legal-dictionary.thefreedictionary.com/Civil+Rights.
Williams, P. S. (2010). *Archaeology and the historical reliability of the New Testament. [Revised July 2012]*. Retrieved April 6, 2019 from https://www.bethinking.org/is-the-bible-reliable/archaeology-and-the-historical-reliability-of-the-new-testament.

Supplementary reading recommendations

American Jewish University. (2018). *Torah resource center*. Retrieved May 21, 2019 from https://www.aju.edu/ziegler-school-rabbinic-studies/torah-resource-center.
Barnett, P. (1990). *Jesus & the rise of early Christianity: A history of New Testament times*. Downers Grove, IL: InterVarsity Press.
Barrett, J. (2016, June 17). *Women: Equality in the Book of Acts?* Retrieved May 21, 2019 from https://trinitysem.edu/women-equality-book-acts/.
Bible Study Tool, Dictionaries, Retrieved March 23, 2019 from https://www.biblestudytools.com/dictionaries, 2019, (i) Baker's Evangelical Dictionary of Biblical Theology (1996); (ii) Condensed Biblical Encyclopedia (1896); (iii) ISBE International Standard Bible Encyclopedia (1915); (iv) Nave's Topical Bible (1896); (v) Smith's Bible Dictionary (1901).
Cain, J. (2018, October 1). *The Lord of the sidelined*. Retrieved May 21, 2019 from https://www.intouch.org/read/magazine/faith-works/the-lord-of-the-sidelined.
Edwards, G. (1974). Preaching from the Book of Acts. *Southwestern Journal of Theology*, *17*, Retrieved May 21, 2019 from http://preachingsource.com/journal/preaching-from-the-book-of-acts/.

Halloran, K. (2014). *A list of sermons in the Book of Acts*. https://www.leadershipresources.org/blog/list-of-sermons-in-acts/.
Holman, S. R. (2014). *The project on lived theology*. Retrieved May 23, 2019 from http://www.livedtheology.org/susan-r-holman/.
Stanley, C. (Host) (2018, November 3). In Touch Ministries [Television program]. The Attractive Quality of Kindness, Expressing Godly Character, Pt. 5. Atlanta, GA. Retrieved May 21, 2019 from https://www.intouch.org/watch/expressing-godly-character/the-attractive-quality-of-kindness.
Woodley, M., & Wright, M. (2018, June 22). *Different together: An honest conversation about race and the church*. Retrieved May 21, 2019 from https://www.intouch.org/read/magazine/features/different-together.

Definitions

Abrahamic Religions generally, a genealogical foundation that harkens back to Abraham (Genesis 12–18; Acts 3:13; Acts 3:24–25).
Abrahamic Seed Two covenants; distinguishes Isaac born from the promise of a child to Abraham and Sarah from Ismael born of Abram and Hagar who is not the seed of promise (Rom 9:6–9; Gen 16–18; Gal 4:21–31).
Apostles Peter, John, James and Andrew; Philip and Thomas, Bartholomew and Matthew; James son of Alphaeus and Simon the Zealot, and Judas son of James (Acts 1:13) and later joined by Matthias (Acts 1:26) after Judas betrayed Jesus and committed suicide.
Circumcision a sacrificial cutting of the foreskin of males foreshadowing the shedding of blood for the pardon of sin; privileges of circumcision were extended to foreigners who submit to the sacramental operation in order to partake in religious privileges (Exodus 12:48, Ezekiel 44:9).
Civil Rights nonpolitical or personal rights/liberties of a citizen or resident of a particular country.
Covenant generally an agreement between God and man or among men.
Covenant of Circumcision an agreement to honor God by the people of Israel to circumcise males on the 8th day; a visible mark of inclusion.
Discrimination when an individual or group is subject to different treatment because of race color, religion, sex, age, handicap, or national origin; an inequality of opportunity for basic needs such as housing, education, and employment.
Epistles Letters written by Apostles through divine inspiration for God's people in cities across the world at that particular point in time to address misunderstandings and to direct proper behavior.
Gentile often references as referenced as Greeks
Faith receiving the gift of salvation by belief in Jesus Christ
Grassroots (grass roots) associated with a movement or social change by ordinary people, not by existing leadership or the elite members of society.
Gospel(s) The first four Books of the New Testament (Matthew, Mark, Luke, and John) presenting the life and ministry of Jesus as a way to salvation for both Jew and Gentile.
New Covenant linked to the willingness of Jesus Christ to be a substitute for the sins of all people unto salvation.
Pharisees minority members of the Sanhedrin that broke from Mosaic written law; created an oral history of divinely inspired elder traditions that extended written Mosaic Law

distinguishing them from the Sadducees; Jesus and His followers were in agreement with some of these concepts.

Sadducees majority ruling members of the Sanhedrin (Acts 5:17); distinguished by their unbelief in the resurrection of Jesus Christ (Act 4:2) and the Holy Spirt (Acts 23:8); belief in the literal interpretation of Old Testament written scripture containing pertinent law; resisted new concepts brought forth by Jesus and His followers because they would diminish their position in society.

Salvation scripture defines multiple definitions but prominent is to save, delivered, or set free from bondage

Sanhedrin Judicial system from Moses that represents provincial government; primarily comprised of Sadducees and also Pharisees.

Scripture term inclusive of religious writings such as the Judeo-Christian text found in the Old and New Testament

Testimony twofold in nature, (1) representing a statement before legal representatives of a court system, (2) statement before others to demonstrate your beliefs, a declaration of your personal faith in God

A woman's (unpaid) work: Global perspectives on gender, healthcare, and caregiving

Yara Asi[a], *Cynthia Williams*[b]

[a]Department of Health Management and Informatics, University of Central Florida, Orlando, FL, United States, [b]Brooks College of Health, University of North Florida, Jacksonville, FL, United States

Abstract

Women make up most of the global unpaid labor force in healthcare, performing work worth trillions of dollars annually. However, this work often limits their personal and professional opportunities and perpetuates gender disparities. The outsized role of women caregivers is attributed to historical, cultural, and social perspectives on gender and caregiving that perpetuate the gender inequalities in unpaid care work. In this chapter, we analyze women in unpaid work considering two regions: The United States and the Middle East and North Africa (MENA). We compare these regions along social, political, and historical contexts, including use of two vignettes. As a result of the comparisons, we make several key recommendations, framed by Bronfenbrenner's ecological model, which include: (1) generating awareness for US women caregivers to receive compensation, and (2) developing the job industry to embrace women caregivers in MENA countries.

Keywords: Gender disparity, Informal caregiving, Healthcare, United States, Middle East and North Africa (MENA).

Chapter outline

Three Facets of Public Health and Paths to Improvements. https://doi.org/10.1016/B978-0-12-819008-1.00010-9

Valuing women in global health services

It's true that women deliver health, but men are paid for it.
Werner Obermeyer, Executive Director, WHO, March 20, 2019

Women make a significant contribution to health services, comprising 70% of the global healthcare workforce (World Health Organization, 2019, p. 5). In addition to an outsized role in paid work, women are also responsible for more than double the unpaid caregiving work of men. The McKinsey Global Institute (2015, p. 2) estimates that unpaid labor performed by women would be worth "as much as $10 trillion of output per year, roughly equivalent to 13 percent" of global gross domestic product (GDP). Every day, women are providing physical and mental health care services for children and adults. The expectation that caregiving is inherently "women's work" seems to justify both the lack of compensation and the belief that women should bear this duty with minimal physical or mental burden (Morgan, Williams, Trussardi, & Gott, 2016). These expectations persist, regardless of a nation's GDP or other economic indicators, demonstrating that there are social factors framed by culture which allow gender inequalities to endure. While the global economic gender gap narrowed in 2018, the income gap remains at 51%, and only 34% of leaders across industries worldwide are women (World Economic Forum, 2018a). The Global Gender Gap Report from 2018 found that the burden of unpaid work on women was one of the reasons for these gaps (World Economic Forum, 2018b). Although gender gaps exist across industries, they are particularly problematic in healthcare because adequate provision of quality care is a key to a country's stability and development.

A sustainable and diverse workforce is critical to achieve any significant goals in the health sector, and health is a critical part of achieving sustainable development around the world. Sustainable Development Goal (SDG) 3, part of the United Nations global development agenda for 2015–30, recognizes the value of and seeks to promote health as a foundation for sustainable progress among societies (Acharya, Lin, &

Dhingra, 2018). The health status of a nation contributes to the degree of social parity, economic growth, and civic engagement in that nation. The attainment of sustainable health is quite complex due to the interplay of social, environmental, and behavioral factors that must be considered at various levels of government and social structures (Lomazzi, 2016; Smith & Henderson-Andrade, 2006). A vibrant workforce is needed to ensure progress of laudable, yet multifaceted goals. While workforce challenges are primarily considered at the national and global levels in relation to economic status, the national impact of an unbalanced workforce affects families and individuals regardless of stage of development. However, developing countries are most ill-equipped to deal with such challenges and require coordinated response from the global and local communities to mitigate the unfavorable outcomes. Equal access and opportunity must be fully realized among all people to achieve sustainable progress.

The SDGs established goals to build and sustain healthy communities for families, communities and countries that are best realized through gender parity. As a result, gender-related targets are present throughout the SDGs as well as with SDG 5, which calls for gender equality and empowerment of women and girls across all sectors (UN WOMEN, 2019). Gender is a key stratifying feature in social systems and a determinant of health and well-being in many countries (Acharya et al., 2018), and like most other social sectors, the healthcare workforce (paid and unpaid) is influenced by gender and social norms. Unfortunately, many of these norms prove to be disadvantageous for women. Women are often expected to ensure the health and security of others at the expense of their own health and security. These gender biases persist not just in the home, but are translated through health care policy and planning for women as both patients and providers (Sen & Ostlin, 2009).

The role of unpaid caregiving requires careful consideration if SDGs are to be achieved. We are defining unpaid care as all "unpaid services provided within a household for its members, including care of persons, housework and voluntary community work. These activities are considered work, because theoretically one could pay a third person to perform them," (Ferrant, Pesando, & Nowacka, 2014, p. 1). Unpaid healthcare work is largely determined by gender, particularly in developing countries. Women provide significantly more hours of care, perform more tasks when in a caregiving role, and deliver more personal forms of care (Pinquart & Sorenson, 2006). In many cases, cultural factors such as familism, filial piety, and family cohesion are more important than an individual's aspirations. Daughters, either married or unmarried, are more likely to assume caregiving roles than sons and with other factors being equal, they are *expected* to provide more care than sons (Grigoryeva, 2017). One of the few familial factors that predicts lower caregiver burden for women is the presence of sisters, with unmarried sisters providing more care than married sisters (Wolf, Freedman, & Soldo, 1997). Women absorb the burden of care when it is not provided by other public or private entities. Most states could not afford to shoulder the responsibilities of providing constant care to all disabled or sick children and adults. As a result, unpaid care work is a fundamental component of any state's economic and healthcare systems.

While there are a host of social issues that lead to gender disparities, perhaps one of the most entrenched is that of cultural attitudes of caregiving work, some of which are unconscious and implicit and others are overt and purposefully oppressive. In this chapter, we discuss the disproportionate obligation that falls on women and the challenges posed by the gender disparities of unpaid caregiving. We reflect on the economic, political, and personal factors associated with caregivers to provide a richer understanding of women's unpaid caregiving practices. We highlight two regions that compare and contrast cultures, economic development, and the status of women: North America, the region with the lowest levels of unpaid caregiving by women (with a focus on the United States), and the Middle East and North Africa (MENA) region, which reports the highest global levels. Lastly, we provide recommendations that may mitigate gender disparity in unpaid caregiving and provide a framework where women are socially and economically empowered.

Challenges of overreliance on unpaid women's caregiving

The decision to care for family members or loved ones with complex illnesses or disabilities is not easy. Caregivers perform tasks such as activities of daily living (e.g., feeding, bathing, dressing), instrumental activities of daily living (e.g., housekeeping, grocery shopping), and medical care (i.e., monitoring medication). While these are beneficial to the care recipient, the caregiver may suffer from poor mental and physical health. Due to the intensity of time required to care for another person, there may be significant loss of wages, decreased employability, and social deprivations. Despite the significant contribution of men and women to their respective communities, the lack of compensation and recognition of women has increased their pay disparity and vulnerability.

The increase in structural familial changes, such as divorce and remarriage, has left gaps in elder care that women are expected to fill (Silverstein & Giarrusso, 2010). Globally, health issues that increase mortality among young and middle-aged adults, such as the AIDS epidemic, have left child caregiving responsibilities to grandmothers, a trend seen widely in sub-Saharan Africa (Oppong, 2006). Additionally, due to increased life expectancy and decreased birth rates, the global population is aging. This leads to a larger population of older adults who require nonmedical and medical care. While multigenerational households were once common around the world, today many young adults leave their hometowns for jobs, schooling, or relationships, often postponing marriage and children (Pesando, Castro, Andriano, Behrman, & Billari, 2018). When women are unmarried or don't have children, they are usually the expected providers of needed healthcare services. Placing them in these positions, sometimes as early as adolescence, further stifles their ability to achieve their personal aspirations and they may be pigeonholed in caregiving roles for their entire lives. Unfortunately, these gender inequities are acceptable to many families who see these differences as "natural" (Sen, 1996). In this section, we will discuss the drawbacks of an unpaid caregiver workforce.

Lack of training and consequences to patient care

Although providing health and wellness services is quite labor-intensive and intellectually demanding, caregiving skill development is usually not a priority. Caregivers are often not counseled on their roles and report feeling insecure about their ability to provide adequate care as well as their level of healthcare knowledge. They often feel unsupported by the patient's doctors and other professional caregivers and may be at odds with their recommendations. Physicians have an important yet often underutilized role in communicating to patients, demonstrating empathy for family emotions and encouraging advanced planning for a patient's family (Rabow, Hauser, & Adams, 2004). Further, informal caregivers may not know they need help or where to find help if they need it. This dynamic can be detrimental for the patient's care quality. Research shows that a tenuous time for patients is the transition between hospital discharge and home, as many caregivers are ill-equipped to provide support at home. Medical professionals report that their directions are not being followed or they are not being kept adequately updated, while caregivers say they weren't prepared for the responsibilities of providing care for an ill or recovering patient (Reinhard, Given, Petlick, & Bemis, 2008).

If a patient suffers a complication or injury, or if an attendant makes a medical mistake, an untrained caregiver will not have the skills to quickly remedy a situation or to recognize when someone may be deteriorating. In the United States, for example, less than 1 in 10 caregivers received any nursing or medical training related to their caregiving responsibilities (Whitman, 2016). Additionally, when parents are not available to provide care for their children, care often falls to less able providers, such as elderly grandparents or other children. In rural parts of Ethiopia, more than half of 5–8 years old girls provide unpaid care every day (Samman et al., 2016, p.10). The expectation of quality care under the ethical dilemma of placing individuals unprepared for such positions of responsibility is untenable yet commonplace.

The health of the caregivers

Caregiving may be physically, mentally, and socially taxing. Informal caregivers are more likely to experience emotional difficulty, physical difficulty, financial stress, social isolation, and loss of work productivity (Riffin, Van Ness, Wolff, & Fried, 2019; Wolff, Spillman, Freedman, & Kasper, 2016). Life strain to the caregiver can occur because their other responsibilities do not cease even when they love the person for whom they are providing care. They may have other people dependent on them, like children or grandchildren, and they are often responsible for doing the bulk of the chores and other housework. Consequently, they may feel like they can never get a "break" and become resentful of the care recipient and those they feel are outsourcing the care to them (Day, Anderson, & Davis, 2014). These women may feel uncertain about their ability to manage someone else's health and have doubts about the care they are able to provide. This may lead to the caregiver becoming compassion-fatigued, leading to declines in mental and physical health.

Caregiver mental health

Caregivers do not provide care in isolation from their other responsibilities. Their personal lives, spouses, children, friends, and job responsibilities intersect with their caregiving roles. Ideally, one might be able to balance these roles; however, this is often not the case. Unpaid caregivers who are engaged with care for a family member that involves more personal care, such as bathing or dressing, report the highest level of restriction in their ability to work or engage socially with family/friends or attend religious services (Committee on Family Caregiving for Older Adults et al., 2016). The role of the caregiver is dynamic and often changes depending on the needs of the care recipient. The burden on the caregiver expands as the needs of the care recipient increase, demonstrating a linear increase in reliance on the caregiver concurrently with mental illness and functional decline.

Informal caregiving comes at a cost to the caregiver in the form of mental stress and physical exhaustion that negatively affect a person's well-being and health. Existing studies suggest that informal caregivers had lower well-being and mental health as indicated by long-term illness than noncaregivers. Caregivers who provide more than 20 hours a week of care show lower levels of health (Legg, Weir, Langhorne, Smith, & Stott, 2013). One community health worker (CHW) reported that she "give[s] out her health for the health of the community," (Raven et al., 2015, p.8). Caregivers are also at higher odds for depression and sleep problems by 1.54 and 1.37 times, respectively (Koyanagi et al., 2018).

These indicators suggest that caregiving takes a mental toll on caregivers. While both genders experience stress from caregiving, women respond to caregiver stress differently than men, making women more likely to suffer from caregiving burdens. Male coping strategies tended to include disassociation, separation, and distancing as opposed to the internalization strategies that women often employ (Sharma, Chakrabarti, & Grover, 2016). Thus, women's response may lead to the increased mental and health distress. Unfortunately, there are limited studies that assess how socioeconomic, familial, and culture-related factors may mediate the relationships between caregiver burden, gender, and mental health (Sharma et al., 2016).

Caregiver physical health

The physical health of caregivers is not as solidified in the literature as the toll on mental health. What is known often varies in terms of gender, cultures, and age groups. Caregiving has a perceived association with higher physical and mental health burdens; however, Musich, Wang, Kraemer, Hawkins, and Wicker (2017) report that healthcare utilization is not higher than their noncaregiving counterparts. A similar study in Germany also suggested that there are no short- or medium-term effects of caregiving on physical health; however, mental health effects remain over time (Schmitz & Westphal, 2015). It is also noted that while poor mental health is seen among caregivers, physical health is generally better; however, the risks of poor physical health are significant given the impact that mental health has on physical well-being (Trivedi et al., 2014). Other studies suggest that caregivers were more likely to self-report health status as fair/poor along with diagnoses of diabetes, arthritis, anxiety, and depression (Stacey, Gill, Price, & Taylor, 2019). Caregivers are also more likely to report household food insecurity (Horner-Johnson, Dobbertin, Kulkarni-Rajasekhara, Beilstein-Wedel, & Andresen, 2015). These studies may be difficult to reconcile due to the diverse nature of caregivers and the duties of caregiving, but nevertheless point to the association between increased stress of caregiving and negative health impacts of caregivers.

Pay disparities

Women's labor force participation and potential for economic empowerment have improved significantly over time. However, while women can engage in meaningful work, their contributions are not equally valued, regardless of the mode of compensation (i.e., familial duty or monetary payment). Health and social services are the fastest growing industries for women's employment, yet over 50% of their contribution to global health is unpaid (Dhatt, Thompson, & Keeling, 2018, p.3).

Women are considered the "hidden resource" in many countries as they make significant contributions toward economic development and better health (De Giusti & Kambhampati, 2016). In many countries, the proportion of women in the healthcare workforce is over 75% (Dhatt et al., 2018, p. 3), yet they fall among the lowest paid (Friedemann, Newman, Buckwalter, & Montgomery, 2014; Newman, 2014). In China, approximately 1/3 of the GDP was attributed to unpaid care work—more than 70% of the work was performed by women, but the value of unpaid work is overlooked by policymakers (Qi & Dong, 2016, p. 144). For many women who work outside of the home as many hours as men, the primary responsibilities of the home traditionally still belong to the woman. This suggests that there is a trade-off in time/productivity between unpaid caregiving work and paid work. Evidence suggests that the amount of time spent on unpaid care work has a negative effect on wages for women (Qi & Dong, 2016).

Healthcare workforce distribution

The gender imbalance in the workforce distribution is seen in pay disparities and the types of work they can acquire. The inequalities are seen in countries across the socioeconomic spectrum. Jobs that are perceived as "less skilled" are more likely to be filled by women. Women are also compensated less for similar work and promoted less often than men (George, 2008). In Kenya, jobs with low educational requirements and wages were noted to be "female jobs," such as nurses, nutritionists, and CHWs (Newman, 2014). Women make up 90% of the CHW workforce, yet only 40% of the medical doctor workforce (Newman, 2014, p.4). In Zimbabwe, CHWs reported that the $14 per month stipend was not regularly paid, or if paid, they spent an entire day without food waiting for payment (while paying $6 in transportation cost to retrieve their wages). One woman reported having to "wait three days to receive payment," (Raven et al., 2015, p. 7). In 2001, Uganda introduced CHWs to promote health; however, these were largely unpaid women who could not read and write

(Morgan et al., 2018). Uganda's initiative to promote CHWs negatively affected a woman's ability to seek a skilled job. In the Democratic Republic of Congo, Uganda, and Zimbabwe, CHWs were more likely to be women over 30 years old. This was due to their already assumed caregiving roles in the family. In Ghana, CHWs were more likely to be men due to the travel requirements, while women were expected to stay home with family (Raven et al., 2015).

Poverty/economic progress

Gender influences the ability to obtain education or experience economic progress. Largely seen as "women's work," caregiving often stifles a women's ability to pursue paid work; informal caregivers are less likely to be employed outside the home. In Europe, the probability of working outside the home was 5% lower if engaged in caregiving work (Bauer & Sousa-Poza, 2015). Thus, political efforts or initiatives designed to promote women's labor force participation are likely ineffective for women who have caregiving responsibilities (Bauer & Sousa-Poza, 2015). If labor force participation negatively impacts workers in caregiving, given the rising life expectancy and the growing need for informal caregivers, seeking a remedy for unpaid care work becomes more urgent. Given these circumstances, labor force participation may not increase unless educational opportunities are available for unskilled workers (Nguyen & Connelly, 2014).

While men and women may serve as unpaid workers, men are less likely to substitute paid work for informal caregiving, thus negatively affecting women. Women perform more caregiving at a higher intensity (Bauer & Sousa-Poza, 2015), yet the unpaid work that women perform is largely seen as unskilled and less valuable. This renders women to be financially and socially dependent upon men, restricting their autonomy (Sepulveda Carmona & Donald, 2014). The intense burden of caregiving leaves many women and girls entrenched in poverty. In South Africa, 80% of caregivers are women and live in poverty. This limits their ability to participate in advanced educational opportunities that may advance their income earning activities (Sepulveda Carmona & Donald, 2014). Unpaid care time provides a barrier to financial security in retirement and for educational progress even in countries affiliated with the Organization for Economic Cooperation and Development (OECD) where income opportunities are more readily available. This has a significant negative impact on human rights and fails to uphold principles of gender equality and women's rights: right to education, decent work, and right to health (Sepulveda Carmona & Donald, 2014).

Cultural perspectives on caregiving

Decades of evidence suggest that across religious affiliations, socioeconomic standing, and many other individual- and state-level characteristics, women are more likely to be engaged in unpaid care work than men. This suggests that some factors that contribute to unpaid caregiving work by women are beyond institutional or governmental control. Ultimately, gender inequalities in the workforce are tolerated because they are accepted and internalized by both men and women. At the same time, unpaid caregiving supplies a significant amount of free health care labor. The combination of these factors means that policymakers have little incentive to evaluate their cultural biases in caregiving practices and policies because attending to them would require significant shifts in the healthcare workforce. In this section, we introduce the global challenges for caregivers through analysis of the region with the highest global rates of unpaid caregiving by women—the Middle East and North Africa (MENA), and the United States in North America, the region with the lowest rates of unpaid women's caregiving work.

Caregiving policy and practice in the United States

Approximately 34.2 million informal caregivers (60%, or 20,500,000, are women) in the United States provide care to a loved one, saving the long-term care system $41.5 billion annually (Plichta, 2018, p.1282). This significantly outweighs the 4.4 million persons who serve as paid caregivers across the United States (Caregiving in the U.S., 2015). No wonder, unpaid caregivers are considered the "cornerstone in the long-term care system" (Plichta, 2018, *para* 1). While the number of male caregivers in the United States is increasing, the ill effects of caregiving are more likely to affect women (Lee & Tang, 2013). Although studies suggest that culture and altruism drive the continued trend in women as caregivers, there is little research done in this area (Cruz, 2016; Pavalko & Wolfe, 2016). Sociological theorists suggest that traditional expectations are ingrained in women at an early age which leads them to approach caregiving differently (Sharma et al., 2016).

Caregivers may work either full- or part-time. Of caregivers who work, 69% worked at least 35 hours per week (Committee on Family Caregiving for Older Adults et al., 2016, p. 127). This has a profound effect on their ability to stay in the workforce and, many times, individuals must adjust their work responsibilities or leave the labor force. Unpaid work adversely affects their job security, retirement savings/benefits, and career opportunities (Committee on Family Caregiving for Older Adults et al., 2016). Women caregivers who are employed often decrease their work hours and face a 3% lower wage as compared to noncaregivers, while persons who provide more intense caregiving (more than 20 hours per week) are less likely to be employed altogether (Jacobs, Van Houtven, Tanielian, & Ramchand, 2019). Although many people expect to work longer, surveys suggest that there is a strong association between caregiving and reduced levels of employment which adversely affects women's pay and retirement security. Women who leave the workforce early found it difficult to return after their caregiving duties ceased (Skira, 2015). However, women who are engaged in paid work are less likely to provide caregiving than their nonemployed counterpart, simply due to opportunity costs. While cross-sectional studies may indicate that the higher the wages earned, the less one will work as a caregiver, this does not seem true in longitudinal studies (Pavalko & Wolfe, 2016).

While other options exist, such as home- or community-based care and residential care, family caregiving is often the economically feasible option as residential care can be costly. In 2018, the annual cost of services for a full-time home maker was $48,000, a home health aid was $50,300, adult day services were $18,700, an assisted living facility was $48,300, and nursing home care was $100,400 (Genworth, 2018). The needs of an aging society suggest that the demand for informal caregivers will become greater; they fill a gap that society cannot physically or economically support. However, this "free labor" to the United States comes at personal costs. In a survey by the National Alliance for Caregiving and the AARP Public Policy Institute, they indicate that 36% of caregivers of older adults reported "high levels of financial strains," (Committee on Family Caregiving for Older Adults et al., 2016, p. 124). Financial strain increased concurrently with caregiving needs.

History and legislation in the United States

The history of the United States portrays women as mainly domesticated members of society fulfilling their roles as wives and mothers. During World War II, women entered the work force to fill jobs that men vacated as they left for war. However, after World War II and during the Cold War, men came home from the war, returned to their prior occupations, and women resumed their primary roles in the home. Society embraced a return to the domesticated roles; many felt that it assisted in the rebuilding of society to prewar levels (Rutherford & Bowes, 2014). However, the shift in workforce demographics continued. While some women returned to domesticity, other women remained in the labor force outside of the home. By the 1950s, some women were balancing a work and home life (Rutherford & Bowes, 2014). This "balancing act" sometimes included women delaying children until they were firmly established in their careers. As the 1970s approached, more women (with children) were entering the labor force and societal beliefs about the role of women in the workplace began to change.

Aside from changes in personal opinions, demographic change helped facilitate society's attitudes toward women. Young adults were more likely to obtain college degrees which liberalized viewpoints about women in the workplace. Children who saw their own mothers enter the workforce were more likely to have egalitarian views, as were men whose wives worked. Viewpoints were changing—although women were changing quicker than men (Brewster & Padavic, 2004). Due to the heightened awareness of the woman's perspective, the convergence of psychology and feminism continued to tackle the issue of the role of women in society. By the 1980s, policy makers and thought leaders were more proactive in changing public opinion on the roles of women in the family and in society.

While relatively more men are taking on caregiving responsibilities and more women are entering the workforce, overall women continue to dominate the caregiving role. Many women continue to handle a majority of household and caregiving responsibilities (Bureau of Labor Statistics, 2015). Aside from external policies that may not support working caregivers, women generally feel a greater obligation to family. The priority on family support for some women is driven by self-interest and not sex role values or work/family conflict (Wayne & Casper, 2016). Studies suggest that the greater a person valued family, the more likely family became a priority over other interests (Li, Butler, & Bagger, 2018). The higher percentage of women caregivers is a complex issue that includes societal norms, workplace policies, and personal interests.

The United States promulgated federal action to deal with the need for caregiving. The Family Medical Leave Act (FMLA) of 1993 provided flexibility for caregivers to work and care for family members (Yang & Gimm, 2013). While well-intentioned, this act fails to keep pace with the aging population and only applies to "qualified employees" (Bailey, 2017; Yang & Gimm, 2013). The FMLA provides basic caregiving benefits to employees defined as persons who have worked at least 1250 hours and to organizations that have at least 50 employees. Since the passage of FMLA, states have taken extended measures to assist caregivers. Wisconsin and Vermont have allowed employees to use sick or vacation time in addition to FMLA time to care for a family member. Massachusetts and Rhode Island extended the time of FMLA beyond the traditional 12 weeks, the District of Columbia and Connecticut reduced the time required to be eligible to 1000 hours, and 3 other states have enlarged the scope of employers to include organizations with at least 15 or more employees (Yang & Gimm, 2013).

Despite well-intentioned public policies and government interventions, the initiatives fail to fully meet the needs of caregivers. Less than 12% of employees have access to FMLA, and many cannot afford to go without pay (Lang & Carlson, 2018). Forty-five percent lack access to any paid sick leave and 7 out of 10 low wage workers do not have access to sick leave (National Partnership for Women & Families, 2017, p. 1). In California, where FMLA is six paid weeks in addition to the 12 weeks Federal Initiative (up to 55% of wages) (Gemilli, 2018), less than 25% are aware of the FMLA and one-third of these individuals could not afford to take FMLA due to the low wages (Kornfeld, 2018, p. 168). In New York, where FMLA is funded by state payroll taxes, the FMLA passed in 2016 will allow up to 12 weeks of paid leave and 67% of salary at a predetermined cap when fully implemented in 2021 (Kornfeld, 2018, p. 169). Similar solutions may be considered which allow caregivers to take paid leave with proper funding mechanisms for sustained viability.

Caregiving policy and practices in MENA

The situation for women on the other side of the globe is generally different than the role of a woman who provides unpaid care in the US. Proliferation of unpaid caregiving work by women, especially within the Middle East and North African (MENA) region (Table 10.1), is common due to a combination of factors including gendered social expectations and cultural values in some communities that are inherently discriminatory. Cultural and societal gender expectations in this region are more likely to influence organizational and structural policies (Tlaiss, 2013). While we are presenting a regional analysis, it is important to remember that the MENA region is not a monolith in terms of religion, ethnicity, language, or state governance. Differences in practices between states, communities, and individual families is going to be much more significant in a woman's life than any overall regional trends. As the global region with the highest levels of unpaid caregiving, it is important to consider

Table 10.1 Countries in MENA region and population (% female), 2018 (World Bank Data, 2019 [database]).

Nation	Population (thousands)	% female
Algeria	42,228.43	49.5
Bahrain	1569.44	36.6
Djibouti	958.92	49.8
Egypt	98,423.60	49.4
Iran	81,800.27	49.7
Iraq	38,433.60	49.4
Israel	8883.80	50.3
Jordan	9956.01	49.4
Kuwait	4137.31	42.6
Lebanon	6848.93	49.8
Libya	6678.57	49.6
Malta	483.53	49.8
Morocco	36,029.14	50.5
Oman	4829.48	33.8
Qatar	2781.68	25.1
Saudi Arabia	33,699.95	42.7
Syria	16,906.28	49.5
Tunisia	11,565.20	50.6
United Arab Emirates	9630.96	27.9
West Bank and Gaza	4569.09	49.3
Yemen	28,498.69	49.5

some broad themes that may help explain the disparities and lead to policy solutions. However, this chapter is not meant to be wholly descriptive of every state, nor any specific religion or culture in the very diverse MENA region.

On average, women in the MENA region spend 4.5 times more hours per day performing unpaid care work than men, the largest regional gender disparity worldwide (OECD, 2014, p. 2). Women's labor force participation in MENA is the lowest in the world. The data shows that in countries where women spend more time on unpaid care work, the women's unemployment rate is higher (OECD, 2014). Interestingly, women performing unpaid care work are counted in regional employment numbers by the World Bank. They found that of all employed women in MENA, 56% are unpaid family helpers, 31% are wage earners, and about 12% are self-employed (World Bank, 2009, p. 49). Women in rural areas are more likely to be unpaid care workers than women in urban areas; most of the women who earn a wage are employed in the public sector (predominantly in education and health) due to family-friendly work schedules and the possibility to receive benefits (World Bank, 2009). Gender labor disparities are not due to lack of education, as is the case in some other environments; women in MENA are literate, and many women pursue secondary and tertiary education.

Today, the rates of paid women's labor force participation in the MENA region are not only the lowest in the world, but are significantly lower than would be expected based on factors such as fertility rates, education, and age structure (World Bank, 2004, p. 1). If current trends persist, MENA countries won't reach global averages for women's labor force participation for 150 years. Aside from the social impacts, the economic impacts are significant. Simulations show that if women's participation in the labor force increased just from the actual levels to predicted levels, average household earnings would increase by 25% and per capita GDP growth in the region would have been 0.7% higher per year (World Bank, 2004, p. 4). The knowledge that women in the MENA region today are educated at rates comparable to their male counterparts and may even perform better (especially in the Gulf states, Jordan, and the Palestinian Territories) and be more likely to pursue higher education may be surprising to some. While efforts have succeeded in pushing women in the MENA region to be literate and educated, societal norms seem to discourage some women from actually working (El-Swais, 2016). One report found that Arab women with postsecondary education were more likely to be unemployed than their counterparts with no postsecondary education (Momani, 2016). This may be because some women with the opportunity to attend college already come from more affluent backgrounds, mitigating the need to work once they start their families. Additionally, at the higher echelons of employment, many of the positions are attained through professional networks and personal relationships, which are easier for men to curate over the course of their schooling and work life.

Like many other countries, the states of the MENA region are also experiencing an aging population. As in much of the developing world, declining fertility rates have led to a "health transition phase." When coupled with changes in traditional family support systems, there will be a significant need for elder care over the coming decades. Unfortunately, many of the same countries experiencing the highest demographic changes are also the most fragile, with low economic growth, inadequate public services, and less developed health care infrastructure. Cultural traditions in the MENA region dictate high levels of respect and care for older family members, indicating that the need for unpaid caregiving for the older adults and sick will continue to increase (Abyad, 2006). The MENA region is facing one of the world's fastest growing rates of Alzheimer's disease, as the WHO estimates a 125% increase in prevalence rate by 2050 (WHO, 2012, p. 13), yet culture dictates that many families would never consider placing their older family members in a care facility even if one was accessible (Al-Olama & Tarazi, 2017). At the same time, more women in MENA states are seeking educational and career opportunities and delaying marriage and children (Krafft & Assaad, 2017). Policymakers in the MENA region must reconcile these trends, while maintaining the role of the family in providing care and allowing women to pursue lives of their own choosing.

History and legislation in MENA

The history of the MENA region predating Islam and the subsequent spread of Islam are important components in the life of most of the region. The population is approximately 383 million with an annual growth rate of 1.4% (Pew Research

Center, 2011, *para* 3). In the MENA region, more than 91% of the population is Islamic, yet they only make up about 20% of the global Muslim population (Pew Research Center, 2011, *para* 1–2). The distinct historical influences of the region distinguish it from other representations of Islam and Muslims around the world. This is an important distinction to make when considering factors like labor participation and fertility rates because, while these are trends that may be influenced in some part by religion and culture, aspects of these and other trends are more reflective of political and social changes in the fragile MENA region. The complexities of the Middle East's past and present cannot be reduced to a single war, dictator, or religion. It is a vast region with many religious and ethnic minorities as well as historical political resentments that play out in various arenas, and gender politics is but one factor to consider. However, to understand what can be done to support and empower women today, it is important to recognize some of the broad reasons that gender roles have evolved as they have.

In general terms, there is a belief that the origins of the broadly patriarchal societies present throughout MENA are a culmination of some Arabic tribal customs and practices instilled before the introduction of Islam in the early 600s. These practices placed a woman as property of a male, subject to marriage or divorce at will, and dowries were paid to the woman's male family members, ensuring her no economic independence (Oxford Islamic Studies, 2019). Furthermore, some of the misogynistic traditions (Eltahawy, 2012; Fisher, 2012) associated with some MENA societies may have been introduced or entrenched as part of a "patriarchal bargain" set into motion by colonization practices of the British, Turkish, and French (Kandiyoti, 1988). The established standard was that the main responsibility of women was to have children. As a result, women's employment can be a provocative topic even today, especially in more conservative villages and states. However, while the global Muslim population is growing, MENA fertility rates are expected to decrease to an average of about three children per woman by 2030–35, with slightly higher fertility rates in the Middle East over North Africa, down from approximately five children per woman in 1990–95 (Pew Research Center, 2011, *para* 15). This trend, along with higher rates of women's educational attainment and more outspoken campaigns against discrimination and violence against women, suggests that more women are willing to counter that "patriarchal bargain" established in their society generations ago.

While male-centered practices have been prevalent for centuries, predating the founding of Islam, the role of conservative interpretations of Islam in some communities cannot be denied. Traditionally, Islam is centered on a patriarchal unit that is viewed as the primary earner, the authority figure, and the public representative of the family. However, many who believe in this model do not necessarily view the woman in the relationship as lesser or oppressed in any way; on the contrary, they believe she has the significant role of guiding the family and being responsible for the adequate raising and educating of children. As Fereshteh Hashemi, an Iranian intellectual, wrote: "God, therefore, absolves the woman from all economic responsibilities so that she can engage herself in [procreation and rearing a generation] with peace in mind." In fact, Islam differed from its monotheistic predecessors (Judaism and Christianity) by affording women economic rights, such as retaining control of their wealth, signing contracts with their own name, and receiving inheritances, even though it was within the framework of a woman still primarily dependent on male patriarchs, like a father, brothers, or a spouse (Moghadam, 2004). Azizah al-Hibri, a notable scholar of Islam, law, and women, describes two viewpoints on this: one that believes that "Islam as it is today is fair and just to women" and another that proposes that "Islam *as it is practiced today* is utterly patriarchal, but that *true* Islam is not," (Al-Hibri, 1982).

Scholar Lila Abu-Lughod famously asked "Do Muslim Women Need Saving?" where she attempted to describe the disparity between the famously depicted oppressed Muslim woman and the more complex observations of Muslim woman in the real world. Ultimately, she argues that history's rife with poverty and authoritarian governance plays a larger part in propagating gender inequalities than religion, and that religion may be used as a convenient backdrop for justifying intervention into Islamic societies (Abu-Lughod, 2013). Yet, religious interpretations have also been used within these societies to justify policy and practice. There are many examples where Islamic law that would promote gender parity is ignored and where the harshest interpretations of Islamic law are used to place women in a submissive status. Some contend that "patriarchy…should not be conflated with Islam, but rather should be understood in social-structural and developmental terms," (Moghadam, 1992, p. 11). Other scholars argue that Islam is inherently patriarchal, and while the belief system provides women certain rights, these rights did not place them at equal status with men in practical terms. The purpose of this chapter is not to negotiate the tension between these viewpoints, but it is important to note their presence when considering how unpaid caregiving work by women in this region has developed and is expressed today.

Structural and legal barriers to women in the workforce are predominant throughout MENA countries. While it is often tempting to blame Islam for these disparities, the Islamic holy book (the *Quran*) does not mandate that women are responsible for all caretaking work. In most cases, directives are given through social policies mandated by the state, sometimes with justifications from strict interpretations of traditional Islamic Sharia law. Some countries that follow the legal Sharia framework, such as Iran, have codified laws that state that women are obligated to perform care for family members (Minguez, 2012). Other structural factors, such as harassment of women, limitations on women's mobility, lower salaries, and industries closed to women workers, have limited their desire and potential to enter the wage-earning labor force. Morocco and Djibouti are the only MENA countries with laws that address gender discrimination in hiring (Lemmon, 2017).

Global models for gender labor protections were incorporated in the mid-20th century, but they were not as influential in the MENA region where social norms suggested that men were fiscally responsible for their families. Even men struggling to support their families may be hesitant to have their wife work as that reflected poorly on a man's ability to care for their family. In the 20th century, these existing traditions combined with the growing conflicts in the Middle East and burgeoning financial support for conservative Islamic movements. These factors thus ensured that the strictest interpretations of the religion, including misogynistic practices, persisted and grew in some areas (Chamlou, 2017). Approximately 86% of Muslims completely or mostly agree that a wife should always obey her husband (Pew Research Center, 2013, Table 2). This reflects a deeply ingrained sense of gender norms on the part of both men and women.

Vignettes illustrating cultural perspectives of caregiving

Vignettes are commonly used in social sciences research. They provide context that allows for clarification and serves as a tool to explore potentially delicate scenarios from a less personal perspective (Barter & Renold, 1999). In this section, we will introduce two short composite vignettes to illustrate the various experiences of caregivers in the United States and the MENA region. These vignettes are not meant to be entirely representative of all women across the full geographic spectrum of each region. They convey a perspective of a typical caregiver in their respective regions. The vignettes were developed from a review of the literature and case studies based on women for an accurate portrayal (Hughes & Huby, 2004). The questions we consider in these vignettes are: What is the experience of unpaid women caregivers in these two environments? How does the political, social, and cultural context influence their roles as caregivers? After exploring these ideas, we will discuss the stories. How are their daily lives the same? How are they different? Importantly, what do they tell us about women and caregiving today?

Julia

Life was hectic in the Midwest, but Julia could not have asked for more. Although Julia completed high school, she never finished college. After entering college, she quickly found the love of her life; 2 years later, they were married. She left college so that her husband could finish his college degree in secondary education in another state. Soon after his graduation, he landed a job as a high school history teacher and they were expecting their first child. Julia had often thought about going back to finish her degree, but family responsibilities became greater, especially after eventually having three children. Julia found steady work as an office manager in a neighborhood child daycare center. Between home and work, her responsibilities were great but manageable, even as her husband took on more responsibilities at work and spent less time at home.

Julia became the primary caregiver for her 72-year-old mother when her father passed away. She loves her mother, but faces the daunting task of caring for her due to early onset of dementia and several comorbidities (e.g., macular degeneration, diabetes, high blood pressure, and heart failure). Her mother lives 20 minutes away from her job at the daycare center; this originally seemed to be convenient as her mom's health was steadily declining with age and her mother became more forgetful. Julia would stop by during her lunch hour 3–4 days a week and spend a few hours on Sunday helping her mom with larger tasks such as grocery shopping, cleaning the house, laundry, and preparing meals for the week.

After a fall where her mother fractured her right hip, the care for her mother intensified. Her mother needed total assistance with bathing and dressing. Julia discussed obtaining paid help, but her mother refused to have a stranger in her home. Julia's labor of love intensified, and she became exhausted. The physical requirements of taking care of her mother began to wear on her as her lower back pain began to increase. As a wife and mother of three, Julia continued to work her full-time job at the day care center, but the obligations of home and her mom prevented her full commitment to the director position that became available. She had little reprieve from her work in her own home. After making dinner, cleaning the house, and helping her children with their homework, Julia crawled into bed—only to start the process again the next day. Julia's marriage became strained as she and her husband began to argue due to her fatigue and being overwhelmed by the responsibilities of work and caring for her family and mother. She found herself with increased anxiety and depression which would often lead to bouts of crying and irritability. She began to take sleeping pills to help her rest at night, but often awoke feeling more tired than before.

Nadia

Even as a child, Nadia helped her mother prepare food, dispense medications, and provide personal care for her elderly grandparents who lived with them. The only girl of six children, Nadia's caregiving workload was always higher than her brothers who were encouraged to go outside and play soccer while she was washing dishes or preparing meals. As her mother aged and was unable to perform her duties, Nadia's caregiving load extended to her aging father, a retired engineer. Nadia helped her father manage his diabetes by checking his blood sugar and assisting with medication management. She happily accepted the responsibility to assist her family as a child, seeing it as an almost spiritual purpose in line with the Quran's view of respecting the elderly, especially one's parents. As a high school student, Nadia wanted to pursue an English degree in college. She studied hard in her minimal spare time for the comprehensive exams administered at the end of high school. She excelled at the exams, but was dismayed when her parents asked her to go to the local college instead of the large university that was further away. In the meantime, her family started bringing suitors to their home in an effort to find Nadia a husband.

Nadia was able to convince her parents that she wanted to wait until after college to find a husband, to which they reluctantly acquiesced. However, her father's health deteriorated and surgeons eventually had to amputate his foot due to diabetic complications. This meant he needed continuous care and her aging mother was starting to exhibit early signs of Alzheimer's, intensifying her own need for care. By this point, most of Nadia's brothers were married with families of their own. Nadia, as the unmarried aunt, was also expected to help care for them. Nadia learned of an opportunity for girls from her country to receive scholarships to the United Kingdom to attend a specialized English graduate program, but she knew she couldn't leave her parents without a caretaker. Upon graduation, she married a man from a neighboring village, and they built a second floor on her parent's house for them to live in. She accepted a job as an English teacher in a local girl's middle school and continued to provide care for her elderly parents, her husband's parents, her nieces and nephews, and after a few years, her own children.

Similarities and differences in vignette profiles

Although both Nadia and Julia were primary caregivers for aging parents, their trajectories were very different based on many culturally derived factors. Julia had a well-established life with her own children once her caretaking duties began, while Nadia began her life with the assumption that she would provide a caretaking role in the family. This may have made the transition more difficult for Julia, who felt herself pulled in multiple directions with no personal time. She found herself feeling both physically and mentally drained. Nadia, on the other hand, felt a lifelong sense of filial duty that was tied up in both her cultural and religious life, which made the expectation of caretaking less of a mental burden. However, while Julia had the time and freedom to pursue her career and make her own decisions about marriage and children, Nadia was expected to stay close to home, limiting her educational and career opportunities. From a younger age, Nadia was expected to get serious about finding a husband and starting a family. Nadia was also subject to distinct gender roles as a young child, watching her brothers grow up with a different set of rules and expectations than she received. Nadia was educated with a college degree and underemployed. In the United States, Julia's education would have likely served as a stepping stone to a better life; this was not the case for Nadia.

Despite the women's differences, some aspects of caregiving cross cultures. Both women were providing care to mentally and physically impaired adults without any formal medical or social work training. Julia's caregiving responsibilities began later in her life and she still felt the same duty to care for her mother that Nadia felt. Neither woman fulfilled their professional potential as both chose jobs that allowed them time to be caregivers. Although their journey to caregiving differed, the physical and mental tolls of caregiving are similar; neither woman was offered assistance or support.

Julia had greater access to pharmaceutical solutions to her physical pains due to the nature of sleep and pain medication offered in the United States. She may have had to turn to these measures especially as her marriage experienced strain. However, Nadia's family only accepted her husband's marriage proposal after extended conversations between both religiously adherent families about their expectations for the marriage. This limited Nadia's freedom in some respects, but also provided an understanding between her and her husband about their roles in the marriage. Julia, unlike Nadia, had the option of placing her mother in an assisted living facility. Although their careers differed, both women were engaged in work that also required some level of caregiving. Julie was a day care employee and Nadia taught young children. These are jobs that are often held by women and are typically underpaid and undervalued. Both women were, like many others, under financial burdens that limited their ability to care for their families and themselves.

Discussion and recommendations: Can cultural perspectives on gender roles in caregiving change?

Although every woman's story is different, these vignettes display circumstances that are commonly experienced by women from these regions. Around the world, many women are trying to balance the demands of the family, work, and caregiving, often sacrificing their own empowerment and fulfillment. However, without informal caregivers, many children and adults in need of care would have no other option. Unpaid caregiving is a necessary component of health care systems, but current belief systems may limit the caregivers and care recipient. Is it possible to change cultural perspectives on caregiving to ensure gender equity and support for all caregivers? Importantly, is it ethical to critique a culture's approach to caregiving if it is composed of sincerely held beliefs about family, duty, and tradition, even if it may be detrimental to some in the process? These are complex questions without obvious answers. However, ignoring the challenges posed by these issues does not make demographic shifts less destabilizing, nor minimize the increased need for caregiving.

Current gender disparities pose a threat to the future of caregiving services. Logically, as more women enter the workforce, they will no longer be able to shoulder the burden of caregiving. This poses a significant challenge to healthcare systems. However, institutional change is not likely without changing the cultural and social perspectives. Cultural change must be structural and personal. Gender beliefs may be held on a personal level, but together these individual beliefs manifest as structures at different levels, from the individual identity of a person, to organizational behaviors, and ultimately to distribution of resources (Ridgeway & Correll, 2004). As a result, we must consider changes at each level to bring lasting societal change. The issue is not eliminating unpaid caregiving by women, but to provide support for those who choose to engage in care provision, offer alternatives for families who need care services, and ensure women and girls have the freedom to choose what path works best for them. We address policy recommendations at four levels based on the Bronfenbrenner (1979) ecological model of human development.

The original model views the place of an individual in the context of increasing social interaction. The microsystem is positioned closest to the individual, reflecting personal identify and motivation. The individual has certain characteristics, such as gender or age, which influences development. At the macrosystem level, which is the farthest away from the individual, the model describes the various conditions in society that can impact the individual. In between these two extreme levels is the mesosystem level that engages family, peers, and work environments, while the exosystem level extends to the larger sense of community. A fifth level, the Chronosystem demonstrates changes in society during a lifespan that can also affect an individual. A period of transition occurs when an individual makes a major decision to follow a certain path. Changing social norms influence an individual (Bronfenbrenner, 1979). Following the model of increasing external impact, we examine the problem of women's caregiving at four levels: individual, community, government and policy, and culture (Fig. 10.1). The model represents the dynamic

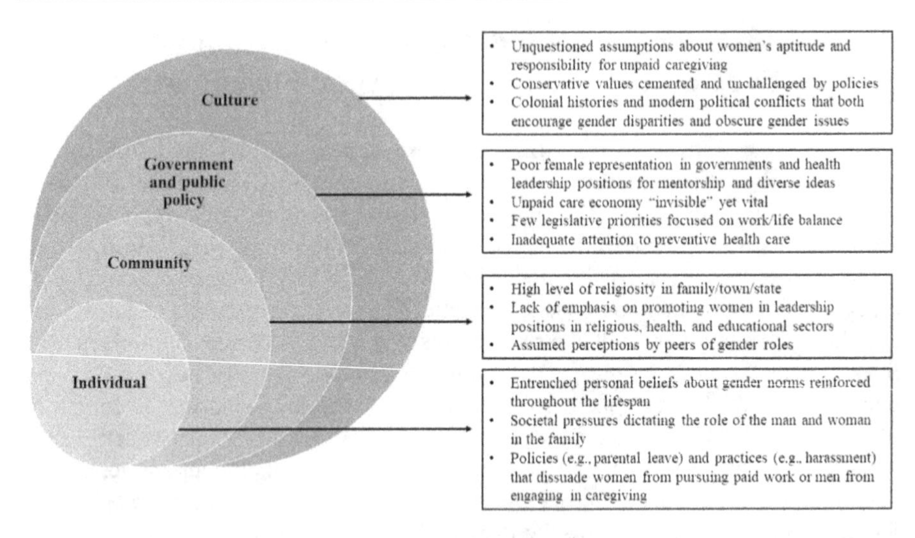

Fig. 10.1 Interpretation of Bronfenbrenner ecological model of human development: Factors that facilitate unpaid women's caregiving work.
Credit: The authors.

interaction between an individual and their environment and proposes that levels closer to the individual will have a higher level of influence in the process of decision-making process.

Individual

Cultural beliefs are only permitted to exist because they are held by a wide margin of a given population, including the men and the women. As a result, any widespread cultural shift must include education and outreach on a personal level. Modifying attitudes about gender equality in unpaid caregiving may lead to changes in behavior at the individual level, but may still perpetuate at the institutional and societal levels. Because attitudes and behaviors are often engrained by adulthood, such efforts must begin as early as childhood. Behaviors such as segregating children by sex as well as guiding children's interests toward gender-normative roles and activities are early ways that cultures indoctrinate gender beliefs in children (Martin & Ruble, 2010). While the bulk of this chapter has focused on the plight of women who perform unpaid care work, there are congruent solutions that must be spearheaded by men. Men that grow up in financially stable homes are able to receive an education that can position them to witness equitable behavior. Men with these experiences are then more likely to have egalitarian attitudes about caregiving. Evidence shows that men are capable of changes in attitude about gender issues and that definitions of masculinity are not static (Levtov, Barker, Contreras-Urbina, Heilman, & Verma, 2014). When gender equality is not seen as a zero-sum game, not only can empowered women achieve more, but men do not fear that they may lose when women are successful.

In addition to providing support for women caregivers, organizations such as UN WOMEN have recognized that men must be willing to perform more of the unpaid caregiving work, especially in cultures where this is not the norm (UN WOMEN, 2018). In the MENA region, evidence suggests that men who saw their fathers perform caregiving work are more likely to do so. We also see that the same gender norms that depress women's labor participation are the same ones that push men into overwork and burnout. Some men may prefer to perform more caregiving tasks, but cannot do so because they feel pressure to work more as they are the primary or only wage earner. Parental leave policies are highly variable across and within countries, and well-intentioned policies do not always translate into desired practice. In some instances where parental leave may exist for fathers, there may be societal pressure that prevents men from utilizing this benefit due to disparagement from their peers. This is a prime example of where policy solutions can change cultural expectations in the workforce, opening opportunities for everyone.

To alleviate personal mental stress and improve outcomes for the care recipient, caregivers must be encouraged to engage in self-care as well as self-education. Caregivers with the means to access computers or the internet can engage in self-education to aid in their ability to provide quality care. In the United States, websites offer free resources and information for caregivers. Next Step in Care (https://www.nextstepincare.org/), created by New York's United Hospital Fund, offers free videos on topics like Caring for and Maintaining Ostomy Bags and Preparing Your Home for Safe Mobility. Other organizations such as the Caregiver Action Network (https://caregiveraction.org/), Family Caregiver Alliance (https://www.caregiver.org/), and the Well Spouse Association (https://wellspouse.org/) also offer educational resources, information, and caregiver support groups. There are few analogues for these services in MENA and much of the developing world, indicating a clear need for additional tools of support. Nevertheless, existing tools can serve as viable examples for policymakers and organizations in MENA countries.

Community

As discussed earlier in this chapter, there is a significant interaction between culture and religion. This interaction forms the foundation for a large portion of the community and civil society. As a result, cultural attitudes on gender may come at least in part from an interpretation of a common religious text. Despite theories that suggest that the religion dictates how a society will treat women, competing ideas suggest that the more nonreligious people are, the more likely they are to have egalitarian perspectives. While primarily Christian populations do appear to be more egalitarian than mostly Muslim nations, these differences are minimal when development is considered. In general, the least gender equal countries have the fewest nonreligious people of any denomination—in other words, level of religiosity in a society is more of an indicator than the religion itself (Schnabel, 2016). This demonstrates that, at times, the focus on a state's particularly religion as a factor in social norms can obscure other indicators, such as socioeconomic status, quality of governance, and who gains when a particular religion is maligned.

With all that said, the interaction between culture and religion is complex. While religion may influence gender inequalities in some nations, it is neither realistic nor ethical to attempt to disconnect a society's culture from their religion, or vice versa. Evidence from the MENA region suggests that most dimensions of religiosity serve "patriarchal socialization," but this is less likely in two populations: women and the highly educated (Glas, Spierings, & Scheepers, 2018). This provides some direction in terms of guiding cultural change in more egalitarian directions, while maintaining religious adherence in those who choose to practice. First, promotion of education is vital in any development agenda. Education serves two outcomes in this regard: providing girls and marginalized populations with opportunities along with informing men about the contribution women make to society. Additionally, promoting women in roles of leadership within religious and cultural sectors may lead to more gender equal interpretations with practices that maintain a society's values. Within the MENA region, it may also ease the restrictions on women and assumptions about their place in the community.

Sometimes environmental circumstance provides a push for social change that might otherwise not be possible. For example, conflict and insecurity in MENA states, such as Syria, the West Bank and Gaza, and Lebanon, have forced women to enter the work force when their husbands are killed or imprisoned; similarly, as the men in these states live in areas of extremely high unemployment and economic vulnerability, they may be more likely to take on domestic tasks (UN WOMEN, 2018). War was also a large impetus for woman to enter the workforce in the early- and mid-20th century in the United States. This is an important reminder that unintended consequences of political and policy decisions across sectors ripple into individual beliefs and behaviors about health provision, labor participation, and gender roles. Context-specific policy solutions that consider local and regional trends and events can be developed by grassroots organizations. Donors and policymakers must take into account the incomparable knowledge of community-based organizations to develop solutions that improve outcomes in vulnerable populations.

Governments and public policy

Evidence suggests that a primary cause for the outsized role of women in the caregiving space is not just "traditional" cultural expectations about gender roles, but that these beliefs become codified into institutional policies that perpetuate the cultural norms. Investing in women and girls is a pathway for economic, educational, and political stability. To do so, the unpaid care economy must be made more visible through policies that consider gender roles and norms, and in some instances, challenge them. While valuable policies may be supported, lack of measurement may lead to good intentions being absolved for poor outcomes. Actionable policies should consider the needs of the healthcare system and the needs of the healthcare workforce. In other words, how does the healthcare system respond to its people, and how do the people respond to the system? Unfortunately, many governments operate under the viewpoint that most caregiving duties will be supplied by families and focus on policies where government expertise is needed (Samman et al., 2016). As caregiving needs in countries around the world will increase, this viewpoint is not sustainable.

States can begin by valuing promotion of preventative care in order to reduce the burden of the need for care by encouraging healthier lifestyles. This includes promoting physical activity, alleviating tobacco, alcohol, and drug abuse, and encouraging healthy eating habits. Such actions could reduce incidence of diseases like Type 2 diabetes and Alzheimer's, which will allow more of the adult population to maintain autonomy (Al-Olama & Tarazi, 2017). Spending money on marketing and outreach events today will reduce the money spent later paying for chronic disease treatment, as well as increase the employability of potential caregivers. Additionally, increased automation in low-skill jobs will push many workers out of the labor force. At the same time, demand for caregiving will increase, but the increase in working women will make them less available to provide such care. The McKinsey Global Institute report estimates that jobs related to caregiving will grow globally by 50–85 million by 2030 (Manyika et al., 2017). This presents a prime opportunity for states to train displaced workers, both men and women, for the caregiving roles their communities will need.

Nations that are more gender egalitarian on measures of women's professional opportunities and political power tend to divide unpaid work more equally than nations that are less egalitarian, even within developed nations like Italy and Japan (World Economic Forum, 2018b). As a result, there is a clear role for public policies in labor regulations, work/life balance, and gender equality. For example, policies that limit working time may make men more available to pursue unpaid caregiving tasks, while policies offering publicly funded childcare and expanding parental leave to both parents could provide women more economic opportunities outside the home (Lachance-Grzela & Bouchard, 2010). Preliminary evidence in OECD countries suggests that favorable changes are noted in areas where pay equity exists (Buchan & Black, 2011). Improving wage settings may provide an effective alignment between efficiencies and performance.

Public policy has a significant impact on caregivers' ability to provide care and participate in the labor force (Lang & Carlson, 2018). Many MENA states recognize that changes in gender parity are needed for the social and economic advancement. Bahrain, Egypt, Jordan, Lebanon, Morocco, the Palestinian Authority, Tunisia, Yemen, and the United Arab Emirates have issued national gender equality strategies in the past decade. However, while a positive step in recognizing gender disparities and empowering women, the strategies outlined in some of these agendas still focus on protecting women within their roles as mothers and caretakers (OECD, 2014). This suggests that a foundational shift is needed in how such policies might be presented. It is not enough to take existing assumptions and maintain them with financial support and programming. We must be willing to challenge assumptions about women's labor to avoid unintended consequences that perpetuate gender inequalities.

In the United States, there are several initiatives that were developed to assist caregivers, but are inadequate in some areas. The Medicaid Home and Community-Based Waiver Services allows for "personal care services" to be provided that "serves the best interest" of the care recipient (Friedman & Rizzolo, 2016; Lang & Carlson, 2018). This initiative permits individuals who will best serve as the caregiver to the recipient to be reimbursed for services rendered. While this alleviates the burden of a nonpaid caregiver, the Medicaid reimbursement is less than a paid caregiver would earn and less than an unpaid caregiver would obtain if employed (Lang & Carlson, 2018). The funding mechanism for parental leave policies varies among OECD countries and should be considered in light of other social funding policies and organizational policies that allow a mix of unpaid and paid leave (Yang & Gimm, 2013). Subsequent calls for gender equality and paid leave are criticized for potential "moral hazards"—due to social norms, women are more likely to take paid leave, costing employers more money and leaving women less attractive for employment or more likely to be paid less than their male counterparts (Yang & Gimm, 2013). Given the limitations of the current legislation to meet income needs in retirement, we recommend providential solutions that extend into retirement. Retirement income is sacrificed when a woman does not participate in the labor force. Women have long been able to claim the husband's retirement benefit (social security) when her work history is inadequate. However, social security caregiver credits are proposed as one way to offset the financial viability in retirement for caregivers (Lang & Carlson, 2018). We also recommend that states consider their own paid caregiving leave (Lang & Carlson, 2018).

Culture

We cannot discount the role of political and legal reforms in changing the nature and status of women's work. However, as one Brookings report noted, "enhanced employment and empowerment of women in Arab economies and societies can be achieved through political, legal, and regulatory reforms. However, without larger changes in societal and cultural attitudes, top-down reforms will have a limited impact," (Momani, 2016, p. 8). Controlling for socioeconomic variables, cultures with higher levels of conservative values report a higher gender gap, while countries that value autonomy are more likely to be gender egalitarian (Yeganeh & May, 2011). Some of these cultural changes will be generational, long-term shifts that will only come when changes at the individual level are internalized, and some societies will be more resistant to these changes than others. We have addressed many of the cultural issues throughout this chapter, but it is clear that a mix of policy and education and outreach is needed to move how communities view women and their roles in the workforce and the household.

Conclusion

In this chapter, we have described the role of unpaid caregiving by women in global health delivery, as well as the challenges posed by the entrenched structural barriers that culture and tradition place on women performing unpaid care work. For women working in MENA countries, we recommend developing the job industry to embrace unpaid women in these roles. The United States and other more developed nations must reconsider their policies in areas such as parental leave and caregiver reimbursement in order to account for unintended consequences as well as inequity of access to these programs. Women in caretaking roles across the world would benefit from prioritizing personal care to offset burnout and other negative health outcomes associated with long-term unpaid healthcare. We suggest that if culture has led to these current trends in caregiving, adapting culture using an inclusive approach to health caregiving can aid the caregiver, patient, and social structure as a whole. Overall findings suggest that a move toward compensated healthcare will value women in this role and provide the balance between caregiving, women's health, and opportunity required to meet sustainable goals in gender equality.

References

Abu-Lughod, L. (2013). *Do Muslim women need saving?* Cambridge, Massachusetts; London, England: Harvard University Press.

Abyad, A. (2006). Health care services for the elderly in the Middle East. *Middle East Journal of Business, 2*(2) Retrieved August 1, 2019 from http://www.mejb.com/mejb_iss2_vol2/Healthcare%20for%20the%20Elderly.html.

Acharya, S., Lin, V., & Dhingra, N. (2018). The role of health in achieving the sustainable development goals. *Bulletin of the World Health Organization. 96*(9). https://doi.org/10.2471/BLT.18.221432. 591–591A.

Al-Hibri, A. (1982). A study of Islamic herstory: Or how did we ever get into this mess? *Women's Studies International Forum, 5*(2), 207–219. https://doi.org/10.1016/0277-5395(82)90028-0.

Al-Olama, M., & Tarazi, F. (2017, October 12). *Middle East cultures treasure the elderly, making Alzheimer's a complex scourge.* World Economic Forum. Retrieved August 1, 2019 from http://www.weforum.org/agenda/2017/10/alzheimers-MENA/.

Bailey, K. S. (2017). The FMLA and psychological support: Courts care about "care" (and employers should, too). *Michigan Law Review, 115*(7), 1213–1237. Retrieved August 1, 2019 from http://michiganlawreview.org/wp-content/uploads/2017/05/115MichLRev1213_Bailey.pdf.

Barter, C., & Renold, E. (1999). The use of vignettes in qualitative research. *Social Research Update, 25,* Retrieved August 1, 2019 from http://sru.soc.surrey.ac.uk/SRU25.html.

Bauer, J., & Sousa-Poza, A. (2015). Impacts of informal caregiving on caregiver employment, health, and family. *Journal of Population Aging, 8,* 113–145. https://doi.org/10.1007/s12062-015-9116-0.

Brewster, K., & Padavic, I. (2004). Change in gender-ideology, 1977–1996: The contributions of intracohort change and population turnover. *Journal of Marriage and Family, 62*(2), 477–487. https://doi.org/10.1111/j.1741-3737.2000.00477.x.

Bronfenbrenner, U. (1979). *The ecology of human development: Experiments by nature and design.* Cambridge, MA: Harvard University Press.

Buchan, J., & Black, S. (2011). *The impact of pay increases on nurses' labour market: A review of evidence from four OECD countries.* OECD Health Working Papers, No. 57 Paris: OECD Publishing. https://doi.org/10.1787/5kg6jwn16tjd-en.

Bureau of Labor Statistics. (2015). Charts by topic: Household activities. In *American Time Use Survey.* Retrieved August 1, 2019 fromhttp://www.bls.gov/tus/charts/household.htm.

Caregiving in the U.S. (2015). National Alliance for Caregiving. Retrieved from https://www.caregiving.org/wp-content/uploads/2015/05/2015_CaregivingintheUS_Final-Report-June-4_WEB.pdf.

Chamlou, N. (2017, October 3). Women's rights in the Middle East and North Africa. *Global Policy*, Retrieved August 1, 2019 from https://www.globalpolicyjournal.com/blog/03/10/2017/women's-rights-middle-east-and-north-africa.

Committee on Family Caregiving for Older Adults, Board on Health Care Services, Health and Medicine Division, & National Academies of Sciences, Engineering, and Medicine. (2016 Nov 8). *Economic impact of family caregiving*. In R. Schulz & J. Eden (Eds.), *Families caring for an aging America*. Washington, DC: National Academies Press (US). (chapter 4). Retrieved April 30, 2019 from https://www.ncbi.nlm.nih.gov/books/NBK396402/.

Cruz, M. (2016). Healthcare, policy implementation, and culture: What cultural values influence unpaid primary caregivers to provide care to older adults? *International Journal of Public Administration*, *40*(2), 176. https://doi.org/10.1080/01900692.2015.1089444.

Day, J., Anderson, R., & Davis, L. (2014). Compassion fatigue in adult caregivers of a parent with dementia. *Issues in Mental Health Nursing*, *35*(10), 796–804. https://doi.org/10.3109/01612840.2014.917133.

De Giusti, G., & Kambhampati, U. (2016). Women's work choices in Kenya: The role of social institutions and household gender attitudes. *Feminist Economics*, *22*(2), 87–113. https://doi.org/10.1080/13545701.2015.1115531.

Dhatt, R., Thompson, K., & Keeling, A. (2018). Gender equality = smart global health: Our story, women in global health. *World Medical Journal*, *64*(1), 7–9.

El-Swais, M. (9 March 2016). *Despite high education levels, Arab women still don't have jobs*. The World Bank. Retrieved August 1, 2019 from http://blogs.worldbank.org/arabvoices/despite-high-education-levels-arab-women-still-don-t-have-jobs.

Eltahawy, M. (23 April 2012). Why do they hate us? *Foreign Policy*, Retrieved August 1, 2019 from https://foreignpolicy.com/2012/04/23/why-do-they-hate-us/.

Ferrant, G., Pesando, L. M., & Nowacka, K. (2014). *Unpaid care work: The missing link in the analysis of gender gaps in labour outcomes*. Organization for Economic Cooperation and Development, OECD. Retrieved August 3, 2019 from https://www.oecd.org/dev/development-gender/Unpaid_care_work.pdf.

Fisher, M. (25 April 2012). The real roots of sexism in the Middle East (It's not Islam, race, or 'hate'). *The Atlantic*, Retrieved August 1, 2019 from https://www.theatlantic.com/international/archive/2012/04/the-real-roots-of-sexism-in-the-middle-east-its-not-islam-race-or-hate/256362/.

Friedemann, M. L., Newman, F. L., Buckwalter, K. C., & Montgomery, R. J. (2014). Resource need and use of multiethnic caregivers of elders in their homes. *Journal of Advanced Nursing*, *70*(3), 662–673. https://doi.org/10.1111/jan.12230.

Friedman, C., & Rizzolo, M. C. (2016). Un/Paid labor: Medicaid home and community based services waivers that pay family as personal care providers. *Intellectual and Developmental Disabilities*, *54*(4), 233–244. https://doi.org/10.1352/1934-9556-54.4.233.

Gemilli, K. D. (2018, March 7). *Navigating the interaction between the FMLA and California leaves: Proceed with caution*. Retrieved August 3, 2019 from http://dmec.org/2018/03/07/navigating-interaction-fmla-california-leaves/.

Genworth. (2018). *Genworth 2018 cost of care survey*. Retrieved April 30, 2019 from https://www.genworth.com/aging-and-you/finances/cost-of-care.html.

George, A. (2008). Nurses, community health workers, and home carers: Gendered human resources compensating for skewed health systems. *Global Public Health*, *3*(s1), 75–89. https://doi.org/10.1080/17441690801892240.

Glas, S., Spierings, N., & Scheepers, P. (2018). Re-understanding religion and support for gender equality in Arab countries. *Gender & Society*, *32*(5), 686–712.

Grigoryeva, A. (2017). Own gender, sibling's gender, parent's gender: The division of elderly parent care among adult children. *American Sociological Review*, *82*(1), 116–146. https://doi.org/10.1177/0003122416686521.

Horner-Johnson, W., Dobbertin, K., Kulkarni-Rajasekhara, S., Beilstein-Wedel, E., & Andresen, E. (2015). Food insecurity, hunger, and obesity among informal caregivers. *Preventing Chronic Disease*, *12*, E170. https://doi.org/10.5888/pcd12.150129.

Hughes, R., & Huby, M. (2004). The construction and interpretation of vignettes in social research. *Social Work and Social Sciences Review*, *11*(1), 36–51.

Jacobs, J. C., Van Houtven, C. H., Tanielian, T., & Ramchand, R. (2019). Economic spillover effects of intensive unpaid caregiving. *PharmacoEconomics*, 1–10. https://doi.org/10.1007/s40273-019-00784-7.

Kandiyoti, D. (1988). Bargaining with patriarchy. *Gender and Society*, *2*(3), 274–290. https://www.jstor.org/stable/190357.

Kornfeld, S. K. (2018). A need not being met: Providing paid family and medical leave for all Americans. *Family Court Review*, *56*(1), 165–179. https://doi.org/10.1111/fcre.12329.

Koyanagi, A., DeVylder, J. E., Stubbs, B., Carvalho, A. F., Veronese, N., Haro, J. M., et al. (2018). Depression, sleep problems, and perceived stress among informal caregivers in 58 low-, middle-, and high-income countries: A cross-sectional analysis of community-based surveys. *Journal of Psychiatric Research*, *96*, 115–123. https://doi.org/10.1016/j.jpsychires.2017.10.001.

Krafft, C., & Assaad, R. (2017). *Employment's role in enabling and constraining marriage in the Middle East and North Africa*. Economic Research Forum. Retrieved from August 1, 2019 from http://erf.org.eg/publications/employments-role-in-enabling-and-constraining-marriage-in-the-middle-east-and-north-africa/.

Lachance-Grzela, M., & Bouchard, G. (2010). Why do women do the lion's share of housework? A decade of research. *Sex Roles*, *63*(11 – 12), 767–780. https://doi.org/10.1007/s11199-010-9797-z.

Lang, K., & Carlson, E. (2018). Making ends meet (or not): How public policy affects caregivers' income. *Generations*, *42*(3), 90–96. Retrieved August 1, 2019 from https://www.questia.com/library/journal/1P4-2188100646/making-ends-meet-or-not-how-public-policy-affects.

Lee, Y., & Tang, F. (2013). More caregiving less working: Caregiving roles and gender difference. *Journal of Applied Gerontology*, *34*(4), 465–483. https://doi.org/10.1177/0733464813508649.

Legg, L., Weir, C., Langhorne, P., Smith, L., & Stott, D. (2013). Is informal caregiving independently associated with poor health: A population based study. *Journal of Epidemiology and Community Health*, *67*(1), 95–97. https://doi.org/10.1136/jech-2012-201652.

Lemmon, G. (2017, February 27). *Improving women's economic participation in MENA nations*. Council on Foreign Relations. Retrieved August 1, 2019 from https://www.cfr.org/blog/improving-womens-economic-participation-mena-nations.

Levtov, R., Barker, G., Contreras-Urbina, M., Heilman, B., & Verma, R. (2014). Pathways to gender-equitable men: Findings from the International Men and Gender Equality Survey in eight countries. *Men and Masculinities*, *17*(5), 467–501. https://doi.org/10.1177/1097184X14558234.

Li, A., Butler, A., & Bagger, J. (2018). Depletion or expansion? Understanding the effects of support policy use on employee work and family outcomes. *Human Resource Management Journal*, *28*(2), 216–234. https://doi.org/10.1111/1748-8583.12174.

Lomazzi, M. (2016). A global charter for the public's health—the public health system: Role, functions, competencies and education requirements. *European Journal of Public Health, 26*(2), 210–212. https://doi.org/10.1093/eurpub/ckw011.

Manyika, J., Lund, S., Chui, M., Bughin, J., Woetzel, J., et al. (2017, November). *Jobs lost, jobs gained: What the future of work will mean for jobs, skills, and wages*. McKinsey & Company. McKinsey Global Institute Report. Retrieved August 1, 2019 from https://www.mckinsey.com/featured-insights/future-of-work/jobs-lost-jobs-gained-what-the-future-of-work-will-mean-for-jobs-skills-and-wages.

Martin, C., & Ruble, D. (2010). Patterns of gender development. *Annual Review of Psychology, 61*, 353–381. https://doi.org/10.1146/annurev.psych.093008.100511.

McKinsey Global Institute (MGI) report. (2015). *The power of parity: How advancing women's equality can add $12 trillion to global growth*. Accessed August 2, 2019 from https://www.mckinsey.com/~/media/McKinsey/Featured%20Insights/Employment%20and%20Growth/How%20advancing%20womens%20equality%20can%20add%2012%20trillion%20to%20global%20growth/MGI%20Power%20of%20parity_Full%20report_September%202015.ashx.

Minguez, A. (2012). Gender, family and care provision in developing countries: Towards gender equality. *Progress in Development Studies, 12*(4), 275–300. https://doi.org/10.1177/146499341201200402.

Moghadam, V. (1992). *Development and patriarchy: The Middle East and North Africa in economic and demographic transition*. WIDER Working Papers (1986-2000) 1992/099 Helsinki: UNU-WIDER.

Moghadam, V. (2004). Patriarchy in transition: Women and the changing family in the Middle East. *Journal of Comparative Family Studies, 35*(2), 137–162.

Momani, B. (2016). *Equality and the economy: Why the Arab world should employ more women*. The Brookings Institution. Retrieved August 1, 2019 from https://www.brookings.edu/wp-content/uploads/2016/12/bdc_20161207_equality_in_me_en.pdf.

Morgan, R., Ayiasi, R. M., Barman, D., et al. (2018). Gendered health systems: Evidence from low- and middle-income countries. *Health Research Policy and Systems, 16*(58) https://doi.org/10.1186/s12961-018-0338-5.

Morgan, T., Williams, L., Trussardi, G., & Gott, M. (2016). Gender and family caregiving at the end-of-life in the context of old age: A systematic review. *Palliative Medicine, 30*(7), 616–624. https://doi.org/10.1177/0269216315625857.

Musich, S., Wang, S., Kraemer, S., Hawkins, K., & Wicker, S. (2017). Caregivers for older adults: Prevalence, characteristics, and health care utilizations and expenditures. *Geriatric Nursing, 28*, 9–16. https://doi.org/10.1016/j.gerinurse.2016.06.017.

National Partnership for Women & Families. (2017). *Paid sick days: Busting common myths with facts and evidence*. Retrieved May 1, 2018 from http://www.nationalpartnership.org/our-work/resources/workplace/paid-sick-days/busting-the-myths-about-paid-sick-days.pdf.

Newman, C. (2014). Time to address gender discrimination and inequality in the health workforce. *Human Resources for Health, 12*(1), 25. https://doi.org/10.1186/1478-4491-12-25.

Nguyen, H. T., & Connelly, L. B. (2014). The effect of unpaid caregiving intensity on labour force participation: Results from a multinomial endogenous treatment model. *Social Science & Medicine, 100*, 115–122. https://doi.org/10.1016/j.socscimed.2013.10.031.

Olson, D. H., Russell, C. S., & Sprenkle, D. H. (1983). Circumplex model of marital and family systems: VI. Theoretical update. *Family Process, 22*(1), 69–83.

Oppong, C. (2006). Familial roles and social transformations: Older men and women in sub-Saharan Africa. *Research on Aging, 28*, 654–668. https://doi.org/10.1177/0164027506291744.

Organization for Economic Cooperation and Development (OECD). (2014). *Women in public life: Gender, law and policy in the Middle East and North Africa.* https://doi.org/10.1787/9789264224636-en.

Oxford Islamic Studies. (2019). In J. L. Esposito (Ed.), *"Women" in the Islamic world: Past and present.* Retrieved August 3, 2019 from http://www.oxfordislamicstudies.com/print/opr/t243/e370.

Pavalko, E. K., & Wolfe, J. D. (2016). Do women still care? Cohort changes in US women's care for the ill or disabled. *Social Forces, 94*(3), 1359–1384. https://doi.org/10.1093/sf/sov101.

Pesando, L. M., Castro, A., Andriano, L., Behrman, J., & Billari, F. (2018). *Global family change: Persistent diversity with development.* University of Pennsylvania Population Center Working Paper (PSC/PARC), 2018-14. Retrieved August 3, 2019 from https://repository.upenn.edu/psc_publications/14.

Pew Research Center. (2011). *The future of the global Muslim population.* Region: Middle East-North Africa. Retrieved August 3, 2019 from https://www.pewforum.org/2011/01/27/future-of-the-global-muslim-population-regional-middle-east/.

Pew Research Center (2013). *Women in society.* In *The world's Muslims: Religion, politics and society.* (chapter 4). Retrieved August 3, 2019 from https://www.pewforum.org/2013/04/30/the-worlds-muslims-religion-politics-society-women-in-society/.

Pinquart, M., & Sorenson, S. (2006). Gender differences in caregiver stressors, social resources, and health: An updated meta-analysis. *Journals of Gerontology-Series B, Psychological Sciences and Social Sciences, 61*(1), 33–45. https://doi.org/10.1093/geronb/61.1.P33.

Plichta, S. B. (2018). Paying the hidden bill: How public health can support older adults and informal caregivers. *American Journal of Public Health, 108*(10), 1282–1284. https://doi.org/10.2105/AJPH.2018.304670.

Qi, L., & Dong, X. (2016). Unpaid care work's interference with paid work and the gender earnings gap in China. *Feminist Economics, 22*(2), 143–167. https://doi.org/10.1080/13545701.2015.1025803.

Rabow, M., Hauser, J., & Adams, J. (2004). Supporting family caregivers at the end of life: "They don't know what they don't know" *Journal of the American Medical Association, 291*(4), 483–491.

Raven, J., Akweongo, P., Baba, A., Olikira, S., Sall, M., Buzuzi, S., et al. (2015). Using a human resource management approach to support community health workers: Experiences from five African countries. *Human Resources for Health, 13*, 45–58. https://doi.org/10.1186/s12960-015-0034-2.

Reinhard, S., Given, B., Petlick, N., & Bemis, A. (2008). Supporting family caregivers in providing care. In R. Hughes (Ed.), *Patient safety and quality: An evidence-based handbook for nurses.* Rockville: Agency for Healthcare Research and Quality.

Ridgeway, C., & Correll, S. (2004). Unpacking the gender system: A theoretical perspective on gender beliefs and social relations. *Gender & Society, 18*(4), 510–531. https://doi.org/10.1177/0891243204265269.

Riffin, C., Van Ness, P., Wolff, J., & Fried, T. (2019). Multifactorial examination of caregiver burden in a national sample of family and unpaid caregivers. *Journal of American Geriatric Society, 67*, 277–283. https://doi.org/10.1111/jgs.15664.

Rutherford, A., & Bowes, A. (2014). Networks of informal caring: A mixed-methods approach. *Canadian Journal on Aging/La Revue Canadienne du Vieillissement, 33*(4), 473–487. https://doi.org/10.1017/S0714980814000361.

Samman, E., Presler-Marshall, E., Jones, N., Bhatkal, T., Melamed, C., Stavropoulou, M., et al. (Check for Year). *Women's work: Mothers, children and the global childcare crisis.*

Overseas Development Institute. Retrieved August 3, 2019 from https://www.odi.org/sites/odi.org.uk/files/odi-assets/publications-opinion-files/10333.pdf.

Schmitz, H., & Westphal, M. (2015). Short- and medium-term effects of informal care provision on female caregivers' health. *Journal of Health Economics*, *42*, 174–185. https://doi.org/10.1016/j.jhealeco.2015.03.002.

Schnabel, L. (2016). Religion and gender equality worldwide: A country-level analysis. *Social Indicators Research*, *129*(2), 893–907. https://doi.org/10.1007/s11205-015-1147-7.

Sen, A. (1996). Gender inequality and theories of justice. In M. Nussbaum & J. Glover (Eds.), *Women, culture, and development: A study of human capabilities* (pp. 259–270). New York: Oxford University Press.

Sen, G., & Ostlin, P. (2009). Gender as a social determinant of health: Evidence, policies, and innovations. In *Gender equity in health: The shifting frontiers of evidence and action*. New York: Routledge.

Sepulveda Carmona, A., & Donald, K. (2014). What does care have to do with human rights? Analyzing the impact on women's rights and gender equality. *Gender and Development*, *22*(3), 441–457.

Sharma, N., Chakrabarti, S., & Grover, S. (2016). Gender differences in caregiving among family-caregivers of people with mental illnesses. *World Journal of Psychiatry*, *6*(1), 7–17. https://doi.org/10.5498/wjp.v6.i1.7.

Silverstein, M., & Giarrusso, R. (2010). Aging and family life: A decade review. *Journal of Marriage and Family*, *72*(5), 1039–1058. https://doi.org/10.1111/j.1741-3737.2010.00749.x.

Skira, M. M. (2015). Dynamic wage and employment effects of elder parent care. *International Economic Review*, *56*(1), 63–93. https://doi.org/10.1111/iere.12095.

Smith, M., & Henderson-Andrade, N. (2006). Facing the health worker crisis in developing countries: A call for global solidarity. *Bulletin of the World Health Organization*, *84*(6), 426. https://apps.who.int/iris/handle/10665/269659.

Stacey, A. F., Gill, T. K., Price, K., & Taylor, A. W. (2019). Biomedical health profiles of unpaid family carers in an urban population in South Australia. *PLoS One*. *14*(3). https://doi.org/10.1371/journal.pone.0208434.

Tlaiss, H. (2013). Women in healthcare: Barriers and enablers from a developing country perspective. *International Journal of Health Policy Management*, *1*(1), 23–33. https://doi.org/10.15171/ijhpm.2013.05.

Trivedi, R., Beaver, K., Bouldin, E. D., Eugenio, E., Zeliadt, S. B., et al. (2014). Characteristics and well-being of informal caregivers: Results from a nationally-representative US survey. *Chronic Illness*, *10*(3), 167–179. https://doi.org/10.1177/1742395313506947.

UN WOMEN (2019). *SDG 5: Achieve gender equality and empower all women and girls*. Retrieved August 30, 2019 from https://www.unwomen.org/en/news/in-focus/women-and-the-sdgs/sdg-5-gender-equality.

United Nations Women (UN WOMEN). (2018). *Promoting men's caregiving to advance gender equality*. Retrieved August 3, 2019 from https://promundoglobal.org/wp-content/uploads/2018/10/F-Understanding-How-to-Promote-Mens-Caregiving-to-Advance-Gender-Equality-1.pdf.

Wayne, J., & Casper, W. (2016). Why having a family-supportive culture, not just policies, matters to male and female job seekers: An examination of work-family conflict, values, and self-interest. *Sex Roles*, *75*(9–10), 459–475. https://doi.org/10.1007/s11199-016-0645-7.

Whitman, E. (2016, December 17). *Daughter, student, caregiver, nurse: Unpaid caregivers are finding their place in the healthcare system*. Modern Healthcare. Retrieved May 25, 2019 from https://www.modernhealthcare.com/article/20161217/MAGAZINE/312179843/daughter-student-caregiver-nurse-unpaid-caregivers-are-finding-their-place-in-the-healthcare-system.

WHO (2019). *Gender equity in the health workforce: Analysis of 104 countries*. Health Workforce Working paper 1. Retrieved August 30, 2019 from https://apps.who.int/iris/bitstream/handle/10665/311314/WHO-HIS-HWF-Gender-WP1-2019.1-eng.pdf?ua=1.

Wolf, D., Freedman, V., & Soldo, B. (1997). The division of family labor: Care for elderly parents. *The Journals of Gerontology*, *52B*, 102–109.

Wolff, J. L., Spillman, B. C., Freedman, V. A., & Kasper, J. D. (2016). A national profile of family and unpaid caregivers who assist older adults with health care activities. *JAMA Internal Medicine*, *176*(3), 372–379. https://doi.org/10.1001/jamainternmed.2015.7664.

World Bank (2004). *Gender and development in the Middle East and North Africa: Women in the public sphere*. Retrieved August 3, 2019 from documents.worldbank.org/curated/en/976361468756608654/pdf/281150PAPER0Gender010Development0in0MNA.pdf.

World Bank (2009). *The status and progress of women in the Middle East and North Africa*. Retrieved August 3, 2019 from http://documents.worldbank.org/curated/en/911511503553406149/pdf/116732-WP-MENA-Status-and-Progress-2009-PUBLIC.pdf.

World Bank Data (2019). *Population, total*. Retrieved August 28, 2019 from https://data.worldbank.org/indicator/SP.POP.TOTL?name_desc=false.

World Economic Forum (2018a). *108 years: Wait for gender equality gets longer as women's share of workforce, politics drops*. Retrieved August 30, 2019 from http://reports.weforum.org/global-gender-gap-report-2018/press-release/.

World Economic Forum (2018b). *The global gender gap report*: (p. 2018). Retrieved August 3, 2019 from http://www3.weforum.org/docs/WEF_GGGR_2018.pdf.

World Health Organization (WHO). (2012). *Dementia: A public health priority*. Retrieved from https://apps.who.int/iris/bitstream/handle/10665/75263/9789241564458_eng.pdf;jsessionid=7B68CD6AEFF9ADD45AACC0C354964EEA?sequence=1.

Yang, Y. T., & Gimm, G. (2013). Caring for elder parents: A comparative evaluation of family leave laws. *The Journal of Law, Medicine & Ethics*, *41*(2), 501–513. https://doi.org/10.1111/jlme.12058.

Yeganeh, H., & May, D. (2011). Cultural values and gender gap: A cross-national analysis. *Gender in Management*, *26*(2), 106–121. https://doi.org/10.1108/17542411111116536.

Supplementary reading recommendations

Ar, Y., & Karanci, A. (2019). Turkish adult children as caregivers of parents with Alzheimer's disease: Perceptions and caregiving experiences. *Dementia*, *18*(3), 882–902.

Ayalong, L. (2004). Cultural variants of caregiving or the culture of caregiving. *Journal of Cultural Diversity*, *11*(4), 131–138.

Booth, A. (2016). Women, work and the family: Is Southeast Asia different? *Economic History of Developing Regions*, *31*(1), 167–197. https://doi.org/10.1080/20780389.2015.1132624.

Brussevich, M., et al. (2018). *Gender, technology, and the future of work*. International Monetary Fund. Staff Discussion Notes No. 18/07. Retrieved from https://www.imf.org/en/Publications/Staff-Discussion-Notes/Issues/2018/10/09/Gender-Technology-and-the-Future-of-Work-46236.

Carmeli, E. (2014). The invisibles: Unpaid caregivers of the elderly. *Frontiers in Public Health*. *2*(91). https://doi.org/10.3389/fpubh.2014.00091.

Community Health Workers, Retrieved August 16, 2019 from https://www.apha.org/apha-communities/member-sections/community-health-workers.

Dumit, N., Abboud, S., Massouh, A., & Magilvy, J. (2015). Role of the Lebanese family caregivers in cardiac self-care: A collective approach. *Journal of Clinical Nursing*, *24*, 3318–3326.

Ferrant, G., & Thim, A. (2019). *Measuring women's economic empowerment: Time use data and gender inequality.* OECD. OECD Development Policy Papers No. 16. Retrieved September 6, 2019 from http://www.oecd.org/dev/development-gender/MEASURING-WOMENS-ECONOMIC-EMPOWERMENT-Gender-Policy-Paper-No-16.pdf.

Friedemann-Sanchez, G., & Griffin, J. (2013). *Economic and health outcomes of unpaid caregiving: A framework from the health and social sciences.* Minnesota Population Center Working Paper No. 2013-10. Retrieved from https://assets.ipums.org/_files/mpc/wp2013-10.pdf.

Jung, A., & O'Brien, K. (2019). The profound influence of unpaid work on women's lives: An overview and future directions. *Journal of Career Development*, *46*(2), 184–200.

Kabitsi, N., & Powers, D. (2002). Spousal motivations of care for demented older adults: A cross-cultural comparison of Greek and American female caregivers. *Journal of Aging Studies*, *16*, 383–399.

Lahoti, R., & Swaminathan, H. (2016). Economic development and women's labor force participation in India. *Feminist Economics*, *22*(2), 168–195.

Magana, I., Martinez, P., & Loyola, M. (2018). Health outcomes of unpaid care workers in low-income and middle-income countries: A protocol for a systematic review. *BMJ Open*, *8*, e018643. https://doi.org/10.1136/bmjopen-2017-018643.

Manandhar, M., Hawkes, S., Buse, K., Nosrati, E., & Magar, V. (2018). Gender, health and the 2030 agenda for sustainable development. *Bulletin of the World Health Organization*, *96*, 644–653.

Maybud, S. (2015). *Women and the future of work—Taking care of the caregivers.* International Labor Organization. ILO's Work in Progress. Retrieved from https://www.ilo.org/wcmsp5/groups/public/—ed_protect/—protrav/—travail/documents/publication/wcms_351297.pdf.

OECD. (2011). Cooking and caring, building and repairing: Unpaid work around the world. In *Society at a glance 2011: OECD social indicators.* Paris: OECD Publishing Retrieved September 6, 2019 from https://www.oecd-ilibrary.org/docserver/soc_glance-2011-3-en.pdf?expires=1567280222&id=id&accname=guest&checksum=EE4ABEF81DA00AB79BA12B886BC1A575.

Pew Research Center (2017). *The future of the global Muslim population, Region: Middle East-North Africa.* Retrieved August 3, 2019 from https://www.pewforum.org/2011/01/27/the-future-of-the-global-muslim-population/.

Ruiz, I., & Nicolas, M. (2018). The family caregiver: The naturalized sense of obligation in women to be caregivers. *Enfermeria Global*, *49*, 434–447. https://doi.org/10.6018/eglobal.17.1.292331.

Tabatabaei, M., & Mehri, N. (2019). Gender inequality in unpaid domestic housework and childcare activities and its consequences on childbearing decisions: Evidence from Iran. *Journal of International Women's Studies*, *20*(2), 26–42.

United Nations Economic and Social Commission for Asia and the Pacific (ESCAP) (2015). *Gender equality and women's empowerment in Asia and the Pacific.* Retrieved from https://www.unescap.org/sites/default/files/publications/B20%20Gender%20Equality%20Report%20v10-3-E.pdf.

WHO (2016). Global strategy on human resources for health: Workforce 2030. Retrieved August 2, 2019 from https://apps.who.int/iris/bitstream/handle/10665/250368/9789241511131-eng.pdf;jsessionid=7DA1D5D5899A018CF6E1EFD88B98C0C6?sequence=1.

World Health Organization (2019). Delivered by women, led by men: A gender and equity analysis of the global health and social workforce. In *Human Resources for Health Observer Series No. 24.* Retrieved from https://apps.who.int/iris/bitstream/handle/10665/311322/9789241515467-eng.pdf.

Videos

UCLA Caregiver Training (series of videos—English and Spanish): <https://www.uclahealth.org/dementia/caregiver-education-videos.

How to Manage Compassion Fatigue in Caregiving, Patricia Smith, TEDxSanJuanIsland: https://www.youtube.com/watch?v=7keppA8XRas.

RCIL Caregiver Training Videos—Arabic: https:/www.youtube.com/watch?v=-95vOXMYD4o&list=PL0ER9jI5iKEJSFyrgEQ9-QwEkNOMqsoNe.

UN WOMEN—What is the real value of unpaid work?: https://www.youtube.com/watch?v=fcqt0QzgUFU

Institute of Development Studies—Who cares: Unpaid care work, poverty and women's/girl's human rights: https://www.youtube.com/watch?v=VVW858gQHoE

Books

Antonopoulos, R., & Hirway, I. (Eds.), (2010). *Unpaid work and the economy: Gender, time use and poverty in developing countries*. London: Palgrave Macmillan.

Chast, R. (2016). *Can't we talk about something more pleasant?: A memoir*. Brooklyn, NY: Bloomsbury USA.

Gates, M. (2019). *The moment of lift: How empowering women changes the world*. New York, NY: Flatiron Books.

Mies, M. (2014). *Patriarchy and accumulation on a world scale: Women in the International Division of Labor*. Chicago, IL: University of Chicago Press.

Murray, P. (Ed.), (2014). *Women and gender in modern Latin America*. New York and London: Routledge.

Solati, F. (2017). *Women, work and patriarchy in the Middle East and North Africa*. London: Palgrave Macmillan.

Stettinius, M. (2012). *Inside the dementia epidemic: A daughter's memoir*. Horseheads, NY: Dundee-Lakemont Press.

van Nederveen Meerkerk, E. (2019). *Women, work and colonialism in the Netherlands and Java: Comparisons, contrasts, and connections, 1830–1940*. London: Palgrave Macmillan.

Waring, M. (1999). *Counting for nothing: What men value and what women are worth*. Toronto, CA: University of Toronto Press.

Definitions

Community health worker (CHW) American Public Health Association defines a Community Health Worker (CHW) as a frontline public health worker who is a trusted member of and/or has an unusually close understanding of the community served. This trusting relationship enables the CHW to serve as a liaison/link/intermediary between health/social services and the community to facilitate access to services and improve the quality and cultural competence of service delivery.

Familism a cultural value that emphasizes family relationship prioritization over the needs of any individual within the family.

Family cohesion the emotional bonding that family members have toward one another (Olson, Russell, & Sprenkle, 1983).

Filial piety the virtue and primary duty of respect, obedience, and care for one's parents and elderly family members.

Gender gap difference between men and women concerning a variety of public and private issues.

Gender parity concerns relative to equality in terms of numbers and proportions of women and men, girls and boys, and is often calculated as the ratio of female-to-male values for a given indicator.

Global Domestic Product (GDP) total market value of the goods and services produced by a country's economy during a specified period of time; includes all final goods and services—that is, those that are produced by the economic agents located in that country regardless of their ownership and that are not resold in any form; global measure of output and economic activity.

Misogynistic strong, negative emotions about women, sometimes characterized by hatred.

Patriarchy A social formation…where property, residence, and descent proceed through the male line. In classic patriarchy, the senior man has authority over everyone else in the family, including younger men, and women are subject to distinct forms of control and subordination (Moghadam, 1992).

Social parity concerns relative to equality in terms of numbers and proportions of racial, gender, and socio-economic levels for a given indicator.

Unpaid care work refers to all unpaid services provided within a household for its members, including care of persons, housework, and voluntary community work. These activities are considered work, because theoretically one could pay a third person to perform these items (Ferrant et al., 2014).

Criminal justice research: Incorporating a public health approach

Lynette Feder, Samantha Angel
University of Central Florida, Department of Criminal Justice, Orlando, FL, United States

Abstract

Criminal justice, a social science discipline intersecting law, psychiatry, psychology, sociology, and social work, continues to work toward developing effective solutions to crime control and reduction. Increasing numbers of criminologists are engaging an applied evidence-based strategy used widely in medical and public health research. These rigorous research methods more validly assess the effects of mandated programs and policies in criminal justice. This chapter provides a foundational understanding of the circumstances that birthed the traditional approach to research taken by criminologists and how several factors are moving the field to using an evidence-based approach. Two overarching recommendations from this review are the need for (1) more classes in experimental research at the graduate level, and (2) mentoring young criminologists on ways to run experiments within criminal justice agencies to ensure continued use of these methods by the next generation of criminologists.

Keywords: Criminal justice, Criminology, Experimental research, Evidence-based research

Chapter outline

Three Facets of Public Health and Paths to Improvements. https://doi.org/10.1016/B978-0-12-819008-1.00011-0

A brief history of criminal justice research

Ironically, just as culture impacts public health, public health can impact culture. Over the last 25 years, several factors, such as (1) the rise of the use of experiments in criminal justice research, (2) the start of the Campbell Collaboration, and (3) the increase in interdisciplinary research often due to funding requirements, have coalesced to move research in criminal justice more in line with the approach used by those within public health. While this may at first seem incompatible, a closer look indicates that this gradual shift can be explained in that both disciplines regularly deal with marginalized populations who may fall under shared domains (e.g., domestic violence, drug addiction, mentally disturbed offender, etc.). The focus of this chapter looks at several factors which have led to a slow but steady convergence as increasing numbers of criminologists adopt some of the most salient elements of public health research. Understanding how research was traditionally conducted within the field of criminal justice helps to explain its original theoretical approach to research. We therefore start with a brief history of criminal justice research.

Before there was criminal justice research, there was crime which led to the criminal justice system. Cesare Beccaria, a young Milanese aristocrat living during the European Renaissance, is credited with being the first person to study the criminal justice system in terms of its impact on crime. At that time, laws criminalizing behaviors did not exist leading to crime being ill-defined yet extensive. Additionally, the criminal justice system punished people often based on who they were instead of what they did (Beccaria, 1963). Beccaria's ideas on how to reform this corrupt system were guided by his thinking on why individuals committed crime and this, in turn, was heavily influenced by the enlightened philosophies of his day. In 1764, he laid out his reasons why the then brutal and capricious criminal justice system had to be reformed and the ways in which these changes should be made in his now famous treatise ***On Crimes and Punishments***.

According to Beccaria, individuals were egotistical and would always endeavor to pursue their own pleasure without regard for the rights of others. However, as man was also a rational being, governments could control these selfish impulses by ensuring that the pain for infringing on the rights of others (through punishment) outweighed the advantages obtained from the illegality. Beccaria's prescription for lowering the crime rate was to make the laws public as well as creating a rational criminal justice system. According to Beccaria, this would allow individuals to know what acts were illegal and, should they choose to engage in these behaviors, the punishment they could expect to receive if caught (Beccaria, 1963). Given man's rational calculations, Beccaria believed that this would lessen his likelihood of committing crime.

Many of the ideas laid out by Beccaria and others in the Classical School of Criminology quickly took hold in western Europe. Though far from perfect, laws began to be written criminalizing specific behaviors. Punishments also became more rational in that judges now were to look at the harm caused when deciding sentences. Despite these reforms, crime rates continued to climb leading to people turning away from classical criminology and its prescribed reforms of the criminal justice system.

An alternative view of the criminal came about with Charles Darwin's publication of the ***Origin of the Species*** in 1864 (Darwin, 2010). In it, he set forth his theories of natural selection and survival of the fittest. A few years later in 1871, Darwin published ***The Descent of Man*** in which he argued that some individuals were less evolved than others and, as such, were closer to their ape-like ancestors (Darwin, 1998). As before, people's world view influenced their understanding of why individuals became criminal. While Classicalists saw crime as a logical choice made by a rational individual when the benefits of the misdeed outweighed the pain for getting caught, a new view of the criminal now emerged. Criminals were instead viewed as individuals whose behavior was biologically driven rather than rationally decided.

Cesare Lombroso, an army physician, became one of the first proponents of biological positivist criminology. In autopsying criminals, he said that he had noticed physical differences that set them apart from noncriminals. In 1876, Lombroso published ***The Criminal Man*** in which he set forth his thesis that the development of individuals followed the same path as the development of society as enumerated by Darwin. Specifically, the physical differences he had noted, what he called stigmata, were indicators that the criminal was an atavistic being. That is, that the criminal was a throwback to an earlier era in our evolutionary history. As such, Lombroso argued that criminals had a biological basis for committing crime, and therefore, could not be expected to peacefully coexist with modern man (Lombroso-Ferrero, 2012).

While many continued to pursue a biological basis for crime well into the early 1900s, biological positivism eventually fell out of favor with researchers and the public alike. However, researchers in criminal justice have continued to search for differences between criminals and noncriminals to better understand why some people break the law, while the largest percentage of individuals do not engage in frequent or serious offending. This search for differences between criminals and noncriminals continues today, though the focus is now on sociological differences, such as those existing in parenting practices, education, neighborhood, peers, or socioeconomic status.

Some might argue that the age-old problems associated with the foundation of the criminal justice system, development of legal definitions, and generalized perception of the criminal are a significant justification for an enlightened approach to criminal justice research. Further, they demonstrate the need to create objective methods based on scientifically gathered evidence free of bias. Given the persistence of these problems, there is an understandable need to demonstrate how historical patterns of the criminal justice system reflect existing gaps in factual understanding of the criminal, their activities, and how best for sovereign entities to address crime reduction for those who commit crime, those impacted by criminal activities, and communities that struggle with the consequences of crime related to the overall quality of life. We seek to illuminate the problem and foster solutions to elicit positive changes.

This chapter largely explains the original theoretical approach to criminal justice research by highlighting some of the most noteworthy differences between how research in criminal justice traditionally proceeded when compared to the approach used by public health researchers. Given the discipline's recent move toward a public health approach, the chapter provides an overview of the salient factors leading to the progression toward evidenced-based methods into criminal justice research. We will then discuss some of the specific methods that criminologists have adopted by providing examples in the criminal justice research literature. Finally, we support sustaining the discipline's continued movement toward an evidence-based approach and recommend that criminal justice pedagogy includes more information on experimental research on the graduate level along with mentorship of young criminologists so as to be able to continue providing the most valid assessments of what works to policymakers.

Two different approaches to research and why who is at the table makes a difference

Criminology's traditional approach to crime and criminals gives rise to some important differences in the way in which the field has customarily conducted research when compared to the methods used by those in public health. These include differences in the purpose for the research, the approaches to the research, and who is brought into the research process.

Differences in the purpose of research

A major difference in the two fields' approach deals with what is viewed as the purpose of the research. Given how the study of crime and criminals evolved, criminal justice researchers use a theoretical approach which has traditionally had less application to program development.

As was true of Lombroso, criminologists continue to search for differences between criminals and noncriminals. This identification is used to either develop or test theories seeking to explain why individuals commit crime. An example of this type of research can be seen in Wolfgang, Figlio, and Sellin's seminal study, ***Delinquency in a Birth Cohort*** (1972). The researchers tracked all official delinquencies of males born in 1945 and living in Philadelphia between the ages of 10 and 18. They found that 6% of the delinquents accounted for 52% of all the juvenile contacts with the police (Wolfgang et al., 1972, p. 204). In other words, a small percentage of individuals accounted for both the most serious and the largest number of delinquencies. While this finding has been important in developing theories on crime causation, it has not provided any information on how to curtail or lessen the likelihood that an adolescent would become involved in delinquent activities. In other words, the study sought to explain delinquency in the community rather than to change the outcomes.

This difference in focus gives rise to one of the largest distinctions between the two fields in their approach to conducting research. While criminology traditionally had a theoretical approach to research, public health's focus is on applied research. That is, research conducted to solve a specific problem rather than to develop or test a theory. Some of the most well-known and successful public health initiatives have begun with research to develop programs seeking to change individual behaviors, for instance, to avoid drunk driving, wear seat belts when driving, or curtail cigarette smoking. In each of these endeavors, public health's focus was on changing the behavior of individuals rather than developing or testing theories as to why individuals drive drunk, fail to wear seat belts, or smoke cigarettes.

The point at which each field typically intervenes is connected to the difference noted above. Traditionally, criminal justice exclusively looked to intervene only after the bad behavior had been committed. (There is one notable exception to this in that the original intent of the juvenile justice system as developed in the late 1800s was to prevent criminality rather than to react to it at a later date.) This is in stark contrast to public health's focus on prevention. Though preventing bad behavior may be more effective than intervening once it has taken root, there is an important reason for criminal justice's backward-looking approach that lies with how our Founding Fathers viewed the criminal justice system and, therefore, originally wrote the laws of the land.

As a colony of England, the Founders felt they had been subjected to a criminal justice system that was used as a tool to oppress them when their views departed from that of the King's. In writing our Constitution, these men, therefore, sought to ensure that the laws they were establishing could not be used to tyrannize its citizens. As such, our criminal justice system can typically only be used reactively (rather than proactively) to punish individuals for behavior that: (a) has first been established as criminal (this is the idea of "no crime without first there being law," also known as principle of legality); and that (b) an individual must first be convicted of having engaged in this criminal conduct before the government can punish (Neubauer & Fradella, 2019). Once convicted, criminal justice personnel look at stopping the offender from immediately re-offending (this is the idea of incapacitation), lessening his likelihood of re-offending in a similar manner in the future (this is the idea of specific deterrence), or even trying to stop the offender from engaging in all bad behavior in the future (more commonly known as rehabilitation).

In contrast, public health does not look to punish the individual, but instead seeks to make him less likely to engage in the harmful behavior in the future. In order to accomplish this, public health researchers look at the underlying antecedents (more commonly known as "risk factors") that are thought to have led to this behavior (Moore, Prothrow-Stith, Guyer, & Spivak, 1994). They will then seek to prevent the development of these antecedents via primary or secondary prevention. While primary prevention typically applies preemptive efforts to a wide population before the negative outcome has occurred, secondary prevention looks to direct its efforts to a select group who have already shown that they are at increased risk for the negative outcome (Welsh, Braga, & Sullivan, 2012).

Differences in the research approach

Linked to the above difference is the way the individual is viewed in criminal justice versus public health. Deriving from the discipline's historical roots in classical criminology, criminal justice personnel regarded the offender as having made an affirmative choice to commit the crime. Consequently, he was thought to be deserving of punishment for the harms he caused to the victim and the larger community. As the criminal justice system seeks to individualize justice, once convicted, the focus would now be on the offender and his background when the courts were called upon to make decisions about how best to handle him. Not unusually, the punishment was incarceration or community corrections, though the system could still introduce rehabilitation in the form of educational or occupational programs with the hope of reducing the likelihood that he would re-offend in the future.

Importantly, most criminal justice interventions have not been founded upon evidence-based research (Sherman, 2003). Instead, the way in which criminal justice deals with crime and criminals has typically been a political issue. In practical terms, this means that policies and programs in criminal justice are often based on the political ideologies of the time (Farrington, 2003). Just as problematic, even fewer of these implemented interventions will then be rigorously evaluated to test whether they are effective in their goal of reducing the individual's likelihood of re-offending in the future (Sherman, 2003).

Public health's approach is quite different. Instead of thinking of the individual as morally unfit, public health looks to determine the risk factors preceding the behavior. To accomplish this, public health researchers will look to find the antecedents among individuals exhibiting this behavior (Moore, 1993). Researchers in public health will then develop interventions seeking to counter or minimize these antecedents with the expectation that this will stop or lessen the bad behavior in the future (Moore et al., 1994). And after the intervention is put into place, public health typically will evaluate whether the specific program implemented achieved its intended outcome (Moore et al., 1994). Therefore, public health researchers use an evidence-based approach to policy and program development. Public health's scientific approach to prevention is diagramed below (Fig. 11.1).

The differences in each discipline's view of why individuals behave as they do have ripple effects for how both the research and the researcher are viewed by community members when they conduct their investigations in the field. It is a truism that areas where crime is high are also typically communities with more criminals. Unfortunately, the criminal justice view of criminals as morally unfit can make it difficult to gain the trust, support, and buy-in from community members. Family, friends, and neighbors may feel like they are being viewed as responsible for the crime and the criminal in such communities. They may also feel that the researcher is judging them and thinking that had the family taught the individual right from wrong, had the community been more cohesive, or had the school provided more direction, the individual would not be a criminal. Where communities feel blamed for the problem, implicitly or explicitly, they may be less trusting of the researcher and therefore less likely to participate in the study.

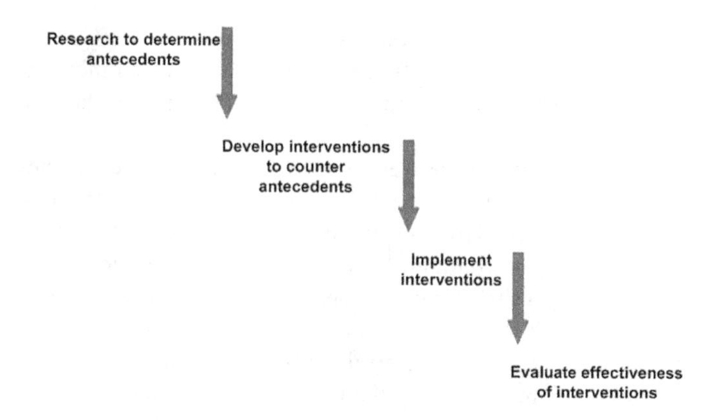

Fig. 11.1 Public health's approach to applied research.

In contrast, public health views the individual's bad behavior as attributable to factors outside of the individual. According to the public health view, crime may be due to such factors as a lack of economic opportunity, drug-infested and gang-ridden neighborhoods, or lack of legitimate opportunities. One can easily see how this latter approach, viewed as less blaming of the individual and those around him, would be more acceptable to individuals. They may, therefore, be more trusting of the researcher and with that more likely to fully engage in the research.

A final distinction in the research approach provides another large and significant difference between the two fields. Specifically, criminal justice personnel rely on the law to change behavior, while public health typically prefers to use education to modify the bad behavior. Public health's hesitancy in using the law to effect change is due to the recognition that criminalizing behavior drives it underground, making it even more difficult to access and then successfully change.

Differences in who is brought into the research process

The differences in the way in which criminal justice versus public health researchers view their subjects may explain why community stakeholders are less comfortable and trusting of both criminal justice research and the individuals conducting these studies. As researchers know, conducting research in a community setting is never easy. But the research process becomes more problematic when there is suspicion and distrust among those who host the study or are the targets of the research. This lack of trust between criminologists and those in the communities where they work might explain why, in the past, criminal justice has not typically sought to include members of the community in their research endeavors.

The same is not true in public health research. We speculate this is probably driven by the fact that the mission of public health research is to gather data from the community and then use that information to solve problems that exist in that community. Necessarily, this means that the first phase of their research includes speaking with those in the community to develop a fuller understanding of the problem they are seeking to target.

Another important difference between the two disciplines lies in their view of agency personnel. [C]riminal justice traditionally used research to develop theories on why some individuals turned to crime while others did not. However, even as they conducted their studies in the community, criminologists did not look to bring criminal justice personnel into the research endeavor. Traditionally, these researchers viewed agency personnel as merely the source of their data. As the researcher controlled the research, they therefore determined both what to do with that data and, consequently, what the results from their analyses indicated in terms of recommended programs and policies. Given this perspective, it is possible that agency personnel may have felt used by criminal justice researchers in the past. It was not uncommon for these individuals to feel that there was little reason for them to open their doors to researchers as it just meant more work for them in pulling together the data with little or no gain in terms of solving the immediate issues their clients or agency personnel were regularly facing (Feder, Jolin, & Feyerherm, 2000). Once again, this process is in direct contradiction to the way in which public health researchers approach their research. Public health researchers seek to bring agency personnel, who are already familiar with the problem, on board from the start of their endeavors (Moore, 1993). As they also typically seek to include community stakeholders, public health research has traditionally brought many more people to the table than have criminologists when conducting research and developing programs and policies.

Undoubtedly, there are some concrete reasons for these differences that go beyond pure preference on the part of criminal justice and public health researchers. Given public health's connection to the medical community, there is a stronger scientific tradition in the training of public health personnel. This is not the case for criminal justice personnel such as police, parole, and correctional officers. To the extent that they have not been educated about research, criminal justice personnel may not have the same level of appreciation or support in comparison to those working within public health agencies. Additionally, the administration of agencies of criminal justice parallels those of a military organization with a very rigid hierarchy. Therefore, criminal justice personnel may not be familiar or comfortable with providing feedback. Finally, criminal justice personnel typically do not have the luxury of being proactive about a problem. The situation is already established as a problem by the time the criminal justice system is engaged. Though public health's charge can at times include issues that must be immediately addressed (as in the spread of a highly contagious disease), they also deal with issues where they can be proactive in implementing policies and interventions.

All told, there are many differences between how criminologists and public health researchers approach their studies. Some of these are due to the contingencies of the work and differences in personnel and clientele which form the parameters that each discipline must work within.

The evolution of criminal justice research

Growing numbers of criminologists have begun a slow but steady movement to adopt a more scientific approach to research. To be sure, not everyone is in full agreement with these changes (Cornish & Clarke, 1983). Additionally, it is understood that research consistent with these changes will not always be applicable to the range of issues encompassed under the heading of criminal justice. But findings from past studies, the increase of funding agencies requesting interdisciplinary teams of researchers, as well as cross-fertilization from other disciplines, have led to this slow growing movement within the research community in criminal justice.

While there are typically a multitude of both small and large factors that contribute to any specific change, we address a few of the major events moving criminologists to adopt a different methodological framework for their research. We present each event as a separate entity for the sake of clarity even as these factors are largely interrelated. Disruption to the status quo began in the 1970s with the "finding" that programs in criminal justice were ineffective to the gradual recognition that, though much of what was being done was not helpful, there were some well-run programs that could help. Simultaneously, there was an increasing awareness that, in our efforts to help, we may have inadvertently caused more harm. This recognition led to the realization that criminal justice research required the most rigorous methods to ensure that evaluations are capturing the full range of outcomes of any intervention that is offered or mandated. That includes outcomes showing an intervention to be helpful as well as those indicating harmful effects.

Nothing works in criminal justice

In 1974, Martinson's article entitled *What Works: Questions and Answers in Prison Reform* was published. After reviewing over 200 studies conducted on programs designed to rehabilitate offenders, Martinson concluded, "with few and isolated exceptions, the rehabilitation efforts that have been reported so far have had no appreciable effect on recidivism," (Martinson, 1974, p. 25). This conclusion led the National Research Council, the research arm of the National Academy of Sciences, to create a panel of highly respected criminologists to ascertain the validity of Martinson's findings. In 1979, the panel reported that Martinson's conclusions were largely correct (Weisburd, Farrington, & Gill, 2017). Additionally, given the steady increase in crime that Americans were experiencing during this time, the belief that nothing was effective in decreasing offenders' likelihood of recidivating rang true with the public. Simultaneously, many researchers were also turning away from the idea that rehabilitative treatment could produce positive behavioral change among offenders.

While large numbers of policies and programs did not make an appreciable difference on recidivism, there were some interventions that did have a significant and positive effect on juvenile and adult offenders that began surfacing two decades later. Specifically, Lipsey conducted a metaanalysis based on 443 studies of interventions aimed at juveniles. His research found tremendous variability in results based on both the research methods used and the treatments implemented. The results indicated that when all the studies were taken together, they led to the conclusion that nothing worked. Lipsey found that, when looked at separately, one could find some programs that were effective in lessening delinquents' likelihood of re-offending. This finding was contrary to Martinson's conclusion (Lipsey, 1992).

A few years later, Sherman and his colleagues at the University of Maryland were funded by the United States Congress to determine "what works" in criminal justice. They reviewed 308 intervention studies and found that, contrary to earlier reviews concluding that nothing worked, there were some interventions that were successful in lowering the likelihood of recidivism among adult offenders (Weisburd, Lum, & Petrosino, 2001). Importantly, Sherman and his colleagues also found that evaluations using more rigorous research methods were less likely to incorrectly find an intervention effective when it was not (Sherman, 2003).

Good intentions can harm

For many years, and still to a large extent today, those developing programs and policies in criminal justice start and end with the idea that a given intervention might help some and not others. This assumption hides a large and important omission. Specifically, that there may be interventions that are not only unhelpful, but may in fact be harmful. While most individuals would not take a pill that was handed to them without first knowing that it had been rigorously tested with the full range of both intended and unintended outcomes known and reported, the same does not generally hold true when it comes to social science interventions. Today, there are many programs and policies in criminal justice which have been implemented and even mandated without first being carefully and thoroughly tested to ensure that they do not cause more harm than good. Indeed, several large programs have in recent years come to the public's attention which, though well-intended, caused more harm than good (Dishion, McCord, & Poulin, 1999).

For example, the Cambridge Somerville Youth Study, conducted by Joan McCord, is believed to be one of the first large experiments conducted in criminology and provides an excellent example of the need to conduct rigorous research before declaring an intervention a success. Starting in 1939, 506 boys from two communities in Massachusetts were recruited into the study. The researchers paired the boys, matching them on such variables as their age, social background, somatotype (body type), and temperament, and then flipped a coin to determine whether they would be placed in the treatment or control (no-treatment) groups. Those in the treated group received an array of services including having one social worker attached to them and their family who provided case management services as well as giving guidance to the boys and counseling services to their parents. Additionally, boys in the treatment group received academic tutoring, job training, medical services, referral to specialists when necessary, and summer camp over the 5.5 years that the study ran (McCord, 2003). In the end, the boys in the treated group were quite sure that the program had been of great help to them. Initial statistics backed up this assessment with two-thirds of the treated boys avoiding involvement in delinquent activities or demonstrating a significant decline in these activities over the years (Oakley, 2000).

Based on these facts, most would have declared the Cambridge Somerville Youth Study a great success. But a more careful investigation of the boys' outcomes indicated a contrary conclusion. Specifically, McCord returned 30 years later to Massachusetts and was able to survey 98% of these now middle-aged men. The men in the treatment group continued to speak positively about the program, noting how helpful it had been in their lives. However, McCord found that these men, when compared to their no-treatment counterparts, were significantly more likely to have been convicted of a crime, less likely to be happily married, more likely to have received a medical diagnosis of alcoholism, less likely to have been steadily employed, and more likely to have had psychiatric problems. Overall, the men in the treated group rated themselves as less pleased with their lives when compared to those who had not been given the program (McCord, 2003). Additionally, McCord found what looked to be a negative dose response in that those in the treatment group who received more services and had been in the program longer were likely to have suffered more adverse life outcomes (McCord, 2003). In other words, her research indicated that these services were not only unhelpful, they were, in fact, harmful.

We need not go back in time to find examples of programs thought to be beneficial which demonstrate iatrogenic effects. Scared Straight provides a more recent example of a well-intended and widely implemented program that, when rigorously evaluated, was found to be harmful. Scared Straight brings delinquent adolescents into a prison where inmates spend 3 hours terrifying them about the horrors of prison life to "scare them straight." The program was videotaped and made into a short documentary. Included in this film were scenes of the adolescents exiting the prison seemingly shaken up by the time spent inside, with most saying that they did not want to end up like the individuals they had just heard from. Some studies also confirmed, as shown in the movie, the positive effects of the Scared Straight program (Feinstein, 2005). Many communities adopted the program given these seemingly positive results coupled with the low costs of implementing the program. However, later experimental research indicated that the program was not helpful, but, in fact, was harmful. Specifically, those who went through the program were more likely to subsequently become involved in criminal activity than those who had not gone through the Scared Straight program (Petrosino, Petrosino, & Buehler, 2007). Despite these results, many communities continue to run their Scared Straight program.

The basis for more rigorous research

The aforementioned studies, and the later findings of the harm they caused, point to an important conclusion. Specifically, without a comparison group one would have incorrectly concluded that a program was beneficial when, in fact, it was found to be harmful. But having a comparison group is not sufficient. The no-treatment control group must also be comparable to the group receiving the intervention when studying a program's effectiveness. The way to ensure equivalency between groups is by using an experimental design. This is because the cornerstone of an experiment is the use of random assignment to treatment and control groups. Probability theory tells us that when large numbers of persons are randomly assigned prior to the implementation of an intervention being studied, within the limits of slight statistical variation, the resulting groups should be equivalent.

Randomized controlled trials (RCTs), therefore, provide three advantages not available when using nonexperimental designs. First, RCTs provide every person an equal chance of selection into the experimental and control groups. As such, this equal probability of selection method (EPSEM) uses chance variability to control for the effect that chance plays on outcomes. Additionally, studies that use random assignment minimize bias in the selection of individuals into experimental and control groups, thereby increasing the likelihood that there is equivalence between groups prior to implementing the intervention. Finally, in order to achieve a fair comparison, the groups cannot differ in any way that may affect the outcome prior to introducing the intervention being investigated. Random assignment is, therefore, the truest method for ensuring equivalency between the groups prior to the implementation of the intervention under question. This is because random assignment safeguards the even distribution of factors, both those that are known and those that are unknown, which may be related to the outcome (Feder, 2019).

As the groups are equivalent prior to implementing the intervention, changes observed in the groups postimplementation can only be ascribed to the one difference between the groups—the introduction of the intervention. Therefore, RCTs allow researchers to make an unambiguous link between the intervention under investigation and the outcome observed. Said another way, experiments provide the most unequivocal evidence of an intervention's effectiveness. This explains why experiments are considered the "gold standard" in evaluation research as its results are most likely to accurately reflect the intervention's effectiveness (more commonly referred to as its "internal validity" by researchers) when contrasted to nonexperimental methods (Farrington, 2003).

Quasiexperiments use statistical controls and matching to make the groups equivalent. However, there is always the possibility that some unknown, and therefore, uncontrolled factor(s) might affect the outcome. In these situations, the researcher will not be able to rule out the possibility that the differences observed posttest might reflect nothing more than the differences that existed between the groups pretest. Given this, researchers find that less rigorous research methods are associated with a higher likelihood of falsely finding treatment effectiveness and a lower likelihood of finding harmful effects when compared to findings from an experimental design (Feder & Wilson, 2005; Weisburd, Petrosino, & Lum, 2003).

Move toward evidence-based research

The above has led to a call for evidence-based practices within criminal justice. Instead of using politics, intuition, or beliefs about what should work, evidence-based practice relies on scientific support when developing and implementing interventions. The idea of using science to inform decisions in policy and practice began somewhat unexpectedly with the development of "evidence-based medicine," first established at McMaster University in the 1980s (McGrath, 2004; Smart & Marwick, 2004).

As noted by Sherman (2003, p. 7):

> *The idea of evidence-based medicine was itself somewhat shocking, since many consumers of medicine assumed that most medical practice was based on solid evidence of what works. Yet in medicine as in government, much of what is done proceeds from theory, conjecture, and untested new ideas.*

The most widely accepted definition of evidence-based medicine (EBM) comes from Sackett (1997, p. 3) who indicates, "the conscientious, explicit, and judicious use of current best evidence in making decisions about the care of individual patients. The practice of evidence-based medicine means integrating individual clinical experience with the best available external clinical evidence from systematic research." EBM has come about because of an increasing awareness that many well-intended treatments sometimes offered by highly respected experts in the field have produced more harm than good. An excellent example of this comes from the internationally recognized pediatrician, Benjamin Spock, who advised caregivers to place infants on their stomach when setting them down for sleep. As innocent as this advice may have seemed, it turned out to be lethal leading to the death of thousands of babies. As we now know, positioning babies on their backs is recommended because putting babies to sleep on their stomachs increases the likelihood of sudden infant death syndrome (SIDS). "If advice as apparently innocuous and theoretically sound as recommending a baby's sleeping position can be lethal, there is clearly no room for complacency among professionals about their potential for harming those whom they purport to help," (Chalmers, 2003, p. 24).

This example points to another argument for evidence-based practice. There are many interventions that are implemented without first undergoing rigorous testing. Therefore, their effectiveness cannot be known. In other words, mandating an intervention which has not first been thoroughly tested is as much an experiment as advising that babies be put to sleep on their stomach. The only difference is that, in these cases, the "experiment" is not controlled and therefore we learn nothing from it. "There is thus far more experimenting going on in the social world than in laboratories, although most of it is wrapped in secrecy and thus fails to make a contribution to knowledge," (Oakley, 2000, p. 316).

Evidence-based medicine has led to the establishment of evidence-based practices in a multitude of disciplines including evidence-based criminology. The idea of basing criminal justice policies and practices on the results from scientific research is gaining momentum among practitioners, researchers, and scholars in criminal justice. This would seem to be intuitive as at the heart of all evidence-based practices is the desire to improve outcomes by using the process of obtaining the best available evidence when formulating recommendations in each discipline. Clear (2010, p. 20) eschewed criminologists in the Presidential address to the American Society of Criminology, "How will we evaluate our role in the production of justice when we look back at this sudden invitation by policy makers to bring evidence to bear on the world of policy?" This is certainly a worthwhile question to be answered as evidence-based practices continue to gain traction in many of the social sciences including criminal justice.

Application of evidenced-based methods in studies of the criminal and the criminal justice system

Integration between disciplines is not unusual (Akers & Lanier, 2009). Toward those ends, some researchers have suggested the integration between criminology and public health to form a new subfield of "epidemiological criminology" (Potter & Akers, 2010). While many researchers working in criminal justice have not taken on this title, the factors that have been previously discussed are leading to significant changes in the way in which criminologists are now approaching their inquiries. We illuminate this by providing a brief introduction into this research through the lens of two different studies (one representing the study of the criminal and the other the criminal justice system) and noting their research methodology which clearly demonstrates the migration of public health's approach into the field of criminal justice. Beyond that, these same factors have also led to significant discipline-wide changes. Following these examples, we discuss the impact that the public health approach has had on the discipline.

The criminal using a public health approach

The Seattle Social Development Project (SSDP) by Hawkins and his collaborators studied the antecedents of early conduct problems and found that low achievement and low commitment to school, rejection by prosocial peers, along with high levels of family stress and conflict were related to early drug use and delinquency among children (Peterson, Hawkins, & Catalano, 1992). Therefore, they developed a three-pronged experimental intervention designed to reduce the risk factors identified above. This intervention was then provided to children in grades one through four. A nonrandomized controlled trial, using 598 children in 8 urban elementary schools in high-crime neighborhoods in Seattle, Washington, was used to assess the effectiveness of the experimental intervention.

Teachers of children in the treatment group were trained on positive classroom management skills, interactive teaching, and cooperative learning to reduce academic failure, peer rejection, and conduct disorders among students. Parents of children in the treatment group were also offered parent training classes (though only a minority regularly took advantage of this training). Finally, children in the treatment group received social competence training. This complement of interventions was designed to reduce the identified risk factors for drug abuse and delinquency, thereby keeping the child attached to his school, bonded to his parents, and relating to pro-social peers (Peterson et al., 1992).

By fifth grade, students in the treatment group reported feeling more attached to their schools, more involved with their families, and experienced more peer interactions than those in the control condition. Additionally, students in the treatment group reported significantly lower rates of alcohol use and delinquent activities (Hawkins, Catalano, Kosterman, Abbott, & Hill, 1999).

By age 18, fully 6 years posttreatment, students in the treatment group reported significantly stronger commitment and attachment to school, improvement in self-reported achievement, and a tendency for achieving improved grade performance when compared to the no-treatment control group. Additionally, Hawkins and his team found that those in the control group reported more frequent drinking and engaging in more risky sexual behaviors including having more sexual partners and experiencing more pregnancies. Students in the control group also committed significantly more violent delinquent acts than students in the treatment group. However, no significant effects were found for drug use or heavy cigarette or marijuana use between the two groups (Hawkins et al., 1999).

Fifteen years after the intervention ended (when the subjects were now 27 years of age), those who had been in the treatment group reported significantly better mental health, sexual health, and higher socioeconomic status when compared to the no-treatment group. At this follow-up, no significant differences were obtained between the groups in terms of variables related to crime or substance use (Hawkins, Kosterman, Catalano, Hill, & Abbott, 2008). Though this last finding was disappointing, it must be remembered that the intervention was not very intensive or expensive and yet it was still demonstrating significant positive effects 15 years after it had ended.

In all, the Seattle Social Development Project indicates that preventive interventions targeting risk factors can succeed in having long-term positive effects on an individual's life trajectory. Very much in line with the public health approach, the underlying antecedents of drug use and delinquency were identified, a program was developed, and then implemented to address those risk factors, followed by a rigorous evaluation of the intervention's effectiveness in achieving its goals of lowering drug use and delinquency. Hawkins and his team even continued to follow these individuals many years after the original intervention had been given to see if any positive effects were maintained.

The criminal justice system using a public health approach

Turning our attention to research on a criminal justice intervention, Braga and his colleagues looked at an evaluation of a criminal justice intervention, "focused deterrence," to study its effects on the prevention of gun violence. The study took place in Boston, Massachusetts, where the city was experiencing a high rate of gun violence. Like the public health approach, the focused deterrence method identified the risk factors at the heart of the problem. This, in turn, helped to guide the development of a response to eliminate or lessen these risk factors. And then, in line with the public health approach, the focused deterrence method evaluated the effectiveness of the implemented intervention. As Braga and Weisburd note, the public health approach comprised three elements, "(a) a focus on prevention, (b) a focus on scientific methodology to identify risks and patterns, and (c) a multidisciplinary collaboration to address the issue," (Braga & Weisburd, 2015, p. 58).

Braga and his colleagues' investigation into the underlying causes of gun violence found that much of it was related to vendetta-like conflicts between members in neighborhood gangs. Though gang members represented less than 1% of the city's youthful population between the ages of 14 and 24, they were responsible for more than 60% of the gun homicides (Braga & Weisburd, 2015, p. 57). Therefore, researchers sought to develop an intervention that would specifically reduce gang-related homicides. Working with the Boston police department, they developed what has now become known as the *pulling levers* approach. Specifically, gangs were continuously notified, and in very explicit terms, that violence would no longer be tolerated in Boston. This message was then backed up by a swift and severe police response when gang violence broke out. Specifically, the police used every lever legally available to them in a coordinated criminal justice response aimed at disrupting all gang activities. When any gang violence occurred, police concentrated their efforts on ensuring that gang members paid the price for any and all criminal activities they engaged in. This included large and small infractions like trespassing, public drinking, and drug activities. Additionally, police used the criminal justice system to harass gang members by serving outstanding warrants, providing stricter probation and parole enforcement, seizing drug and other assets, and ensuring stiffer plea-bargaining negotiations. The message to gang members was a promise that gang violence would ensure an immediate, intense, and negative response by the police solely designed to make things as unpleasant as possible for them (Braga, Weisburd, & Turchan, 2018).

It is notable that the focused deterrence approach was not implemented to eliminate gangs or to stop gang activity, but to exclusively deter gang members' use of violence. Using a nonrandomized quasiexperimental design, Braga and his colleagues tested the effectiveness of this approach in lowering gun violence and found that it led to significantly fewer gun violence behaviors when compared to gangs in Boston that weren't exposed to this approach (Braga, Hureau, & Papachristos, 2014).

Dissemination of public health's influence on criminal justice

The influence of public health's research methodologies has led to large and small discipline-wide changes in criminal justice. This includes the formation of a multinational collaboration for systematic reviews of policies and programs in criminal justice, the creation and establishment of a new experimental division within the discipline's professional association, and the formation of a new journal. Below each change is discussed in turn.

A significant component of an evidence-based system is the need for systematic reviews of research that has already been conducted to determine which interventions are effective. These studies must rely on thorough literature reviews encompassing all available evaluations meeting carefully specified criteria, including the use of experimental and high-quality quasiexperimental evaluations. Systematic reviews are necessary tools in that they reduce the importance of any one specific study by placing it within the context of all other available studies on the topic (Sherman, 2003). The Cochrane Collaboration, a worldwide partnership of researchers, was begun by Sir Ian Chalmers in 1992. Since then this database has become a leading source of systematic reviews of medical treatments and is credited with leading to significant advances in medicine (Weisburd et al., 2001).

Given Cochrane's success, Sir Chalmers convinced Dr. Robert Boruch, from the University of Pennsylvania, to create a similar repository for reviews in the fields of criminal justice, social welfare, and education. The international Campbell Collaboration, named after the famous social psychologist Donald Campbell, was begun in 2000. Akin to Cochrane, its mission is to synthesize evidence from high quality evaluations with the intent of making these results more comprehensible and therefore user-friendly to practitioners and policy makers. The Campbell Collaboration (https:/campbellcollaboration.org/) is an "international social science research network that produces high quality, open and policy-relevant evidence syntheses, plain language summaries and policy briefs."

Additionally, there is now a Division of Experimental Criminology (DEC) founded in 2009. DEC is one of eight divisions within the American Society of Criminology (ASC), the professional international association for those pursuing scholarly and scientific knowledge in criminal justice. Though housed in the ASC, the Division's genesis began within the Academy of Experimental Criminology which came about due to an international group of highly respected criminologists who hoped to draw increased attention to more rigorous research methods, including the use of experiments in criminal justice. DEC was also envisioned as a method for encouraging and mentoring young researchers to engage in experimental research, thereby helping to grow the next generation of experimental criminologists. In line with this mentorship, the DEC also offers seminars on different aspects of running an experimental study.

The *Journal of Experimental Criminology* (https:/www.springer.com/journal/11292) was started in 2005 to focus on high-quality experimental and quasiexperimental research in criminal justice. The quarterly peer-reviewed academic journal is committed to the advancement of experimental research, systematic reviews, and metaanalyses to continue to advance evidence-based practices within the discipline.

Finally, the dissemination of the public health research methodologies is expanding as increasing numbers of practitioners are now reaching out to criminal justice researchers to collaborate on specific problems. A recent example is provided by Project Safe Neighborhoods (PSN). This federal government initiative provided funds for each of the 94 US Attorneys' districts to hire an academic researcher to work with them to address serious gun violence within their jurisdiction (Welsh et al., 2012). An evaluation of PSN found that those jurisdictions which demonstrated high implementation of the intervention showed decreases in their rates of violent crime (McGarrell, Corsaro, Hipple, & Bynum, 2010). Another example is in the Centers for Disease Control and Prevention's Academic Centers for Excellence (ACE). The goal of these centers is, in part, to foster collaboration between academic researchers and communities to promote strategies to prevent youth violence (Vivolo, Matjasko, & Massetti, 2011). Additionally, many federal funding agencies are now letting it be known that they want interdisciplinary teams encompassing both practitioners and researchers on grant applications. All of these are leading to an increase in collaborations and partnerships between criminologists and practitioners along with others from diverse disciplines.

Summarily, the dissemination of the idea of evidence-based research is important to unraveling some of the most serious social problems that have been called "wicked problems." That is, problems that are too large and complex for any one discipline to be able to solve on their own.

Discussion

The evidence-based approach used in public health whereby the underlying factors leading to the targeted behavior are identified, an intervention is developed and implemented to counter these underlying factors, and the researcher then rigorously evaluates the effectiveness of this intervention is beginning to be more widely used in criminal justice research. What has been more difficult to employ is the use of experimental research in criminal justice agencies. Happily, research indicates that randomized experiments are feasible and that they can be conducted in a variety of criminal justice settings (Petrosino, 1998). However, implementing and running experiments in criminal justice settings continues to be more difficult when compared to engaging in nonexperimental research. Unlike public health personnel, those in criminal justice are less likely to have been schooled in research, and therefore, are less trusting of experiments. Criminal justice personnel have voiced concerns that randomly assigning some individuals to treatment and others to a no-treatment control group is unfair and possibly an illegal denial of services (Feder et al., 2000). Of course, their fear that it is an unfair denial of services assumes that the treatment has previously been rigorously tested and found to be beneficial. Where this has already occurred, there would be no need to evaluate the intervention. As for the concern that an experiment may be illegal, the United States Supreme Court's research arm, the Federal Judicial Center, endorsed the use of experimental research where researchers can ensure that there is equal treatment of subjects (which does not translate to all subjects receiving the same treatment) and that the study guarantees basic respect for all involved individuals (Feder, 2019).

The medical community's experience in implementing experiments provides a good deal of hope for experimental criminologists. Specifically, when medicine increased its use of experiments, practitioners eventually grew more comfortable with this methodology, making it easier to conduct experimental research. And this surge in the use of experiments led to a significant increase in the number of cures for a wide variety of ills which, in turn, led to medical personnel feeling even more comfortable with experimental research (Oakley, 2000). One can envision it one day being the same within the criminal justice community.

As noted by Feder et al. (2000, p. 398):

> *There is one thing that will help get experiments more widely accepted and used [in criminal justice]. That is, doing more experiments. Most researchers already understand the advantages that randomized experiments offer over other types of research designs. As more randomized experiments are published, researchers will see that they are possible and perhaps they will think more seriously about implementing them. Also, as researchers increasingly use this methodology, agencies will become more comfortable with experiments.*

Conclusion

In this chapter, we've introduced some of the practical, scientific, and political factors that have led to a gradual evolution of how research is conceived and conducted in criminal justice. We have distinguished how differences in the purpose, research approach, and collaborative potential have shifted the criminal justice research objective from theoretical (developing theories as to why individuals commit crime) to applied research (developing methods to stop individuals from committing crimes). We have noted other important contributions of the public health methodologies which emphasize prevention instead of intervention and the need to collaborate with agency personnel and community stake holders. Together, these factors have led to this relatively new discipline's gradual maturation. However, addressing problems through science also requires changes in criminal justice pedagogy to include experimental research and to find ways to provide mentorship to the next generation of researchers on how to implement and run an experiment in a criminal justice setting. Armed with evidenced-based results, it is hoped that logic and sound judgment will prevail and that legislators will allow science, rather than politics, to take precedence in discussions on criminal justice policies and programs.

References

Akers, T. A., & Lanier, M. M. (2009). Epidemiological criminology: Coming full circle. *American Journal of Public Health, 99*(3), 397–402. https://doi.org/10.2105/ajph.2008.139808.

Beccaria, C. (1963). *On crimes and punishments.* Indianapolis, IN: Bobbs-Merrill.

Braga, A. A., Hureau, D. M., & Papachristos, A. V. (2014). Deterring gang-involved gun violence: Measuring the impact of Boston's operation ceasefire on street gang behavior. *Journal of Quantitative Criminology, 30*(1), 113–139. https://doi.org/10.1007/s10940-013-9198-x.

Braga, A. A., & Weisburd, D. L. (2015). Focused deterrence and the prevention of violent gun injuries: Practice, theoretical principles, and scientific evidence. *Annual Review of Public Health, 36*(1), 55–68. https://doi.org/10.1146/annurev-publhealth-031914-122444.

Braga, A. A., Weisburd, D. L., & Turchan, B. (2018). Focused deterrence strategies and crime control. *Criminology & Public Policy, 17*(1), 205–250. https://doi.org/10.1111/1745-9133.12353.

Chalmers, I. (2003). Trying to do more good than harm in policy and practice: The role of rigorous, transparent, up-to-date evaluations. *The Annals of the American Academy of Political and Social Science, 589*, 22–40.

Clear, T. R. (2010). Policy and evidence: The challenge to the American Society of Criminology: 2009 Presidential address to the American Society of Criminology. *Criminology, 48*(1), 1–25. https://doi.org/10.1111/j.1745-9125.2010.00178.x.

Cornish, D. B., & Clarke, R. V. (1983). *Crime control in Britain: A review of policy research.* Albany: State University of New York Press.

Darwin, C. (1998). *The descent of man.* Amherst, NY: Prometheus Books.

Darwin, C. (2010). *The origin of the species.* Pacific Publishing Studio.

Dishion, T. J., McCord, J., & Poulin, F. (1999). When interventions harm: Peer groups and problem behavior. *American Psychologist, 54*(9), 755–764. https://doi.org/10.1037//0003-066x.54.9.755.

Farrington, D. P. (2003). British randomized experiments on crime and justice. *The Annals of the American Academy of Political and Social Science*, *589*(1), 150–167. https://doi.org/10.1177/0002716203254695.

Feder, L. (2019). Experimental criminology. In R. Morgan (Ed.), *Encyclopedia of criminal psychology* (pp. 436–439). Thousand Oaks, CA: Sage Publications.

Feder, L., Jolin, A., & Feyerherm, W. (2000). Lessons from two randomized experiments in criminal justice settings. *Crime & Delinquency*, *46*(3), 380–400. https://doi.org/10.1177/0011128700046003007.

Feder, L., & Wilson, D. B. (2005). A meta-analytic review of court-mandated batterer intervention programs: Can courts affect abusers' behavior. *Journal of Experimental Criminology*, *1*, 239–262.

Feinstein, S. (2005). Another look at scared straight. *Journal of Correctional Education*, *56*(1), 40–44.

Hawkins, J. D., Catalano, R. F., Kosterman, R., Abbott, R., & Hill, K. G. (1999). Preventing adolescent health-risk behaviors by strengthening protection during childhood. *Archives of Pediatrics & Adolescent Medicine*, *153*(3). https://doi.org/10.1001/archpedi.153.3.226.

Hawkins, J. D., Kosterman, R., Catalano, R. F., Hill, K. G., & Abbott, R. D. (2008). Effects of social development intervention in childhood 15 years later. *Archives of Pediatrics & Adolescent Medicine*, *162*(12), 1133. https://doi.org/10.1001/archpedi.162.12.1133.

Lipsey, M. (1992). Juvenile delinquency treatment: A meta-analytic inquiry into the variability of effects. In T. Cook, et al. (Eds.), *Meta-analysis for explanation* (pp. 83–127). New York, NY: Russell Sage Foundation.

Lombroso-Ferrero, G. (2012). Criminal man. In J. Jacoby, T. Severance, & A. Bruce (Eds.), *Classics of criminology* (pp. 183–198). Long Grove, IL: Waveland Press.

Martinson, R. (1974). What works? Questions and answers about prison reform. *The Public Interest*, *35*, 22–54.

McCord, J. (2003). Cures that harm: Unanticipated outcomes of crime prevention programs. *The Annals of the American Academy of Political and Social Science*, *587*, 16–30.

McGarrell, E., Corsaro, N., Hipple, N., & Bynum, T. (2010). Project safe neighborhoods and violent crime trends in the US: Assessing violent crime impact. *Journal of Quantitative Criminology*, *26*, 165–190.

McGrath, P. (2004). Review: Exercise training in patients with heart failure is safe. *Evidence-Based Medicine*, *9*(6), 174. https://doi.org/10.1136/ebm.9.6.174.

Moore, M. H. (1993). Violence prevention: Criminal justice or public health? *Health Affairs*, *12*(4), 34–45. https://doi.org/10.1377/hlthaff.12.4.34.

Moore, M., Prothrow-Stith, D., Guyer, B., & Spivak, H. (1994). Understanding and preventing violence. In A. Reiss & J. Roth (Eds.), *Consequences and control* (pp. 167–216). Washington, DC: National Academy Press.

Neubauer, D., & Fradella, H. (2019). *America's courts and the criminal justice system*. Australia: Cenage Learning.

Oakley, A. (2000). A historical perspective on the use of randomized trials in social science. *Crime and Delinquency*, *46*(3), 315–329.

Peterson, P. L., Hawkins, J. D., & Catalano, R. F. (1992). Evaluating comprehensive community drug risk reduction interventions. *Evaluation Review*, *16*(6), 579–602. https://doi.org/10.1177/0193841x9201600601.

Petrosino, A. (1998). A survey of 150 randomized experiments in crime reduction: Some preliminary findings. *The Forum*, *16*(1), 1–8.

Petrosino, A., Petrosino, C.-T., & Buehler, J. (2007). Scared straight and other juvenile awareness programs. In *Preventing crime* (pp. 87–101). https://doi.org/10.1007/1-4020-4244-2_6.

Potter, R. H., & Akers, T. A. (2010). Improving the health of minority communities through probation-public health collaborations: An application of the epidemiological criminology framework. *Journal of Offender Rehabilitation*, *49*(8), 595–609. https://doi.org/10.1080/10509674.2010.519674.

Sackett, D. (1997). Evidence-based medicine. *Seminars in Perinatology*, *21*(1), 3–5.

Sherman, L. W. (2003). Misleading evidence and evidence-led policy: Making social science more experimental. *The Annals of the American Academy of Political and Social Science*, *589*(1), 6–19. https://doi.org/10.1177/0002716203256266.

Siegel, L. (2018). *Criminology: Theories, patterns, and typologies.* United States: Cenage Learning.

Smart, N., & Marwick, T. H. (2004). Exercise training for patients with heart failure: A systematic review of factors that improve mortality and morbidity. *American Journal of Medicine*, *116*, 693–706.

Vivolo, A. M., Matjasko, J. L., & Massetti, G. M. (2011). Mobilizing communities and building capacity for youth violence prevention: The National Academic Centers of Excellence for Youth Violence Prevention. *American Journal of Community Psychology*, *48*(1–2), 141–145. https://doi.org/10.1007/s10464-010-9419-5.

Weisburd, D., Farrington, D. P., & Gill, C. (2017). What works in crime prevention and rehabilitation? *Criminology & Public Policy*, *16*(2), 415–449. https://doi.org/10.1111/1745-9133.12298.

Weisburd, D., Lum, C., & Petrosino, A. (2001). Does research design affect study outcomes in criminal justice? *The Annals of the American Academy of Political and Social Science*, *578*, 50–70. https://doi.org/10.1177/000271620157800104.

Weisburd, D., Petrosino, A., & Lum, C. M. (2003). Preface. *The Annals of the American Academy of Political and Social Science*, *587*(1), 6–14. https://doi.org/10.1177/0002716202250882.

Welsh, B. C., Braga, A. A., & Sullivan, C. J. (2012). Serious youth violence and innovative prevention: On the emerging link between public health and criminology. *Justice Quarterly*, *31*(3), 500–523. https://doi.org/10.1080/07418825.2012.690441.

Wolfgang, M., Figlio, R., & Sellin, T. (1972). *Delinquency in a birth cohort.* Chicago, IL: University of Chicago Press.

Supplementary reading recommendations

Allen, P., Jacob, R. R., Lakshman, M., Best, L. A., Bass, K., & Brownson, R. C. (2018). Lessons learned in promoting evidence-based public health: Perspectives from managers in state public health departments. *Journal of Community Health*, *43*(5), 856–863. https://doi.org/10.1007/s10900-018-0494-0.

Braga, A., & Weisburd, D. (2013). Editors' introduction: Advancing program evaluation methods in criminology and criminal justice. *Evaluation Review*, *37*(3–4), 163–169. https://doi.org/10.1177/0193841x14524208.

Boruch, R. F. (1997). *Randomized experiments for planning and evaluation: A practical guide. Vol. 44.* Thousand Oaks, CA: Sage.

Campbell, D. (1988). The experimenting society. In D. Campbell & S. Overman (Eds.), *Methodology and epistemology for social science: Selected papers* (pp. 290–314). Chicago, IL: University of Chicago Press.

Campbell, D. T. (2017). Reforms as experiments. In *Research design* (pp. 79–112). https://doi.org/10.4324/9781315128498-9.

Farrington, D. P. (2003). A short history of randomized experiments in criminology. *Evaluation Review*, *27*(3), 218–227. https://doi.org/10.1177/0193841x03027003002.

Hawkins, J. D., Smith, B. H., Hill, K. G., Kosterman, R., Catalano, R. F., & Abbott, R. D. (2007). Promoting social development and preventing health and behavior problems during the elementary grades: Results from the Seattle Social Development Project. *Victims & Offenders*, *2*(2), 161–181. https://doi.org/10.1080/15564880701263049.

Shepherd, J. P. (2003). Explaining feast or famine in randomized field trials. *Evaluation Review*, *27*(3), 290–315. https://doi.org/10.1177/0193841x03027003005.

US Department of Justice, National Institute of Corrections (NIC). (2013). *Evidence-based practices in the criminal justice system: An annotated bibliography* [Last update 2017]. Accession No. 026917. Aurora, CO: NIC Information Center (NICIC). https://nicic.gov/library/026917.

Van Ness, V. N., III. (2018). Streetwise community policing to inform United States national policy. In B. A. Fiedler (Ed.), *Translating national policy to improve environmental conditions impacting public health through community planning* (pp. 285–304): Springer International Publishing. Editor's recommendation.

Welsh, B. C., & Farrington, D. P. (2001). Toward an evidence-based approach to preventing crime. *The Annals of the American Academy of Political and Social Science*, *578*(1), 158–173. https://doi.org/10.1177/000271620157800101O.

Definitions

Criminology the scientific study of the nature, extent, cause and control of criminal behavior," (Siegel, 2018, p. 580)

Criminal justice a field of study that focuses on law enforcement, the legal system, corrections, and other agencies of justice involved in apprehension, prosecution, defense, sentencing, incarceration, and supervision of those suspected of or charged with criminal offenses," (Siegel, 2018, p. 580).

Iatrogenic side effects and other negative outcomes as a result of medical intervention, care, or procedure (e.g., health-acquired infections, allergic response to vaccination, drug interaction).

Incapacitation criminal justice intervention to help prevent a convicted criminal from committing a second offense.

General Deterrence application of punishment to the specific offender to decrease the probability of others, who see that offender getting punished, engaging in the bad act

Recidivism when a previously convicted criminal returns to the criminal justice system

Rehabilitation criminal justice intervention to stop a convicted criminal from engaging in bad behavior which could prevent a repeat offense.

Specific Deterrence application of punishment to the specific offender to decrease the probability of that offender engaging in that bad act again

Section C

Environmental impact on public health

Environmental perspectives from the ground up: The cost of poor environmental health on human health

Beth Ann Fiedler
Independent Research Analyst, Jacksonville, FL, United States

Abstract

Risk exposure to environmental contaminants and occupational hazards are prevalent in certain factors of the social determinants of health (e.g., economic status, gender, geography) impacting opportunity and poor health outcomes. Using a case study methodology and data from the Global Burden of Disease 2017 Study, we compare relative environmental risks between and within the regions of Europe and North America to demonstrate the variance in leading causes of death, mortality rate, and contributing risks by stratifying data by levels of income. We find consistent evidence of environmental health disparities, provide the economic and human costs associated with these inequalities, and consider how scientific innovation and policy response can generate improved efficiencies. We conclude that the role of income and economic development in the global scheme of health systems, environmental, and public health to reconcile inequalities must change to halt rapid development unless accompanied by planned action to address environmental health.

Keywords: Environmental management, Health risks, Health systems, Wellness, Europe, North America.

Chapter outline

Three Facets of Public Health and Paths to Improvements. https://doi.org/10.1016/B978-0-12-819008-1.00012-2

Particulate matters

We kill all the caterpillars, then complain there are no butterflies.[a]

John Marsden

The association between environmental and public health is of topical interest throughout the literature, and rightly so, because there is an inherent cost to Earth and all her inhabitants (e.g., plants, animals, and people) in the interaction between man and nature. The strain on human body organs (e.g., heart, lungs) and the places where we live and work are of particular concern. But, how poor environmental health (e.g., tiny air- or waterborne pollutants, microscopic biohazards, environmental strain sustained from development, and synthetic products) attacks the planet and human body thus limiting human development is often lost in grizzly, but important, statistics on heart, respiratory, and other diseases which cause reductions in quality of life, disability, and premature death. To answer this question, (1) we establish a broad overview of global health status in relation to environmental and occupation risks using the regions of Europe and North America as case studies, (2) present the economic cost of a poor environment, and (3) discuss the role of health systems in addressing divergent and complex problems associated with environmental and human health. There is, of course, remarkable consensus that poor environments impact human health, but the literary challenge herein is addressing the financial, human, and destructive cost of a number of causal perspectives spanning climate change, environmental injustice, and environmental inequalities. Ultimately, understanding health inequalities and available associated costs will inform this chapter about methods of achieving health equity; but, because of the diversity in contributory causes, there will be some unfortunate reliance on the hard facts of mortality and morbidity. The objective here is to demonstrate the broader view of human interaction with Earth and the obvious medical impact on human health. We define common terms herein to set the stage for the balance of the chapter.

Healthy environmental quality is measured by the lack or minimal presence of harmful toxins in the air we breathe, water we drink, and soil in which we play, live, and grow our food. Poor environmental quality globally accounts for "23% of all deaths and 26% of deaths among children under age 5," (Office of Disease Prevention and Health Promotion (ODPHP), 2014a, 2014b, 2014c, *para* 4–5) in two ways: first, by increasing risk for those with a compromised health status, and second, by introducing toxins to children in the early stages of growth, when protective immunities have not fully developed and/or development is impaired. Most

[a] All initial section quotes in this chapter from Juma (2019).

environmental hazards have two opposing common features: they are preventable, but they are often difficult to escape.

"Environmental health consists of preventing or controlling disease, injury, and disability related to the interactions between people and their environment," (ODPHP, 2014a, 2014b, 2014c, *para* 1). Thus, many would agree that environmental health is directly attributable to negative human behavior (e.g., overconsumption) and offset by changing how we interact with the environment with positive human behavior (e.g., recycling, using mass transit, or water conservation). The supporting evidence for negative human behavior is apparent in pesticide agricultural runoff, food waste, and shorter product life cycles that place "obsolete" technologies at a record pace into the garbage heaps. In turn, the environment acts upon the human body in a vicious cycle through air pollution, water pollution, and mounting waste on land and in water. Adverse chronic health conditions and deadly outcomes, such as cancer, heart disease, stroke, respiratory illnesses and others (So, Mamary, & Shenoy, 2018), are exasperated or caused by these and other unhealthy environmental conditions. The Global Burden of Disease Study (GBD 2017 Risk Factor Collaborators, 2018) parse environmental risk factors into two general categories: (1) environmental, and (2) occupational. Environmental factors consist of air and water quality, unsafe sanitation, and ambient particulate matter pollution. Common occupational environmental risks include exposure to carcinogens found in chemical emissions through combustion, processing, or handling of raw materials. For example, occupational risk hazards include asbestos, diesel engine exhaust, and formaldehyde but also noise, injuries, ergonomic factors, and other asthmagens.

Finally, "environmental health inequalities refer to general differences in environmental health conditions" measured in various factors such as "socioeconomic and demographic inequalities in exposure to environmental hazards [that] exist everywhere and can be expressed in relation to factors that may affect the risk of being exposed, such as income, education, employment, age, sex, race/ethnicity and specific locations or settings," (World Health Organization (WHO) Regional Office for Europe, 2012, p. xvi). Further, the World Health Organization (WHO) Regional Office for Europe (2012) considers that environmental health inequalities are (1) caused by vulnerability to certain risks as a result of social or demographic differences, and (2) generate variance in the level of exposure to harmful environmental conditions, contaminants, or other environmental health inequalities.

This chapter provides an overview of global environmental health status focusing on regions in North America and Europe by distinguishing environmental health impact in high-income regions compared to similar counterparts within the larger regional distinction and between the two major continental designations. The case studies rely on data reported by the Global Burden of Disease Study 2017 (GBD 2017 Risk Factor Collaborators, 2018) and were selected to demonstrate how certain social determinants of health (SDOH) impact subgroups within and across seemingly similar SDOH and other environmental conditions. We report the economic and public health costs of poor environmental health on heart and lung health followed by the role of health systems to ameliorate the effect of poor environmental health on the public and/or specific subpopulations. We find consistent evidence of environmental health disparities, provide the economic and human costs associated with these inequalities, and consider how scientific innovation and policy response can generate improved efficiencies. We concur with current literature that reconciling environmental inequalities is a path to progress and must be evident in health system revision. First and foremost, we must reconsider the role of income in the global scheme of health systems, the environment, and public health from new perspectives, such as the systematic conduct of soil composition and the role of microorganisms relative to human health.

Human and environmental health status: Europe and North American overview

In nature there are neither rewards or punishments-there are consequences.
Robert Green Ingersoll

The impact of environmental health on public health is persistent across the globe with a number of research organizations addressing various aspects of the persistent battle between economic development and safe environments. However, we discover in this section that there are environmental conditions in regional environments that exact a greater toll on certain groups of people who experience unstable economic situations and lack of education which cumulatively leads to shorter life spans.

Organizations, such as the US National Institute of Environmental Health Sciences (NIEHS), look at diseases (e.g., respiratory ailments, cancer, cardiovascular, waterborne, vector-borne, and zoonotic) through the lens of the health impact of climate change (NIEHS, n.d.-a, n.d.-b). On the other hand, the ODPHP through the Healthy People 2020 Environmental Health "objectives focus on 6 themes: 1) outdoor air quality, 2) surface and ground water quality, 3) toxic substance and hazardous wastes, 4) homes and communities, 5) infrastructure and surveillance, and 6) global environmental health," (ODPHP, 2014a, *para* 2). The ODPHP focus on similar diseases, such as cancer, diabetes, maternal, infant, and child health, occupational safety and health, and respiratory diseases, but include environmental conditions that are localized at the community level. These environmental conditions include access to health services, genomics, public health infrastructure, and health communication and health information technology (ODPHP, 2014c). All are important and contribute to some extent to public health.

Global organizations, such as the World Health Organization (WHO) Commission on the Social Determinants of Health, established three primary recommendations in 2008 and subsequent research continues to build on the premise to overcome the various inputs at home and work that result in poor health outcomes. They suggest a focus on (1) improving daily living conditions; (2) tackling the inequitable distribution of power, money, and resources; and (3) measuring and understanding the problem and assessing the impact of action (CSDH, 2008, p. 2). In similar fashion, the Global Burden of Disease Study has been consistently addressing the factors associated with behavior, environment, and metabolic risks (The GBD 2017 Risk Factor

Collaborators, 2018), while others address various problems associated with growing up unequal in SDOH factors such as gender or socioeconomic status (Inchley et al., 2016) or the cumulative effect of environmental conditions from childhood to conception (NIEHS, 2016). Regardless of whether the research institution focuses on any specific part of the human and environmental interaction, environmental conditions, or the cumulative interaction of human activity or Earth's response, the negative impact on the environment is linked to poor human health, which is clearly demonstrated in poor health outcomes (Fig. 12.1).

North America and Europe serve as regional comparisons on the environmental risks experienced in these areas that contribute to poor health. Using data from the Global Burden of Disease Study (GBD 2017 Risk Factor Collaborators, 2018) and supporting information, we provide an overview of the human and environmental health status in these regions. Both regions offer subgroups of high-income nations to demonstrate the "relationship between development and risk exposure by modelling the relationship between the Socio-demographic Index (SDI) and risk-weighted exposure prevalence and estimated expected levels of exposure and risk-attributable burden by SDI," (GBD 2017 Risk Factor Collaborators, 2018, p. 1923; Global Burden of Disease Collaborative Network. Global Burden of Disease Study 2017 (GBD 2017), 2018b). In addition, there are some marked differences between environmental risk, mortality and morbidity, and gender.

European region

In 2012, WHO Europe baselined 14 specific problems in 3 inequality categories in relation to housing, work injuries, and the environment. A second assessment report (World Health Organization (WHO) Regional Office for Europe, 2019a) added five more total problems and two new categories encompassing basic service inequalities and inequalities related to urban environments and transportation there. The follow-up evaluation indicates that 15% of deaths in this region are attributable to unsafe environmental conditions leading to poor health and diminished well-being (World Health Organization (WHO) Regional Office for Europe, 2019a, 2019b, p. x). These conditions include lack of a flush toilet and/or shower (World Bank Group, 2019b [database]), energy poverty (World Health Organization (WHO) Regional Office for Europe, 2019a), poor sanitation (World Bank Group, 2019a [database]; World Health Organization (WHO) Regional Office for Europe, 2019a), exposure to hazardous chemicals and air pollutants (Attina et al., 2016; Chanel, Perez, Künzli, & Medina, 2016; Rotter et al., 2018), work-related injury and risk exposures (GBD 2017 Risk Factor Collaborators, 2018; World Health Organization (WHO) Regional Office for Europe, 2019a, 2019b), and fatal accidents/injuries (GBD 2017 Risk Factor Collaborators, 2018; World Health Organization (WHO) Regional Office for Europe, 2019a, 2019b).

Similar to the 2012 baseline report, the 2019 evaluation of 53 countries in 4 subregions demonstrates a persistent "uneven distribution of environmental risks within societies and the related impacts on health and health equity are therefore of increasing concern," (World Health Organization (WHO) Regional Office for Europe, 2019a,

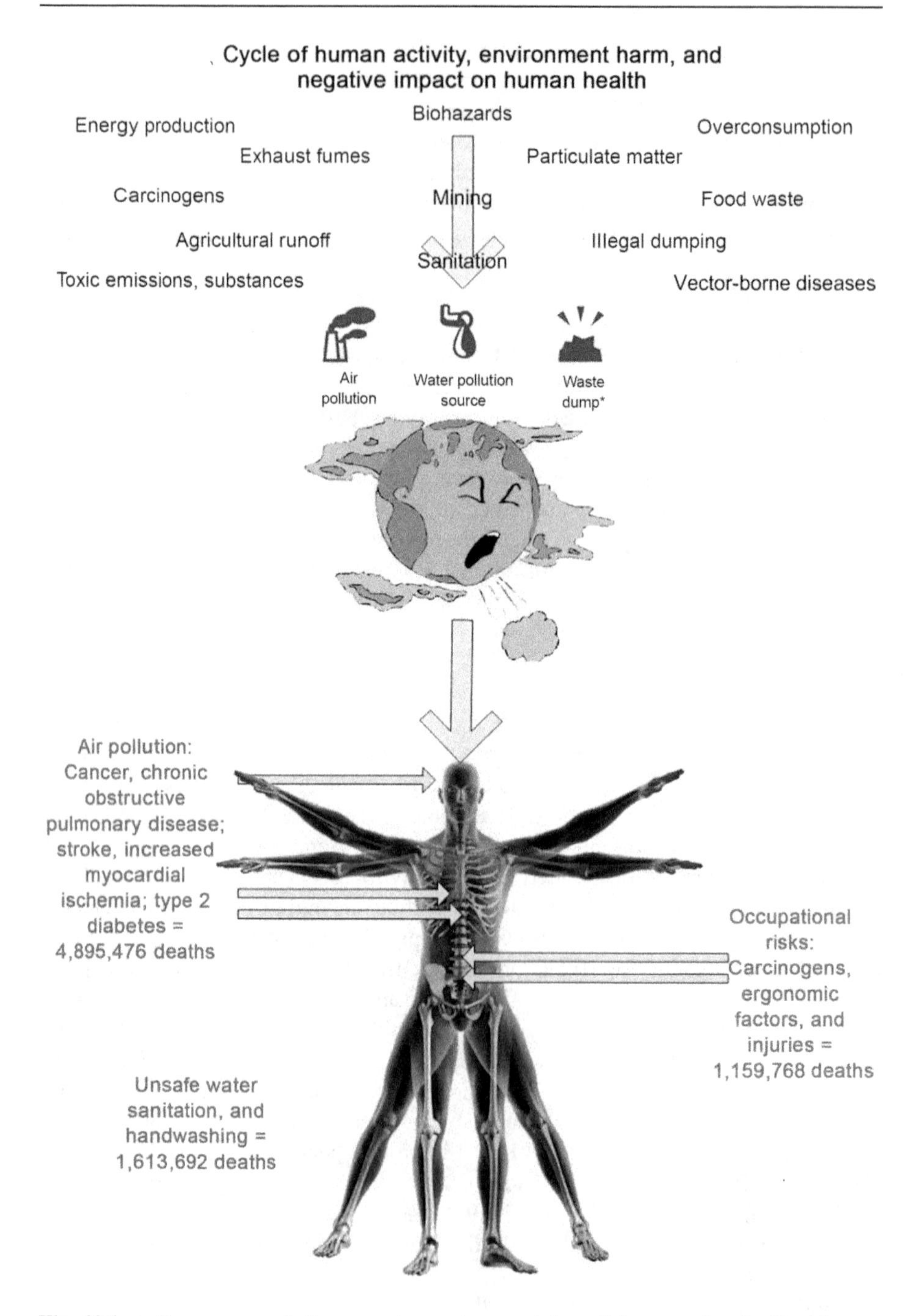

Fig. 12.1 Anthropocene activity generates environmental conditions causing the Earth's environmental health to deteriorate and, in turn, impacts human health; the image is not intended to represent a passive linear flow but to suggest the cyclical nature of human and earth interaction eliciting negative environmental and health outcomes (mortality data from GBD 2017 Risk Factor Collaborators, 2018; Earth and human image courtesy of http://insertmedia.office.microsoft.com).

2019b, p. x). Metrics (e.g., socioeconomic status, demographic and spatial variables) from international databases bring to the fore variance demonstrated within subgroups of the same nation and between nations in terms of environmental impact on public health. Differential exposure and vulnerabilities combined with a lack of institutional/societal support (World Health Organization (WHO) Regional Office for Europe, 2019a, 2019b) leave some subgroups five times more likely to be exposed to environmental hazards and injury when compared to other subgroups who are categorized as "advantaged" by comparison (2019, p. 130). Achieving equity is a prime objective of the persistent research and status monitoring by WHO and other research organizations.

The GBD 2017 Risk Factors Collaborators (2018) have reported specific environmental and occupational risks leading to poor health demonstrating a consistent pattern in which SDOH factors, such as socioeconomic status, play a role in environmental inequalities. One of the most alarming environmental concerns is the extreme health impact of ambient particulate matter pollution that acts on the lungs, brain, vasculature, blood, and heart causing ischemic heart disease, stroke (ischemic and hemorrhagic), Chronic Obstructive Pulmonary Disease (COPD), lung cancer, acute lower respiratory infection, and Type-II diabetes. Unsafe water sources produce preventable deaths from diarrheal disease and lead exposure continues to create intellectual disability and IQ loss due to early childhood exposure. We view various environmental risks and mortality in Central, Eastern and High-Income Western Europe (Table 12.1) to bring forth the divergent challenges these areas face.

Comparison results within the European region drawn from Table 12.1 demonstrate some similarities in the leading types of environment/occupational risks leading to mortality in relation to income and gender. While particulate matter is the leading E/O risk in all three European regions, the mortality rate in Western Europe (35.48) is less than half of Central (71.54) and Eastern Europe (76.34). More men succumb to the cumulative impact of particular matter on various organs (e.g., lungs, heart, kidneys). However, the presence of occupational carcinogens (21.6) in Western Europe is more than two times this risk in Central Europe (8.98) and five times more than Eastern Europe (4.00). There are some obvious differences in the top five leading E/O risks in Western, Central, and Eastern Europe. For example, occupational particulates impact Western Europe, but radon is prevalent in the other sections. Also, occupational injury is unique to Eastern Europe. In all cases for these leading E/O risks, men are more adversely impacted as evidenced by a higher mortality rate for every listed category.

North American region

When most people are confronted with the term "North America," they often default to the United States or Canada due to their relatively high socioeconomic status that seemingly differentiates them from other nations such as Mexico, Cuba, or Greenland, but who also have a relatively high levels of income depending on

Table 12.1 Leading environmental/occupational (E/O) risks[a] (http://ihmeuw.org/4ygd) in 2017 with aggregated gender mortality rate[b], cause of death attributable to E/O leading risk factors, and the disaggregated gender mortality rate attributable to rate of deaths gender impacting European regions[c–e] based on Socio-demographic Index (GBD 2017 Risk Factor Collaborators, 2018, p. 1972; Global Burden of Disease Collaborative Network; Global Burden of Disease Study 2017 (GBD 2017), 2018a, 2018b); Institute for Health Metrics and Evaluation, 2019 [database]).

	Leading E/O risk factor[a]	Aggregated mortality rate[b] attributable to risk factor, both genders	Cause of death attributable to E/O leading risk factor[a]	Disaggregated mortality rate attributable rate of deaths by gender
High-income Western Europe[c]	Particulate matter air pollution	35.4 (27.63–43.50)[f]	Diabetes and kidney disease, chronic respiratory diseases, cardiovascular diseases, neoplasms, respiratory infections, and tuberculosis	Male 39.54 (30.88–48.76) Female 31.58 (24.72–38.84)
	Occupational carcinogens	21.06 (17.08–24.87)	Neoplasms	Male 35.29 (27.71–42.69) Female 7.35 (5.90–8.84)
	Lead	8.84 (4.47–14.14)	Cardiovascular diseases	Male 10.02 (5.32–15.25) Female 7.84 (3.66–13.12)
	Ozone	5.21 (1.92–8.65)	Chronic respiratory diseases	Male 5.73 (2.12–9.60) Female4.72 (1.70–7.91)
	Occupational particulates	3.74 (2.76–4.85)	Chronic respiratory diseases	Male 5.75 (3.85–7.90) Female 1.81 (1.11–2.61)

Central Europe[d]	Particulate Matter	71.54 (61.04–82.22)	Diabetes and kidney diseases, chronic respiratory diseases, cardiovascular diseases, neoplasms, respiratory infections and tuberculosis	Male 80.34 (68.19–93.05) Female 63.18 (54.03–72.49)
	Lead	12.46 (4.51–21.70)	Cardiovascular diseases	Male 16.43 (7.52–26.05) Female 8.69 (1.74–17.88)
	Occupational carcinogens	8.98 (6.73–11.66)	Neoplasms	Male 15.51 (11.14–20.52) Female 2.78 (2.06–3.56)
	Ozone	4.14 (1.51–6.79)	Chronic respiratory diseases	Male 5.18 (1.88–8.47) Female 3.16 (1.14–5.23)
	Radon	4.10 (2.29–6.46)	Neoplasms	Male 6.13 (3.46–9.55) Female 2.17 (1.21–3.43)
Eastern Europe[e]	Particulate matter	76.34 (63.71–90.83)	Diabetes and kidney diseases, chronic respiratory diseases, cardiovascular diseases, neoplasms, respiratory infections and tuberculosis	Male 86.85 (72.60–102.87) Female 67.23 (55.65–80.28)
	Lead	7.08 (0.85–17.12)	Cardiovascular diseases	Male 8.66 (1.36–19.38) Female 5.72 (0.40–15.15)

Continued

Table 12.1 Leading environmental/occupational (E/O) risks (http://ihmeuw.org/4ygd) in 2017 with aggregated gender mortality rate, cause of death attributable to E/O leading risk factors, and the disaggregated gender mortality rate attributable to rate of deaths gender impacting European regions[c–e] based on Socio-demographic Index (, p. 1972; Global Burden of Disease Collaborative Network; Global Burden of Disease Study 2017 (GBD 2017), 2018a, 2018b); [database])—cont'd

	Leading E/O risk factor	Aggregated mortality rate attributable to risk factor, both genders	Cause of death attributable to E/O leading risk factor	Disaggregated mortality rate attributable rate of deaths by gender
	Occupational injury	4.64 (4.10–5.23)	Unintentional injuries, transport injuries	Male 9.24 (8.07–10.51) Female 0.66 (0.77–0.56)
	Occupational carcinogens	4.00 (3.00–5.13)	Neoplasms	Male 6.88 (4.90–9.17) Female 1.50 (1.22–1.81)
	Radon	2.76 (1.52–4.45)	Neoplasms	Male 4.87 (2.63–7.79) Female 0.95 (0.51–1.53)

[a] This subset of risk factors and explanations can be found at http://ghdx.healthdata.org/record/ihme-data/gbd-2017-cause-rei-and-location-hierarchies or the parent page of http://ghdx.healthdata.org/gbd-2017.
[b] Mortality rate is the number of deaths per 100,000.
[c] Western European countries: Andorra, Austria, Belgium, Cyprus, Denmark, Finland, France, Germany, Greece, Iceland, Ireland, Israel, Italy, Luxembourg, Malta, Netherlands, Norway, Portugal, Spain, Sweden, Switzerland, and United Kingdom.
[d] Central European countries: Albania, Bosnia and Herzegovina, Bulgaria, Croatia, Czech Republic, Hungary, Macedonia, Montenegro, Poland, Romania, Serbia, Slovakia, Slovenia.
[e] Eastern European countries: Belarus, Estonia, Latvia, Lithuania, Moldova, Russian Federation, and Ukraine.
[f] 95% Confidence Interval (CI) with lower and upper limits in parenthesis.

a particular organization's criteria for inclusion. However, there are also two dozen other nations generally located in North America, but often designated as the Caribbean or Central Latin America in the region facing the challenge of environmental health. We draw attention to the environmental impact on public health when high-income nations in the same general region present markedly different health outcomes.

In the United States, the ODPHP Healthy People 2020 initiative has focused on 68 environmental health objectives to decrease a number of adverse health effects caused by airborne toxins (Environmental Health EH-3.2), hazardous sites (EH-9), or improving items such as access to safe drinking water (EH-4) through policy and regulation (ODPHP, 2014b). Canada, like the United States, is a high-income nation and member of the Organization for Economic Cooperation and Development (OECD). The Frasier Institute has compiled a list of 17 indicators, drawn from overarching concern for protecting human health (7) and ecosystems (10), to create a comparative Index of Environmental Performance ranging from 0 to 100 for Canada and 32 other OECD nations to assess the level of environmental performance that is reflected in higher index scores (McKitrick, Aliakbari, & Stedman, 2018). (Note, Mexico and Turkey were not included in the McKitrick et al., 2018, study because they are not designated as high-income per OECD even though they are OECD nations.) In the United States and Canada, progress has been made toward targeted reductions and proactive response in terms of both environmental and public health, but still demonstrates that pockets of low performance persist.

The United Nations Development Programme (2017) Human Development Index (HDI) classifies Barbados as "very high" on the HDI scale placing them in the category of high-income based on components such as life expectancy, level of education, and gross national income (GNI) per capita. These components are in line with United Nations Sustainable Development Goals (SDG) such as SDG 3 to ensure health and well-being for all; SDG 4.3 (equal access to tertiary education, technical, or vocational training for men/women) and 4.6 (youth and adults, regardless of gender, to be literate); and SDG 8.5 (equal employment opportunities with equal pay for men/women). To allow for a basis of comparison, the composite HDI scale for the OECD nations is 0.895, the United States 0.924, Canada is 0.926, and Barbados is at 0.800. Other nations outside of North America, such as Ireland and Germany, lead the HDI scale with 0.938 and 0.936, respectively. Notable is that a number of countries in North America have achieved a "high" HDI status including Costa Rica 0.794, Panama 0.789, Trinidad & Tobago 0.784, and nine others, including Belize 0.708, which is the lowest ranking nation in the "high" sector of North America. While country affiliations vary and there is overlap, we provide some general country affiliations with North America to include the Caribbean and Central Latin American nations designated by the GBD 2017 Risk Factor Collaborators (2018). Data from the GBD 2017 Risk Factor Collaborators (2018) distinguish the high-income environmental and occupational risks in North America as shown in Table 12.2. Notable are the obvious differences in the contributing risks, cause of death, and mortality rates reported for high-income North America as opposed to those with less economic production, education, or life expectancy.

Table 12.2 Leading environmental/occupational (E/O) risks[a] (http://ihmeuw.org/4ygd) in 2017 with aggregated gender mortality rate[b], cause of death attributable to E/O leading risk factors, and the disaggregated gender mortality rate attributable to rate of deaths gender impacting North American regions[c–e] based on Socio-demographic Index (GBD 2017 Risk Factor Collaborators, 2018, p. 1972; Global Burden of Disease Collaborative Network; Global Burden of Disease Study 2017 (GBD 2017), 2018a, 2018b); Institute for Health Metrics and Evaluation, 2019 [database]).

	Leading E/O risk factor[a]	**Mortality rate[b] attributable to risk factor, both genders**	**Cause of death attributable to E/O leading risk factor**	**Mortality rate attributable rate of deaths by gender**
High-income North America[d]	Particulate matter pollution	25.65 (18.27–32.87)	Diabetes and kidney diseases, chronic respiratory diseases, cardiovascular diseases, neoplasms, respiratory infections, and tuberculosis	Male 28.01 (20.2–35.74) Female 23.36 (16.27–30.32)
	Occupational carcinogens	13.85 (10.93–16.72)	Neoplasms	Male 22.88 (17.24–28.46) Female 7.06 (2.58–11.56)
	Lead exposure	8.18 (3.75–13.07)	Cardiovascular diseases	Male 9.89 (4.96–15.22) Female 6.53 (2.49–11.15)
	Ambient ozone pollution	7.08 (2.6–11.62)	Chronic respiratory diseases	Male 7.1 (2.63–11.73) Female 5.1 (3.94–6.33)
	Occupational particulate matter, gases, and fumes	4.76 (3.81–5.84)	Chronic respiratory diseases	Male 6.34 (4.62–8.12) Female 3.22 (2.24–4.31)

Caribbean/North America[e]	Particulate matter pollution	48.68 (39.4–57.04)	Diabetes and kidney diseases, chronic respiratory diseases, cardiovascular diseases, neoplasms, respiratory infections, and tuberculosis	Male 50.6 (40.35–60.78) Female 46.81 (37.58–55.17)
	Lead exposure	18.73 (12.94–24.87)	Cardiovascular diseases	Male 23.03 (16.57–30.31) Female: 14.54 (9.23–20.3)
	Unsafe water source	8.56 (5.51–12.07)	Enteric infections	Male 9.71 (5.53–14.98) Female 7.45 (4.58–10.51)
	No access to handwashing facility	6.96 (4.52–9.59)	Respiratory infections and tuberculosis, enteric infections	Male 7.7 (4.86–11.16) Female 6.24 (4.1–8.49)
	Unsafe sanitation	5.76 (4.03–8.11)	Enteric infections	Male 6.90 (6.19–7.68) Female 4.98 (3.38–7.02)
Central Latin America/North America[f]	Particulate matter pollution	31.59 (26.15–36.67)	Diabetes and kidney diseases, chronic respiratory diseases, cardiovascular diseases, neoplasms, respiratory infections, and tuberculosis	Male 34.29 (28.43–39.93) Female 29.00 (23.74–33.56)
	Lead exposure	11.24 (7.73–14.82)	Cardiovascular diseases	Male: 13.66 (9.76–17.76) Female 8.91 (5.83–12.19)

Continued

Table 12.2 Leading environmental/occupational (E/O) risks (http://ihmeuw.org/4ygd) in 2017 with aggregated gender mortality rate, cause of death attributable to E/O leading risk factors, and the disaggregated gender mortality rate attributable to rate of deaths gender impacting North American regions[c–e] based on Socio-demographic Index (, p. 1972; Global Burden of Disease Collaborative Network; Global Burden of Disease Study 2017 (GBD 2017), 2018a, 2018b); [database])—cont'd

	Leading E/O risk factor	Mortality rate attributable to risk factor, both genders	Cause of death attributable to E/O leading risk factor	Mortality rate attributable rate of deaths by gender
	Occupational injuries	3.69 (3.36–4.03)	Unintentional injuries, transport injuries	Male 6.92 (6.26–7.64) Female 2.44 (1.07–3.64)
	Occupational particulate matter, gases, and fumes	2.71 (2.14–3.32)	Chronic respiratory diseases	Male 4.35 (3.21–5.57) Female 2.14 (0.81–3.55)
	Unsafe water source	2.52 (1.15–3.65)	Enteric infections	Male 2.59 (1.17–3.82) Female 1.34 (0.84–1.84)

[a] This subset of risk factors and explanations can be found at http://ghdx.healthdata.org/record/ihme-data/gbd-2017-cause-rei-and-location-hierarchies or the parent page of http://ghdx.healthdata.org/gbd-2017.

[b] Mortality rate is the number of deaths per 100,000.

[c] High-income North American countries according to the Global Burden of Disease Study 2017 are Canada, Greenland, and the United States. (However, high-income North American countries with "very high" designation on the United Nations Human Development Index (HDI): Canada, United States (and Puerto Rico which is not listed separately), and Barbados.)

[d] Caribbean North American countries according to the Global Burden of Disease Study 2017 are Antigua and Barbuda, The Bahamas, Barbados, Belize, Bermuda, Cuba, Dominica, Dominican Republic, Grenada, Guyana, Haiti, Jamaica, Puerto Rico, Saint Lucia, Saint Vincent and the Grenadines, Suriname, Trinidad and Tobago, Virgin Islands, United States. Territories such as Puerto Rico and US Virgin Islands are not considered part of US data. Also some nations, such as Saint Kitts and Nevis, are not distinguished in the Global Burden of Disease Study 2017. (However, Caribbean North American countries with "high" UN HDI designations are Costa Rica, Panama, Trinidad & Tobago, Antigua & Barbuda, Saint Kitts and Nevis, Cuba, Mexico, Grenada, Saint Lucia, Jamaica, Dominican Republic, Saint Vincent and the Grenadines, Dominica, and Belize.)

[e] Central Latin American countries according to the Global Burden of Disease Study 2017 indicate these nations are designated in Columbia, Costa Rica, El Salvador, Guatemala, Honduras, and Mexico. (However, other Central Latin North American countries with other HDI designations such as "medium" are El Salvador, Nicaragua, Guatemala, and Honduras or "low": Haiti. General note: the Turks and Caicos, though a nation in North America, are part of the United Kingdom Overseas Territory and do not have a separate HDI designation.)

[f] 95% Confidence interval with lower and upper limits.

Comparison results within the North American region drawn from Table 12.2 demonstrate some similarities in the leading types of environment/occupational risks leading to mortality in relation to income and gender. While particulate matter is the leading E/O risk in all three North American regions, the mortality rate in high-income North America (25.65) is nearly half of the Caribbean (48.68) and about 80% of Central Latin America (31.59). More men succumb to the cumulative impact of particular matter on various organs (e.g., lungs, heart, kidneys), but in the North American region the mortality rate gap is smaller between men and women in this and all leading categories. However, high-income North Americans struggle with occupational carcinogens and ambient ozone, while the Caribbean and Central Latin America must contend with unsafe water. In addition, the Caribbean has limited access to handwashing facilities and unsafe sanitation, leading to diseases of the small intestine and respiratory infections. Central Latin America struggles with injuries and chronic respiratory diseases as a consequence of occupational injuries and occupational particular matter.

European vs North American environmental risk regional observations

Comparison results between the three regions of Europe and North America drawn from Tables 12.1 and 12.2 demonstrate several similarities and differences in leading E/O risks in relation to income and gender. Mortality from particulate matter, which topped the list in all six categories, was observed less in high-income North America and high-income Western Europe, while occupational injuries were observed in regions where development is a factor: Central Latin America and Eastern Europe. Radon, an odorless and colorless gas that is radioactive, is prominent in Central and Eastern Europe, but unsafe water is problematic for the Caribbean and Central Latin America. Lead exposure is the second or third leading E/O mortality risk in all six regions.

In terms of the E/O risk on mortality, high-income Western European men (39.54) and women (31.58) are at greater risk of impact to organs and tissue from particulate matter than high-income North American men (28.01) and women (23.36). Central European men (80.34) and women (63.18) and Eastern European men (86.85) and women (67.23) are at greater health risk compared to Caribbean men (50.06) and women (46.81) and Central Latin American men (34.29) and women (29.00). Notable is that particulate matter overshadows by nearly double and up to 10 time more in all regions except for high-income Western Europe where PM is 35.48 and the second leading E/O risk factor is occupational carcinogens (21.06).

European vs North American overall nonenvironmental risk regional observations

Additional analysis between high-income Europe and high-income North America demonstrates how behavioral choices because of access to higher income lead to different types of illnesses and mortality for the national population as a whole and in part in special pockets where various SDOH (i.e., income, gender) dramatically differ from national averages. We briefly summarize the overall behavioral and metabolic risks.

Overall behavior risks, such as smoking or alcohol consumption, represent the top two risks for males in high-income Western Europe and Eastern Europe, while these risks are one and three for Central Europe (GBD 2017 Risk Factor Collaborators, 2018, p. 1972). Behavior choices for females, according to the same study, list smoking as the number one risk for high-income Western Europe, third for Central Europe, and fifth for Eastern Europe. In both genders, metabolic rates such as high body mass index (BMI), high fasting plasma glucose (FPG), and high systolic blood pressure (SBP) are present in the balance of the top five. Poor choices lead to further health impairment and metabolic indicators, such as BMI, FPG or SBP, could be indicative of sedentary behavior, poor eating habits/overconsumption, and/or the invasive nature of pollutants on the human body.

The behavioral risk of using illicit drugs such as opioids, cocaine, amphetamines, cannabis, and others is obvious in the top five overall risks for high-income North America for both males at number three and females at number four (GBD 2017 Risk Factor Collaborators, 2018, p. 1972). The top two overall risk factors for women in high-income North America are high BMI and smoking, while the Caribbean nations and Central Latin America are experiencing high rates of FGP, BMI, and SBP. Further, females in the Caribbean nations are still struggling with short gestation and low birthweight (numbers 4 and 5), while Central Latin America has problems with kidney function impairment and short gestation. For high-income North American males, high BMI and smoking are the leading risks, while the leading risk for Caribbean males are high SBP and FPG. For Central Latin American, high FPG and alcohol are leading risks.

Environmental economics and public health

At one time in the world there were woods that no one owned.

Cormac McCarthy

Place is important as this is the link between the universal desire to meet daily requirements of food, clothing, and shelter and the means to access them through employment. Thus, the topic of environmental health in relation to public health boils down to the topic of land use and several questions about soil and proper use (Fiedler, 2018a, 2018b, 2018c, 2018d), resource distribution (Cook & Fiedler, 2018; Costantini & Mele, 2018), environmental and social justice (Fiedler, 2018e; Walig, Canencia, & Fiedler, 2018; Webster, 2018), and balancing development (Onel & Fiedler, 2018) against the impact of same on public health (Gonzales & Sawyer, 2017; Howard, 2018; Panwhar & Fiedler, 2018; Rao & Ross, 2018), while facing social challenges of immigration (Van Ness III, 2018) and other factors related to land such as energy and forest research (Ferretti & Fischer, 2013; Matyssek et al., 2013; Percy, 2012). The applied sciences struggle with soil composition and political science with ownership, while social, geographical, and other sciences carefully consider the end point of soil extraction, application, and as a foundation for the built environment. At stake is how to address the cost of disease, especially for those who are least able to manage health in low income nations (Roche, Broutin, & Simard, 2018) or where pockets of disease exist in poverty-riddled areas within nations that exhibit higher standards of living. Therefore, understanding the complex impact of the environment, health, and associated costs is important when considering viable solutions, which encompass the responsibility of resource allocation in terms of equitable distribution and development advantages to meet needs with least impact to public health, while addressing social and environmental justice. We provide an overview of the cost burden for European and North American health in relation to the impact on human tissue and organs. Note that some information is expressed in costs for the European Union (EU), which may differ slightly from the previous expression of Western, Central, and Eastern Europe information. While there are a number of European nations not in the EU such as Albania, Bosnia-Herzegovina, and Ukraine among others (Mercy, 2018), the cost information provides a meaningful overview despite the absence of certain national information that may be unavailable or limited based on level of priority of tracked diseases as in the case with the US data. Further, the definition of direct cost varies and may exclude important information such as prescription medicine costs or other items. These differences in data tracking and parameters make direct national comparisons difficult. Therefore, the objective is to provide some meaningful costs with associated diseases while noting differences, as necessary.

We focus on lung and heart health, but note the impact of blood, brain, and vascular health of pollutants, especially particulate matter, acting on the human body. According to the Health and Environment Alliance (HEAL, 2013), typical blood health involves such problems as reduced oxygen saturation, peripheral thrombosis or clotting, and altered blood flow; brain health disorders consist mainly of increased areas of the brain that are affected by bleeding (i.e., cerebrovascular ischemia); and vascular health when fatty material deposits block blood flow in the inner layer of arteries (i.e., atherosclerosis) or burst causing blood clots; and failure of lining of blood and organ cavities (i.e., endothelial dysfunction), constriction of blood vessels (i.e., vasoconstriction), and hypertension. Notable is that there were roughly 336 million people in Europe in 2010 compared to 310 million in the United States, 34 million in Canada, and about 119 million in Mexico (PopulationPyramid.net, n.d.). This may be helpful when assessing the problem of certain diseases in consideration of overall population.

Lung health

"More than 600,000 people in the EU die of respiratory disease," (European Respiratory Society, European Lung White Book, 2019, p. 5); moreover, respiratory diseases are responsible for 1/8 deaths within the European Union (European Respiratory Society, European Lung White Book, 2019, p. 7). The 2011 cost of respiratory disease due to inflammation and a reduction of lung function in the EU converted to the 2019 valuation of the Eurodollar exceeds €413 billion annually, which includes the costs of direct care (i.e., primary, hospitalization) accounting for about €59.85 billion, about €45.06 billion in lost production, and €308.23 billion in lost disability-adjusted-life-years (DALYs) (Euro Inflation Calculator, 2019; European Lung Foundation, 2019, pp. 9–11). "The annual costs of healthcare and lost productivity due to COPD [in 2011 converted to 2019 Eurodollars] are estimated as €48.4 billion [52.68] and those due to asthma at €33.9 billion [36.90]," (European Respiratory Society, European Lung White Book, 2019, *para* 2) (Table 12.3). Finally, major risk factors of respiratory disease include early-life events, outdoor/indoor environments (including energy production), hereditary factors, smoking, and occupational risk factors.

In the United States, 6.7% of all deaths (in 2015) were due to chronic respiratory diseases, which were the fifth leading cause of death (IHME, 2017, *para* 2) and "between 1980 and 2014, the rate of death from chronic respiratory diseases, such as COPD, increased by nearly 30 percent overall in the US," (IHME, 2017, *para* 2).

Table 12.3 Aggregated annual direct and indirect costs and the value of disability-adjusted life-years (DALYs) lost for EU countries 2011 by disease (billions of Eurodollars[a] at 2011 values converted to 2019 equivalent purchasing power) (Euro Inflation Calculator, 2019; European Respiratory Society, European Lung White Book, 2019).

Disease	**Direct costs[b] € bn 2011/2019**	**Indirect costs[c] € bn 2011/2019**	**Monetized value of DALYs lost € bn 2011/2019**	**Total costs € bn 2011/2019**
Chronic Obstructive Pulmonary Disease (COPD)	23.3/25.36	25.1/27.32	93.0/101.22	141.4/153.90
Asthma	19.5/21.22	14.4/15.67	38.3/41.69	72.2/78.58
Lung cancer	3.35/3.65	NA	103.0/112.10	106.4/115.80
Tuberculosis (TB)	0.54/0.59[d]	4	5.37/5.41	5.9/6.42
Obstructive Sleep Apnea Syndrome (OSAS)	5.2/5.66	1.9/2.07	NA	7.1/7.73
Cystic fibrosis (CF)	0.6/0.65	NA	NA	0.6/0.65
Pneumonia/Acute Lower Respiratory Infections (ALRI)	2.5/2.72	NA	43.5/47.34	46.0/50.07
Total	**55.0/59.85**	**41.4/45.06**	**283.2/308.23**	**379.6/413.15**

NA: Not available.
[a] One Eurodollar is equivalent to 1.11560 US Dollars as of October 2019 (Exchangerate.com).
[b] Primary care, hospital outpatient and inpatient care, drugs, and oxygen.
[c] Lost production including work absence and early retirement.
[d] Indirect costs included with direct costs.

Table 12.4 Aggregated annual direct and indirect costs for select respiratory disease for the United States 2008 by disease (billions of Dollars in 2008 values converted to 2019 equivalent purchasing power) (Inflation Calculator, 2019; National Institutes of Health, National Heart, Lung, and Blood Institute, 2012, pp. 19–20).

Disease	Direct costs[a] $ bn 2008/2019	Indirect costs[b] $ bn 2008/2019	Total costs[b] $ bn 2008/2019
Chronic Obstructive Pulmonary Disease (COPD)/Asthma	53.7/64.04	14.3/17.05	68.0/81.09
Pneumonia	14.0/16.70	6.4/7.63	20.4/24.33

[a] Direct costs include hospital inpatient stays, emergency room visits, outpatient or office-based providers, prescription medicines, and home healthcare.

[b] Indirect costs of lost productivity due to premature mortality are based on mortality data from the National Center for Health Statistics and the initial 2008 value of lifetime earnings estimates by age and sex from the Institute for Health and Aging, University of California at San Francisco.

Theoretically, the data shown in Table 12.4 could be increased by as much as 1/3 to reflect this long-term jump in respiratory disease. Notable is that, by comparison, the EU spends about €232 billion ($258 billion) annually on COPD and asthma for roughly the same population as the United States who spends about $81 billion or three times less. Canada, with about 1/10th the population of the United States, spends C$2.34 billion ($1.79 billion) annually on COPD and asthma or more than 45 times less. (While there is some speculation in supporting literature that estimates the number of undiagnosed patients for certain diseases, there is little clarity on the actual numbers of people who are dying as a result of lack of healthcare for various diseases. Improved estimates over a greater number of diseases may further enhance the understanding of costs related to treatment and associated deaths when treatment is not provided.)

In Canada in the period from 1999 to 2012, lung health showed improvements based on national incident rates of asthma and COPD (Hamm et al., 2019). In 2010, the cost of respiratory healthcare was 6.53 billion Canadian Dollars (C$) or the inflation-adjusted 2019 monetary value of C$7.6 billion and costs were basically equal across gender (Table 12.5). Leading respiratory illnesses impacting Canadians were asthmas and other diseases of the upper respiratory tract including strep, rhinoviruses, whooping cough, and diphtheria (excluding influenza listed separately). Healthcare costs are low which may be attributed to lack of prescription medicine data in the compilation of direct costs in Canada. However, the numbers are still remarkably low.

Heart health

Seventeen million deaths per year globally are attributed to preventable cardiovascular disease (CVD) and "80 percent of the deaths from CVD occur in low- and middle-income countries," (Nelson, 2016, *para* 2). However, the European Society of

Table 12.5 Economic burden of diseases of the respiratory system in Canada[a] aggregated annual direct and indirect costs for Canada 2010 by disease (Canadian Dollars[b] in billions at 2010 values converted to 2019 equivalent purchasing power) (CAD Inflation Calculator, 2019; Public Health Agency of Canada, 2017a).

Disease	Male C$ bn 2010/2019	Female C$ bn 2010/2019	Total costs[c,d] C$ bn 2010/2019
Asthma[e]	0.61/0.72	0.63/0.73	1.24/1.45
Bronchiectasis[f]	0.09/0.113	0.11/0.126	0.20/0.24
Chronic diseases of tonsils and adenoids	0.10/0.12	0.12/0.14	0.22/0.26
Chronic Obstructive Pulmonary Disease (COPD)	0.38/0.45	0.39/0.44	0.77/0.89
Influenza	0.02/0.03	0.03/0.030	0.050/0.06
Other acute lower respiratory infections	0.16/0.19	0.18/0.21	0.34/0.40
Other acute upper respiratory infections	0.29/0.34	0.38/0.44	0.67/0.78
Other diseases of the respiratory system	0.50/0.58	0.37/0.43	0.87/1.01
Other diseases of upper respiratory tract	0.69/0.81	0.63/0.73	1.32/1.54
Pneumonia	0.42/0.49	0.41/0.47	0.83/0.96
Totals	**3.29/3.83**	**3.24/3.77**	**6.53/7.60**

[a] Data are grouped by International Classification of Disease 10th Revision (ICD-10) chapter and diagnostic categories based on the International Short List of Hospital Morbidity Tabulation (ISHMT). Please refer to the Economic Burden of Illness in Canada (EBIC) 2010 Report for complete data limitations. EBIC 2010 is availabe at https://www.canada.ca/en/public-health/services/publications/science-research-data/economic-burden-illness-canada-2010.html.
[b] One Canadian Dollar (C$) is equivalent to 0.763829 US Dollars as of October 2019 (Exchangerate.com).
[c] Direct costs include hospital care including inpatient, day surgery, other ambulatory, outpatient and emergency care; and physician care. Drug costs for the territories (Alberta, British Columbia, Manitoba, New Brunswick, Newfoundland and Labrador, Northwest Territories, Nova Scotia, Nunavut, Ontario, Prince Edward Island, Quebec, Saskatchewan, and Yukon) were not included in this data due to lack of availability.
[d] Indirect costs are associated with premature mortality.
[e] Table interpretation example: Male C$ bn 2010 so 0.61 billion is 610 million.
[f] Table interpretation example: Male C$ bn 2010 so 0.09 billion is 90 million.

Cardiology (2019, *para* 1) reports a decline in CVD in countries with high income such as Western Europe with <30% of all deaths attributed to CVD compared to >50% in middle income nations. In the same way, there are notable inequalities in hypertension with just about half reporting an increase in prevalence in middle income countries compared to small but persistent declines in higher income countries. Statistically, slightly more men (30.4%) than women (25.3%) succumb to CVD in Europe. Finally, DALYs for ischemic heart disease in middle income nations are three times higher than high-income nations of the 57 ESC members. Risk factors here are obesity, diabetes, smoking, mean cholesterol levels, and high blood pressure (European Society of Cardiology, 2019).

The EU nations experience less annual cases of CVD (>6 million), but across Europe the instance of CVD is almost double (>11 million), while the CVD cause of death in Europe is 45% compared to the EU rate of 37% (European Society of

Cardiology, 2017, *para* 1, 6). "With almost 49 million people living with the disease in the EU, the cost to the EU economies is high at €210 billion a year," (European Society of Cardiology, 2017, *para* 1). Notable is that this figure is limited to EU costs and does not include non-EU members.

The CDC Foundation in Atlanta reports that "one in every three deaths" in the United States annually (nearly 800,000) are attributed to some form of cardiovascular disease, heart disease, or stroke at the cost of "one in every six U.S. healthcare dollars," (Stinson, 2015, *para* 1). "By 2030, annual direct medical costs associated with cardiovascular diseases are projected to rise to more than $818 billion, while lost productivity costs could exceed $275 billion," (Stinson, 2015, *para* 1). Major outcomes of cardiovascular disease besides death include the hazard of debilitating heart attacks and strokes that impact 1.5 million people annually and account for $320 billion of the costs associated with cardiovascular disease (Stinson, 2015, *para* 3).

The Public Health Agency of Canada (2017b) reports' analysis of the 2012/2013 data from the Canadian Chronic Disease Surveillance System indicates that heart disease is second only to cancer as the leading cause of death, but accounts for the greatest amount of DALYs due to the cumulative impact of premature mortality, costs associated with hospitalization, and long-term requirements for the most prevalent form of heart disease—ischemic heart disease. More than 10% of Canadians (approximately 3 million) either live with ischemic heart disease or have reported a heart attack, while "about 669,600 (3.6%) Canadian adults aged 40 years and older live with diagnosed heart failure," (Public Health Agency of Canada, 2017b, p. intro). However, heart health reportedly improved in Canada during the period from 1999 to 2012 based on national incident rates of hypertension, ischemic heart disease, and stroke (Hamm et al., 2019) (Table 12.6). Many of the diseases are similar across gender. However, there are noticeable differences in gender in circulatory system illnesses related to acute myocardial infarction, other diseases of the circulatory system, and ischemic heart diseases. Some examples of other circulatory system diseases not listed separately (i.e., angina, arrhythmia, atherosclerosis, and hypertension) include congenital heart defects, cardiomyopathy, hypercholesterolemia, peripheral vascular system, and rheumatic heart disease.

Heart health in other parts of North America, such as Mexico, is demonstrating the ill health effects of modern economic development referred to as the epidemiological transition from agricultural/rural populations to industrial/urban populations already experienced in the United States and Canada (Blair, Huffman, & Shah, 2013; GBD 2017 Risk Factor Collaborators, 2018). The government has responded to the threat of CVD by implementing policy that focuses on early detection and access to trained healthcare professionals with proper medication for treatment. In addition, legislation has limited advertising to youth and enacted taxes and new labelling information on items (i.e., food, beverages with high sugar content) that place patients at risk for poor health and these preventive measures help to reduce the long-term strain on health systems (Nelson, 2016).

Table 12.6 Economic burden of diseases of the circulatory system by disease in Canada[a] aggregated annual direct and indirect costs for Canada 2010 by disease (Canadian Dollars[b] in billions at 2010 values converted to 2019 equivalent purchasing power) (CAD Inflation Calculator, 2019; Public Health Agency of Canada, 2017a).

Disease	Male C$ bn 2010/2019	Female C$ bn 2010/2019	Total costs[c,d] C$ bn 2010/2019
Acute but ill-defined stroke[e]	0.11/0.13	0.10/0.11	0.21/0.24
Acute myocardial infarction	0.90/1.05	0.41/0.47	1.31/1.52
Angina pectoris	0.22/0.26	0.11/0.12	0.33/0.38
Atherosclerosis[f]	0.09/0.10	0.09/0.10	0.18/0.20
Cerebral infarction	0.19/0.23	0.17/0.19	0.36/0.42
Conduction disorders and cardiac arrhythmias	0.60/0.69	0.44/0.52	1.04/1.21
Essential hypertension	0.26/0.31	0.30/0.35	0.56/0.66
Heart failure	0.33/0.39	0.32/0.37	0.65/0.76
Intracerebral hemorrhage	0.14/0.17	0.11/0.13	0.25/0.30
Other cerebrovascular diseases	0.08/0.09	0.06/0.07	0.14/0.16
Other diseases of the circulatory system	0.76/0.88	0.50/0.58	1.26/1.46
Other hypertensive diseases	0.14/0.16	0.10/0.11	0.24/0.27
Other ischemic heart diseases	0.90/1.04	0.30/0.36	1.20/1.40
Pulmonary heart disease and diseases of pulmonary circulation	0.08/0.10	0.10/0.12	0.18/0.22
Subarachnoid hemorrhage	0.04/0.5	0.07/0.80	0.11/0.13
Varicose veins of lower extremities	2.84/3.31	2.23/2.60	5.07/5.91
Totals	**7.70/8.96**	**5.42/6.31**	**13.12/15.27**

[a] Data are grouped by International Classification of Disease 10th Revision (ICD-10) chapter and diagnostic categories based on the International Short List of Hospital Morbidity Tabulation (ISHMT). Please refer to the Economic Burden of Illness in Canada (EBIC) 2010 Report for complete data limitations. EBIC 2010 is availabe at https:/www.canada.ca/en/public-health/services/publications/science-research-data/economic-burden-illness-canada-2010.html.
[b] One Canadian Dollar (C$) is equivalent to 0.763829 US Dollars as of October 2019 (Exchangerate.com).
[c] Direct costs include hospital care including inpatient, day surgery, other ambulatory, outpatient and emergency care; and physician care. Drug costs for the territories (Alberta, British Columbia, Manitoba, New Brunswick, Newfoundland and Labrador, Northwest Territories, Nova Scotia, Nunavut, Ontario, Prince Edward Island, Quebec, Saskatchewan, and Yukon) were not included in this data due to lack of availability.
[d] Indirect costs are associated with premature mortality.
[e] Table interpretation example: Male C$ bn 2010 so 0.11 billion is 110 million.
[f] Table interpretation example: Male C$ bn 2010 so 0.09 billion is 90 million.

Health systems

Nature is not a place to visit. It is home.

Gary Snyder

Where we live is critical to our health, but so, too, is where we can't live. "As a fundamental component of a comprehensive public health system, environmental health works to advance policies and programs to reduce chemical and other environmental exposures in air, water, soil, and food to protect residents and provide communities with healthier environments. Environmental health protects the public by tracking environmental exposures in communities across the United States and potential links with disease outcomes," (National Environmental Health Partnership Council, 2017, p. 1). Other US organizations, besides those already mentioned throughout the text, such as the Environmental Protection Agency (EPA), provide access to regional and local health and environmental agencies (U.S. Environmental Protection Agency (EPA), 2017). In the same way, WHO Europe, the European Environmental Agency, and others track important information relating to their goals of "delivering services that improve, maintain or restore the health of individuals and their communities," (World Health Organization (WHO) Regional Office for Europe, 2019b, *para* 1) and support the concept that "a clean environment is essential for human health and well-being," (EEA, n.d., *para* 1). A variety of local environmental stressors, such as noise pollution, distribution of vector-borne diseases, biodiversity loss, health, and workforce planning, contribute to the human condition.

A number of models demonstrate different approaches to reconciling environmental and public health as discussed by Graham and White (2016). Eventually, the authors point to methods of addressing the high-income or increasing income lifestyles and associated behaviors by focusing damage control on overconsumption which is particularly harmful to nature (Graham & White, 2016). But poverty or moving from poverty is not the only problem.

Working collaboratively to address transboundary problems, such as air pollution emanating from development, is also important to global health (HEAL, 2013), various assessments and trends of environmental and other factors impacting public health (Eckelman & Sherman, 2016; GBD 2017 Risk Factor Collaborators, 2018), frameworks that seek to balance the complex nature of the interaction between man and nature (Bales & Lindland, 2014; Château, Dellink, & Lanzi, 2014; Frameworks Institute, 2014; Graham & White, 2016), and persistent attention to data collection and new methodologies that could change projected public health impact to preventive measures. For example, the "OECD ENV-Linkages Computational General Equilibrium (CGE) model is an economic model that describes how economic activities are interlinked across several macroeconomic sectors and regions. It links economic activity to environmental pressure, specifically to emissions of greenhouse gases (GHGs)," Château et al., 2014, *para* 2). The American Public Health Association (AHPA, 2017, p. 1) recommends that all levels of government "should adopt standard approaches to ensuring environmental health equity, protections, and access for all, particularly vulnerable and at-risk populations." They suggest an environmental health monitoring system in which a variety of data (e.g., diseases, hazards, exposures) are measured in relation to health outcomes over time to (1) aid in decision-making for quick response to emergency situations, and (2) identify any potential problems that must be taken into consideration for program planning. Further, health systems that disentangle politics from environmental concerns seem best likely to lead to healthier environments and healthier people. There are, of course, multiple standards and approaches available and the real question of what type of data would meet these needs.

Various tracking mechanisms continue to collect data and compile information on multiple hazardous agents, such as the toxicology reports by the Agency for Toxic Substances and Disease Registry (ATSDR), and associated health outcomes to help communities address environmental toxins. The "ATSDR Brownfields/Land Revitalization Action Model…is a grassroots, community-level model designed to foster dialogue, communication, and vision," (Agency for Toxic Substances and Disease Registry (ATSDR), Division of Community Health Investigations, n.d., p. 1) for the development community and their diverse members. Brownfield sites (funded by the EPA) and Land Reuse sites (not funded by the EPA Brownfield Program) provide communities with the opportunity to identify community issues with these lands, determine how development can resolve open issues, foresee the community health benefits from conversion, and determine what data are needed to measure change according to these four steps of the ATSDR Action Model. While this model has been applied successfully to many urban areas, the problem is that not all infectious diseases or toxic chemicals are limited to abandoned manufacturing areas that are readily visible. Instead, emerging infectious diseases, zoonotic pathogens, and environmental toxins have natural pathways (e.g., natural disasters, habitat destruction through development, globalization) to spread across borders. These transmission pathways are (1) less obvious than dilapidated buildings and rusting metal on commercial sites with a history of working with or producing toxic chemicals as waste, and (2) less predictable without specific information on existing conditions and potential challenges in certain regions.

Nevertheless, strategies that encompass the ATSDR Action Model have allowed researchers to develop strategies and compile important information to assist specific communities in advancing green space, eliminating harmful byproducts of productions, and clearing waterways. They have also discovered that some of the data details of their substance toxicology reports, such as the profile and further data needs of arsenic, should be supplemented with information about the organic interaction of certain chemicals with the environment (U.S. Department of Health and Human Services, Public Health Service, Agency for Toxic Substances and Disease Registry (ATSDR), 2007). A microbiological approach is also advocated in other sciences and interpretation of environmental and population health (Sniehotta et al., 2017). This suggests another level of analysis research for understanding and researching microorganism level interactions in soil as a preventive measure versus a reactive response to local or global pandemics. The method must also stretch beyond the limited analysis of local areas targeted for redevelopment to a systematic approach to identifying harmful substances in the soil. Clearly, there is a need to view land, particularly soil, from the ground up.

Discussion and recommendations

Like music or art, love of nature is a common language that can transcend political or social boundaries.

Jimmy Carter

How science responds to emerging virulent and morphing influenza viruses, pandemics, bioterrorism, development, persistent poverty, and naturally occurring weather emergencies that disturb microbes is important to understand the cumulative impact of these activities on environmental and public health. Draining swamps may be great for political reformation, but not necessarily the best thing when these natural filters are lost to development or move mosquitoes carrying West Nile Virus to other locations without protective barriers. The same can be said for the loss of green space, such as the deforestation in the Amazon, and other areas experiencing strain on natural resources. At the same time, we recognize the need for heating fuel, alleviating overcrowding, and addressing long-standing illnesses associated with poverty. We understand the complex nature of the need for economic development, scientific innovation, and loss experienced in local communities as progress in science and policy for human development is slow to keep pace with the push for growth. Given these opposing perspectives, we concur that resource allocation remains a large policy, ethics, and social problem (Resnik, MacDougall, & Smith, 2018).

Tackling long-standing inequalities in economic decisions that lead to various environmental, health, and economic disparities (Fiedler & Costantini, 2018) is not a new problem (Fig. 12.2), but one that must be solved or, at minimum, balanced. Perhaps

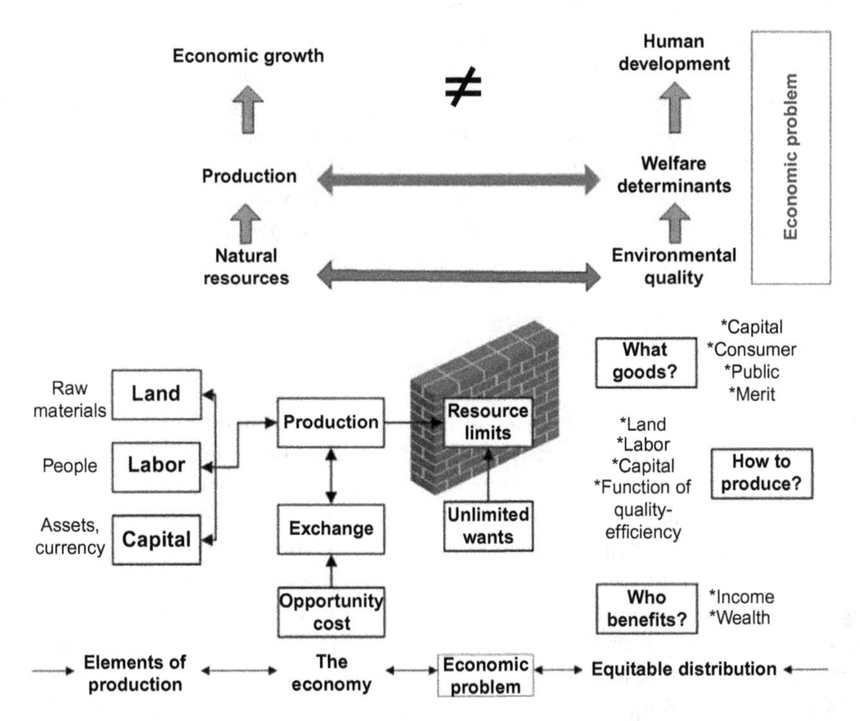

Fig. 12.2 Both sides of the economic problem represent a basic concern—The determination of asset distribution to produce items addressing consumer wants while weighing the cost of asset depletion, limited opportunity, and equitable distribution of products (reproduced with permission from Fiedler, 2009; Fiedler and Costantini, 2018, p. 5).

policy dilemmas and data collection can be funneled toward a better understanding of less complex but equally important soil status and consequences of poor environmental health which separates subgroups and national populations from prosperity and good health. However, analysis cannot be limited to known commercial areas of manufacturing districts. Understanding the environmental condition of the soil from the ground up, including microbial interaction with the environment and various substrates, requires a new approach to data collection with the influx of other technologies to present a clear picture of differences in environmental exposure and susceptibility to soil and other conditions that diminish human health. Already, we understand that data differ from national, regional, and local health perspectives. Perhaps going even smaller to what is in the soil in more directed research and analysis could be the next scientific innovation to address problems, instead of just shedding light on the inequalities.

While establishing gaps in environmental health and public health is certainly important as demonstrated in the case study comparison of European and North American regions, the battle between economic development and human health may become closer to equilibrium if more attention was focused on revealing and resolving environmental quality (e.g., land, soil, air) that impacts health versus the emphasis on environmental and health inequalities, once established. Energy expended on solutions versus opposing factions is the end goal. Focusing on dirt in a different way may be the answer. We suggest starting with a simple question such as "What's in the soil?" through systematic analysis and simple approaches can lead to a greater understanding of the various interactions leading to meaningful changes in the immediate environments of vulnerable and other populations suffering from the ill effects of poor environmental health. Soil analysis cannot just be a focal point of areas that are destined for revitalization. The same would hold true for water and air. Science should be using investigation, mapping, and other technology to identify hazardous zones as a preventive measure, not always in responsive mode.

Conclusion

In this chapter, we discuss the cycle of human interaction that pollutes the environment and the negative impact to the human body due to poor environmental health. Ironically, solving an important dimension of the social determinants of health—poverty, often results in environmental abuse and an imbalance in the economic problem. Consequently, our case study of three regions in North America and Europe provides some insight into the cost of development from the perspective of high-income regions and gender. We find that particulate matter is a global problem regardless of national income or gender. However, we observe that men across these two regions are more susceptible to the leading causes of death related to environmental/occupational risks than women. Surprisingly, the gender gap in the mortality rate is smaller in the regions of North America compared to the regions in Europe. Gauging the cost of life and environment to the healthcare system provides some basis for reviewing how we look at health and environmental inequalities. Current analysis systems are limited to response to outbreaks or the potential for land development. Before we change topography to develop, build, or reduce natural habitats, we should have better knowledge of existing systems. Thus, recommendations include adapting a systematic approach to assess environmental quality beginning with soil analysis to balance our natural assets.

Acknowledgment

The author would like to thank the researchers with the Global Burden of Diseases, Injuries, and Risk Factors Study and special thanks to Varsha Krish, Chris Odell, and Molly R. Nixon from the Institute for Health Metrics and Evaluation for their assistance in extracting requested data fields for Tables 12.1 and 12.2. Views and interpretation of all data source content and use do not necessarily reflect the opinions of the various agencies. The author accepts sole responsibility for same.

References

Agency for Toxic Substances and Disease Registry (ATSDR), Division of Community Health Investigations (n.d.). ATSDR action model. Retrieved October 22, 2019 from https://www.atsdr.cdc.gov/sites/brownfields/model.html.

American Public Health Association (AHPA). (2017). *Environmental health playbook: Investing in a robust environmental health system.* Retrieved October 24, 2019 https://www.apha.org/~/media/files/pdf/topics/environment/eh_playbook.ashx.

Attina, T. M., Hauser, R., Sathyanarayana, S., Hunt, P. A., Bourguignon, J.-P., et al. (2016). Exposure to endocrine-disrupting chemicals in the USA: A population-based disease burden and cost analysis. *Lancet Diabetes Endocrinology*, *4*, 996–1003. https://doi.org/10.1016/S2213-8587(16)30275-3.

Bales, S. N., & Lindland, E. (2014). *Talking environmental health: A FrameWorks message memo*. Washington, DC: FrameWorks Institute.

Blair, J. E., Huffman, M., & Shah, S. J. (2013). Heart failure in North America. *Current Cardiology, Reviews*, *9*, 128–146. https://doi.org/10.2174/1573403X11309020006.

CAD Inflation Calculator. (2019, October 21). *Inflation calculator-Canadian Dollar*. Retrieved October 21, 2019 from https://www.inflationtool.com/canadian-dollar.

Chanel, O., Perez, L., Künzli, N., & Medina, S. (2016). The hidden economic burden of air pollution-related morbidity: Evidence from the Aphekom project. *European Journal of Health Economics*, *17*, 1101. https://doi.org/10.1007/s10198-015-0748-z.

Château, J., Dellink, R., & Lanzi, E. (2014). *An overview of the OECD ENV-linkages model: Version 3*. OECD Environment Working Papers, No. 65 Paris: OECD Publishing https://doi.org/10.1787/5jz2qck2b2vd-en.

Cook, K., & Fiedler, B. A. (2018). Foundations of community health: Planning access to public facilities. In B. A. Fiedler (Ed.), *Translating national policy to improve environmental conditions impacting public health through community planning* (pp. 107–130): Springer International Publishing.

Costantini, V., & Mele, A. (2018). Green aid flows: Trends and opportunities for developing countries. In B. A. Fiedler (Ed.), *Translating national policy to improve environmental conditions impacting public health through community planning* (pp. 23–40): Springer International Publishing.

CSDH. (2008). *Closing the gap in a generation: Health equity through action on the social determinants of health*. Final Report of the Commission on Social Determinants of Health

Geneva: World Health Organization. Retrieved October 13, 2019 from https://www.who.int/social_determinants/final_report/csdh_finalreport_2008.pdf.

Eckelman, M. J., & Sherman, J. (2016). Environmental impacts of the U.S. health care system and effects on public health. *PLoS One*, *11*(6), e0157014. https://doi.org/10.1371/journal.pone.0157014.

Euro Inflation Calculator. (2019, October 21). *U.S. official inflation data.* Alioth Finance. Retrieved October 21, 2019 from https://www.officialdata.org/Euro-inflation.

European Environmental Agency (EEA). (n.d.). Environment and health. [Last read September 23, 2019]. Retrieved October 24, 2019 from https://www.eea.europa.eu/themes/human/intro.

European Lung Foundation. (2019). *Lung health in Europe—Facts and figures. Licensed: Creative Commons Attribution-NonCommercial 4.0 International license.* Retrieved October 21, 2019 from https://www.erswhitebook.org/files/public/About/Slideset%20for%20White%20Book.pdf.

European Respiratory Society, European Lung White Book. (2019). *The cost of respiratory disease.* Retrieved October 12, 2019 from https://www.erswhitebook.org/chapters/the-economic-burden-of-lung-disease/the-cost-of-respiratory-disease/.

European Society of Cardiology (ESC). (2017). *Fact sheets for press: CVD in Europe and ESC Congress figures.* Retrieved October 21, 2019 from https://www.escardio.org/The-ESC/Press-Office/Fact-sheets.

European Society of Cardiology (ESC) (2019). *Key messages from Atlas.* Retrieved October 21, 2019 from https://www.escardio.org/Research/ESC-Atlas-of-cardiology.

Ferretti, M., & Fischer, R. (2013). *Forest Monitoring: Methods for terrestrial investigations in Europe with an overview of North America and Asia. Vol. 12.* Oxford/Amsterdam/Waltham, MA/San Diego, CA: Elsevier.

Fiedler, B. A. (2009). *Environment, human development and economic growth.* Presentation at the University of Central Florida, PAF7315, 24 March.

Fiedler, B. A. (2018a). A regulatory primer of international environmental policy and land-use. In B. A. Fiedler (Ed.), *Translating national policy to improve environmental conditions impacting public health through community planning* (pp. 79–90): Springer International Publishing.

Fiedler, B. A. (2018b). A regulatory primer of United States multi-sectoral land use and environmental policy. In B. A. Fiedler (Ed.), *Translating national policy to improve environmental conditions impacting public health through community planning* (pp. 91–106): Springer International Publishing.

Fiedler, B. A. (2018c). Food sustainability index report on the United States: The good, the bad, and the ugly. In B. A. Fiedler (Ed.), *Translating national policy to improve environmental conditions impacting public health through community planning* (pp. 41–50): Springer International Publishing.

Fiedler, B. A. (2018d). International changes in environmental conditions and their personal health consequences. In B. A. Fiedler (Ed.), *Translating national policy to improve environmental conditions impacting public health through community planning* (pp. 255–284): Springer International Publishing.

Fiedler, B. A. (2018e). The buzz about restoring Mother Nature at the urban core. In B. A. Fiedler (Ed.), *Translating national policy to improve environmental conditions impacting public health through community planning* (pp. 155–170): Springer International Publishing.

Fiedler, B. A., & Costantini, V. (2018). The challenge of implementing macroeconomic policy in an increasingly microeconomic world. In B. A. Fiedler (Ed.), *Translating national policy to improve environmental conditions impacting public health through community planning* (pp. 1–22): Springer International Publishing.

Frameworks Institute (2014). *Environmental health toolkit: A brief intro to strategic frame analysis.* Retrieved February 18, 2019 from https://www.frameworksinstitute.org/toolkits/environmentalhealth/pdfs/eh_briefintrotoSFA.pdf.

GBD 2017 Risk Factor Collaborators. (2018). Global, regional, and national comparative risk assessment of 84 behavioural, environmental and occupational, and metabolic risks or clusters of risks for 195 countries and territories, 1990–2017: A systematic analysis for the Global Burden of Disease Study 2017. *Lancet, 392*, 1923–1994.

Global Burden of Disease Collaborative Network. Global Burden of Disease Study 2017 (GBD 2017). (2018a) *Results.* Seattle, Washington: Institute for Health Metrics and Evaluation (IHME). Available at: http://ghdx.healthdata.org/gbd-results-tool.

Global Burden of Disease Collaborative Network. Global Burden of Disease Study 2017 (GBD 2017). (2018b). *Socio-Demographic Index (SDI) 1950–2017.* Seattle, Washington: Institute for Health Metrics and Evaluation (IHME).

Gonzales, S., & Sawyer, B. (2017, July 7). How does infant mortality in the U.S. compare to other countries? In *Peterson-Kaiser HealthSystem Tracker.* Retrieved May 20, 2019 from https://www.healthsystemtracker.org/chart-collection/infant-mortality-u-s-compare-countries/#item-start.

Graham, H., & White, P. C. L. (2016). Social determinants and lifestyles: Integrating environmental and public health perspectives. *Public Health, 141*, 270–278. https://doi.org/10.1016/j.puhe.2016.09.019.

Hamm, N. C., Pelletier, L., Ellison, J., Tennenhouse, L., Reimer, K., et al. (2019). Trends in chronic disease incidence rates from the Canadian Chronic Disease Surveillance System. [Last modified June 12, 2019]. *Health Promotion and Chronic Disease Prevention in Canada. 39*(6/7). https://doi.org/10.24095/hpcdp.39.6/7.02.

Health and Environment Alliance (HEAL). (2013). *The unpaid health bill: How coal power plants make us sick.* Retrieved May 20, 2019 from www.env-health.org/unpaidhealthbill.

Howard, J. (2018, January 8). *Among 20 wealthy nations, US child mortality ranks worst, study finds. [CNN].* Retrieved May 20, 2019 from https://www.cnn.com/2018/01/08/health/child-mortality-rates-by-country-study-intl/index.html.

Inchley, J., Currie, D., Young, T., Samdal, O., Torsheim, T., et al. (Eds.), (2016). Growing up unequal: Gender and socioeconomic differences in young people's health and well-being. In *Health Behavior in School-aged Children (HBSC) study: International report from the 2013/2014 survey. Health Policy for Children and Adolescents, No. 7.* Retrieved October 13, 2019 from http://www.euro.who.int/__data/assets/pdf_file/0003/303438/HSBC-No.7-Growing-up-unequal-Full-Report.pdf.

Inflation Calculator (2019). *U.S. official inflation data.* Alioth Finance. Retrieved October 21, 2019 from https://www.officialdata.org/.

Institute for Health Metrics and Evaluation (IHME) (2019). *Global health data exchange. [database]*http://ghdx.healthdata.org/.

Institute for Health Metrics and Evaluation (IHME) (2017, September 26). *Large increase in recent decades in rate of death from chronic respiratory diseases in US.* Retrieved October 21, 2019 from http://www.healthdata.org/news-release/large-increase-recent-decades-rate-death-chronic-respiratory-diseases-us.

Juma, N. (2019). *80 nature quotes about reconnecting with mother Earth. [Everyday Power. com].* Retrieved October 3, 2019 from https://everydaypower.com/nature-quotes/.

Matyssek, R., Clarke, N., Cudlin, P., Mikkelsen, T. N., Tuovinen, J.-P., Wieser, G., & Paoletti, E. (Eds.), *Vol. 13.* (2013). *Climate change, air pollution and global challenges:*

Understanding and perspectives from forest research. In Vol. 13. Oxford/Amsterdam/Waltham, MA/San Diego, CA: Elsevier.

McKitrick, R. R., Aliakbari, E., & Stedman, A. (2018). *Environmental ranking for Canada and the OECD*. Fraser Institute.

Mercy, M. (2018, June 3). *European countries who are not part of the European Union*. Retrieved October 21, 2019 from https://www.worldatlas.com/articles/european-countries-who-are-not-part-of-the-european-union.html.

National Environmental Health Partnership Council (2017). *Environmental Health Playbook: Investing in a Robust Environmental Health System*. Retrieved February 14, 2020 from https://apha.org/topics-and-issues/environmental-health/partners/national-environmental-health-partnership-council.

National Institute of Environmental Health Sciences (NIEHS). (2016, September 21). *NIH awards more than $150 million for research on environmental influences on child health: ECHO program to investigate exposures from conception through early childhood*. Retrieved October 4, 2019 from https://www.nih.gov/news-events/news-releases/nih-awards-more-150-million-research-environmental-influences-child-health.

National Institute of Environmental Health Sciences (NIEHS). (n.d.-a). Climate change and human health literature portal. Retrieved October 4, 2019 from https://tools.niehs.nih.gov/cchhl/index.cfm.

National Institute of Environmental Health Sciences (NIEHS). (n.d.-b). Health impacts: Climate and human health. Retrieved October 4, 2019 from https://www.niehs.nih.gov/research/programs/geh/climatechange/health_impacts/index.cfm.

National Institutes of Health, National Heart, Lung, and Blood Institute. (2012). *Morbidity & mortality: 2012 chart book on cardiovascular, lung, and blood*. Retrieved October 21, 2019 from https://www.nhlbi.nih.gov/files/docs/research/2012_ChartBook.pdf.

Nelson, P. (2016, September 16). *Experts gather at CDC to discuss cardiovascular disease solutions*. Retrieved October 21, 2019 from https://www.cdcfoundation.org/blog-entry/experts-gather-cdc-discuss-cardiovascular-disease-solutions.

Office of Disease Prevention and Health Promotion (ODPHP). (2014a). *Environmental health*. [Last updated 10/04/19] https://www.healthypeople.gov/2020/topics-objectives/topic/environmental-health.

Office of Disease Prevention and Health Promotion (ODPHP). (2014b). *Environmental health*. [Last updated 10/19/2019]. Retrieved October 19, 2019 from https://www.healthypeople.gov/2020/data-search/Search-the-Data#topic-area=3525.

Office of Disease Prevention and Health Promotion (ODPHP). (2014c). *Topics and objectives*. [Last updated October 4, 2019]. Retrieved October 4, 2019 from https://www.healthypeople.gov/2020/topics-objectives.

Onel, N., & Fiedler, B. A. (2018). Green business—Not just the color of money. In B. A. Fiedler (Ed.), *Translating national policy to improve environmental conditions impacting public health through community planning* (pp. 171–202): Springer International Publishing.

Panwhar, S. T., & Fiedler, B. A. (2018). Upstream policy recommendations for Pakistan's child mortality problem. In B. A. Fiedler (Ed.), *Translating national policy to improve environmental conditions impacting public health through community planning* (pp. 203–218): Springer International Publishing.

Percy, K. E. (Ed.), (2012). *Vol. 11. Alberta oil sands*. Oxford/Amsterdam: Elsevier.

PopulationPyramid.net. (n.d.). Population pyramids of the world from 1950 to 2100. Retrieved October 23, 2019 from https://www.populationpyramid.net/.

Public Health Agency of Canada. (2017a). *Economic burden of illness in Canada custom report generator*. [Last date modified October 23, 2019]. Retrieved October 23, 2019 from https://cost-illness.canada.ca/custom-personnalise/national.php.

Public Health Agency of Canada. (2017b). *Heart disease in Canada: Highlights from the Canadian chronic disease surveillance system*. Retrieved October 23, 2019 from https://www.canada.ca/content/dam/phac-aspc/documents/services/publications/diseases-conditions/heart-disease-fact-sheet/heart-disease-factsheet-eng.pdf. © Her Majesty the Queen in Right of Canada, as represented by the Minister of Health, 2017 vert; Cat.: HP35–85/2017E-PDF | ISBN: 978-0-660-08776-4 | Pub.: 170112.

Rao, A., & Ross, C. L. (2018). Planning healthy communities: Abating preventable chronic diseases. In B. A. Fiedler (Ed.), *Translating national policy to improve environmental conditions impacting public health through community planning* (pp. 51–78): Springer International Publishing.

Resnik, D. B., MacDougall, D. R., & Smith, E. M. (2018). Ethical dilemmas in protecting susceptible subpopulations from environmental health risks: Liberty, utility, fairness, and accountability for reasonableness. *The American Journal of Bioethics, AJOB*, *18*(3), 29–41. https://doi.org/10.1080/15265161.2017.1418922.

Roche, B., Broutin, H., & Simard, F. (2018). *Ecology and evolution of infectious diseases: Pathogen control and public health management in low-income countries*. Oxford University Press.

Rotter, S., Beronius, A., Boobis, A. R., Hanberg, A., van Klaveren, J., et al. (2018). Overview on legislation and scientific approaches for risk assessment of combined exposure to multiple chemicals: The potential EuroMix contribution. *Critical Reviews in Toxicology*, *48*(9), 796–814. https://doi.org/10.1080/10408444.2018.1541964.

Sniehotta, F. F., Araújo-Soares, V., Brown, J., Kelly, M. P., Michie, S., & West, R. (2017). Complex systems and individual-level approaches to population health: A false dichotomy? *The Lancet, Public Health*, *2*(9), e396–e397. https://doi.org/10.1016/S2468-2667(17)30167-6.

So, J. Y., Mamary, A. J., & Shenoy, K. (2018). Asthma: Diagnosis and treatment. *European Medical Journal*, *3*(4), 111–121.

Stinson, C. (2015, April 29). *Heart disease and stroke cost America nearly $1 billion a day in medical costs, lost productivity*. CDC Foundation. Retrieved October 21, 2019 from https://www.cdcfoundation.org/pr/2015/heart-disease-and-stroke-cost-america-nearly-1-billion-day-medical-costs-lost-productivity.

U.S. Department of Health and Human Services, Public Health Service, Agency for Toxic Substances and Disease Registry (ATSDR). (2007). *Toxicological profile for arsenic*. Retrieved October 22, 2019 from https://www.atsdr.cdc.gov/substances/toxsubstance.asp?toxid=3.

U.S. Environmental Protection Agency (EPA). (2017, January 19). *Health and environmental agencies of U.S. states and territories*. [Last updated September 19, 2018]. Retrieved October 24, 2019 from https://www.epa.gov/home/health-and-environmental-agencies-us-states-and-territories.

United Nations Development Programme. (2017). *Human Development Index and its components*. Retrieved October 19, 2019 from http://www.hdr.undp.org/en/composite/HDI.

Van Ness, V. N., III. (2018). Streetwise community policing to inform United States national policy. In B. A. Fiedler (Ed.), *Translating national policy to improve environmental conditions impacting public health through community planning* (pp. 285–304): Springer International Publishing.

Walig, A. M. P., Canencia, O. P., & Fiedler, B. A. (2018). Water quality: Mindanao Island of the Philippines. In B. A. Fiedler (Ed.), *Translating national policy to improve environmental conditions impacting public health through community planning* (pp. 219–254): Springer International Publishing.

Webster, R. (2018). Food reservations at the reservation? In B. A. Fiedler (Ed.), *Translating national policy to improve environmental conditions impacting public health through community planning* (pp. 131–154): Springer International Publishing.

World Bank Group: (2019a). *WHO/UNICEF Joint Monitoring Programme (JMP) for water supply, sanitation and hygiene (washdata.org).* License: CC BY-4.0. Retrieved October 19, 2019 from https://data.worldbank.org/indicator/SH.STA.SMSS.ZS.

World Bank Group: (2019b). *World Health Organization, Global health observatory data repository (apps.who.int/ghodata).* License: CC BY-4.0. Retrieved October 19, 2019 from https://data.worldbank.org/indicator/SH.STA.WASH.P5.

World Health Organization: (WHO) Regional Office for Europe (2012). *Environmental health inequalities in Europe: Assessment report.* Retrieved October 4, 2019 from http://www.euro.who.int/en/publications/abstracts/environmental-health-inequalities-in-europe.-assessment-report.

World Health Organization: (WHO) Regional Office for Europe (2019a). *Environmental health inequalities in Europe: Second assessment report.* Retrieved October 4, 2019 from https://www.who.int/social_determinants/publications/environmental-health-inequalities-in-europe/en/.

World Health Organization: (WHO) Regional Office for Europe (2019b). *Health systems.* Retrieved October 24, 2019 from http://www.euro.who.int/en/health-topics/Health-systems.

Supplementary reading recommendations

American Heart Association (2019). *Healthy living. Available at:*https://www.heart.org/en/.

American Lung Association (2019). *Lung health and diseases. Available at:* https://www.lung.org/lung-health-and-diseases/.

American Public Health Association (AHPA) (2019). *Health equity. Available at:* https://www.apha.org/topics-and-issues/health-equity.

Benjamin, E. J., Blaha, M. J., Chiuve, S. E., Cushman, M., et al.on behalf of the American Heart Association Statistics Committee and Stroke Statistics Subcommittee, (2017). Heart disease and stroke statistics—2017 update: A report from the American Heart Association. [published online ahead of print January 25, 2017]*Circulation.* https://doi.org/10.1161/CIR.0000000000000485.

*European Lung White Book. Available at:*https://www.erswhitebook.org/chapters/.

Lung Institute (2014, October 17). *The cost of lung disease. [Blog-Exhale].* Retrieved May 20, 2019 from https://lunginstitute.com/blog/the-cost-of-lung-disease/.

U.S. Department of Health and Human Services, Agency for Healthcare Research and Quality (AHRQ). United States, 1996–2015. Medical Expenditure Panel Survey. [Online interactive]. Available at: https://meps.ahrq.gov/mepstrends/hc_cond/.

U.S. Environmental Protection Agency's National Environmental Justice Advisory Council (NEJAC) (n.d.). Environmental justice. [Last update August 29, 2019]. Available at: https://www.epa.gov/environmentaljustice.

Thorpe, K. E., & Philyaw, M. (2012). The medicalization of chronic disease and costs. *Annual Review of Public Health, 33*(1), 409–423.

World Economic Forum (2016). *Which of the world's four income groups are you in?* Available at: https://www.weforum.org/agenda/2016/05/these-maps-divide-the-world-by-average-income/.

Definitions

Asthmagens any environmental exposure to inhaled substances (e.g., chemical and particulate matter in air) found in the home, work, or as a result of an occupation, that is causally related to the development of asthma.

Brownfield sites "abandoned, idled, or underused industrial and commercial properties where reuse or redevelopment is complicated by real or perceived contamination." These sites have received funding from EPA Brownfield Program (Agency for Toxic Substances and Disease Registry (ATSDR), Division of Community Health Investigations, n.d., p. 1).

Environmental Health consists of preventing or controlling disease, injury, and disability related to the interactions between people and their environment (ODPHP, 2014a).

Gross National Income (GNI) per capita "aggregate income of an economy generated by its production and its ownership of factors of production, less the incomes paid for the use of factors of production owned by the rest of the world, converted to international dollars using PPP rates, divided by midyear population," (United Nations Development Programme, 2017).

Human Development Index (HDI) "a composite index measuring average achievement in three basic dimensions of human development—a long and healthy life, knowledge and a decent standard of living," (United Nations Development Programme, 2017).

Land Reuse sites "any site formally utilized for commercial and industrial purposes complicated by real or perceived contamination" that has not received funding from the EPA Brownfield Program (Agency for Toxic Substances and Disease Registry (ATSDR), Division of Community Health Investigations, n.d., p. 1).

Particulate Matter composition of microscopic organic and inorganic hazardous materials (e.g., dust, soot, liquid, smoke) that can be inhaled from the air causing damage to lungs, blood, and other organs.

Purchasing Power Parity (PPP) "to compare economic statistics across countries, the data must first be converted into a common currency. Unlike market exchange rates, PPP rates of exchange allow this conversion to take account of price differences between countries. In that way GNI per capita (PPP $) better reflects people's living standards uniformly. In theory, 1 PPP dollar (or international dollar) has the same purchasing power in the domestic economy of a country as US$1 has in the US economy. The current PPP conversion rates have been introduced in May 2014. They were based on the 2011 International Comparison Programme (ICP) Surveys, which covered 199 economies from all geographical regions and from the OECD," (United Nations Development Programme, 2017).

Socio-demographic Index (SDI) relationship between level of economic development and the exposure prevalence of risk associated with socioeconomic status. Increasing SDI have an inverse relationship with several decreasing risks such as unsafe water sources, intimate partner violence, and underweight children, while other risks increase directly with increasing SDI such as high red meat consumption, alcohol use, smoking, and discontinued breastfeeding.

Structural racism and social environmental risk: A case study of adverse pregnancy outcomes in Louisiana

Vanessa Lopez-Littleton[a], *Carla Jackie Sampson*[b]
[a]California State University, Monterey Bay, Seaside, CA, United States, [b]Fox School of Business, Temple University, Department of Risk, Insurance, and Healthcare Management, Philadelphia, PA, United States

Abstract

This chapter examines structural racism as a fundamental cause of health inequities by challenging conventional social determinants of health models. Evidence of structural racism exists in regions where high rates of pregnancy-related Black mortality persist at rates three times higher than rates of non-Hispanic White women in the US. Social environmental conditions experienced in racially segregated Black communities contribute to health disadvantages experienced by Black women. This case study explores the social environmental impact of structural racism and a combination of policy artifacts as the underlying cause of health inequities faced by Black women in Louisiana, by comparing maternal and infant mortality between Black and White women and between Louisiana parishes. We recommend the reconceptualization of public health approaches to health inequities through individual-level efforts made by public health professionals and policymakers to create, support, and engage in mechanisms that improve the social environmental deficits experienced by Black women.

Keywords: Social environmental inequality, Infant mortality, Maternal mortality, Residential segregation, Structural racism.

Chapter outline

Three Facets of Public Health and Paths to Improvements. https://doi.org/10.1016/B978-0-12-819008-1.00013-4

Challenging conventional determinants of health

Current, perhaps color-blind, approaches to addressing health inequities which center on social determinants of health (SDOH) models should be reconsidered in light of persistent pockets of preventable mortality. Disproportionately high rates of Black infant and maternal mortality are prime examples of the adverse health outcomes experienced by vulnerable populations in the United States. As an illustration, the US Black infant mortality rate of 10.97 deaths per 1000 live births is nearly twice the overall US rate of 5.79 (Ely & Driscoll, 2019, p. 2). As such, the Black infant mortality rate is on par with lower-income countries such as Libya, Tunisia, and Armenia, each with an infant mortality rate of 11 (World Bank, Global Development Data, 2019a). In much the same way, the US maternal mortality rate for Black women of 42.8 maternal deaths per 100,000 live births is 2.5 times higher than the overall US rate of 17.2 (Centers for Disease Control and Prevention (CDC), 2019, *para* 2). The US rate outpaces rates in lower middle-income countries such as El Salvador and Iraq with rates of 35 and 38, respectively (World Bank, Global Development Data, 2019b). Utilizing data from the Louisiana Vital Records and the US Census Bureau, this chapter illustrates the high rate of adverse pregnancy outcomes (infant and maternal mortality) experienced by Black women, across income and education levels, in the state of Louisiana. Findings suggest structural racism plays an essential role in policymaking that alters the course of preventable mortality in areas where we suggest specific patterns of health inequities exist. As a result, we contend the inclusion of structural racism—a social environmental factor—as a fundamental cause of health inequities in future SDOH models.

Adverse pregnancy outcomes are closely tied to socioeconomics (Kim & Saada, 2013), yet poverty does not fully account for variations in outcomes. According to the America's Health Rankings (2019, *para* 4), the US infant mortality rate of 5.9 is higher than the average rate of 3.9 for other high-income countries. Declines in the United States have not kept pace with those experienced by other high-income countries. In contrast, maternal mortality rates in the United States are increasing, while rates in other countries are decreasing (MacDorman, Declercq, Cabral, & Morton, 2016, p. 447). Between 2000 and 2014, the US maternal mortality rate rose by 27% (MacDorman et al., 2016, *para* 2). Social and economic characteristics vary within and between countries making comparisons difficult, but comparing countries based on income categories (e.g., high, middle, and low income) provides an important level of comparison for benchmarking outcomes.

Adverse birth outcomes are not merely a function of socioeconomics. Social and environmental exposures are also known to contribute to health disadvantages (Padilla et al., 2013, p. 1). Social environmental factors shaped by policies and practices contribute to vulnerabilities experienced by some groups in the United States and internationally (Browne et al., 2016; McCalman, Jongen, & Bainbridge, 2017). In fact, Latina women benefit from the "Latina Paradox" where Latina women tend to have better birth outcomes than their White counterparts despite having lower socioeconomic levels (Zolitschka et al., 2019, p. 1). The level of protection varies by country and outcome. In the United States, Latina women from Cuba, Central and South America, and Mexico tend to have better birth outcomes than Latinas from Puerto Rico whose rates are higher than White women (Mathews, MacDorman, & Thoma, 2015, p. 5). Yet, the pregnancy outcomes of Black women in the United States across all income and education levels are worse than those outcomes of Latina women from countries at any income strata.

The SDOH are closely associated with the distribution of resources, social policies, and politics and, resultantly, health inequities (Healthy People 2020, 2019a). As such, health outcomes emerge from complex interactions between and among health determinants at multiple levels—biological to societal level of individuals and amalgamated within their social networks—and are affected by location and influenced over time (Fink, Keyes, & Cerdá, 2016). Environmental factors, such as poor air, water, and soil quality, negatively affect both individual and community health, but an increasing body of literature points to social environmental factors, through more indirect routes, as entrenched contributors to inequities. However, there are marked differences in health outcomes for Black populations that persist despite increasing education or income when compared to similar White populations (Braveman, Cubbin, Egerter, Williams, & Pamuk, 2010). Consequently, neither socioeconomics nor poverty explains the persistent differential outcomes. These differences suggest there are other factors, perhaps structural, that result in different outcomes for Black populations yielding to an inherent vulnerability. Further, the seemingly entrenched nature of differential outcomes would also suggest conventional SDOH models should explicitly include the proposed variable of structural racism.

The SDOH model used in Healthy People 2020 includes five key domains (determinants): economic stability, education, social and community context, health and healthcare, and neighborhood and built environment (Healthy People 2020, 2019a). The model uses discrimination, within the domain of social and community context, as a proxy for "socially structured action that is unfair or unjustified and harms individuals and groups," (Healthy People 2020, 2019b, *para* 1). While the authors of Healthy People 2020 acknowledge the lack of a standard definition or perspective related to the concept of discrimination, they purport "discrimination on the basis of race (commonly referred to as racism) has been linked to disparities in health outcomes for racial/ethnic minorities," (Healthy People 2020, 2019b, *para* 10). Yet, there is a fundamental distinction between racism and discrimination. Discrimination in the broadest sense "refers to all means of expressing and institutionalizing social relationships of dominance and oppression," (Krieger, 2014, p. 650). Removing the critical aspect of race creates an overly broad generalization applicable across numerous domains. Racism, by contrast, provides a lens for understanding oppressive systems which undergird a society that seemingly tolerates persistent inequities by failing to name past and contemporary insults experienced by specific racial and ethnic groups, Blacks in particular.

Note that while this chapter focuses on race in the United States, there are many examples of papers regarding racial inequities in health outcomes in other countries (Ben, Cormack, Harris, & Paradies, 2017; Bostos, Harnois, & Paradies, 2018). While these resources focus on individual-level issues relating to patient perception of care and racial bias that may go unreported or impact decision-making, the notion of race as a universal metric is plausible. Race is more of a social construct than a purely biologic phenomenon having direct and indirect implications for health and health outcomes. Public health's examination of race and racism, by extension, is a population-based concern well within the purview of the field. Racism has long been identified as a contributor to health inequities with physical, mental, and behavioral effects (Pachter & Garcia Coll, 2009, p. 256). "Racism produces rates of morbidity, mortality, and overall well-being that vary depending on socially assigned race," (Ford & Airhihenbuwa, 2010, p. S30). The rarity of including racism in SDOH models suggests a marked deficit in approaches to promoting equity and specific acknowledgment of the role and function of racism in producing differential health outcomes (Graham, Brown-Jeffy, Aronson, & Stephens, 2011). The American Public Health Association (APHA) explicitly acknowledges racism—intentional or unintentional—as a barrier to health equity (APHA, 2019, *para* 1). Public Health Critical Race theory posits public health approaches should include solutions that address inequities in housing, health, employment, and other related factors that contribute to the disadvantages experienced by vulnerable groups (Ford & Airhihenbuwa, 2010). Together, these perspectives represent clear recognition of the limitations of current SDOH models.

Health, in the United States and Europe, has been shown to follow a social gradient and is strongly influenced by socioeconomic factors (Braveman & Gottlieb, 2014). Poverty, at the end of the socioeconomic scale, is an important predictor of poor health outcomes. While race and class are intertwined, concentrated poverty is a function of the economic conditions where a person lives and the socioeconomic characteristics of a person's neighbors. Racial residential segregation is pervasive and may concentrate poverty, yielding exposure to social risk factors and environmental pollutants (Gee & Ford, 2011). Areas of concentrated poverty serve as a source of stress with an associated decrease in social support. Neighborhoods with high concentrations of poverty may cause residents additional stress or reduce the amount of time spent outdoors due to a fear of crime with deleterious effects on their overall health (Egerter, Barclay, Grossman-Kahn, & Braveman, 2011). As a consequence of location, residents in these neighborhoods have less access to networks that connect them to information and resources that prevent social exclusion (Carpiano, 2006; Kawachi, Kennedy, & Glass, 1999). Included among these limitations is the lack of access to information necessary to improve health literacy and subsequently health outcomes.

This chapter focuses on the long-term structural barriers that establish social environmental obstacles to care leading to preventable infant and maternal mortality. We use infant and maternal mortality outcomes in Louisiana as a case study to argue for the inclusion of structural racism—a social environmental risk factor—as (1) a determinant of health, and (2) an essential component for any comprehensive public health approach. As the social environmental context is thought to shape and sustain behavioral and cultural dimensions, we contend structural racism is a critical SDOH, whose effects are intertwined in a variety of policies at the state and local level, contributing to an intractable public health problem. As such, we argue for changes to education and other policy approaches as a public health imperative to achieving health equity. Strategies for public health researchers, practitioners, and policymakers are presented to reduce place-based risk associated with structural racism.

Defining structural racism and policy artifacts of structural racism

Many are familiar with the vocabulary associated with SDOH. However, to understand the impetus for this research, we introduce several key definitions that surround the concept of structural racism. Then, we examine the combination of policy artifacts contributing to the unequal levels of access to resources and important opportunities that impact population health.

Defining structural racism

Racism is a stratification system that, through various practices and arrangements, allocates resources and creates opportunities in favor of those in the privileged position, while unfairly disadvantaging those in unprivileged positions (Jones, 2002). As a broader concept, structural racism includes numerous ways in which societies perpetuate systems of "racial discrimination through mutually reinforcing systems of housing, education, employment, earnings, benefits, credit, media, healthcare, and criminal justice," (Bailey et al., 2017, p. 1453). Racism, first appearing in public health research in 2002 when Camara Jones began arguing that racism was a fundamental cause of health inequities, has a diminutive history in public health research (Jones, 2002). Researchers disagree on the role racism plays in determining health outcomes (Krieger, 2014), but there is an increasing acknowledgment of structural racism as a "wicked problem."

The pervasiveness of racism shapes health and health outcomes through deeply ingrained social norms and practices. Structural racism operates at the macro-level and is constantly reconstituting the conditions necessary to ensure reinforcing systems remain in place (Gee & Ford, 2011) perpetuating further adaptations of a "White racial frame" in public health and potentially other US systems. These narratives posit perspectives that support and protect social inequities (Castle, Wendel, Kerr, Brooms, & Rollins, 2019) that are reported to favor Whites and become apparent in statistical comparisons. For example, while White women at the lower end of the socioeconomic scale have worse health outcomes than those at the higher end, White women with less than a high school education have better health outcomes than college-educated Black women (Peterson et al., 2019; Singh & Yu, 2019). Specifically, the maternal mortality rate for college-educated Black women is 5.2 times the rate of White women with a college degree and 1.6 times that of White women with less than a high school diploma (Peterson et al., 2019). These inequities persist for Black women without regard to location, income, education, or insurance status, thereby making the policy context in which inequities occur particularly meaningful.

Policy artifacts of structural racism

Structural racism operates through a system of interrelated policies with intergenerational effects. No single policy creates structural racism. As a system, structural racism unfairly disadvantages those who are not in positions of power or authority. The following examples demonstrate how social policies have contributed to the systemic disadvantage of Black and other marginalized groups.

Social security

The **Social Security Act of 1935** prevented domestic and agricultural workers from receiving benefits. This agreement was viewed as a compromise in favor of southern states to ensure they would be able to pass on their wealth to their heirs, while intentionally excluding the people who worked in the homes and on the lands of White southerners. This policy contributed to the persistent and continuing economic disadvantage experienced by racial groups in the United States (Bailey et al., 2017).

Education

In 1868, the 14th Amendment created equal protection under the law, yet legal segregation remained pervasive throughout America. Education policies are closely tied to segregation as children in the United States are educated in the neighborhoods where they reside. The US Supreme Court overturning **Plessy v. Ferguson** in the landmark **Brown v. Board of Education** decision concluded that separate educational facilities are inherently unequal. Yet, public schools across the United States remain segregated. Even though the *Brown* decision sought to ensure students, regardless of race, would have access to a level of education comparable to those of White students, 80% of Black students today attend school in segregated schools (Orfield, Ee, Frankenberg, & Siegel-Hawley, 2016). One of the hallmarks of the US education system is that Black students routinely underperform compared to their White counterparts. The net result has been continual differences in education outcomes where

Black children are more likely to drop out of high school, to be suspended or expelled, to not go on to college and to enter the prison system (Alexander, 2012; National Center for Education Statistics, US Department of Education, 2019a, 2019b).

Mass incarceration

The US War on Drugs and stringent anticrime policies ushered in a new era of incarceration. Many of these policies enacted in the 1970s and 1980s disproportionately targeted Black and other marginalized groups. The consequences of these policies contributed to conditions in racially segregated communities by stripping them of potential employees and disrupting family life for countless families and communities. Young Black men are more likely to be searched and imprisoned for nonviolent offenses. Once they have a record, they lose their voting rights. For most, finding adequate employment is a challenge due to background check requirements. While many states have worked to reform sentencing laws, the effect of the harsh sentencing under laws (e.g., three strike laws, heavy penalties for crack cocaine, stop and frisk laws) has long term consequences for those who were sentenced as well as for their families and communities (Alexander, 2012). Notable is that Blacks are incarcerated at a rate five times greater than Whites, and one in every three Black males in the US population is expected to go to prison during their lifetime (The Sentencing Project, 2013). Thus, mass incarceration of Black males has a deleterious effect on their health and the health of their communities.

Healthcare

Despite the passage of the Civil Rights Act of 1964, healthcare delivery remained segregated in the Jim Crow South. Segregation in healthcare delivery continued blatantly with separate entrances and underresourced wards/units, or at wholly different Black hospitals (Smith, 2005). Discrimination also meant that Black patients were asked to pay fees in advance of treatment when they could ill afford to do and that admitting privileges to hospitals were denied to Black physicians (Chandra, Frakes, & Malani, 2017; Smith, 2005). These practices continued until the implementation of Medicare in 1966, which led to an investigation forcing the necessary change that required integration for receiving Federal funds (Smith, 2005). Still, former "Blacks only" hospitals attracted a majority-Black patient base and served to train minority clinicians, but remained severely underresourced (Zheng, 2016). The "better" hospitals were found in predominantly White suburbs, some conducting their own "White Flight" by moving from majority-minority inner-city areas (Institute of Medicine, 2003). There is abundant literature which confirms that health providers continue to hold negative biases against people of color, which continues to affect access, quality of care, and health outcomes (Hall et al., 2015; Institute of Medicine, 2003; Maina, Belton, Ginzberg, Singh, & Johnson, 2018; Ramaswamy & Kelly, 2015).

Housing

In retrospect, no other set of discriminatory policies have been as blatant as housing policies. Often the foundation of intergenerational wealth transfer, Black people were systematically excluded via several mechanisms, not limited to the Jim Crow South, which directly or indirectly affected health outcomes for generations (Mendez, Hogan, & Culhane, 2011). The exclusions from homeownership manifested in the behavior of landlords and brokers who would not show available homes to persons of color (or show available homes in Black neighborhoods to White clients). Lenders applied spurious rules and requirements only for clients of color. A landmark study by Smith and Cloud (1996) found patterns of unfair practices experienced by Blacks and Latinos, such as required credit checks, restrictive qualifications, and stringent documentation for preappointments. They also found Whites benefited from practices such as exceptions to criteria, lower escrow and reserve payments, and constructive application advice. Blacks also faced additional challenges such as restrictions based on property location—redlining and the Federal Housing Administration (FHA) denial of backing for mortgages on properties in or in proximity to Black neighborhoods. In 1933, New Deal housing construction in the new suburbs explicitly excluded persons of color, who instead were steered toward inner-city housing projects (Massey & Denton, 1993). These practices contributed to patterns of segregation still evident today.

Racial residential segregation

Racial residential segregation (residential segregation, hereafter) remains one of the most enduring features of neighborhoods in the United States (Kollmann, Marsiglio, & Suardi, 2018). Residential segregation, the degree to which racial groups live apart from one another, can be measured in different dimensions (Massey & Denton, 1993). Each measure is associated with a distinct health risk for marginalized populations (e.g., geographic isolation, constraints on social and economic resources, reduced social support and cohesion, substandard housing) (Mehra, Boyd, & Ickovics, 2017). The hypersegregated (>90% from one racial group) urban areas where Blacks reside are often characterized by a high degree of clustering, concentration, and isolation (Acevedo-Garcia, Lochner, Osypuk, & Subramanian, 2003). Hypersegregation crosses all five geographic patterns and is closely associated with racism and adverse pregnancy outcomes, including low birth weight and preterm births (Mehra et al., 2017). While fewer US cities are considered hypersegregated (segregated across multiple measures of segregation) than in the previous decades, residential segregation is an archetypal characteristic of neighborhoods throughout the United States and the foundation of structural racism. Conditions experienced in segregated communities include concentrated poverty, limited employment opportunities, high crime rates, economic oppression, and spatially concentrated disadvantage (Massey & Tannen, 2015).

Residential segregation has been described as a root cause of health inequities with evidence suggesting a detrimental effect on the health of Blacks (Acevedo-Garcia et al., 2003; Bailey et al., 2017; Mehra et al., 2017). Studies examining the relationship between structural racism and health outcomes at the individual-level have identified discrimination and bias affects Blacks in the clinical encounter, throughout daily life, and across the lifespan (Geronimus, Hicken, Keene, & Bound, 2006; Krieger, 2014; Lu et al., 2010). More recent studies focusing on residential segregation have found various factors, including incarceration rates and median household incomes, to contribute to adverse pregnancy outcomes (Mehra et al., 2017; Wallace, Crear-Perry, Richardson, Tarver, & Theall, 2017; Wallace, Green, Richardson, Theall, & Crear-Perry, 2017).

The disadvantages experienced in segregated neighborhoods yield negative consequences for Black women residing in them. By virtue of their location and the limited access to healthcare providers and health-related networks, Black women in segregated neighborhoods have limited access to health-promoting factors (Mehra et al., 2017). High crime and incarceration rates in segregated Black neighborhoods reduce the pool of marriageable Black men (Passel, Wang, & Taylor, 2010). Other factors related to segregated Black neighborhoods include greater access to unhealthy foods, increased availability of guns and subsequent exposure to gun violence, and a lack of safe places for physical activity (Institute of Medicine and National Research Council, 2013). These factors coalesce to yield health and social disadvantages for those residing in segregated Black neighborhoods.

Case study: Disproportionate Black infant and maternal mortality in the United States and Louisiana

Black women in Louisiana face health risks associated with race and class (socioeconomic status), both of which are exacerbated by social environmental risk factors. Current trends in infant and maternal mortality point to intractable inequities between racial groups, particularly in the state of Louisiana. This case study of adverse pregnancy outcomes demonstrates how the problem disproportionately impacts Black women and likely forecasts little positive change in health outcomes for this population.

Adverse pregnancy outcomes and Black women in the United States

The effect of structural racism on pregnant women has consequences for Black mothers and their infants, who, by birth, are at an increased risk of early death. Adverse pregnancy outcomes are a broad classification of pregnancy-related outcomes affecting the mother or infant. While low birth weight and other morbidity conditions are also considered adverse pregnancy outcomes, infant and maternal mortality are notably the most severe type of pregnancy outcome. Miscarriages or medically induced terminations are not included as adverse pregnancy outcomes unless the event leads to the death of the mother. Individual risk factors for adverse pregnancy outcomes include chronic health conditions, maternal infections, biological, and genetic factors.

The health and well-being of infants are closely linked to community health status, which is a function of social, economic, political, and environmental factors. These factors contribute to the quality of, access to, and the delivery of healthcare services, as well as how individuals access health information and use health information to improve health and health outcomes. Neighborhood conditions serve as powerful predictors of morbidity and mortality. As a result, social environmental factors are an increasingly important factor in determining health and health outcomes for vulnerable populations. In the United States, the infant mortality rate for non-Hispanic Black infants is more than two times the rate for non-Hispanic Whites (Ely & Driscoll, 2019, p. 2). In 2016, the US infant mortality rate for non-Hispanic Blacks was 10.97 deaths per 1000 live births, yet the rate for non-Hispanic Whites was 4.67 (Ely & Driscoll, 2019, p. 2). While maternal socioeconomic and behavioral risk factors account for some of the variation in birth outcomes, no permutation of these factors fully account for the persistence in differentials in birth outcomes experienced by US Black women.

Women in the United States die from pregnancy-related complications at a higher rate than women in comparable countries. As used in this chapter, maternal mortality is defined as "the death of a woman while pregnant or within 42 days of termination of pregnancy, irrespective of the duration and site of the pregnancy, from any cause related to or aggravated by the pregnancy or its management, but not from accidental or incidental causes," (World Health Organization, 2019, *para* 2). Only two countries, the United States and Serbia, have experienced increases (MacDorman, Declercq, & Thoma, 2017) in maternal mortality rates, while rates in other developed/high-income countries have steadily decreased. The exact cause of the increase in the United States is not known, but recent changes in the way deaths are reported or errors in reporting pregnancy status may contribute to some of the overestimation of maternal deaths (CDC, 2019a).

Each year, approximately 700 women in the United States die of delivery or pregnancy-related complications, of which nearly 60% of these deaths are considered preventable (CDC, 2019, *para* 1). Chronic health conditions account for some of the variation, but most women die due to severe bleeding (mostly after childbirth), infections (mainly after childbirth), high blood pressure during pregnancy, delivery complications, or unsafe terminations (CDC, 2019a, *para* 11). Maternal mortality rates for non-Hispanic Black women in the United States are nearly three times the rate of non-Hispanic White women. Between 2011 and 2015, the maternal mortality rate for Black women was 42.8 deaths per 100,000 live births compared to 17.2 deaths per 100,000 live births for all US women and 13.0 deaths per 100,000 births for White women (CDC, 2019a, *para* 7). Various factors (e.g., social, economic, environmental, behavioral, or genetic) contribute to the differential outcomes experienced by Blacks, but none of the known factors fully account for the differential in outcomes between Black women and other racial and ethnic groups.

Black infant mortality in Louisiana

Louisiana persistently has one of the highest rates of infant mortality in the United States. Each year, nearly 504 infants in Louisiana die before the age of one (Louisiana Department of Health (LDH), Office of Public Health (OPH), Bureau of Family Health, 2017, p. 3). Fig. 13.1 demonstrates that the overall infant mortality in Louisiana rate has shown a small decline since the early 2000s, but remains higher than the US rate (Mathews, MacDorman, & Thoma, 2015). While the United States met the Healthy People 2020 infant mortality goal of 6.0 deaths per 1000 live births, Louisiana lags at 7.1, the third highest in the United States (CDC, 2019b). One of the main drivers of the higher rate experienced in Louisiana is the persistently high rate of Blacks infant deaths. Black infants in the United States face a 122% greater chance of dying within the first year of life than their White counterparts (Singh & Yu, 2019). In 2016, the non-Hispanic Black infant mortality rate in Louisiana was 11.4 deaths per 1000 live births compared to the non-Hispanic White rate of 4.9 deaths per 1000 live births (CDC, 2019c).

The leading causes of infant death in Louisiana were perinatal period conditions (including preterm-related causes such as low birth weight), sudden unexplained infant death, congenital anomalies, and other medical conditions (Louisiana Department of Health (LDH), Office of Public Health (OPH), Bureau of Family Health, 2017). The health of the mother during the perinatal period (0–28 days) and the period before conception account for nearly 45% of infant deaths (LDH-OPH, 2017, p. 7). Some of the contributors to infant deaths during the perinatal period relate to stress, inadequate healthcare services, and maternal health conditions such as diabetes, depression, hypertension, and infection (LDH-OPH, 2017). Access to prenatal care has previously been thought to reduce pregnancy risks, but more recent studies have shown that

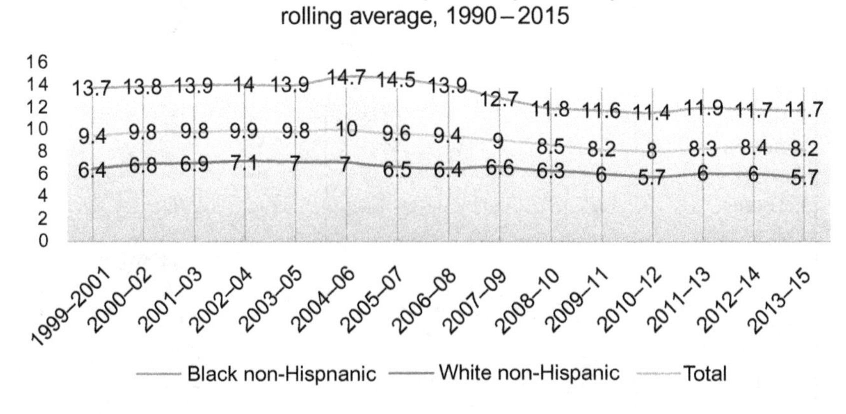

Fig. 13.1 Louisiana infant mortality rates, by race/ethnicity, 1999–2015, 3-year rolling average.
Data from U.S. Department of Health and Human Services (2019).

increasing access to prenatal care does not reduce inequities or improve outcomes for all women, racial and ethnic minorities in particular (Alexander, Wingate, Bader, & Kogan, 2008; Healy et al., 2006; MacDorman & Mathews, 2008).

Health inequities are often summarized in comparison to other populations. In this case, Black infant mortality rates (the population with the highest infant mortality rates in Louisiana) are compared to White infant mortality rates (the population with the lowest infant mortality rates in Louisiana) in which the data was stable enough for comparison. Using disparity ratios, a ratio is calculated by dividing the lowest health outcome rate by the highest health outcome rate (Healthy People 2020, 2019c). Rate ratios are expressed numerically with values equal to or larger than 1.00. Rates closer to 1.00 show less statistical significance. Rates larger than or equal to 2.00 are statistically significant and mean the difference is greater than 100% of the lower rate.

Using death certificate data from Louisiana Vital Records prepared by the Louisiana Vital Records (2019), disparity ratios were calculated for infant mortality rates for Black and White infants. The Black/White disparity ratio for the state of Louisiana increased from 2.13 to 2.28 (Table 14.1). Among parishes where data counts

Table 14.1 Black/White infant disparity ratios for Louisiana and Louisiana Parishes, 2010–12, 2013–15 (data from Louisiana Vital Records, 2019).

	Black/White infant disparity ratio 2010–12	Black/White infant disparity ratio 2013–15
Louisiana	2.13	2.28
Acadia	3.15	4.04
Ascension	2.32	2.85
Bossier	2.68	3.95
Caddo	2.43	2.74
Calcasieu	2.40	2.60
East Baton Rouge	2.27	2.82
Iberia	2.24	–
Jefferson	–	3.22
Jefferson Davis	4.08	–
Lafayette	–	2.91
Lafourche	2.13	–
Livingston	–	2.84
Orleans	5.99	2.19
Rapides	2.10	2.91
St. Bernard	–	2.38
St. Martin	–	2.37
St. Tammany	2.47	–
Terrebonne	2.68	2.32
Vernon	–	2.72
Washington	–	2.85

– denotes nonreportable data.

were greater than 5 and therefore reportable, 20 unique parishes experienced disparity ratios greater than 2.00. The highest disparity ratio occurred in Orleans Parish (2010–12), which included a non-Hispanic Black infant mortality rate of 10.70 deaths per 1000 live births and a White rate of 1.79 deaths per 1000 live births. This Black/White disparity gap meant Black infants in Orleans Parish were nearly six times more likely to die before their first birthday than their White counterparts. During the subsequent period, the disparity ratio in Orleans Parish decreased by 63%, declining to 2.19 deaths per 1000 live births. Of the five most populous parishes (East Baton Rouge, Jefferson, Orleans, Caddo, St. Tammany), the Black/White disparity ratio increased in East Baton Rouge and Caddo by 24.2% and 12.8%, respectively.

Black maternal mortality in Louisiana

According to the Louisiana Maternal Mortality Review Report, there were 187 identified maternal deaths in Louisiana between 2011 and 2016, but only 47 maternal deaths met the criteria described above (Louisiana Department of Health (LDH), Office of Public Health (OPH), Bureau of Family Health, 2018). During the review period, the Louisiana maternal mortality rate increased from 9.7 (deaths per 100,000 live births) to 77.6, roughly 34% per year. Fig. 13.2 demonstrates that the maternal mortality rates (not verified through a maternal mortality review process) in Louisiana and the United States have increased in recent years. In 2016, the Louisiana rate was nearly 3.6 times the US rate. The leading causes of maternal deaths in Louisiana were related to hemorrhage, cardiomyopathy, and cardiovascular and coronary conditions. Sixty-two percent of pregnancy-related deaths occurred to women who had Medicaid for their insurance (LDH-OPH, 2018, p. 21).

While maternal mortality affects US women from all backgrounds, non-Hispanic Black women bear a disproportionate burden. In Louisiana, non-Hispanic Black women experienced 68% of all maternal deaths, while only accounting for 37% of

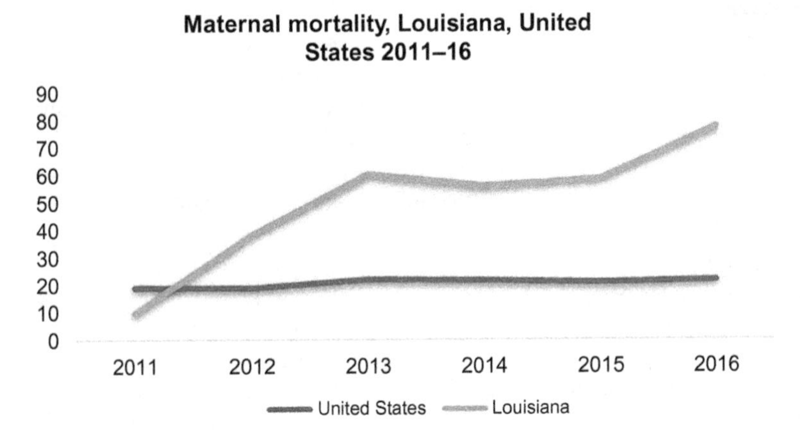

Fig. 13.2 Louisiana maternal mortality rates, Louisiana and United States 2011–16. Data from Louisiana Department of Health, Office of Public Health (2018).

all women who gave birth (LDH-OPH, 2018, p. 22). The Black maternal mortality rate of 22.8 deaths per 100,000 live births is more than four times the rate of 5.6 deaths per 100,000 live births for White women.

In Louisiana, nearly half of all maternal deaths occurred between 24 hours and 42 days after delivery. Of all maternal deaths reviewed, 45% were considered preventable (LDH-OPH, 2018, p. 19). Some of the contributing factors related to maternal deaths included provider and facility-level factors, such as failure to screen/inadequate assessment of risk, lack of standardized policies and procedures, lack of referral or consultation, poor communication, and lack of case coordination or continuity of care (LDH-OPH, 2018, p. 20). It is worth noting that most of the maternal deaths, about 57%, occurred while the women were hospital patients. Of these deaths, one in five women needed transfer to a different facility for higher acuity obstetric or medical care. Roughly 21% of maternal deaths were to women who presented to an emergency department or outpatient facility. The remaining maternal deaths, to pregnant or postpartum women, occurred in a nonmedical setting (LDH-OPH, 2018, p. 3).

Demographic trends in Louisiana

Louisiana is the 25th most populous state in the United States, with a population of 4.65 million. The demographic make-up includes 62.9% White, 32.7% Black or African American, 5.2% Hispanic/Latina/o/x, and less than 2% Asian, Other races, Native American, Native Hawaiian, or Pacific Islander (US Census Bureau, 2018). In 2005, Hurricane Katrina caused the loss of nearly half of New Orleans's residents (254,502 people) (Plyer, 2015, *para* 5). Post-Katrina, New Orleans has become one of the fastest-growing US cities. Yet, recent population estimates show Louisiana, as one of 10 US regions (including DC and Puerto Rico), projected to experience a population loss between 2017 and 2018 (US Census Bureau, n.d.-a).

According to the United Health Foundation (2019), Louisiana is ranked as the least healthy state in the United States, while Hawaii, Massachusetts, and Connecticut have the best rankings. The ranking is a composite index of state comparisons developed from over 30 metrics to derive an annual snapshot of health status. The healthiest states' high rankings are partly attributed to low rates of obesity, uninsured, children in poverty, smoking, and other factors. Louisiana's overall poor ranking is negatively affected by the disproportionately high rates of smoking, children in poverty, obesity, mental and physical distress, low birth weight births, high school graduation rates, premature deaths, violent crime, infectious diseases, preventable hospitalizations, infant mortality, and other factors.

According to the 2010 Census, the top five most populous cities in Louisiana are New Orleans, Baton Rouge, Shreveport, Lafayette, and Lake Charles. Blacks represent 32.7% of Louisiana's population, and more than half the population in Baton Rouge (54.8%), New Orleans (59.8%), and Shreveport (56.7%) (US Census Bureau, 2018). Hispanics/Latina/o/x account for 5.2% of the population in Louisiana in comparison to 18.3% of the US population. The median household income in each of the most populous parishes was lower than the overall state median income of $46,710 with the exception of Lafayette ($48,533). The rate of children in poverty in Shreveport (Caddo Parish) of 38.1% outpaces other populous cities levels (US Census Bureau, n.d.-b). Other less populous parishes have higher rates of childhood poverty, including East Carroll Parish (62.3%) and Claiborne Parish (47.6%). Tensas Parish, the least populous parish in the state, has a child poverty rate of 55.1% (US Census Bureau, n.d.-b).

Education outcomes in Louisiana are varied. High school graduation rates in Louisiana and each of the most populous cities are less than the US rate (US Census Bureau, n.d.-b). While 78.1% of people in the United States over 25 years of age have a high school diploma, only 66.1% in Shreveport and 67.6% in Baton Rouge achieve this milestone. In the United States, only 30.9% of persons 25 years of age and over have a Bachelor's degree. The US rate is outpaced by New Orleans (36.5%) and Baton Rouge (32.4%) (US Census Bureau, 2018).

Disproportionate impact of policy on Black women in Louisiana

Public health plays a critical role in considering how social policies adversely affect health outcomes for various groups. In Louisiana, health inequities experienced by Black women are ominous with the lives and livelihoods of Black families being affected for generations, even though inequities are considered preventable. This section presents three relevant social environmental policies and their disproportionate impact on Black women in Louisiana.

Sex education

Louisiana has some highest rates of sexually transmitted infections in the United States. In 2017, Louisiana ranked second in reported cases of chlamydia (Louisiana rate 799.8 per 100,000 of the population/US rate 528.8 per 100,000 of the population), third in reported cases of gonorrhea (256.7 rate per 100,000 of the population/171.9 US rate per 100,000 of the population), and third in reported cases of primary and secondary syphilis (14.5 Louisiana per 100,000 of the population/9.5 US rate per 100,000 of the population) (CDC, 2018). A fundamental barrier to improving sexually transmitted infection rates relates to the lack of health literacy related to sexual health, particularly for young women and teens. Louisiana is one of the 27 states in the United States that does not mandate sex education in the K-12 system (National Conference on State Legislators, 2019). Policy changes to allow education related to preconception care may help to address the high rate of infection as well as unintentional pregnancies, now at around 60% of all pregnancies in the state (Maternal Health and Rights Initiative, 2018).

Criminal justice

Louisiana has one of the highest imprisonment rates in the nation. In June 2017, Louisiana's imprisonment rate of 712 per 100,000 was 1.8 times higher than the US rate of 450 per 100,000 at the end of 2016 (Gelb & Compa, 2018, *para* 2). Blacks are overrepresented in Louisiana's prison population. Black males make up for 68.2%

of those in the correctional system, while accounting for only 32.7% of the general population (Louisiana Commission on Law Enforcement and Administration of Criminal Justice. Statistical Analysis Center, 2019, p.17). In 2017, 18.1% of the prison population was serving 20-year fixed-term sentences, and 14.9% was serving a life sentence (Louisiana Commission on Law Enforcement and Administration of Criminal Justice. Statistical Analysis Center, 2019, p.14). While the state has made some progress in addressing mandatory minimums and restrictive parole policies (Louisiana Department of Public Safety and Corrections and LA Commission on Law Enforcement, 2018), the incarceration of such a large proportion of Black males has an undue negative influence on Black women related to early employment decisions and socioeconomics (Mechoulan, 2011).

Residential segregation and education

As a result of historical patterns in housing, access to wealth, and socioeconomic positioning, Blacks in Louisiana reside in highly segregated communities (Weldon Cooper Center for Public Service, 2013). The conditions experienced in Black segregated neighborhoods have been shown to contribute to poor health outcomes (Massey & Denton, 1993; Popescu, Duffy, Mendelsohn, & Escarce, 2018; Williams & Collins, 2001). The Racial Dot Map, a map created from data reported on the 2010 decennial census, demonstrates that segregation by race is observable in all the major cities in Louisiana.

Race and poverty create educational challenges for Blacks in segregated neighborhoods. Segregation is a significant determinant of school quality, educational attainment, and access to opportunities (Orfield et al., 2016; Walters, 2001) for students isolated in disadvantaged neighborhoods. Segregation in the public school system contributes to vast disparities between Blacks and Whites and entrenches a two-tiered system whereby students who attend intensely segregated minority schools (90%–100% non-White) have worse educational outcomes than those who do not attend these schools (Frankenberg, Ee, Ayscue, & Orfield, 2019). In recent decades, the number of White intensely segregated non-White schools has more than tripled, rising from 5.7% to 18.6% between 1988 (the peak of desegregation) and 2016 (Orfield et al., 2016, p. 3). Black and Latina/o/x students, on average, attend public schools where the White student population is relatively low, about 12% (Frankenberg et al., 2019, p. 32). In 2016, 40% of Black students attended schools where more than 90% of the students were Black or Hispanic/Latina/o/x (Frankenberg et al., 2019, p. 4). In comparison, White students are much more likely to attend public schools with a higher percentage of White students, 45% White on average (Frankenberg et al., 2019, p. 32). The number of intensely segregated minority schools has decreased over the past several decades, dropping from 38.9% in 1988 to 16% in 2016 (Frankenberg et al., 2019, p. 21). Within school segregation, the degree to which students are placed in advanced placement or accelerated classes is also divided down racial lines (Rodriguez & McGuire, 2019) with Black students having less access to these types of courses. Education is critical to the promotion of health equity as "potentially avoidable factors associated with lower educational status account for almost half of all deaths among working-age adults in the U.S.," (Jemal et al., 2008 p. 1).

Over the past several years, Louisiana has made harsh divestment decisions regarding higher education budgets. Between 2008 and 2017, Louisiana cut the budget for colleges and universities by 44.9% or nearly $9 billion (Mitchell, Leachman, & Masterson, 2017, p. 5). Louisiana public schools are now more segregated than they were in the 1970s (Kotok, Beabout, Nelson, & Rivera, 2018). Louisiana leads the United States in per-student funding decreases with a funding reduction of $5340 per student in just under a decade. These decreases are particularly harsh for Black students who are more likely to incur student loan debt and experience difficulty in paying back loans (Addo, Houle, & Simon, 2016; Grinstein-Weiss, Perantie, Taylor, Guo, & Raghavan, 2016). Louisiana has two historically Black colleges and universities (HBCU), Grambling State University and Southern University. While HBCUs traditionally cost less than other 4-year universities, students who attend there also tend to have higher rates of default compared to students who attend predominantly White institutions (Mitchell & Fuller, 2019). More often, Black students have different starting points than their White counterparts, in that they are often unable to afford tuition without supplemental loans. Also, Black students often lack the resources to increase their out-of-pocket expenses as tuition costs rise. These factors also place Blacks at a disadvantage in accessing health information, social and professional networks, and skills and abilities to effectively navigate the clinical encounter and subsequent healthcare interactions.

Discussion

Social environmental factors are complex societal norms that are deeply embedded in every level of society. The case study of Louisiana highlights the impact of social environmental influences and the critical need to address structural racism as a root cause of health inequities. Dismantling systems that have for centuries contributed to disparate outcomes warrant broad multitiered approaches capable of driving structural change. Medical and behavioral interventions alone cannot eliminate health inequities (Bailey et al., 2017; Woolf, 2019). Social environmental factors such as structural racism, in particular, play an important role in increasing risk and producing health disadvantage for Blacks and other marginalized groups. The harmful effects of structural racism experienced by Blacks, and other marginalized racial and ethnic groups, are deeply embedded in various systems through a series of interweaving policies. The consequences of racial stratification are dire and intergenerational, particularly for Black women and infants (Gee & Ford, 2011), who face a greater risk of death than their White counterparts. "Intense racial separation and concentrated poverty in schools that offer inferior opportunities fundamentally undermine the American belief that all children deserve an equal educational opportunity. Segregated schools build and sustain a segregated society," (Orfield et al., 2016, p. 9). Thus, the challenge for public health, as well as the broader society, is overcoming resistance to the role race and racism play in producing health inequities. Despite reform and varied legislation, the United States can be characterized as a racialized capitalist society built with unpaid labor by generations of Blacks and indigenous groups making acknowledgment of the entrenchment of racism within various US systems a difficult task. This becomes blatantly obvious in attempting to reconcile the decree to achieve health equity, while the inclusion of "color-blind" or race-neutral policies remains off the public radar. Health disparities will continue along with the fallacy of a level playing field where everyone has an equal chance of success, health, and well-being until the United States truly becomes a postracial society (Williams, Lawrence, & Davis, 2019).

Recommendations

Improving conditions for Black women involves addressing social environmental factors across a variety of settings that challenge the bounds of public health research, practice, and policymaking. We propose that the remedy process requires a fundamental shift in the public health role and bridging of the racial schisms that undergird traditional SDOH. This multifaceted approach involves policymakers and practitioners integrating critical consciousness and structural competency practices to inform policy development by centering on the margins. Utilizing the Black Mamas Matter policy agenda as a guideline, we recommend a hybrid between activism and public health to address social environmental stressors that affect Black populations as illustrated in the case study. These recommendations have the potential to affect Black women positively by establishing new social norms in treatment protocols, policymaking, and equality of opportunity.

This systemic change calls for an inclusion of structural racism as a fundamental contributor to health disadvantage and the development and implementation of targeted social policies, research, and provider-patient interactions that result in widespread systemic change for Black women across various socioeconomic levels (Brown et al., 2019). The following section describes mediations and how they lay the foundation for addressing structural racism in Louisiana and potentially in other areas where social environmental risks contribute to the health disadvantage of marginalized populations.

Context and critical social interventions

Multifaceted approaches, essential to address deeply rooted societal norms that contribute to health disadvantages, must include structural interventions to remedy deeply entrenched policy and social norms that span individual, community, and societal levels. Addressing these issues cannot be singularly confined to geographic location as Black women experience inequities without regard to place. As such, policies targeting stress, quality of life, well-being, and healthcare experiences are needed for Black women in high and middle-income brackets who may not be isolated in segregated Black communities. Historical patterns of residential segregation in Louisiana have concentrated poverty and deprivation with intergenerational effects that have contributed to additional structural barriers for Black women residing in economically depressed communities. Research studies (e.g., mixed-methods longitudinal, small area analysis) that take into consideration structural disadvantages and toxic environmental stressors are needed to illuminate variations that occur within cities and parishes and examine the health outcomes of Black women across the lifespan.

Disease-agnostic interventions

The health disadvantage experienced by Black women in Louisiana is fueled by long-standing structural barriers. Addressing deeply entrenched intergenerational health disadvantages in Louisiana requires rejecting race-neutral policies in favor of policy solutions that center on equity. These approaches should seek to improve living conditions in segregated communities, create opportunities and access for vulnerable populations, and reduce risk associated with racism, discrimination, and bias. Local and statewide initiatives that improve well-being, promote equity, enhance social mobility, and create equality of opportunity may serve to reduce some of the risks for Black women residing outside of racially segregated communities.

Using critical consciousness to support structural competency

As we have shown, Louisiana faces the difficult task of overcoming the intractable vestiges of slavery (e.g., discrimination, patterns of incarceration, student experiences in the education system, socioeconomic barriers) still present there today. Recognizing the historic discord between racial groups in the segregated south, as well as persistent social stressors of being Black in the United States, health care professionals must be able to examine their social positioning and integrate lessons from social structuring, some of which contribute to stigma, bias, and mistreatment. This approach will allow public health officials and healthcare providers to demonstrate their ability to reject the dominant world view and develop the level of critical consciousness needed to drive incremental change. This consciousness will involve the incorporation of an inclusive approach to marginalized citizens and provide a broader perspective on what metrics are important.

Centering on the margins

In Louisiana, much like the rest of the United States, the clinical encounter for Black women is marked by mistreatment, provider bias, and discrimination. Narratives that reveal the experiences of Black women in their communities and during their interactions with healthcare providers are needed to inform their care and the services received. One approach to shifting these patterns is by centering on the margins, a concept that requires human-centered approaches to policymaking which moves beyond the bounds of typical policy and decision-making (Ford & Airhihenbuwa, 2010; hooks, 2000). The inclusion of Black women's voices must become a central focus of any policy agenda aimed at improving Black health outcomes. One such exemplar is Black Mamas Matter (2018), a reproductive justice alliance whose policy agenda seeks to engage and prioritize the maternal health needs of Black women. At the core of the Black Mamas Matter agenda is the notion that the success of these initiatives is predicated on the need for projects to be led by Black women. The level of disruption codified in the Black Mamas Matter agenda has the potential for incremental policy changes.

Conclusion

Reducing health inequities, in this case adverse pregnancy outcomes, requires systematic approaches that include culturally appropriate, high-quality clinical care, and concerted efforts to improve the experiences of Black women throughout the healthcare system by naming and actively working to dismantle structural racism. Moreover, the systematic approaches must intentionally include strategies to address community- and individual-level structural barriers to eliminate risks to the SDOH exacerbated by and grounded in structural racism. Black women in Louisiana are at a health disadvantage, as indicated in the case study, particularly those residing in segregated Black communities. The entangled relationships between the health determinants may distract from the root cause of health inequity, mainly because these conversations are challenging to engage and sustain. For public health policymakers, measures of success will continue to be hampered by the paucity of data on maternal mortality, differing definitions for the phenomena, as well as correlations among the health determinants that are not easily understood. Thus, these approaches should, therefore, include improved efforts to collect maternal mortality data, the development of suitable proxy measures for the interventions, and authentic engagement with the affected communities. Without an intentional disruptive shift in ideology, health inequities will remain an enduring feature of US society.

References

Acevedo-Garcia, D., Lochner, K. A., Osypuk, T. L., & Subramanian, S. V. (2003). Future directions in residential segregation and health research: A multilevel approach. *American Journal of Public Health*, *93*(2), 215–221. https://doi.org/10.2105/ajph.93.2.215.

Addo, F. R., Houle, J. N., & Simon, D. (2016). Young, black, and (still) in the red: Parental wealth, race, and student loan debt. *Race and Social Problems*, *8*(1), 64–76. https://doi.org/10.1007/s12552-016-9162-0.

Alexander, M. (2012). *The new Jim Crow: Mass incarceration in the age of colorblindness*. New York: The New Press.

Alexander, G. R., Wingate, M. S., Bader, D., & Kogan, M. D. (2008). The increasing racial disparity in infant mortality rates: Composition and contributors to recent US trends. *American Journal of Obstetrics and Gynecology*, *198*(1), 51.e1–51.e9. https://doi.org/10.1016/J.AJOG.2007.06.006.

America's Health Rankings. (2019). *Annual report: Louisiana*. Retrieved September 8, 2019 from https://www.americashealthrankings.org/explore/annual/measure/Overall/state/LA.

American Public Health Association. (2019). *Racism and health*. Retrieved September 8, 2019 from https://www.apha.org/topics-and-issues/health-equity/racism-and-health.

Bailey, Z. D., Krieger, N., Agénor, M., Graves, J., Linos, N., & Bassett, M. T. (2017). Structural racism and health inequities in the USA: Evidence and interventions. *The Lancet, 389* (10077), 1453–1463. https://doi.org/10.1016/S0140-6736(17)30569-X.

Ben, J., Cormack, D., Harris, R., & Paradies, Y. (2017). Racism and health services utilization: A systemic review and meta-analysis. *PLoS One. 12*(12). https://doi.org/10.1371/journal.pone.0189900.

Black Mamas Matter. (2018). *Advancing holistic maternal care for black women through policy.* Retrieved September 8, 2019 from https://blackmamasmatter.org/wp-content/uploads/2018/12/BMMA-PolicyAgenda-Digital.pdf.

Bostos, J. L., Harnois, C. E., & Paradies, Y. C. (2018). Health care barriers, racism, and intersectionality in Australia. *Social Science & Medicine, 199*, 209–218.

Braveman, P. A., Cubbin, C., Egerter, S., Williams, D. R., & Pamuk, E. (2010). Socioeconomic disparities in health in the United States: What the patterns tell us. *American Journal of Public Health. 100*(Suppl. 1) https://doi.org/10.2105/AJPH.2009.166082.

Braveman, P., & Gottlieb, L. (2014). The social determinants of health: It's time to consider the causes of the causes. *Public Health Reports (Washington, D.C.: 1974), 129*(Suppl. 2), 19–31. https://doi.org/10.1177/00333549141291S206.

Brown, A. F., Ma, G. X., Miranda, J., Eng, E., Castille, D., et al. (2019). Structural interventions to reduce and eliminate health disparities. *American Journal of Public Health, 109*(S1), S72–S78. https://doi.org/10.2105/AJPH.2018.304844.

Browne, A. J., Varcoe, C., Lavoie, J., Smye, V., Wong, S. T., Krause, M., et al. (2016). Enhancing health care equity with indigenous populations: Evidence-based strategies from an ethnographic study. *BMC Health Services Research, 16*(544), 1–17. https://doi.org/10.1016/j.scitotenv.2013.03.027.

Carpiano, R. M. (2006). Toward a neighborhood resource-based theory of social capital for health: Can Bourdieu and sociology help? *Social Science and Medicine, 62*(1), 165–175. https://doi.org/10.1016/j.socscimed.2005.05.020.

Castle, B., Wendel, M., Kerr, J., Brooms, D., & Rollins, A. (2019). Public health's approach to systemic racism: A systematic literature review. *Journal of Racial and Ethnic Health Disparities.* https://doi.org/10.1007/s40615-018-0494-x. Springer International Publishing.

CDC. (2019a). *Pregnancy-related surveillance system.* Retrieved September 8, 2019 from https://www.cdc.gov/reproductivehealth/maternalinfanthealth/pregnancy-mortality-surveillance-system.htm.

CDC. (2019b). *Infant mortality rates by state.* Retrieved September 8, 2019 from https://www.cdc.gov/nchs/pressroom/sosmap/infant_mortality_rates/infant_mortality.htm.

CDC. (2019c). *Infant mortality.* Retrieved September 15, 2019 from https://www.cdc.gov/reproductivehealth/maternalinfanthealth/infantmortality.htm.

Centers for Disease Control and Prevention. (2018). *Sexually transmitted disease surveillance 2017.* Atlanta, GA: US Department of Health and Human Services.

Centers for Disease Control and Prevention (CDC). (2019). *Pregnancy-related deaths.* Retrieved September 8, 2019 from https://www.cdc.gov/reproductivehealth/maternalinfanthealth/pregnancy-relatedmortality.htm.

Chandra, A., Frakes, M., & Malani, A. (2017). Challenges to reducing discrimination and health inequity through existing Civil Rights Laws. *Health Affairs, 36*(6), 1041–1047. https://doi.org/10.1377/hlthaff.2016.1091.

Egerter, S., Barclay, C., Grossman-Kahn, R., & Braveman, P. (2011). *Violence, social disadvantage and health.* Princeton, NJ: Robert Wood Johnson Foundation (RWJF). Retrieved September 8, 2019 from www.rwjf.org/content/dam/farm/reports/issue_briefs/2011/rwjf70452.

Ely, D. M., & Driscoll, A. K. (2019). Infant mortality in the United States, 2017: Data from the period linked birth/death data file. *National Vital Statistics Report*, *68*(10), 1–20. https://www.cdc.gov/nchs/data/nvsr/nvsr68/nvsr68_10-508.pdf.

Fink, D. S., Keyes, K. M., & Cerdá, M. (2016). Social determinants of population health: A systems sciences approach. *Current Epidemiology Reports*, *3*(1), 98–105. https://doi.org/10.1007/s40471-016-0066-8.

Ford, C. L., & Airhihenbuwa, C. O. (2010). The public health critical race methodology: Praxis for antiracism research. *Social Science & Medicine*, *71*(8), 1390–1398. https://doi.org/10.1016/J.SOCSCIMED.2010.07.030.

Frankenberg, E., Ee, J., Ayscue, J. B., & Orfield, G. (2019). *Harming our common future: America's segregated schools 65 years after Brown.* Retrieved September 10, 2019 from https://www.civilrightsproject.ucla.edu/research/k-12-education/integration-and-diversity/harming-our-common-future-americas-segregated-schools-65-years-after-brown/Brown-65-050919v4-final.pdf.

Gee, G. C., & Ford, C. L. (2011). Structural racism and health inequities: Old issues, new directions. *Du Bois Review*, *8*(1), 115–132. https://doi.org/10.1017/S1742058X11000130.

Gelb, A., & Compa, E. (2018). *Louisiana no longer leads nation in imprisonment rate.* The Pew Charitable Trusts. Retrieved September 19, 2019, from https://www.pewtrusts.org/en/research-and-analysis/articles/2018/07/10/louisiana-no-longer-leads-nation-in-imprison ment-rate.

Geronimus, A. T., Hicken, M., Keene, D., & Bound, J. (2006). "Weathering" and age patterns of allostatic load scores among Blacks and Whites in the United States. *American Journal of Public Health*, *96*(5), 526–833. https://doi.org/10.2105/AJPH.2004.060749.

Graham, L., Brown-Jeffy, S., Aronson, R., & Stephens, C. (2011). Critical race theory as theoretical framework and analysis tool for population health research. *Critical Public Health*, *21*(1), 81–93. https://doi.org/10.1080/09581596.2010.493173.

Grinstein-Weiss, M., Perantie, D. C., Taylor, S. H., Guo, S., & Raghavan, R. (2016). Racial disparities in education debt burden among low- and moderate-income households. *Children and Youth Services Review*, *65*, 166–174. https://doi.org/10.1016/J.CHILD YOUTH.2016.04.010.

Hall, W. J., Chapman, M. V., Lee, K. M., Merino, Y. M., Thomas, T. W., et al. (2015). Implicit racial/ethnic bias among health care professionals and its influence on health care outcomes: A systematic review. *American Journal of Public Health*, *105*(12), e60–e76. https://doi.org/10.2105/AJPH.2015.302903.

Healthy People 2020. (2019a). *Social determinants of health.* Retrieved September 8, 2019 from https://www.healthypeople.gov/2020/topics-objectives/topic/social-determinants-of-health.

Healthy People 2020. (2019b). *Discrimination.* Retrieved September 8, 2019 from https://www.healthypeople.gov/2020/topics-objectives/topic/social-determinants-health/interventions-resources/discrimination.

Healthy People 2020. (2019c). *Health disparities tool: A user's guide.* Healthy People 2020. Retrieved from https://www.healthypeople.gov/2020/disparities-user-guide.

Healy, A. J., Malone, F. D., Sullivan, L. M., Porter, T. F., Luthy, D. A., et al. (2006). Early access to prenatal care. *Obstetrics & Gynecology*, *107*(3), 625–631. https://doi.org/10.1097/01.AOG.0000201978.83607.96.

hooks, b. (2000). *Feminist theory: From margin to center.* Chicago, IL: Pluto Press.

Institute of Medicine. (2003). In B. D. Smedley, A. Y. Stith, & A. R. Nelson (Eds.), *Unequal treatment: Confronting racial and ethnic disparities in health care.* Washington, DC: National Academy Press. https://doi.org/10.17226/10260.

Institute of Medicine and National Research Council. (2013). *US health in international perspective: Shorter lives, poorer Health.* Washington, DC: The National Academies Press https://doi.org/10.17226/13497.

Jemal, A., Thun, M. J., Ward, E. E., Henley, S. J., Cokkinides, V. E., & Murray, T. E. (2008). Mortality from leading causes by education and race in the United States, 2001. *American Journal of Preventive Medicine*, *34*(1), 1–8. e7. https://doi.org/10.1016/j.amepre.2007.09.017.

Jones, C. P. (2002). Confronting institutionalized racism. *Phylon (1960-)*, *50*(1/2), 7–22. https://doi.org/10.2307/4149999.

Kawachi, I., Kennedy, B. P., & Glass, R. (1999). Social capital and self-rated health: A contextual analysis. *American Journal of Public Health*, *89*(8), 1187–1193. https://doi.org/10.2105/AJPH.89.8.1187.

Kim, D., & Saada, A. (2013). The social determinants of infant mortality and birth outcomes in Western developed nations: A cross-country systematic review. *International Journal of Environmental Research and Public Health*, *10*(6), 2296–2335.

Kollmann, T., Marsiglio, S., & Suardi, S. (2018). Racial segregation in the United States since the Great Depression: A dynamic segregation approach. *Journal of Housing Economics*, *40*, 95–116. https://doi.org/10.1016/j.jhe.2018.03.004.

Kotok, S., Beabout, B., Nelson, S. L., & Rivera, L. E. (2018). A demographic paradox: How public school students in New Orleans have become more racially integrated and isolated since Hurricane Katrina. *Education and Urban Society*, *50*(9), 818–838. https://doi.org/10.1177/0013124517714310.

Krieger, N. (2014). Discrimination and health inequities. *International Journal of Health Services*, *44*(4), 643–710. https://doi.org/10.2190/hs.44.4.b.

Louisiana Commission on Law Enforcement and Administration of Criminal Justice. Statistical Analysis Center (2019). *2018 status of state and local corrections facilities and program report (R.S. 15:1204.1).* http://lcle.la.gov/programs/uploads/2018%20Status%20of%20State%20and%20Local%20Corrections%20Facilities%20and%20Program%20Reportrd.pdf.

Louisiana Department of Health (LDH), Office of Public Health (OPH), Bureau of Family Health. (2017). *Louisiana child death review annual report 2014–2016.* Retrieved October 1, 2019 from http://ldh.la.gov/assets/oph/Center-PHCH/Center-PH/maternal/CDR_Report_2014-2016.pdf.

Louisiana Department of Health (LDH), Office of Public Health (OPH), Bureau of Family Health. (2018). *Louisiana maternal mortality review report 2011–2016.* Retrieved October 1, 2019 from http://ldh.la.gov/assets/oph/Center-PHCH/Center-PH/maternal/2011-2016_MMR_Report_FINAL.pdf.

Louisiana Department of Public Safety & Corrections and LA Commission on Law Enforcement. (2018). *Louisiana 's justice reinvestment reforms first annual performance report.* Retrieved September 8, 2019 from http://gov.louisiana.gov/assets/docs/JRI/LA_JRI_Annual_Report_FINAL.PDF.

Louisiana Vital Records (2019). Adverse pregnancy outcomes. In *Infant and maternal mortality.* Unpublished data.

Lu, M. C., Kotelchuck, M., Hogan, V., Jones, L., Wright, K., & Halfon, N. (2010). Closing the Black-White gap in birth outcomes: A life-course approach. *Ethnicity & Disease*, *20* (1 Suppl. 2), 62–76.

MacDorman, M. F., Declercq, E., Cabral, H., & Morton, C. (2016). Recent increases in U.S. maternal mortality rate: Disentangling trends from measurement issues. *Obstetrics & Gynecology*, *128*(3), 447–455. https://doi.org/10.1097/AOG.0000000000001556.

MacDorman, M. F., Declercq, E., & Thoma, M. E. (2017). Trends in maternal mortality by sociodemographic characteristics and cause of death in 27 states and the District of Columbia. *Obstetrics & Gynecology, 129*(5), 811–818. https://doi.org/10.1097/AOG.0000000000001968.

MacDorman, M. F., & Mathews, T. J. (2008). *Understanding racial and ethnic disparities in U.S. infant mortality rates* (pp. 1–8). National Center for Health Statistics. 74.

Maina, I. W., Belton, T. D., Ginzberg, S., Singh, A., & Johnson, T. J. (2018). A decade of studying implicit racial/ethnic bias in healthcare providers using the implicit association test. *Social Science & Medicine, 199*, 219–229. https://doi.org/10.1016/J.SOCSCIMED.2017.05.009.

Massey, D. S., & Denton, N. A. (1993). *American apartheid: Segregation and the making of the underclass.* Cambridge, MA: Harvard University Press.

Massey, D. S., & Tannen, J. (2015). A research note on trends in Black hypersegregation. *Demography, 52*(3), 1025–1034.

Maternal Health and Rights Initiative. (2018). *Maternal health in Louisiana.* Retrieved August 20, 2019 from https://reproductiverights.org/sites/default/files/documents/MHRI/USPA-MHRI-LA-FS-Final.pdf.

McCalman, J., Jongen, C., & Bainbridge, R. (2017). Organizational systems' approaches to improving cultural competence in healthcare: A systematic scoping review of the literature. *International Journal for Equity in Health*, 16–78. https://doi.org/10.1186/s12939-017-0571-5.

Mechoulan, S. (2011). The external effect of Black male incarceration on Black females. *Journal of Labour Economics, 29*(1), 1–35. https://doi.org/10.1086/656370.

Mehra, R., Boyd, L. M., & Ickovics, J. R. (2017). Racial residential segregation and adverse birth outcomes: A systemic review and meta-analysis. *Social Science and Medicine, 191*, 237–250.

Mendez, D. D., Hogan, V. K., & Culhane, J. (2011). Institutional racism and pregnancy health: Using Home Mortgage Disclosure act data to develop an index for mortgage discrimination at the community level. *Public Health Reports (Washington, D.C.: 1974), 126* (Suppl. 3), 102–114. https://doi.org/10.1177/00333549111260S315.

Mitchell, J., & Fuller, A. (2019, April 17). The student-debt crisis hits hardest at historically Black colleges. *The Wall Street Journal.* Retrieved August 20, 2019 from https://www.wsj.com/articles/the-student-debt-crisis-hits-hardest-at-historically-black-colleges-11555511327.

Mitchell, M., Leachman, M., & Masterson, K. (2017). *A lost decade in higher education funding: State cuts have driven up tuition and reduced quality.* Center on Budget and Policy Priorities. Retrieved August 20, 2019 from https://www.cbpp.org/sites/default/files/atoms/files/2017_higher_ed_8-22-17_final.pdf.

National Center for Education Statistics, US Department of Education. (2019a). *Status and trends in the education of racial and ethnic groups: Indicator 6: Elementary and secondary enrollment. Retrieved September 9, 2019, from* https://nces.ed.gov/programs/raceindicators/indicator_rbb.asp.

National Center for Education Statistics, US Department of Education. (2019b). *Status and trends in the education of racial and ethnic groups: Indicator 7: High school dropout rates.* Retrieved September 10, 2019, from https://nces.ed.gov/programs/raceindicators/indicator_RDC.asp.

National Conference on State Legislators. (2019). *State policies on sex education in schools.* Retrieved August 20, 2019 from http://www.ncsl.org/research/health/state-policies-on-sex-education-in-schools.aspx.

Orfield, G., Ee, J., Frankenberg, E., & Siegel-Hawley, G. (2016). *Brown at 62: School segregation by race, poverty and state*. Civil Rights Project. 9. Retrieved September 9, 2019 from https://eric.ed.gov/?id=ED565900.

Pachter, L. M., & Garcia Coll, C. (2009). Racism and child health: A review of the literature and future directions. *Journal of Developmental and Behavioral Pediatrics, 30*(3), 255–263. https://doi.org/10.1097/DBP.0b013e3181a7ed5a.

Padilla, C. M., Deguen, S., Lalloue, B., Blanchard, O., Beaugard, C., Troude, F., et al. (2013). Cluster analysis of social and environment inequalities of infant mortality. A spatial study in small areas revealed by local disease mapping in France. *Science of the Total Environment, 454–555*, 433–441. https://doi.org/10.1016/j.scitotenv.2013.03.027.

Passel, J., Wang, W., & Taylor, P. (2010). *One-in-seven new marriages is interracial or interethnic: Marrying out*. Washington, DC: Pew Research Center. Retrieved August 20, 2019 from https://www.pewsocialtrends.org/2010/06/04/marrying-out/.

Peterson, E. E., Davis, N. L., Goodman, D., Cox, S., Syverson, C., Seed, K., et al. (2019). Racial/ethnic disparities in pregnancy-related deaths—United States, 2007-2016. *MMWR, Morbidity and Mortality Weekly Report, 68*(35), 762–765. https://doi.org/10.15585/mmwr.mm6835a3.

Plyer, A. (2015). *Facts for features: Katrina impact*. The Data Center. Retrieved August 20, 2019 from https://www.datacenterresearch.org/data-resources/katrina/facts-for-impact/.

Popescu, I., Duffy, E., Mendelsohn, J., & Escarce, J. J. (2018). Racial residential segregation, socioeconomic disparities, and the White-Black survival gap. *PLoS One, 13*(2), e0193222. https://doi.org/10.1371/journal.pone.0193222.

Ramaswamy, M., & Kelly, P. J. (2015). Institutional racism as a critical social determinant of health. *Public Health Nursing, 32*(4), 285–286. https://doi.org/10.1111/phn.12212.

Rodriguez, A., & McGuire, K. M. (2019). More classes, more access? Understanding the effects of course offerings on Black-White gaps in advanced placement course-taking. *The Review of Higher Education, 42*(2), 641–679. Johns Hopkins University Press https://doi.org/10.1353/rhe.2019.0010.

Singh, G. K., & Yu, S. M. (2019). Infant mortality in the United States, 1915-2017: Large social inequalities have persisted for over a century. *International Journal of MCH and AIDS (IJMA), 8*(1), 19–31. https://doi.org/10.21106/ijma.271.

Smith, D. B. (2005). Racial and ethnic health disparities and the unfinished civil rights agenda. *Health Affairs*. https://doi.org/10.1377/hlthaff.24.2.317.

Smith, S. L., & Cloud, C. (1996). The role of private, nonprofit fair housing enforcement organizations in lending testing. In J. Goering & R. Wienk (Eds.), *Mortgage lending, racial discrimination, and federal policy* (pp. 589–610). Taylor & Francis.

The Sentencing Project. (2013). *Report on the sentencing project to the United Nations Human Rights Committee: Regarding racial disparities in the United States criminal justice system*. Retrieved September 10. 2019 from https://www.sentencingproject.org/wp-content/uploads/2015/12/Race-and-Justice-Shadow-Report-ICCPR.pdf.

United Health Foundation. (2019). *2018 annual report*. Retrieved August 20, 2019 from https://www.americashealthrankings.org/learn/reports/2018-annual-report/findings-state-rankings.

US Census Bureau. (2018). *QuickFacts Louisiana population estimates*. US Census Bureau. Available from: https://www.census.gov/quickfacts/fact/table/LA/PST045218.

US Census Bureau. (n.d.-a). Vintage 2018 population estimates. How does your state stack up? Population changes for states (and Puerto Rico) from July 1, 2017, to July 1, 2018. Retrieved August 20, 2019 from https://www.census.gov/content/dam/Census/library/visualizations/2018/comm/popest-change-2017-2018.pdf.

US Census Bureau. (n.d.-b) Small area income and poverty estimates. Retrieved October 1, 2019 from https://www.census.gov/data-tools/demo/saipe/#/?map_geoSelector=u18_c&s_measures=u18_snc&s_year=2017.

U.S. Department of Health and Human Services (DHSS) (2019). *Infant deaths: Linked birth/ infant death records*. Retrieved September 10, 2019 from https://wonder.cdc.gov/lbd.html.

Wallace, M. E., Crear-Perry, J., Richardson, L., Tarver, M., & Theall, K. (2017). Separate and unequal: Structural racism and infant mortality in the US. *Health & Place*, *45*, 140–144. https://doi.org/10.1016/J.HEALTHPLACE.2017.03.012.

Wallace, M. E., Green, C., Richardson, L., Theall, K., & Crear-Perry, J. (2017). "Look at the whole me": A mixed-methods examination of Black infant mortality in the US through women's lived experiences and community context. *International Journal of Environmental Research and Public Health*, *14*(7), 727. https://doi.org/10.3390/ijerph14070727.

Walters, P. B. (2001). Educational access and the state: Historical continuities and discontinuities in racial inequality in American education. *Sociology of Education*, *74*(Spec. Iss), 35–49. https://doi.org/10.2307/2673252.

Weldon Cooper Center for Public Service. (2013). *The racial dot map. Created by Dustin A. Cable*. Retrieved September 9, 2019 from http://www.coopercenter.org/demographics/Racial-Dot-Map.

Williams, D. R., & Collins, C. (2001). Racial residential segregation: A fundamental cause of racial disparities in health. *Public Health Reports*, *116*, 404–416. https://doi.org/10.1093/phr/116.5.404.

Williams, D. R., Lawrence, J. A., & Davis, B. A. (2019). Racism and health: Evidence and needed research. *Annual Review of Public Health*, *40*, 105–125. https://doi.org/10.1146/annurev-publhealth-040218-043750.

Woolf, S. (2019). Necessary but not sufficient: Why health care alone cannot improve population health and reduce health inequities. *Annals of Family Medicine*, *17*(3), 196–199. http://www.annfammed.org/content/17/3/196.full#ref-5.

World Bank, Global Development Data. (2019a). *Mortality, infant (per 1,000 live births)*. Retrieved September 10, 2019 from https://data.worldbank.org.

World Bank, Global Development Data. (2019b). *Maternity mortality ratio (national estimate, per 100,000 live births)*. Retrieved August 31, 2019 from https://data.worldbank.org.

World Health Organization. (2019). *Health statistics and information systems: Maternal mortality ratio (per 100 000 live births)*. Retrieved October 27, 2019 from https://www.who.int/healthinfo/statistics/indmaternalmortality/en/.

Zheng, J. (2016). Black comfort: A brief history of African American hospitals and clinics in the Mississippi Delta in the early modern south. *The Southern Quarterly*, *53*(3), 176–189. College of Arts and Letters, University of Southern Mississippi. Retrieved August 19, 2019, from Project MUSE database.

Zolitschka, K. A., Miani, C., Breckenkamp, J., Brenne, S., Borde, T., David, M., et al. (2019). Do social factors and country of origin contribute towards explaining a "Latina paradox" among immigrant women giving birth in Germany? *BMC Public Health*, *19*(181), 1–10. https://doi.org/10.1186/s12889-019-6523-9.

Supplementary reading recommendations

Black Mamas Matter Alliance. (2019). *Black Mamas Matter advancing the human right to safe and respectful maternal health care*. Retrieved September 15, 2019 from https://blackmamasmatter.org/.

Byrd, M. W., & Clayton, L. A. (2003). Racial and ethnic disparities in healthcare: A background and history. In B. D. Smedley, A. Y. Stith, & A. R. Nelson (Eds.), *Unequal treatment: Confronting racial and ethnic disparities in healthcare* (pp. 455–527). Washington, DC: The National Academies Press.

Ford, C. L., Griffith, D. M., & Bruce, M. A. (2019). In K. L. Gilbert (Ed.), *Racism: Science & tools for the public health professional.* American Public Health Association. https://ajph.aphapublications.org/doi/book/10.2105/9780875533049.

Hicken, M. T., Kravitz-Wirtz, N., Durkee, M., & Jackson, J. S. (2018). Racial inequalities in health: Framing future research. *Social Sciences & Medicine. 199.* https://doi.org/10.1016/j.socscimed.2017.12.027.

Measurement of Segregation by the U.S. Bureau of the Census in Racial and Ethnic Residential Segregation in the United States: 1980–2000 by Weinberg, Iceland, and Steinmetz. Retrieved September 17, 2019 from https://www.censuc.gov/hhes/www/housing/housing patterns/pdf/massey.pdf.

Louisiana Women's Policy and Research Commission. (2019). *2018 annual report.* Retrieved September 8, 2019 from http://gov.louisiana.gov/assets/docs/LWPRC_2018AnnualReport ONLINE.pdf.

MacDorman, M. F., Mathews, T. J., Mahangoo, A. D., & Zeitlin, J. (2014). International comparisons of infant mortality and related factors: United States and Europe, 2010. *National Vital Statistics Report*, *63*(5), 1–6. https://www.cdc.gov/nchs/data/nvsr/nvsr63/nvsr63_05.pdf.

Massey, D. S., White, M. J., & Phua, V. (1996). The dimensions of segregation revisited. *Sociological Methods & Research*, *25*(2), 172–206. https://doi.org/10.1177%2F004912 4196025002002.

Mathews, T. J., Ely, D. M., & Driscoll, A. K. (2018). State variations in infant mortality by race and Hispanic origin of mother, 2013-2015. National Center for Health Statistics, 1–8, 295.

Mathews, T. J., MacDorman, M. F., & Thoma, M. E. (2015). Infant mortality statistics from the 2013 period linked birth/infant death [Data Set]. *National Vital Statistics Reports*, *64*(9), 1–30 Hyattsville, MD: National Center for Health Statistics.

Muennig, P., Reynolds, M. M., Jiao, B., & Pabayo, R. (2018). Why is infant mortality in the United States so comparatively high? Some possible answers. *Journal of Health Politics, Policy and Law*, *43*(5), 877–895. https://doi.org/10.1215/03616878-6951223.

Pew Research Center. (2014). *Party affiliation by state.* Pew Research Center. Available from: https://www.pewforum.org/religious-landscape-study/compare/party-affiliation/by/state/.

Pew Research Center. (2016). 2. Party affiliation among voters: 1992–2016. Pew Research Center. Available from: https://www.pewresearch.org/topics/political-party-affiliation/.

The Ferguson Commission. (2019). Forward through Ferguson: A path toward racial equity. Retrieved August 20, 2019 from https://forwardthroughferguson.org/.

U.S. Department of Housing, & Urban Development, (1996). Expanding housing choices for HUD-assisted families: First biennial report to congress moving to opportunity for fair housing demonstration.

U.S. Department of Health and Human Services (DHHS), *Infant Mortality Rates by States.* Retrieved September 15, 2019 from https://www.cdc.gov/nchs/pressroom/sosmap/infant_mortality_rates/infant_mortality.htm.

Zapata, A., Herwehe, J., Gruenfeld, R., & Plante, L. (2018). Louisiana maternal mortality review report 2011–2016. Retrieved August 20, 2019 from http://ldh.la.gov/assets/oph/Center-PHCH/Center-PH/maternal/2011-2016_MMR_Report_FINAL.pdf.

Definitions

Brown v. Board of Education the landmark decision of the US Supreme Court in which the Court ruled that American state laws establishing racial segregation in public schools are unconstitutional, even if the segregated schools are otherwise equal in quality.

Latina/o/x inclusive term to signify feminine, masculine, and gender neutral persons of Latin American origin or descent.

People of color Blacks (African American, African, Afro-Caribbean) and Latina/o/x.

Plessy v. Ferguson the landmark 1896 US Supreme Court decision that upheld the constitutionality of racial segregation under the "separate but equal" doctrine. Plessy was arrested in 1892 for taking an empty seat in a "Whites only" train car for a trip from Covington to New Orleans, Louisiana.

Prison industrial complex the intersecting interests of government and industry that use surveillance, policing, and imprisonment as solutions to economic, social, and political problems; the complex extends beyond prisons and criminal justice system (probation service, police, courts) to the companies that benefit from transporting, feeding, and exploiting the inmate population.

Redlining a discriminatory practice by which banks and/or insurance companies refuse or limit loans, mortgages, insurance, and other products; within defined geographic areas such as inner-city neighborhoods.

Structural competency the ability of healthcare providers to appreciate how symptoms, clinical problems, diseases, and attitudes toward patients, populations, and health systems are influenced by "upstream" social determinants of health.

Structural racism systems in which public policies, institutional practices, cultural representations, and other norms work in various, often reinforcing, ways to perpetuate racial group inequity.

White racial frame an overarching White worldview that encompasses a broad and persisting set of racial stereotypes, prejudices, ideologies, images, interpretations and narratives, emotions, and reactions to language accents, as well as racialized inclinations to discriminate

Wicked problems difficult or impossible to solve because of incomplete, contradictory, and changing requirements, often hard to define.

Improving health through a home modification service for veterans

Luz Mairena Semeah[a], *Xinping Wang*[b], *Diane C. Cowper Ripley*[c], *Mi Jung Lee*[d,e], *Zaccheus James Ahonle*[d,e], *Shanti P. Ganesh*[f], *Jennifer Hale Gallardo*[b], *Charles E. Levy*[g], *Huanguang Jia*[b]

[a]Health Services Research and Development, Veterans Health Administration, North Florida/South Georgia Veterans Health System, Research Service, Malcom Randall VAMC, Gainesville, FL, United States, [b]Veterans Health Administration, North Florida/South Georgia Veterans Health System, Research Service, Malcom Randall VAMC, Gainesville, FL, United States, [c]Veterans Health Administration, North Florida/South Georgia Veterans Health System, Malcom Randall VAMC, Gainesville, FL, United States, [d]University of Florida, Department of Occupational Therapy, Gainesville, FL, United States, [e]Veterans Rural Health Resource Center-GNV (VRHRC-GNV), North Florida/South Georgia Veterans Health System, Malcom Randall VAMC, Gainesville, FL, United States, [f]Department of Physical Medicine and Rehabilitation, North Florida/South Georgia Veterans Health System, Malcom Randall VAMC, Gainesville, FL, United States, [g]Physical Medicine and Rehabilitation Service, North Florida/South Georgia Veterans Health System, Malcom Randall VAMC, Gainesville, FL, United States

Abstract

Inaccessible living environments are problematic for some US Veterans, due to injuries sustained in service or attributed to aging, but can be addressed through home modification (HM) afforded by the Home Improvements and Structural Alterations (HISA) program. The retrospective national assessment of HISA, considering a 50% expected increase in Veterans enrollees over a 10-year time span (2013–23) and a 48% increase in dementia patients by 2033, demonstrates the need for attention to major shifts in HM health services demand. Utilizing data from the National Prosthetics Patient Database and the VA National Medical Outpatient Database, the analyses provide key information to support functional independence in the home by providing data about the HISA program. These include filling a data gap (e.g., Veteran demographic and clinical characteristics, modifications cost comparisons, and regional patterns of HISA utilization) for medically prescribed HM and assess underutilization of HM health services.

Keywords: Home modification, Veterans with disabilities, Accessibility, Health services, Housing

Three Facets of Public Health and Paths to Improvements. https://doi.org/10.1016/B978-0-12-819008-1.00014-6

Chapter outline

Veteran housing and health services

US Veterans aged 60 years or older ($N = 11{,}485{,}512$) account for 56.3% of the 20,392,192 total Veteran population (U.S. Department of Veterans Affairs [VA], 2016 [database]). Moreover, this subpopulation is forecasted to have an increased need for long-term support services such as geriatric or palliative care, in-home care, rehabilitation, and community-based care (VA, 2015, 2017). Important to this subpopulation is the desire to age in their homes and communities and slow-down or prevent expensive institutionalization options (Hwang, Parrott, & Brossoie, 2019; Joint Center for Housing Studies [JCHS], 2019a, 2019b), drawing attention to a gap in information on home-based care services (VA, 2015, 2016, 2017). For example, the rate per Veteran spent by Veterans Health Administration (VHA) for nursing home care is $56,353 vs $6570 for home and community-based services (VA, 2017, p. 4). This study provides the important links between Veterans' health care, home modification (HM) services, and health care delivery by quantifying patterns of Home Improvements and Structural Alterations (HISA) utilization and Veteran attributes. The analyses describe the utilization pattern of the national HISA program, define the pattern of type of prescribed HM, and examine sociodemographic and clinical characteristics of Veterans who obtained HM services. The data provide information on HISA's ability to meet the needs of underserved Veteran subpopulations and aid in decision-making to prepare for persistent demand and growth patterns across several demographics including Veterans over 65 with disabilities who are enrolled in VHA services and the utilization potential for recent returning Veterans whose medical condition may demand HMs as they age.

The HISA program provides medically necessary HMs that may improve accessibility, safety, and functional independence in the home while facilitating access to medical services in community settings. In an unpublished retrospective observational study, Ganesh et al. (2019) investigated the impact of obtaining HISA HMs on health services utilization in the form of hospitalizations and outpatient encounters. They found that HM provided by the HISA program was associated with decreased hospitalization and increased outpatient encounters that are considered preventative in nature and part of health maintenance strategies (Ganesh et al., 2019). These results suggest that greater use of HM services through the HISA program may yield better health outcomes and reduce costly institutionalization options such as hospital or nursing home stays.

HM is a remedy to inaccessible living environments to reduce risk for (1) older adults, (2) those with frailty due to declining health, and (3) individuals prone to falling (Chase, Mann, Wasek, & Arbesman, 2012; Davenport et al., 2009; Semeah et al., 2019). Individuals with loss in motor function, reduced levels of activity, pain, depression, and/or the ill effects of poly-pharmaceutical management are also at greater risks (Chase et al., 2012). HM remedies can include installing ramps, replacing steps, and providing grab bars in the bathroom.

Inaccessible living environments can negatively impact the elderly, especially those individuals experiencing social isolation. Of special concern are individuals living alone, living in rural communities, and/or living in substandard housing. Elderly in both rural and urban settings can face scarcity of walkability; lack of reliable and low-cost means of commuting; and suitability of environments to accommodate disabilities; all of which contribute to diminishing independence (U.S. Environmental Protection Agency [EPA], 2011). Sparsely inhabited areas are more likely to experience these challenges to a greater extent and less likely to have diversified health experts than urban areas. Rural communities are home to 5 million Veterans with approximately 2.3 million Veterans served by the VHA (Holder, 2017), while more than 2/5 of rural households will be headed by persons 65 years and older by 2030 (Pendall, Goodman, Zhu, & Gold, 2016). Rural areas are associated with long travel time to medical facilities and distant lifeline services (EPA, 2011; Smith et al., 2018). Travel time is a major barrier that negatively influences the elderly and people with disability to delay or put off required medical care and services (Cowper-Ripley et al., 2009, 2015; HAC, 2012; U.S. Department of Agriculture, 2018). Appropriate HMs can enhance social interactions (Bo'sher, Chan, Ellen, Karfunkel, & Liao, 2015; Maisel, Smith, & Steinfeld, 2008) and health-related quality of life (Jesus, Bright, Kayes, & Cott, 2016).

Tailored for individual needs, skills, and environments, HMs have significantly reduced the risk of falls (Chase et al., 2012; Cowie, Limousin, Peters, Hariz, & Day, 2012; Pynoos & Nishita, 2003; Pynoos, Steinman, & Nguyen, 2010), improved independence (Hwang et al., 2019; Semeah et al., 2017), and lessened hospitalization rates and costs (Ganesh et al., 2019; Pynoos & Nishita, 2003). Thus, increasing Veterans' access to HM has long-term health and health system advantages. The process of investigating data to benchmark existing services and Veteran characteristics provides important information to facilitate identification, engagement, and enrollment of underserved Veterans in the HISA program.

In this chapter, we provide background information linking the problems of Veterans with disabilities to their inaccessible living environments that can further impact their health. These factors inform the research design and methodology of the retrospective study of two database sources: (1) National Prosthetics Patient Database, and (2) the VA National Medical Outpatient Database. A final sample size of 37,307 was used in the analyses which distinguish two comparison classes—Class 1 are Veterans with a service-connected disability (76.7%) and Class 2 are Veterans who have a disability but are not classified as service-connected injured (23.3%). The two classes also differ on the amount of award they are eligible to receive (Class 1 = $6800; Class 2 = $2000). Analyses show that substantially more men obtained HISA awards than women which is consistent with the larger proportion of men enlisted in the military service. Still, further investigation is needed into the housing accessibility needs of female Veterans and how they are faring in relation to inaccessible housing considering their increased participation in combat. More importantly, findings suggest that Class 2 Veterans are unaware of eligibility for HM. The study confirms the need to extend services through advocacy to reach appropriate levels of overall utilization and to marginalized population segments such as those who are female, racially diverse, residing in rural areas, or have returned from recent conflicts.

Safety and rehabilitation at home

Rehabilitation professionals and their clients have long recognized that good home designs, which may include alterations or HMs, are a critical component of successful rehabilitation and recovery (Chase et al., 2012; Jesus et al., 2016). HMs enable performance of activities of daily living by compensating for decline in functional abilities and enabling better interaction with the physical environment. They can be both preventive and corrective in prolonging and promoting functional health, reducing institutionalization, and promoting independent living (Hale-Gallardo et al., 2017; Pynoos et al., 2010). A randomized clinical trial found that an Occupational Therapist (OT) home assessment and implementation of HMs significantly reduced falls (Gillespie et al., 2012). Thus, HMs support independence and enhance safety by reducing fall risks (Keall et al., 2015).

The amount of support a person with a disability requires through HM depends on the level of disability and on the accessibility of the physical environment. A disability can be exacerbated or created when the physical environment presents barriers that inhibit full participation in daily life activities and experiences. Thus, HM can be essential in making the environment supportive and accommodating (Lee, Romero, Hong, & Park, 2018). In one analogy, HMs render the environment as a "mat or a cushion" to offset for physical fragilities (Lee et al., 2018).

In relation to the built environment, *accessibility* refers to an environment that can be approached and used autonomously by persons with diverse disabilities and abilities. Results from the 2011 American Housing Survey reveal that the numbers of homes with prominent accessibility elements do not meet current or forecasted needs (Bo'sher et al., 2015). Accessibility research has focused on HM, visitability legislation, and smart technology as possible solutions (Chase et al., 2012; Maisel, 2006; Maisel et al., 2008; Semeah et al., 2017). Visitability legislation refers to local and state guidelines to promote the construction of single-family dwellings that facilitate visits by friends and relatives who experience mobility challenges or use mobility devices and the ability to visit others or use/maneuver in the physical environment.

Veterans with disabilities and HISA HM

An increasing number of Veterans with disabilities require accessible housing. The number of Veterans with considerable disabilities over the age of 65 seeking VA services is expected to increase by 50% by 2023 (VA, 2017, p. 3). Severely disabled Veterans seek housing options and solutions to improve health-related quality of life and housing satisfaction (Semeah et al., 2019; Semeah, Beamish, Schember, & Cook, 2016). The HISA national HM program allows for the implementation of medically prescribed HM for service members and Veterans (Semeah et al., 2017). HISA-funded HMs promote accessibility within homes that are either owner-occupied or rented. HISA is operated by the VHA, which is the division of the US Department of Veterans Affairs (VA) that provides medical services to qualifying Veterans; other HM programs are offered by the Veterans Benefits Administration (VBA), which administers other services including home loans, compensation, and employment programs. Under HISA, HMs are prescribed by VHA clinical providers. The HISA program also facilitates the discharge of inpatient Veterans as part of home health services. With a financial lifetime award of up to $6800, service-connected injured Veterans can obtain such HMs as installation of handrails, roll-in, barrier-free showers, fencing for visually impaired Veterans entitled to have service dogs, central air conditioning, and plumbing system adjustments to support medical equipment. The awards can be used to pay for labor and materials. In this study, we addressed all allowable HMs using variables addressed under the Methods section. HMs not covered by the HISA program include new home building, elevators, medical equipment for home use such as hospital beds, and powered mobile floor-based lifts. However, some of these items can be obtained through other VHA or VBA channels.

According to the US Government Accountability Office (GAO), there is a lack of information on HM programs such as HISA, Specially Adapted Housing, and Special Housing Adaptation, the utilization of such programs, and the characteristics of the users (U.S. Government Accountability Office [GAO], 2009, 2010). Therefore, Veterans are not well-informed and VA HM programs are underutilized. A database search of PubMed, EBSCO, Academic Search Premier, and ProQuest produced no articles that offered information on HISA users, average annual cost by class or user, types of HMs obtained, or the facility types administering HISA. Furthermore, little is known on the levels of program utilization by Veterans throughout the program history (Ganesh et al., 2019; Hoffman & Livermore, 2012; Semeah et al., 2017).

Potential benefits of research

Data on HISA utilization patterns or effectiveness are not available. Establishing a record of program use, cost, profiles of Veterans users, and profiles of the facilities that administer the program will identify facilities with lower HISA program use and identify Veterans who are not currently using the program but for whom the program could benefit. Given unmet housing accessibility needs and Veterans' lack of awareness of HM services, this study is the first step in ascertaining utilization and how Veterans with disabilities are faring in the housing accessibility market. Understanding HISA program users and the type of HMs they obtain can help inform specific population segments and can enable specific outreach to underserved patients. Identifying areas of underutilization generates the possibility for intervention. Additionally, healthcare providers—*information dissemination agents* for HM-users—can be targeted to increase their awareness of HISA and shrink the knowledge gap. More broadly, a better understanding of the role of medically prescribed HMs in addressing the national housing accessibility gap is necessary to assist in advancing and enhancing these programs.

Program utilization studies are conducted to better understand the needs of patients as evidenced by their patterns of resource use (Jia et al., 2007). Besides providing information about past, current, or gaps in usage patterns, program utilization studies can assist in planning future strategies. The current study provides information on geographic variation of HISA use at VA facilities in different regions.

Policy makers and VA administrators can use findings for deliberations on program administration and funding, especially considering national trends calling for accessible housing. Examples of trends are but not limited to (1) deinstitutionalization movement of persons with disabilities, (2) a growing number of Vietnam Veterans with disabilities, (3) the aging of the housing stock, and (4) a demographic shift to an increasing older population. These issues impact aging-in-place for both the general population and Veterans with disabilities. The latter trend is first-time prevalence in the United States bringing concerns relating to the adequate supply of affordable and accessible housing to the forefront (JCHS, 2014, 2016, 2019a). By 2035, one in five individuals residing in United States will be 65 years old or older and more than 2/5 of rural households will be headed by elderly persons by 2040 (JCHS, 2016, p. 5; Pendall et al., 2016, p. 28). By 2060, more than 98.1 million will be over the age of 65 (Mather, Jacobsen, & Pollard, 2015; Pendall et al., 2016; U.S. Census Bureau, 2014, Table 3). The expected physiological decline that can accompany aging, combined with the aging of the housing stock, suggests a coming housing crisis, since only 1% of homes contain all or most of the prominent accessibility elements (Bo'sher et al., 2015; JCHS, 2013, 2019a; Larson, Lakin, & Hill, 2012).

As the Veteran population ages and disabilities become more prevalent, widespread dissemination about HISA will be needed. The facts below highlight the urgency to examine the utilization of the program and that of the suitability of living environments to sustain Veterans in their home and communities instead of institutional settings:

- Veteran enrollees seeking VA services who are 65 and over are estimated to increase from 4.1 to 4.7 million over the 10-year time span of 2013–23 (VA, 2017, p. 3, Figure 2). These individuals are more like to use long-term support services.
- The quantity of Veterans to enroll for VA services that are significantly disabled (Priority 1a) are expected to increase by 50% over the 10-year time span of 2013–23 (VA, 2017, p. 3, Figure 2).
- VA patients with dementia exceeded 226,000 in 2015 and are estimated to increase to 335,000 in 2033 (VA, 2015, p. 31).
- Average expenditure per Veteran is $56,353 as opposed to $6570 for home and community-based services (VA, 2017, p. 3, Figure 4).

This study on HISA utilization is the first within the VHA HM program. The current literature has established that Veterans with disabilities are unaware of HM programs and the advantages of accessible elements in improving health-related quality of life (GAO, 2009, 2010; Semeah et al., 2016, 2019). Knowing the profiles of HISA users can assist in identifying potential users who can benefit from HISA at each facility to address disparities in HISA utilization.

Link between HMs and health care

HISA is a program that links structural HMs to clinical care since an evaluation by clinicians is required before the approval of the HM application. Although there are numerous studies on HMs (Chase et al., 2012; GAO, 2009, 2010; Pynoos et al., 2010; Pynoos & Nishita, 2003; Semeah et al., 2019), there is limited understanding of medically prescribed HM services such as HISA. This limitation prompted informal interviews with three groups of VHA HISA stakeholders in the North Florida/South Georgia Health System (NF/SG VHS) to build a foundation and understanding for exploring HISA utilization. NF/SGVHS is one of the busiest VHA health systems with 120,000 unique patient encounters monthly consisting of 2 medical centers, 3 large multispecialty outpatient clinics, and 9 community-based primary care outpatient clinics. The informal interviews were conducted with 10 Veterans with disabilities who received a HISA HM; 9 VHA healthcare providers who ordered HMs for Veterans; and HISA Committee members yielding insight on HM utilization and HISA program implementation. Inclusion of the HISA Committee was important because of their role in the review and approval of HISA prescriptions.

Four key takeaways from these stakeholders suggest that an opportunity exists to improve information to Veterans to increase program utilization and enhance quality of life. First, providers indicated that, due to Veterans' health conditions and level of functional limitations, HMs were needed to make their homes accessible and safe. Second, both providers and Veteran users reported a lack of availability of information about VHA HM services and HISA grants, which may have hindered HM use for eligible Veterans. Third, Veterans indicated dissatisfaction with insufficient or inappropriate HMs, which negatively impacted their quality of life and health. Finally, HISA Committee members noted that the understanding of the intent and goals of HM services was inconsistent among providers, Veterans, and Committee members. Results from interviews suggest a lack of program awareness, understanding of the types of

HM items allowed by the program, and utilization in services that may hinder optimal Veteran health. The interviews support the need to assess a greater number of Veterans and providers to determine if our small sample is indicative of concerns in the larger Veteran population and/or other VHA health systems.

Research design and data analysis

The research design was a retrospective data base analysis. Below we present an overview of the two datasets used and variables of interest. Results of a descriptive analysis are provided as well as findings. Descriptive statistics were obtained on (1) HISA utilization scope, mode, and recipient; and (2) HISA utilization patterns (e.g., number of HM prescription, HM types, and cost of HM). Additionally, we compared the national utilization prescription patterns between different VA medical centers and their regions. Within the VHA, the country has been divided into specific regions known as Veterans Integrated Service Networks (VISN). Potential research directions or recommendations relating to the HISA program are discussed. The implications of findings to housing accessibility policy and the field of rehabilitation science are also explored. Consideration is given to the significance of findings to the VHA and its healthcare providers and HISA administrators, Veterans, researchers, and policy makers concerned with accessible housing, both within and outside the VHA.

Methods

This study was approved by VA Research and Development at the NF/SGVHS and the affiliated Institutional Review Board at the University of Florida.

This chapter has three objectives:

(1) Describe sociodemographic and clinical characteristics of HISA users.
(2) Describe HISA utilization patterns in number, types, and cost of HMs.
(3) Compare utilization between the different VA Medical Centers (VAMC) and VISNs.

Data sources

This study utilized two major sources of data: (1) the National Prosthetics Patient Database (NPPD) and (2) the VHA Medical Outpatient Datasets. The NPPD is a national comprehensive administrative database that is part of the VA Prosthetics Work at the VA Corporate Data Warehouse. The database is comprised of patient-specific details on all prosthetic-related items offered to VHA patients and ordered by clinical staff. Such orders include HISA items as well as orthotic and prosthetic devices, sensory aids, durable medical equipment, assistive devices, implantable devices, and home oxygen equipment. The NPPD was used to obtain data on the number and types of HMs. Several variables were obtained from the NPPD including clinical characteristics, facility (administrative parent medical facility) and geographic region, hospital complexity level, and cost.

Clinical characteristics were gleaned from the International Classification of Diseases, 9th Revision (ICD9) and International Classification of Diseases, 10th Revision (ICD10) codes for diagnosis were used to create an eight-level *primary diagnosis* variable. Levels were determined by common diagnoses from the top 20 most common primary diagnoses in the databases grouped into general body systems, including vision and hearing impairment, and an "other" general category. The eight levels of the primary diagnosis variable are: 1 = musculoskeletal, 2 = pulmonary, 3 = neurologic, 4 = endocrine, 5 = cardiovascular, 6 = vision impairment, 7 = hearing impaired, and 8 = other.

The VHA provides health care services to over 9 million Veterans enrolled yearly (VA, n.d.). The VHA is the largest integrated healthcare system in the United States, comprising 18 VISNs, and has 1255 health care facilities, including 170 VAMCs and 1074 outpatient locations of differing size and degree of complexity. The 18 VISNs are categorized as follows: 1, 2, 4, 5, 6, 7, 8, 9, 10, 12, 15, 16, 17, 19, 20, 21, 22, and 23. Existing facility location and geographic regions were identified from VISN designations.

VHA groups VAMCs into five complexity levels for administrative purposes and complexity level is determined by the intricacy of clinical services rendered and the patient population served (Carney, West, Neily, Mills, & Bagian, 2010). The five complexity categories are 1a, 1b, 1c, 2, and 3 based on numerous factors, including the level of care provided by the intensive care unit. Highest complex VAMCs are designated as 1a; additionally, these facilities support substantial education and research activities and extensive areas of medical specialization.

The cost variable is predetermined by Public Law 111–163 and predicated on whether a Veteran: (1) has a service-connected disability and/or a nonservice-connected disability with a 50% or above disability rating and is thus entitled to the HISA maximum of $6800 (Class 1), or (2) a nonservice-connected disability and is thus entitled to a HISA maximum of $2000 (Class 2). We obtained cost data related to items ordered and recorded information on prescriptions made by clinical staff of the VHA Prosthetics and Sensory Aids Service (PSAS). To identify users, codes of the NPPD and the Health Care Financing Administration Coding System of PSAS were used. Codes were sorted and raw data de-identified for analysis. Finally, the VHA Medical Outpatient Datasets administrated by VINCI (VA Informatics and Computing Infrastructure) contain comprehensive, national health services information on outpatient visit and event.

Sample selection and study variables

The study cohort included Veterans who qualified for HISA and received HMs under the HISA program between fiscal year (FY)-2011 and FY-2017. Veterans were included regardless of age, gender, and race/ethnicity. A total of 41,203 individuals were identified as HISA users. After executing exclusionary criteria, the study cohort consisted of 37,307: 2322 had no cost reported; 532 had cost exceeding the $6800 limit; 989 Class 2 had cost exceeding the $2000 limit; and 53 Veterans could not be identified in the outpatient file. Data were merged with the outpatient data by scrambled social security number to retrieve the following study variables:

- *Age*: Calculated based on date of birth and most recent outpatient visit date, then this number was placed into one of the following six categories: $\leq$44, 45–64, 65–69, 70–74, 75–79, and $\geq$80.
- *Race*: Standard categories were utilized, but in instances where race was recorded as *unknown*, the patient was excluded. If multiple races were recorded, the race initially reported was assigned.
- *Ethnicity*: Typically categorized as either Hispanic or non-Hispanic; patients with multiple ethnicities were designated Hispanic.
- *Marital Status*: This variable had six categories in the database that were collapsed into three: unmarried, married, and unknown. For patients with more than one marital status where one was *unknown*, the *unknown* value was eliminated. If a patient's status was both *married* and *unmarried* status, the most recent status was selected.
- *Gender*: The database included only male, female, or *unknown*. For patients with multiple values recorded that included "*unknown*," the *unknown* value was eliminated. If a patient had both male and female status, the female status was selected.
- *Rurality*: VA uses the Rural-Urban Commuting Areas (RUCA) system to define rurality. RUCA was developed by the US Department of Agriculture and the Department of Health and Human Services. This system considers population density and how closely a community is linked socioeconomically to larger urban centers. We used three categories of rurality: rural, urban, and both. If a patient had both rural and urban classifications, "both" was assigned.
- *Comorbidity*: A modified Charlson Comorbidity Index calculated from VHA Medical Outpatient diagnosis was used to assess the burden of comorbid conditions (Chu, Ng, & Wu, 2010; O'Connell & Lim, 2000). The weighted sum score of the Charlson index helped to further describe the impact of patients' other nonprimary medical conditions on HISA utilization. It was important to adjust for the association between utilization and Class (e.g., Veterans with service-connected disability [Class 1] or nonservice-connected disability [Class 2]). The index is based on designating an integer weight from 1 to 6 for specific medical conditions with a weight of 6 constituting maximum illness. The summation of the weighted comorbidity scores indicates that the higher the score, the heavier the burden of medical comorbid conditions. The outpatient visit closest to the HISA award year was used for the comorbidity calculation.

Types of HM

HISA HMs used are as follows, but not limited to: roll-in showers; permanent ramps, installation of central air-conditioning systems, improved lighting, flooring replacement, kitchen/bathroom modifications; and home inspections (see Tables 14.2 and 14.3 for HM items included in this study).

Data analysis

Data was analyzed using SAS, version 9.4. Descriptive statistics were used for sociodemographic, clinical and disease severity, and cost (6800 for class 1 and 2000 for class 2) from the NPPD and Outpatient datasets. *t*-tests were used for continuous variables and Chi-square for categorical variables.

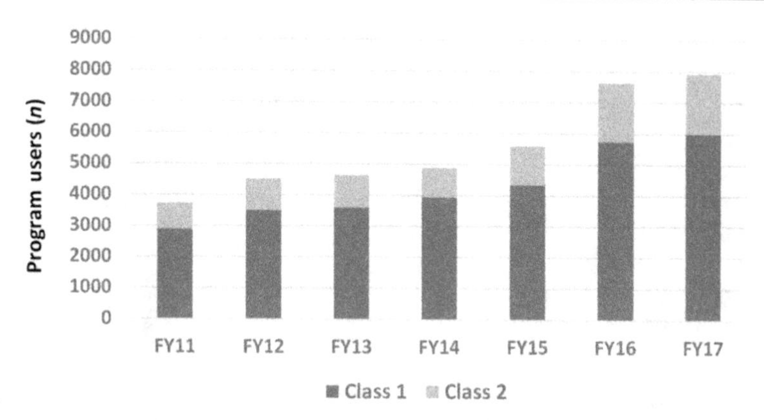

Fig. 14.1 Number of Veterans in the Home Improvements and Structural Alterations (HISA) program of the Veterans Health Administration from fiscal year 2011 to 2017. Class 1 Veterans have a service-connected disability and qualify for maximum lifetime HISA funding of $6800; Class 2 Veterans have a nonservice-connected disability and qualify for $2000.

Results

HISA utilization grew over the study period (Fig. 14.1). In FY-2011, the HISA users totaled 3734 with 77.5% for Class 1 group and 22.5% for Class 2 group; by FY-2017, however, the number of the users almost doubled (N=7929) including 75.6% for Class 1 and 24.5% for Class 2. Thus, from FY-2011 to FY-2017, there was an average percentage change of 112.4%. Examination of year to year provides the following average percentage changes results (Figs. 14.2 and 14.3); between FY-2012 to

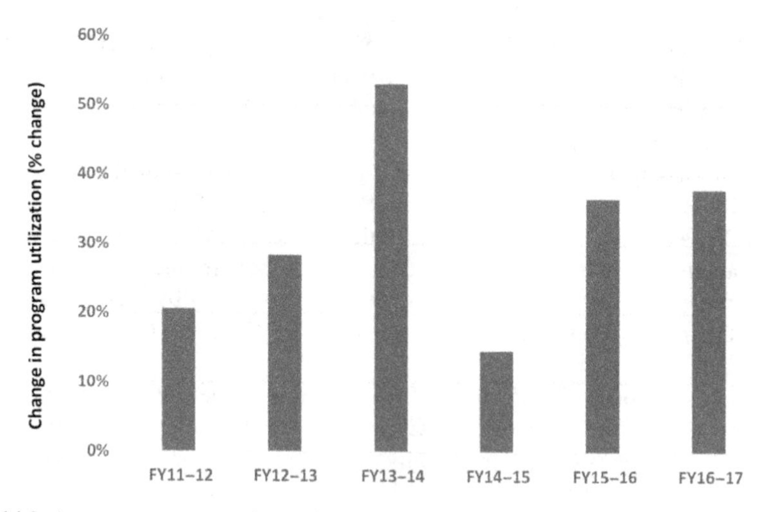

Fig. 14.2 Average percentage change in utilization of the Home Improvements and Structural Alterations (HISA) Program of the Veterans Health Administration by Veterans from fiscal year 2011 to 2017.

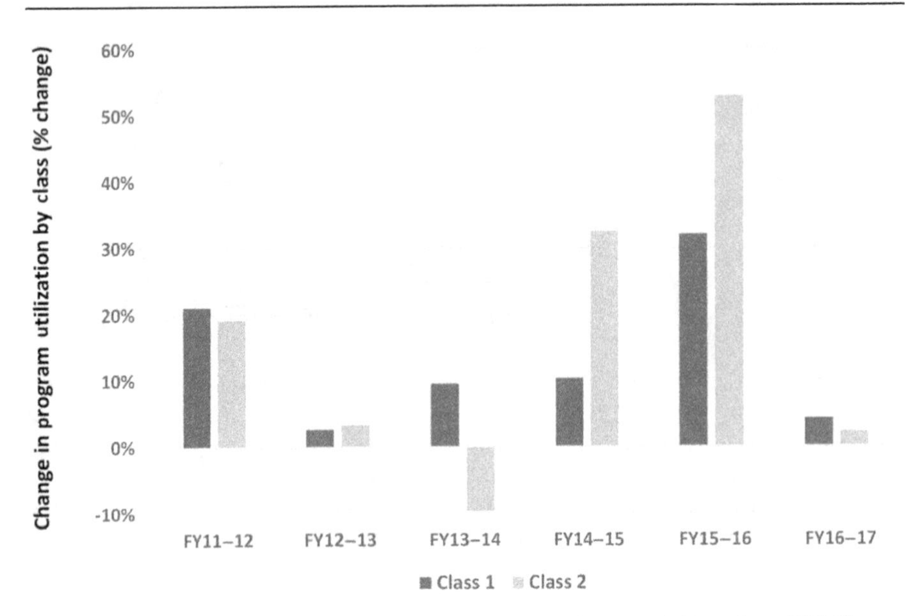

Fig. 14.3 Percentage change in Home Improvements and Structural Alterations (HISA) Program utilization by class from fiscal year 2011 to 2017. Class 1 Veterans have a service-connected disability and qualify for maximum lifetime HISA funding of $6800, whereas Class 2 Veterans have a nonservice-connected disability qualifying for $2000.

FY-2013 and FY-2013 to FY-2014 revealed a 24.7-point percent increase in utilization of the program to 53.0% in FY-2013 to FY-2014. However, there was 38.5-point decrease between FY-2013 to FY-2014 and FY-2014 to FY-2015 in the use of HISA with a 22-point increase yielding a total of 36.6% change between FY-2014 to FY-2015 and FY-2015 to FY-2016. The percentage increases in both FY-2014 to FY-2015 and FY-2015 to FY-2016 for Class 2 were 32.5% and 52.9% and for Class 1, 10.3% and 32.0%, respectively.

Table 14.1 compares sociodemographic, geographic, and clinical characteristics between Class 1 and Class 2 HISA users. For the total HISA cohort: the average age was 72 years and majority were male (95.6%), White (70.7%), married (77.5%), and living in an urban setting (62.1%). The most frequent medical diagnoses were musculoskeletal conditions (31.5%), neurologic disorders (26.6%), and cardiovascular conditions (6.3%). The study cohort Charlson Comorbidity score averaged 2.3.

The two user groups significantly differed in most of the sociodemographic and clinical characteristics: Compared with Class 1 users, Class 2 users were significantly older (77.8 vs 70.6 years of age), marginally more likely to be male (96.4% vs 95.3%), White (76.0% vs 69.1%), unmarried (29.7% vs 19.8%), and from rural areas (30.4% vs 28.9%). Clinically, fewer Class 2 users had musculoskeletal diagnoses (23.9% vs 33.8%) and more had neurologic disorders (29.0% vs 25.9%) and cardiovascular conditions (7.0% vs 6.0%) with fewer burdens of comorbid medical conditions.

Table 14.1 Sociodemographic, clinical, and geographic characteristics of HISA home modification users by Veteran disability Class 1: Service-connected disability and lifetime maximum $6800, and Class 2: Nonservice-connected disability and $2000 lifetime maximum.

Characteristics	Overall (N = 37,307)	Class 1 (n = 28,613)	Class 2 (n = 8694)	P value[a]
Age group				<.0001
≤44	789 (2.11%)	751 (2.62)	38 (0.44)	
45–64	6725 (18.03%)	5669 (19.81)	1056 (12.15)	
65–69	8009 (21.47%)	6991 (24.43)	1018 (11.71)	
70–74	7471 (20.03%)	6356 (22.21)	1115 (12.82)	
75–79	3910 (10.48%)	2709 (9.47)	1201 (13.81)	
≥80	10,403 (27.88%)	6137 (21.45)	4266 (49.07)	
Age, mean yr. ± SD	72.3 ± 11.9	70.6 ± 11.7	77.8 ± 11.0	<.0001
Gender, *n* (%)				<.0001
Male	35,646 (95.6)	27,263 (95.3)	8383 (96.4)	
Female	1661 (4.5)	1350 (4.7)	311 (3.6)	
Race, *n* (%)				<.0001
Black	6941 (18.6)	5741 (20.1)	1200 (13.8)	
White	26,367 (70.7)	19,758 (69.1)	6609 (76.0)	
Others	981 (2.6)	853 (3.0)	128 (1.5)	
Unknown	3018 (8.1)	2261 (7.9)	757 (8.7)	
Marital status, *n* (%)				<.0001
Married	28,913 (77.5)	22,838 (79.8)	6075 (69.9)	
Unmarried	8243 (22.1)	5658 (19.8)	2585 (29.7)	
Unknown	151 (0.4)	117 (0.4)	34 (0.4)	
Rurality, *n* (%)				<.0001
Rural	10,914 (29.3)	8271 (28.9)	2643 (30.4)	
Urban	23,178 (62.1)	17,766 (62.1)	5412 (62.3)	
Both	3214 (8.6)	2575 (9.0)	639 (7.4)	
Primary medical diagnosis				<.0001
Musculoskeletal	10,610 (31.5)	8716 (33.8)	1894 (23.9)	
Pulmonary	1399 (4.2)	1040 (4.0)	359 (4.5)	
Neurologic	8951 (26.6)	6658 (25.9)	2293 (29.0)	
Endocrine	1305 (3.9)	1058 (4.1)	247 (3.1)	
Cardiovascular	2113 (6.3)	1556 (6.0)	557 (7.0)	
Vision impairment	721 (2.1)	484 (1.9)	237 (3.0)	
Hearing impaired	16 (0.1)	12 (0.1)	4 (0.1)	
Other	8565 (25.4)	6237 (24.2)	2328 (29.4)	
Charlson Comorbidity Index mean score ± SD	2.33 ± 1.36	2.31 ± 1.35	2.27 ± 1.36	0.0503

SD—standard deviation.
[a] Chi-Square test for discrete variables and *t*-test for age and comorbidity.

To compare HISA users with VHA enrollees and VHA users, we extracted data from the VHA Service Support Center (2019). A VA enrollee can qualify for VHA services, but not necessarily use services. When compared with VHA enrollees (50.8%) and users (51.6%) in FY-2017, there was a larger proportion of HISA users (79.9%) who were 65 and older. When compared with VHA enrollees (59.5%) and users (29.7%), there were fewer HISA users (18.0%) who were between the ages of 45–64; with a very low use of the HISA program by Veterans with disabilities (2.1%) below the age of 44, when compared with VHA enrollees (19.7%) and users (18.5%). HISA users were marginally less diverse in ethnicity (70.7% for White) than VHA enrollees (64.1% for White) and users (69.0% for White). When compared with VHA enrollees (91.9%) and users (89.0%), the HISA program serves a higher fraction of males (95.6%). There are fewer female Veterans in HISA (4.5%) when compared to VHA enrollees (8.1%) and users (10.8%).

Tables 14.2 and 14.3 show the number and type of HM by FY 2011–17 and user class. Concomitant with the rapid growth of unique HISA recipients over time, the number of HMs funded by the HISA program also showed a sizeable increase. In FY-2011, a total of 4078 HMs were made; by FY-2017, this number more than doubled to 8152 with a 99.9% increase. Bathroom alterations were by far the most common modification throughout the timeframe and comprised nearly 2/3 of all modifications in FY-2011 (65.6%), increasing to 75.5% in FY-2014 and to 80.5% in FY-2017. Among the bathroom alterations, the following five modifications were the most common: (1) bath, (2) shower, (3) grab bar, (4) tub, and (5) toilet. Doorway adjustments and railing placement were distant second and third most common modification across time. The breakdown by Class had similar patterns in modifications by type and over time. Although Jacuzzi and spa modifications are not covered by HISA, these modifications were listed for 13 HISA users and were not included as part of the analysis.

Tables 14.4 and 14.5 demonstrate HISA utilization by Class, VISN, and FY. For the total period under investigation, VISN 17 (Texas) had the largest number of HISA users both for the entire study cohort ($N=4513$) and for Class 1 users ($n=3723$; 82.5%). The region with the largest number of Class 2 users was VISN 8 (Florida excluding the Panhandle, South Georgia, Puerto Rico, and the Virgin Islands) ($n=1266$; 35.2%). The distribution of Class 1 and Class 2 users varied across VISNs. For example, out of the total HISA users in VISN 23 ($n=1537$), located in the Upper Mid-West, 57.9% were Class 1 and 42.1% were Class 2. By contrast, and on the other end of the spectrum, in VISN 21 ($n=1216$), located on the West Coast, 88.4% of recipients were Class 1 and 11.6% Class 2. This finding implies that there may be regional differences in how HISA committees are prioritizing who should receive funding or that there are different distributions of service-connected individuals in local populations.

Table 14.6 lists the top 20 HISA-use VAMCs by Class, facility complexity level, VISN, and percentage of rural patients serve. There is great variability in the use by Class among the facilities: Fayetteville (91.0%) and San Antonio (88.9%) VAMCs have the largest proportions of Class 1 and Minneapolis (52.3%) and Pittsburgh (44.8%) VAMCs have the largest proportions of Class 2.

Table 14.2 Number and type of home modifications by fiscal year (FY) 2011–14 and users by Veteran disability Class 1: Service-connected disability and lifetime maximum $6800, and Class 2: Nonservice-connected disability and $2000 lifetime maximum.

	FY11			FY12			FY13			FY14		
Type	**Total**	**Class 1**	**Class 2**	**Total**	**Class 1**	**Class 2**	**Total**	**Class 1**	**Class 2**	**Total**	**Class 1**	**Class 2**
Bathroom	2676	2187	489	3413	2779	634	3529	2890	639	3849	3279	570
Air conditioning	26	22	4	17	17	0	11	11	0	29	27	2
Doorway	323	199	124	344	216	128	354	219	135	296	189	107
Driveway	42	33	9	49	35	14	34	25	9	23	18	5
Electrical	72	47	25	50	31	19	40	31	9	42	36	6
Kitchen	16	13	3	19	17	2	29	24	5	22	18	4
Not specified	492	401	91	535	435	100	487	385	102	370	278	92
Permits/inspection	42	29	13	18	9	9	17	15	2	15	11	4
Plumbing	18	13	5	16	12	4	8	6	2	18	10	8
Railing	250	158	92	246	140	106	299	167	132	318	174	144
Ramp/cement	60	34	26	54	43	11	55	38	17	55	44	11
Walkway	61	49	12	79	61	18	54	39	15	64	45	19
Total by year	4078	3185	893	4840	3795	1045	4917	3850	1067	5101	4129	972

Table 14.3 Number and type of home modifications by fiscal year (FY) 2015–17 and by Veteran disability Class 1: Service-connected disability and lifetime maximum $6800, and Class 2: Nonservice-connected disability and $2000 lifetime maximum.

	FY15			FY16			FY17			Total by type		
Type	**Total**	**Class 1**	**Class 2**	**Total**	**Class 1**	**Class 2**	**Total**	**Class 1**	**Class 2**	**Total**	**Class 1**	**Class 2**
Bathroom	4454	3626	828	6152	4784	1368	6563	5070	1493	30,636	24,615	6021
Air conditioning	14	9	5	22	18	4	25	21	4	144	125	19
Doorway	320	191	129	373	225	148	370	243	127	2380	1482	898
Driveway	49	40	9	62	51	11	62	51	11	321	253	68
Electrical	61	53	8	88	72	16	59	50	9	412	320	92
Kitchen	20	15	5	18	12	6	22	20	2	146	119	27
Not specified	395	294	101	475	338	137	441	326	115	3195	2457	738
Permits/inspection	16	12	4	9	5	4	3	2	1	120	83	37
Plumbing	13	10	3	9	6	3	6	4	2	88	61	27
Railing	306	157	149	451	249	202	473	282	191	2343	1327	1016
Ramp/cement	39	32	7	51	46	5	32	30	2	346	267	79
Walkway	61	43	18	96	63	33	96	64	32	511	364	147
Total by year	5748	4482	1266	7806	5869	1937	8152	6163	1989	40,642	31,473	9169

Table 14.4 Utilization of home modifications during fiscal year (FY) 2011–14 by region and user by Veteran disability Class 1: Service-connected disability and lifetime maximum $6800, and Class 2: Nonservice-connected disability and $2000 lifetime maximum.

	FY11			FY12			FY13			FY14		
VISN[a]	Total	Class 1	Class 2	Total	Class 1	Class 2	Total	Class 1	Class 2	Total	Class 1	Class 2
1	76	57	19	103	87	16	114	105	9	88	73	15
2	142	93	49	157	105	52	177	126	51	242	171	71
4	125	82	43	228	142	86	252	173	79	273	190	83
5	29	23	6	34	30	4	46	41	5	63	57	6
6	344	291	53	470	401	69	486	381	105	584	496	88
7	235	184	51	271	226	45	389	322	67	378	324	54
8	441	317	124	488	323	165	366	226	140	404	297	107
9	181	133	48	189	151	38	211	173	38	269	217	52
10	142	99	43	183	132	51	206	148	58	315	227	88
12	317	255	62	379	326	53	368	307	61	350	280	70
15	212	137	75	216	143	73	160	115	45	148	120	28
16	246	226	20	287	255	32	263	230	33	203	176	27
17	515	418	97	599	472	127	603	473	130	618	528	90
19	82	60	22	97	64	33	99	74	25	115	84	31
20	169	135	34	165	133	32	170	144	26	125	104	21
21	65	61	4	109	100	9	107	94	13	138	122	16
22	303	262	41	399	344	55	444	383	61	440	389	51
23	110	61	49	131	70	61	172	83	89	126	89	37
Total	3734	2894	840	4505	3504	1001	4633	3598	1035	4879	3944	935

VISN—Veterans Integrated Service Network.
[a] See Fig. 14.4 for location of the regions.

Table 14.5 Utilization of home modifications during fiscal year (FY) 2015–17 by region and user class by Veteran disability Class 1: Service-connected disability and lifetime maximum $6800, and Class 2: Nonservice-connected disability and $2000 lifetime maximum.

VISN[a]	FY15			FY16			FY17			Total by VISN		
	Total	Class 1	Class2	Total	Class 1	Class2	Total	Class 1	Class 2	Total	Class 1	Class 2
1	117	82	35	203	162	41	199	150	49	900	716	184
2	257	187	70	352	259	93	389	269	120	1716	1210	506
4	277	193	84	374	279	95	533	389	144	2062	1448	614
5	39	37	2	402	300	102	395	300	95	1008	788	220
6	809	674	135	859	687	172	930	772	158	4482	3702	780
7	442	379	63	514	417	97	505	405	100	2734	2257	477
8	493	298	195	654	407	247	752	464	288	3598	2332	1266
9	237	195	42	157	118	39	141	120	21	1385	1107	278
10	351	229	122	808	543	265	849	534	315	2854	1912	942
12	353	259	94	215	141	74	202	148	54	2184	1716	468
15	229	168	61	358	240	118	377	284	93	1700	1207	493
16	264	236	28	386	320	66	364	311	53	2013	1754	259
17	605	535	70	821	681	140	752	616	136	4513	3723	790
19	82	68	14	135	110	25	145	121	24	755	581	174
20	292	256	36	378	292	86	378	308	70	1677	1372	305
21	195	179	16	320	276	44	282	243	39	1216	1075	141
22	311	259	52	304	268	36	375	334	41	2576	2239	337
23	238	118	120	399	245	154	361	224	137	1537	890	647
Total	5591	4352	1239	7639	5745	1894	7929	5992	1937	38,910	30,029	8881

VISN—Veterans Integrated Service Network.
[a] See Fig. 14.4 for location of the regions.

Table 14.6 Top 20 HISA utilization VA Medical Centers (VAMCs) and class, complexity level, VISN, and rurality of patients from fiscal years 2011–17.

VAMCs	Total (*n*)	Class 1[a] (*n*, %)	Class 2 (*n*, %)	Complexity level[b]	VISN[c]	Rural patients (%)[a]
San Antonio	2231	1983 (88.9)	248 (11.1)	1a	17	23.1
Cleveland	1401	858 (61.2)	543 (38.8)	1a	10	26.7
Dallas	1249	908 (72.7)	341 (27.3)	1a	17	27.1
North Florida/South Georgia	1031	680 (66.0)	351 (34.0)	1a	8	41.4
Durham	901	731 (81.1)	170 (18.9)	1a	6	44.8
Fayetteville, North Carolina	815	742 (91.0)	73 (9.0)	1c	6	42.8
Tampa	814	534 (65.6)	280 (34.4)	1a	8	7.9
Salisbury, North Carolina	751	604 (80.4)	147 (19.6)	1c	6	34.1
Columbia, South Carolina	746	582 (78.0)	164 (22.0)	1c	7	39.1
Huntington	692	514 (74.3)	178 (25.7)	1c	5	49.3
Texas Valley Coast	691	532 (77.0)	159 (23.0)	3	17	–
Asheville	645	464 (71.9)	181 (28.1)	1c	6	50.9
Birmingham	565	450 (79.6)	115 (20.4)	1a	7	39.7
Phoenix	531	457 (86.1)	74 (13.9)	1b	22	17.6
Gulf Coast HCS	525	460 (87.6)	65 (12.4)	1c	16	30.5
Minneapolis	501	239 (47.7)	262 (52.3)	1a	23	43.2
Pittsburgh	482	266 (55.2)	216 (44.8)	1a	4	32.3
Richmond	464	374 (80.6)	90 (19.4)	1a	6	35.5
Marion, Illinois	457	332 (72.6)	125 (27.4)	2	15	74.8
Philadelphia	451	332 (73.6)	119 (26.4)	1b	4	5.6

[a] Class 1 Veterans have service-connected disability and receive lifetime maximum of $6800 for home modifications (HM); Class 2 Veterans have a nonservice-connected disability and receive a lifetime maximum of $2000 for HM.
[b] Reference: Rural Veterans Health Care Atlas, Available online at: https:/www.ruralhealth.va.gov/docs/atlas/CHAPTER_02_RHRI_Pts_treated_at_VAMCs.pdf.
[c] See Fig. 14.4 for location of the regions/VISN.

All but two facilities (90.0%) have high complexity and all facilities (administrative parent medical facilities) with high HISA utilization numbers are in urban settings. Many urban hospitals have a large portion of their patient population residing in rural areas. For example, the Marion, Illinois, medical center is in an urban area, but approximately 75% of the patients served come from rural areas. Fig. 14.4 depicts the locations of the facilities with high utilization. The high HISA-use facilities are located primarily in the eastern United States.

Table 14.7 shows the cost data of HISA use by class, FY, and region/VISN. On the average, the cost per HISA funding for all years was \$4761.42 (SD = \$2352.5), with \$5688.2 ± \$1858.3 for Class 1 and \$1711.4 ± \$503.0 for Class 2. The overall mean costs did not fluctuate a great deal over time nor did the mean cost by Class. In FY-2011, overall mean cost per HISA award was \$4510.4 and, in FY-2017, the mean was \$4580.6, a percentage increase of 1.56%. For Class 1, the FY-2011 mean cost was \$5348.0 and in FY-2017 it was \$5513.1, a percentage increase of 3.09%. For Class 2, the FY-2011 mean cost was \$1624.6, and in FY-2017 it was \$1696.0, a percentage increase of 4.39%. The mean cost is similar over time, despite an almost doubling of the number of recipients, overall increase in number of HM prescribed, and by Class.

By region, the lowest overall mean cost was \$3545.9 identified in the northern Midwest (see VISN 23 in Fig. 14.4), while the highest overall mean cost is in the southeast (see VISN 17 in Fig. 14.4) at \$5530.1. When we examine the mean cost by Class, we find that portions of Oklahoma and Texas (see VISN 19 in Fig. 14.4) had the lowest mean cost for Class 1 recipients at \$4225.7. The region with the highest mean cost for Class 1 recipients was primarily Texas (see VISN 17 in Fig. 14.4) at \$6310.8. Thus, there is a difference of mean costs for Class 1 recipients across regions. Interestingly, while VISN 17 had the highest mean cost for Class 1 recipients, it also had the lowest mean cost for Class 2 (\$1237.3). The region comprising Washington, DC, Maryland, Virginia, and West Virginia (see VISN 5 in Fig. 14.4) had the highest mean cost for Class 2 (\$1827.7).

As shown in Table 14.8, the five most costly modifications for the overall cohort during the study period were for bathroom renovations (\$5206.6), followed by central air-conditioning installation (\$4988.0), kitchen modifications (\$4665.4), driveway (\$4267.3), and plumbing (\$4249.3). For Class 1, these modifications were also the costliest, but the rank order was slightly different: Bathroom, plumbing, central air conditioning, kitchen, and driveway. For Class 2, the rank order and type of modification for the top five most costly modification were somewhat different with central air conditioning ranking the costliest (\$1900.23) followed by bathroom (\$1819.2), walkway (\$1773.2), doorway (\$1678.25), and driveway (\$1672.3).

Findings

Overall, we find that gender and age are factors in the utilization rates for HISA services. The program is being prescribed at a higher proportion to older male Veterans with service-connected disabilities.

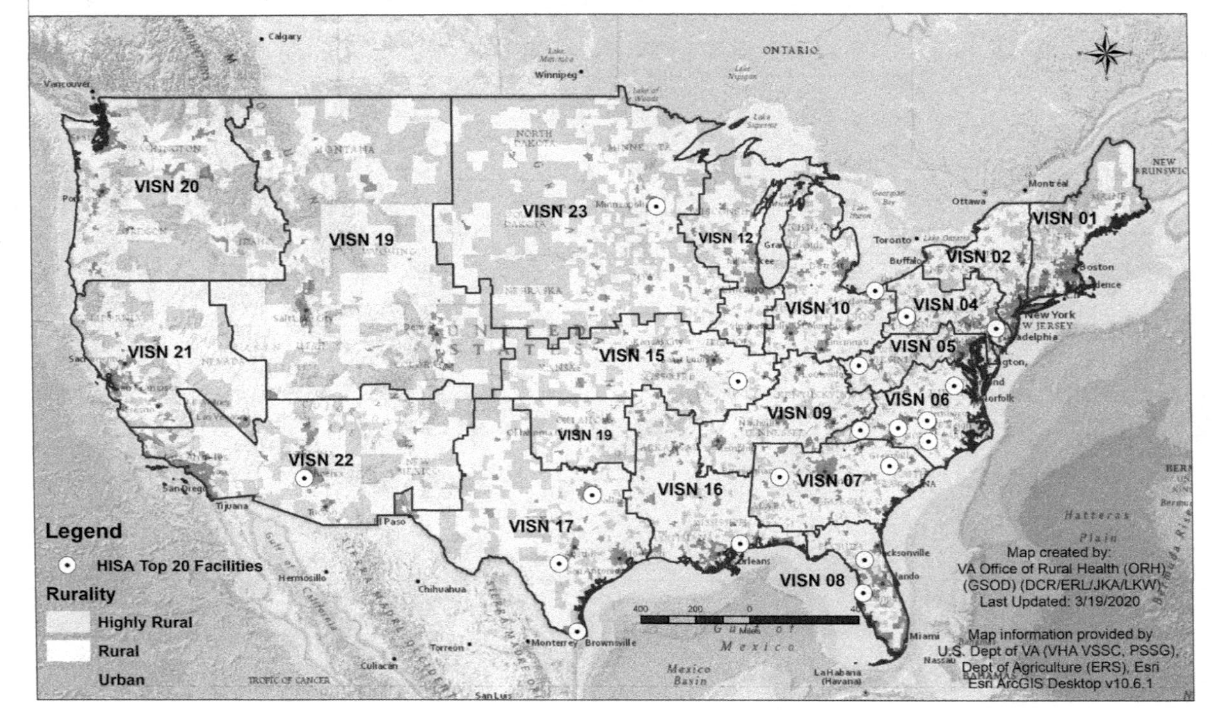

Fig. 14.4 HISA program utilization map shows the 20 Veterans Affairs Medical Centers that have the highest use fiscal year 2011 through 2017.

Table 14.7 Cost data of HISA use by class[a], fiscal year (FY), and region[b].

	Overall (*N* = 37,307)	Class 1 (*n* = 28,613)	Class 2 (*n* = 8694)
Overall cost	**$4761.4 ± $2352.5**	**$5688.2 ± 1858.3**	**$1711.4 ± $503.0**
FY11	*N*=3734 $4510.4 ± $2372.2	*n*=2894 $5348.0 ± $2014.3	*n*=840 $1624.6 ± $538.1
FY12	*N*=4505 $4802.31 ± $2301.4	*n*=3504 $5676.31 ± $1818.15	*n*=1001 $1742.87 ± $478.66
FY13	*N*=4633 $4872.04 ± $2319.6	*n*=3598 $5766.44 ± $1812.75	*n*=1035 $1762.82 ± $460.11
FY14	*N*=4879 $5082.52 ± $2275.4	*n*=3944 $5886.32 ± $1723.16	*n*=935 $1691.97 ± $517.30
FY15	*N*=5591 $4950.77 ± $2328.7	*n*=4352 $5872.18 ± $1749.80	*n*=1239 $1714.30 ± $506.75
FY16	*N*=7639 $4728.89 ± $2354.4	*n*=5745 $5715.08 ± $1836.05	*n*=1894 $1737.53 ± $481.84
FY17	*N*=7929 $4580.61 ± $2377.1	*n*=5992 $5513.11 ± $1957.55	*n*=1937 $1695.99 ± $514.99
VISN			
1	*N*=876 $4590.10 ± $2370.2	*n*=694 $5376.17 ± $2005.39	*n*=182 $1592.66 ± $597.75
2	*N*=1671 $4061.30 ± $2419.2	*n*=1168 $5073.82 ± $2203.03	*n*=503 $1710.17 ± $512.40
4	*N*=2012 $4280.28 ± $2412.2	*n*=1399 $5389.47 ± $2056.61	*n*=613 $1748.87 ± $475.91
5	*N*=1003 $4994.55 ± $2280.8	*n*=783 $5884.35 ± $1734.16	*n*=220 $1827.67 ± $396.74
6	*N*=4405 $5394.21 ± $2094.3	*n*=3632 $6157.01 ± $1402.83	*n*=773 $1810.11 ± $405.80
7	*N*=2564 $4791.93 ± $2340.3	*n*=2102 $5492.99 ± $1971.54	*n*=462 $1602.26 ± $547.12
8	*N*=3380 $4118.06 ± $2391.8	*n*=2178 $5461.05 ± $1914.97	*n*=1202 $1684.58 ± $499.97
9	*N*=1371 $5023.21 ± $2251.6	*n*=1096 $5839.92 ± $1722.43	*n*=275 $1768.22 ± $436.79
10	*N*=2797 $4151.09 ± $2416.9	*n*=1861 $5382.66 ± $2029.33	*n*=936 $1702.42 ± $504.69
12	*N*=1996 $4460.26 ± $2439.2	*n*=1535 $5320.12 ± $2104.31	*n*=461 $1597.19 ± $595.15
15	*N*=1658 $4413.70 ± $2471.9	*n*=1169 $5571.13 ± $1997.29	*n*=489 $1646.74 ± $565.20
16	*N*=1844 $5250.1 ± $2137.2	*n*=1598 $5781.15 ± $1769.44	*n*=246 $1800.4 ± $404.24
17	*N*=4473 $5530.14 ± $2032.9	*n*=3685 $6310.81 ± $1237.30	*n*=788 $1237.3 ± $343.62
19	*N*=728 $3574.95 ± $2464.5	*n*=556 $4225.65 ± $2458.39	*n*=172 $1471.52 ± $616.87

Table 14.7 Continued

	Overall (N = 37,307)	Class 1 (n = 28,613)	Class 2 (n = 8694)
20	N=1626 $5038.81±$2236.3	n=1329 $5771.07±$1770.30	n=297 $1762.12±$461.81
21	N=1146 $5380.90±$2133.1	n=1011 $5866.88±$1767.15	n=135 $1741.48±$465.81
22	N=2340 $5186.52±$2163.0	n=2023 $5729.5±$1787.53	n=317 $1721.43±$503.14
23	N=1475 $3545.86±$2376.3	n=840 $5025.32±$2139.51	n=635 $1588.77±$578.66

[a] Class 1 Veterans have service-connected disability and receive lifetime maximum of $6800 for home modifications (HM); Class 2 Veterans have a nonservice-connected disability and receive a lifetime maximum of $2000 for HM.
[b] See Fig. 14.4 for location of the regions.

Table 14.8 Costliest home modification by class from fiscal year 2011–17.

Modification type	Overall cost	Class 1 cost[a]	Class 2 cost[b]
Bathroom	$5206.6±$2153.6	$6042.6±$1486.8	$1819.2±$410.5
Central air conditioning	$4988.0±$2008.3	$5461.1±$1715.6	$1900.2±$316.5
Kitchen	$4665.4±$2583.1	$5403.9±$2258.4	$1410.8±$753.8
Driveway	$4267.3±$2318.5	$4960.0±$2112.0	$1672.3±$511.3
Plumbing	$4249.3±$2525.7	$5485.3±$2025.6	$1502.6±$655.5
Electrical	$3953.1±$2606.4	$4728.6±$2446.4	$1302.8±$688.4
Walkway	$3795.0±$2202.3	$4618.3±$2101.8	$1773.1±$396.4
Ramp, cement	$3586.0±$2430.2	$4181.9±$2447.4	$1553.9±$533.1
Doorway	$3312.4±$2195.1	$4319.6±$2235.2	$1678.2±$462.7
Permit, inspection	$2193.2±$2335.7	$2820.8±$2544.7	$785.5±$581.9
Railing	$1875.4±$1719.8	$2385.2±$2094.5	$1222.5±$608.3

[a] Class 1 Veterans have service-connected disability and receive lifetime maximum of $6800 for home modifications (HM).
[b] Class 2 Veterans have a nonservice-connected disability and receive a lifetime maximum of $2000 for HM.

HISA users

The lower utilization of HISA by persons between the ages of 45 and 64 (18.0%) and the dismal utilization within the age category of 44 years old (2.1%) may suggest that Veterans from the recent wars, such as Iraq and Afghanistan, are underutilizing the program or that younger Veterans are less impaired (less functionally compromised). However, a large percentage of soldiers in Operation Enduring Freedom (OEF), Operation Iraqi Freedom (OIF), and Operation New Dawn (OND) experience diseases and conditions related to the musculoskeletal system: approximately 62% in 2015, which is one of the primary medical conditions accounting for 31% (Haskell et al., 2012). This calls for a close examination as to why more Veterans of recent wars are not utilizing the HISA program considering the US active engagement in Afghanistan. Nevertheless, musculoskeletal problems, such as low back pain, may be less disabling, than neurological diagnoses such as stroke, Multiple Sclerosis, or Parkinson's.

Special consideration must be given to the increased participation of women in the most recent conflicts and their increased enrollment for VA medical services (Haskell et al., 2012; Meehan, 2006). Women represent 8.1% of VHA enrollees and 10.8% of users, but only make up 4.5% of HISA users. In both groups (Class 1 and Class 2), substantially more men obtained HISA awards than women. Further investigation is needed into the housing accessibility of female Veterans, how they come to know about the HISA benefit, and how they are faring in relation to housing accessibility needs considering their increased participation in combat. Additionally, Haskell et al. (2012) found that the prevalence of musculoskeletal conditions and joint disorders was greatly increased years after deployment for both male and female Veterans using VHA services and that the odds of musculoskeletal conditions were higher for female than male Veterans and this difference increased over time. Considering that as individuals age the impact of their disabilities may exacerbate calling for a loss of independence (Hwang et al., 2019), the need for HMs will only increase for these individuals with musculoskeletal conditions and joint disorders.

The HISA Class 1 (service-connected) users were statistically significantly different from Class 2 (nonservice-connected) users in all sociodemographic and clinical characteristics except for comorbid conditions. Class 2 Veterans were significantly older, more likely to be White and unmarried, and more likely to reside in a rural area. Veterans with nonservice-connected disability (Class 2) are entitled to a lifetime award amount of HISA funding of only $2000 compared with $6800 for Veterans with service-connected disability (Class 1). Due to the limited resources in rural areas and the low funding for Class 2 Veterans, the use of some parts of their dwelling may continue to be inadequate even after implementing a HISA HM. While the quantity of substandard dwellings in the United States has decreased substantially in recent decades, the rate of substandard residences in rural areas is slightly higher than the national rate. In 2013, it was estimated that 63% of rural homes were constructed before 1980 and thus require accessibility alterations, upkeep, and/or repairs (U.S. Department of Housing and Urban Development [HUD], 2017). Furthermore, according to data from the Housing Assistance Council (HAC) reported by the Urban Institute, more than 6.7 million rural households reside in housing units with no or inadequate plumbing or kitchens (Pendall et al., 2016). Low incomes and lesser salaries exacerbate housing cost burdens, thus negatively impacting housing affordability in rural communities (HUD, 2017). This invites discussions of whether the HISA funding is enough to increase those Veterans' independence and ability to engage in activities of daily living such as cooking, bathing, and house cleaning, and furthermore, whether Class 2 Veterans in particular are left to suffer reduced quality of life due to inaccessible dwellings.

According to the American Occupational Therapy Association, clients at all ages with various health conditions such as sensory or movement impairment or cognitive disorders can benefit from HMs (Fagan & Sabata, 2016). HM enables environmental tailoring to individuals' conditions and needs to improve safety and functioning at home (Gitlin, 2015). The American Stroke Association determined HM is a major component of recovery after stroke (American Stroke Association, n.d.), and the Shirley Ryan AbilityLab provides HM guidelines for low vision patients (Shirley Ryan AbilityLab, 2018). This indicates that Veterans with a wide variety of disabilities disorders need support to enhance their safety or activity at home. Further studies should be conducted to investigate how to improve HISA participation by Veterans with various disabling conditions.

HM types and cost

Since the most common diagnoses for both Classes were the same, similar types of HMs were used across the two classifications. The most commonly obtained HISA HMs were related to physical accessibility elements such as bathroom modification, doorway, and railing placement. These components help users' mobility, and navigation at home. Health and mobility issues demand HMs and supportive services for helping residents remain at home. Using the biopsychoecological model as an organizing framework in a qualitative study, Semeah et al. (2019) interviewed a cohort of Veterans with disabilities (Semeah et al., 2019). Veterans, especially those with physical disabilities, acknowledge that when their physical environment presented barriers to mobility and to activities of daily living, financial health was impacted (Ganesh et al., 2019; Lee, Kim, Parrott, Giddings, & Robinson, 2017; Semeah et al., 2019). The lack of appropriate HM leads to early institutionalization and/or out-of-pocket expenses for those residents with disabilities whose environments are not conducive to activities of daily living (Bo'sher et al., 2015; Pynoos et al., 2010; Pynoos & Nishita, 2003; Semeah et al., 2019).

Of note, HMs associated with fine motor activities, such as pinching, grasping, or squeezing, are not noticeably funded by HISA programs. The following are examples of these types of HMs: push button door openers to automate the opening and closing of doors, or easy to use fixtures such as oversized light switches and levers replacing or installed over faucet knobs. These are important since aging is significantly associated with overall motor functioning decline. Fine motor dexterity has been identified as one of the most apparent age-related losses. Future research should focus on why the data does not capture HMs supporting fine motor activities.

We found that the overall mean costs for HMs ($4761.4) did not fluctuate a great deal between 2011 and 2017, or by Class. However, the HMs of the following vary greatly: bathroom, which averaged $5207; kitchen, $4665; and central air conditioning, $4988. These modifications represented some of the most expensive HMs. There was increasing disproportionally larger number of bathroom prescription throughout the timeframe under observation, increasing from 65.6% of the total HMs in FY-2011 to 80.5% in FY-2017. The reason for this is unknown. An examination at the *HISA HM ordering system* in one VAMC reveals that, among other small portions of HMs items, bathroom HM is a forced choice, thus, not providing providers many options to select from. Thus, bathroom HMs may be selected as a default.

The highest mean cost for modifications for Class 1 occurred primarily in Texas (see VISN 17 in Fig. 14.4) at $6310, which had the most HISA users. The lowest overall mean cost was $3545.9 in the northern Midwest (see VISN 23). The lower overall mean cost in the northern Midwest may be due to the larger portion of Class 2 HISA awards (43% of total *N*; $2000 maximum). It is important to note that the Caregivers and Veterans Omnibus Health Services Act of 2010 augmented the lifetime grant limit from $4100 and $6800 for Veterans with service-connected disability (Class 1) and from $1200 to $2000 for Veterans with nonservice-connected disability (Class 2). These increases were much needed since previous award amounts were in effect since 1992 through Public Law 102–405. It is unknown how the price of labor and materials can fluctuate, influence, or impact the sufficiency of the funding or quality of the work performed.

Regional and VAMC differences

This study demonstrates that the utilization of the HISA program is growing in terms of numbers of Veterans accessing this service, especially among those Veterans who are 65 years and older. For example, from FY-2011 to FY-2017, the number of the users nearly doubled ($N=3734$ vs $N=7929$). Furthermore, the number of HISA users is expected to increase as aging Vietnam era Veterans progressively require HMs to permit them to age in place. Although HISA utilization has been noticeably increasing—with Class 2 showing prescription growth, but still lagging behind Class 1 prescriptions segments—considerations must be given to the average percentage changes by FY and by Class (Fig. 14.3), which have fluctuated, and to the overall HISA budgets. In FY-2014, the total national HISA budget of $26,963,262 was about 1.1% of the entire PSAS budget of $2.5 Billion (VHA, Freedom of Information Act Office [FOIAO], 2014). Furthermore, in FY-2017, the total HISA obligation of $44,801,936 was 1.5% of the $3.1 Billion total appropriation for PSAS. From FY-2014 to FY-2017, the total HISA obligation increased by 26.3% (VHA, FOIAO, 2019).

VAMC with high HISA utilization are in urban settings. This does not mean that all HISA users are urban Veterans; for many urban VA hospitals, a large portion of their patient population resides in rural areas and commute for health care services. VHA patients in rural areas often travel long distances to receive services in urban facilities to obtain more technical care than what is accessible near their residence (Cowper-Ripley, Ahern, Litt, & Wilson, 2017).

It was noted earlier that Class 2 users live in rural areas, are older, and less likely to be married. This may mean that Class 2 Veterans are also less likely to receive informal and formal support (e.g., from a spouse or local agency) at the expense of a reduced quality of life. These Veterans might be more likely to be isolated, more vulnerable to depression, and lacking advocacy, all factors which might interfere with completion of the application process and other related processes (e.g., such as the vendorization process). If this is true, unless VAMCs serving these vulnerable Veterans implement strategies to increase knowledge of the program and provide support for completing all the processes involved, it is likely Class 2 Veterans will be underserved. Research to quantify how well VHA and the HISA program are meeting the needs for HM, and then how to bridge the gap if one is detected, is clearly indicated.

The distribution of users by Class fluctuates across regions. This variability by Class also occurs among the top 20 HISA-user VAMCs. These present questions and concerns to be addressed at the organizational level. Whether VAMCs are uniformly implementing HISA (VHA Directive 1173.14) and how HISA Committees in different regions prioritize or make decisions on which classes (Class 1 vs Class 2) of Veterans obtain HISA funds should be examined. For example, VISN 21 reported 88% for Class 1 vs 58% in VISN 23, whereas 76.7% of our total cohort was Class 1.

Discussion

This section discusses some obstacles to achieve higher utilization rates found in our study. They include HISA administrative structure that could block enrollment, policy changes, and the plight of older Veterans with financial challenges. Service locations (VAMC section) also play a role in providing services that could help overcome known obstacles.

HISA administration and major policy changes

The HISA administrative structure, application approval, and vendorization processes are very complex. These may encourage users and providers to seek alternative programs or solutions to avoid the laborious and time-consuming processes (Semeah et al., 2017). Tracking the number of HMs prescribed by key personnel involved in the prescription of HISA (e.g., physicians, OTs, PTs, nurses, social workers) could be part of employee performance assessments and rewards. As *information dissemination agents*, providers play key roles affecting if Veterans know the particulars of the program and what is allowed. For example, through our informal interviews, we gathered that staff knowledge of the HISA program vs what Veterans believe the program is or what kind of HM items are allowed was described as a hindrance to program utilization.

In FY-2015, there was a major policy change with the enactment of 38 CFR Part 17, which impacted HISA. One change that may negatively impact potential HISA users is in the area of contracting with workers/handymen to perform the medically prescribed HMs. Before FY-2015, the VHA vetted bonded contractors who submitted competitive bids for HISA work. After the enactment of the legislation, the responsibility for contracting workers/handymen and weighing affordability of quotes, credentials and quality of contractors, and the work to be performed, all have passed to Veterans. For example, in the case of rentals, a Veteran obtains notarized written permission from a landlord, finds and oversees the work of a contractor, and submits evidence to the VA that HM was done to satisfaction. The VA then releases funding directly to the Veteran, and the Veteran pays the contractor. However, before VA releases funds to the HISA-approved Veteran, the Veterans must become a vendor through a vendorization process. The latter process can take additional months to complete. Many HISA users are 70 years of age or older and some reside in rural areas that have a shortage of skilled construction workers

(Housing Assistance Council [HAC], 2012). Special provisions (e.g., identification and/or partnering with resource agencies) should be implemented to ease challenges (e.g., location disadvantage or low skilled worker pool to select from) and physical decline (e.g., fragility that accompanies old age).

Aging veterans facing additional burdens of financial challenges

Older individuals are likely to experience financial challenges due to decreases in income after retirement; compounding that situation is the inability to maintain property due to a decline in physical and mental health from the aging process and worsening of health conditions. These problems, in turn, create a mismatch between the home environment and the ability to maintain independence, which reduces quality of life, locus of control, and housing satisfaction (Golant, 2008; Lee et al., 2017; Semeah et al., 2016). When faced with these challenges, some forego the performance of the HM until they are unavoidable. As explained above, many rural communities are resource deserts, for example, there is a limited supply of skilled workers to perform HM (HAC, 2012). Based on an analysis of American Community Survey, data analysis conducted by Holder (2017) from 2011 to 2015, a time period also examined by this current study, rural Veterans overall had a greater rate of disability than similar urban Veterans (Holder, 2017). Many HMs will likely be needed to improve safety, independence of daily living, participation in community life, quality of life, and to reduce fall risks (Chase et al., 2012; Cowie et al., 2012; Olij et al., 2019; Semeah et al., 2019) for all Veterans, whether disability is service-connected or not or a factor of the aging process.

Higher complexity VAMCs: Rural HISA patients

Comparatively, some of the VAMCs among the top 20 HISA users are also among the top 10 VAMCs with the highest number of rural patients (FY-2015), for example, the VAMCs in Gainesville, Florida; Minneapolis, Minnesota; Columbia, South Carolina; and Durham, North Carolina (Cowper-Ripley et al., 2017). These facilities also have been designated as complexity level 1a except VAMC-Columbia, South Carolina, which is a 1c facility. This suggests that VAMCs with higher complexity, which are facilities that support substantial education and research activities and that have extensive variation of medical specialization, tend to prescribe HISA to rural Veterans at a greater extent due in part to the larger number of rural patients. Medical centers that emphasize education and research are conducive to and provide a good foundation for what is needed to address the knowledge and utilization gaps in services use. These centers may have resources to develop training modalities and programing by competent researchers, dissemination and implementation agents, and specialists. We focused on the parent medical facilities since the NPPD reports at this level. We encourage future studies to examine potential affiliated sites under each parent medical facility and their role in HISA prescription.

In a recent study by Semeah et al. (2019), Veterans reported that they desired assistance in acquiring HMs and accessible housing, and specifically noted wishing assistance with completing the housing-related application processes, which entailed navigating complex policies, eligibility requirements, and logistics (Semeah et al., 2019). A possible solution is connecting Veterans to their community-based agencies (Hale-Gallardo et al., 2017) such as the following that can provide guidance, resources, referrals, help identifying skilled workers/handymen, or with completing laborious paperwork and processes: Centers for Independent Living, Area Agencies on Aging, Centers for the Visually Impaired, Aging Program at Habitat for Humanity International, and Veteran Serving Organizations (VSOs) such as Paralyzed Veterans of America and the Wounded Warrior Project. This strategy could increase participation of more vulnerable segments of VHA users. This possible solution is supported by the recommendations provided by Veteran respondents in the study by Semeah et al. (2019), which identified VSOs such as Paralyzed Veterans of America as facilitators to obtaining and maintaining housing because VSOs advise Veterans on relevant housing information.

Future studies should focus on the impact of the HISA program on health outcomes and provide information on factors that act as barriers in successfully accessing the program. Further information will aid in developing an educational program for medical providers and inpatient/outpatient care teams to improve patient outcomes, decrease the effect of functional limitations on quality of life, and decrease healthcare costs related to emergent hospitalization due to injuries related to an unsafe home environment.

Sufficiency of lifetime grant limit amounts

An important question remains regarding whether the lifetime grant limit amounts by Class are enough to cover major and more expensive HM expenditures such as kitchens and bathrooms modifications. The JCHS (2019a) found that implementing accessibility enhancements costs more, on average, than remodeling for other motives such as aesthetic improvements. The difference in costs could be as high as 30% for homeowners below the age of 55 and approximately 40% for those 55 and over (JCHS, 2019a). These findings suggest that older adults are more susceptible to paying higher amounts of money for HMs. For example, in 2017, the average expenditures nationwide for renovation of a bathroom was $6362 (Class 1, $6042.6; Class 2, $1819.2); for a kitchen, $12,255 (Class 1, $5403.9; Class 2, $1410.8); for central air conditioning, $5113 (Class 1, 5461.1; Class 2, $1900.2); and for driveways or walkways, $3220 (Class 1, $4960.0; Class 2, $1672.3). HISA does not group driveways and walkways together as reported by the JCHS. According to the current study, the overall cost for a walkway for the HISA cohort was $3795.0 (Class 1, $4618.3; Class 2, $1773.1). In comparing the national averages reported by the JCHS (2019a) to the averages reported in this study on HISA, Class 2 Veterans received what could be considered insufficient funding—ranging from 12% (kitchen) to 51% (driveways or walkways) of the national average reported by JCHS, and Class 1 receiving 44% or exceeding the national average.

Recommendations

In the coming decades, increasing life expectancy, decreasing birth rate, and aging of the baby boom generation will dramatically increase the number of the US population over the age of 65. Those demanding accessible housing should be connected with funding streams and skilled workers/handymen to perform HMs and clinical staff with sufficient knowledge of the program to facilitate the prescriptions and obtainment of HM services. In order to support the need for the current study, we conducted interviews with stakeholders as detailed above, which revealed that a large portion of HISA applications get denied, since HM requests demonstrate a disconnect between what Veterans request to improve function and what clinical staff and program administrators believe is necessary or what the program allows for. VA has HM programs that are offered to Veterans with disabilities, but increasing utilization or awareness by vulnerable (those that live in rural areas and females) and younger Veterans from recent conflicts will be important and will entail strategies that increase knowledge and assistance to both older and younger Veterans across gender.

Class 2 Veterans may be underutilizing and underserved by the HISA program. However, we really don't know the scope of the problem, or whether some populations are underserved. The research design here is only retrospective and depends on what is recorded in different databases. More robust designs using primary data including functional evaluations and home assessments are needed.

The exploration of additional resources and support are necessary to facilitate increased awareness, knowledge, and utilization of HM services among rural healthcare providers (*information dissemination agents*). Additional examination of the extent of marketing efforts of HM services is necessary to address communication needs or challenges of rural vs urban communities. For example, some rural communities are not well-equipped with internet and/or technology services in comparison to urban settings that have this commodity readily available (Smith et al., 2018). Thus, providing marketing collateral electronically may not address lack of access to information on HM services to those with limited or no access to internet.

Conclusion

This chapter provides information about the national usage patterns for the HISA HM program by Veterans from 2011 to 2017. The HISA program has the potential to contribute or influence the VHA's ability to deliver easy-to-access and trusted community-based care by improving the physical environment where Veterans reside and may receive health services. This study offers new knowledge and extensive data on (1) HISA users by demographic and clinical characteristics and how they compare with VA enrollees and VHA users, (2) types of HMs awarded, (3) cost associated with HMs and compares with the average national cost for similar items, and (4) VAMCs and regional patterns of HISA HM prescriptions. Major recommendations include a program analysis of the HISA enrollment procedures to minimize the disparity between what Veterans believe the program to be or allows for and what clinicians believe the program to be or allowed and collaboration with VSOs that may assist Veterans with laborious processes. Information obtained through this work can aid in promoting HM services that may be underutilized by Veterans or underprescribed by providers. Findings also suggest the opportunity to increase overall utilization and utilization by vulnerable populations by using materials to inform/educate segments of Veterans, such as those who are female, racially diverse, reside in rural areas, or have returned from recent conflicts.

References

American Stroke Association. (n.d.). Home modifications. *Recovery after stroke: Home modifications*. Retrieved on June 26, 2019 from https://www.stroke.org/en/life-after-stroke/recovery/home-modifications.

Bo'sher, L., Chan, S., Ellen, I. G., Karfunkel, B., & Liao, H.-L. (2015). *Accessibility of america's housing stock: Analysis of the 2011 american housing survey (AHS)*. Retrieved on June 22, 2019 from https://www.huduser.gov/portal/publications/mdrt/accessibility-america-housingStock.html.

Carney, B. T., West, P., Neily, J., Mills, P. D., & Bagian, J. P. (2010). The effect of facility complexity on perceptions of safety climate in the operating room: Size matters. *American Journal of Medical Quality*, *25*(6), 457–461. https://doi.org/10.1177/1062860610368427.

Chase, C. A., Mann, K., Wasek, S., & Arbesman, M. (2012). Systematic review of the effect of home modification and fall prevention programs on falls and the performance of community-dwelling older adults. *American Journal of Occupational Therapy*, *66*(3), 284–291. https://doi.org/10.5014/ajot.2012.005017.

Chu, Y. T., Ng, Y. Y., & Wu, S. C. (2010). Comparison of different comorbidity measures for use with administrative data in predicting short- and long-term mortality. *BMC Health Services Research*, *10*(140), 1–17. https://doi.org/10.1186/1472-6963-10-140.

Cowie, D., Limousin, P., Peters, A., Hariz, M., & Day, B. L. (2012). Doorway-provoked freezing of gait in Parkinson's disease. *Movement Disorders*, *27*(4), 492–499. https://doi.org/10.1002/mds.23990.

Cowper-Ripley, D. C., Ahern, J. K., Litt, E. R., & Wilson, L. K. (2017). *Rural veterans health care atlas* (2nd ed.). FY-2015. Retrieved July 2, 2019 from https://www.ruralhealth.va.gov/docs/atlas/CHAPTER_00_Introduction.pdf.

Cowper-Ripley, D. C., Kwong, P. L., Vogel, W. B., Kurichi, J. E., Bates, B., & Davenport, C. (2015). How does geographic access affect in-hospital mortality for veterans with acute ischemic stroke? *Medical Care*, *53*(6), 501–509. https://doi.org/10.1097/mlr.0000000000000366.

Cowper-Ripley, D. C., Reker, D. M., Hayes, J., Vogel, B., Wu, S. S., et al. (2009). Geographic access to VHA rehabilitation services for traumatically injured veterans. *Federal Practitioner*, *26*(1), 28–39.

Davenport, R. D., Vaidean, G. D., Jones, C. B., Chandler, A. M., Kessler, L. A., Mion, L. C., et al. (2009). Falls following discharge after an in-hospital fall. *BMC Geriatrics*, *9*, 53. https://doi.org/10.1186/1471-2318-9-53.

Fagan, L. A., & Sabata, D. (2016). *Home modifications and occupational therapy*. Retrieved on June 26, 2019 from https://www.aota.org/~/media/Corporate/Files/AboutOT/Professionals/WhatIsOT/RDP/Facts/HomeMod-Occ-Therapy.pdf.

Ganesh, S. P., Semeah, L. M., Wang, X., Cowper-Ripley, D., Lee, M. J., et al. (2019). Home modification and health services utilization in rural and urban veterans with disabilities. Manuscript submitted for publication.

Gillespie, L. D., Robertson, M. C., Gillespie, W. J., Sherrington, C., Gates, S., Clemson, L. M., et al. (2012). Interventions for preventing falls in older people living in the community. *Cochrane Database of Systematic Reviews, 9.* https://doi.org/10.1002/14651858.CD007146.pub3.

Gitlin, L. (2015). Environmental adaptations for individuals with functional difficulties and their families in the home and community. In I. Söderback (Ed.), *International handbook of occupational therapy interventions* (pp. 165–175). Cham: Springer. https://doi.org/10.1007/978-3-319-08141-0_12.

Golant, S. M. (2008). Commentary: Irrational exuberance for the aging in place of vulnerable low-income older homeowners. *Journal of Aging & Social Policy, 20*(4), 379–397. https://doi.org/10.1080/08959420802131437.

Hale-Gallardo, J., Jia, H., Delisle, T., Levy, C. E., Osorio, V., Smith, J. A., et al. (2017). Enhancing health and independent living for veterans with disabilities by leveraging community-based resources. *Journal of Multidisciplinary Healthcare, 10*, 41–47. https://doi.org/10.2147/jmdh.s118706.

Haskell, S. G., Ning, Y., Krebs, E., Goulet, J., Mattocks, K., Kerns, R., et al. (2012). Prevalence of painful musculoskeletal conditions in female and male veterans in 7 years after return from deployment in Operation Enduring Freedom/Operation Iraqi Freedom. *The Clinical Journal of Pain, 28*(2), 163–167. https://doi.org/10.1097/AJP.0b013e318223d951.

Hoffman, D. W., & Livermore, G. A. (2012). The house next door: A comparison of residences by disability status using new measures in the American Housing Survey. *Cityscape, 14*(1), 5–33.

Holder, K. A. (2017). *Veterans in rural America, 2011–2015: American community survey reports.* Retrieved on June 28, 2019 from https://www.census.gov/content/dam/Census/library/publications/2017/acs/acs-36.pdf.

Housing Assistance Council. (2012). *Taking stock: Rural people, poverty and housing in the 21st century.* Retrieved on July 1, 2019 from http://www.ruralhome.org/storage/documents/ts2010/ts_full_report.pdf.

Hwang, E., Parrott, K., & Brossoie, N. (2019). Research on housing for older adults: 2001 to 2018. *Family and Consumer Sciences Research Journal, 47*(3), 200–2019. https://doi.org/10.1111/fcsr.12295.

Jesus, T. S., Bright, F., Kayes, N., & Cott, C. A. (2016). Person-centred rehabilitation: What exactly does it mean? Protocol for a scoping review with thematic analysis towards framing the concept and practice of person-centred rehabilitation. *BMJ Open, 6*(7), e011959. https://doi.org/10.1136/bmjopen-2016-011959.

Jia, H., Zheng, Y., Reker, D. M., Cowper, D. C., Wu, S. S., et al. (2007). Multiple system utilization and mortality for veterans with stroke. *Stroke, 38*(2), 355–360. https://doi.org/10.1161/01.STR.0000254457.38901.

Joint Center for Housing Studies. (2013). *The U.S. housing stock: Ready for renewal. Retrieved April 11, 2019 from* http://www.jchs.harvard.edu/research/publications/us-housing-stock-ready-renewal.

Joint Center for Housing Studies. (2014). *Housing America's older adults: Meeting the needs of an aging population.* Retrieved April 10, 2019 from https://www.jchs.harvard.edu/research-areas/reports/housing-americas-older-adults—meeting-needs-aging-population.

Joint Center for Housing Studies. (2016). *Projections and implications for housing a growing population: Older households 2015-2035*. Retrieved on June 22, 2019 from https://www.jchs.harvard.edu/research-areas/reports/projections-and-implications-housing-growing-population-older-households-2015.

Joint Center for Housing Studies. (2019a). *Improving America's housing 2019*. Retrieved September 1, 2019 from https://www.jchs.harvard.edu/research-areas/reports/improving-americas-housing-2019.

Joint Center for Housing Studies. (2019b). *The state of the nation's housing 2019*. Retrieved September 30, 2019 from https://www.jchs.harvard.edu/sites/default/files/Harvard_JCHS_State_of_the_Nations_Housing_2019.pdf.

Keall, M. D., Pierse, N., Howden-Chapman, P., Cunningham, C., Cunningham, M., Guria, J., et al. (2015). Home modifications to reduce injuries from falls in the home injury prevention intervention (HIPI) study: A cluster-randomised controlled trial. *The Lancet*, *385*(9964), 231–238. https://doi.org/10.1016/S0140-6736(14)61006-0.

Larson, S., Lakin, C., & Hill, S. (2012). Behavioral outcomes of moving from institutional to community living for people with intellectual and developmental disabilities: U.S. studies from 1977 to 2010. *Research and Practice for Persons with Severe Disabilities*, *37*(4), 235–246. https://doi.org/10.2511/027494813805327287.

Lee, S. J., Kim, D., Parrott, K. R., Giddings, V. L., & Robinson, S. R. (2017). Perceptions on residential environments for urban low-income elderly homeowners aging in place. *Housing and Society*, *44*(1–2), 4–21. https://doi.org/10.1080/08882746.2017.1384992.

Lee, M. J., Romero, S., Hong, I., & Park, H. Y. (2018). Evaluating Korean personal assistance services classification system. *Annals of Rehabilitation Medicine*, *42*(5), 758–766. https://doi.org/10.5535/arm.2018.42.5.758.

Maisel, J. L. (2006). Toward inclusive housing and neighborhood design: A look at visitability. *Community Development*, *37*, 26–34. https://doi.org/10.1080/15575330.2006.10383105.

Maisel, J. L., Smith, E., & Steinfeld, E. (2008). *Increasing home access: Designing for visitability*. Retrieved on June 22, 2019 from http://udeworld.com/documents/visitability/pdfs/IncreasingHomeAccess.pdf.

Mather, M., Jacobsen, L. A., & Pollard, K. M. (2015). *Aging in the United States*. Retrieved June 29, 2019 from https://assets.prb.org/pdf16/aging-us-population-bulletin.pdf.

Meehan, S. (2006). Improving health care for women veterans. *Journal of General Internal Medicine*, *21*(S3), S1–S2. https://doi.org/10.1111/j.1525-1497.2006.00382.x.

O'Connell, R. L., & Lim, L. L. (2000). Utility of the Charlson comorbidity index computed from routinely collected hospital discharge diagnosis codes. *Methods of Information in Medicine*, *39*(1), 7–11. https://doi.org/10.1055/s-0038-1634260.

Olij, B. F., Erasmus, V., Barmentloo, L. M., Burdorf, A., Smilde, D., et al. (2019). Evaluation of implementing a home-based fall prevention program among community-dwelling older adults. *International Journal of Environmental Research and Public Health*, *16*(6), 1–13. https://doi.org/10.3390/ijerph16061079.

Pendall, R., Goodman, L., Zhu, J., & Gold, A. (2016). *The future of rural housing*. Retrieved on September 25, 2019 from https://www.urban.org/sites/default/files/publication/85101/2000972-the-future-of-rural-housing_6.pdf.

Pynoos, J., & Nishita, C. M. (2003). The cost and financing of home modifications in the United States. *Journal of Disability Policy Studies*, *14*(2), 68–73. https://doi.org/10.1177/10442073030140020201.

Pynoos, J., Steinman, B. A., & Nguyen, A. Q. (2010). Environmental assessment and modification as fall-prevention strategies for older adults. *Clinics in Geriatric Medicine*, *26*(4), 633–644. https://doi.org/10.1016/j.cger.2010.07.001.

Semeah, L. M., Ahrentzen, S., Cowper-Ripley, D. C., Santos-Roman, L. M., Beamish, J. O., & Farley, K. (2019). Rental housing needs and barriers from the perspective of veterans with disabilities. *Housing Policy Debate*, *29*(4), 542–558. https://doi.org/10.1080/10511482.2018.1543203.

Semeah, L. M., Ahrentzen, S., Jia, H., Cowper-Ripley, D. C., Levy, C. E., & Mann, W. C. (2017). The home improvements and structural alterations benefits program: Veterans with disabilities and home accessibility. *Journal of Disability Policy Studies*, *28*(1), 43–51. https://doi.org/10.1177/1044207317696275.

Semeah, L. M., Beamish, J. O., Schember, T. O., & Cook, L. H. (2016). The rental housing needs and experiences of veterans with disabilities. *Administration and Society*, *51*(2), 299–324. https://doi.org/10.1177/0095399716666355.

Shirley Ryan AbilityLab. (2018). *Home modification suggestions for low vision*. Retrieved August 23, 2019 from https://www.sralab.org/lifecenter/resources/home-modification-suggestions-low-vision.

Smith, M., Towne, S., Herrera-Venson, A., Cameron, K., Horel, S., et al. (2018). Delivery of fall prevention interventions for at-risk older adults in rural areas: Findings from a national dissemination. *International Journal of Environmental Research and Public Health*, *15*(12), 1–14. https://doi.org/10.3390/ijerph15122798.

U.S. Census Bureau. (2014). *2014 national population projections tables: Table 3 projections of the population by sex and selected age groups for the United States: 2015 to 2060*. Retrieved on June 20, 2019 from https://www.census.gov/data/tables/2014/demo/popproj/2014-summary-tables.html.

U.S. Department of Agriculture. (2018). *Rural america at a alance: 2018 edition*. Retrieved on June 24 from https://www.ers.usda.gov/webdocs/publications/90556/eib-200.pdf?v=5899.2.

U.S. Department of Housing and Urban Development. (2017). *Housing challenges of rural seniors*. Retrieved from June 23, 2019 https://www.huduser.gov/portal/periodicals/em/summer17/index.html.

U.S. Department of Veterans Affairs. (2015). *Uniform geriatrics and extended care services in VA medical centers and clinics*. Retrieved June 26, 2019 from http://www.rstce.pitt.edu/varha/Documents/Part2_03312016Documents/Uniform%20Geriatrics%20and%20Extended%20Care%20Services.pdf.

U.S. Department of Veterans Affairs. (2016). *National center for veterans analysis and statistics: Table 1L, Vetpop2016 living veterans by age group, gender, 2015-2045*. Retrieved August 25, 2019 from https://www.va.gov/vetdata/Veteran_Population.asp.

U.S. Department of Veterans Affairs. (2017). *Gerontology advisory committee*. Retrieved August 25, 2019 from https://www.va.gov/ADVISORY/docs/Minutes%20-%20GGAC%202017%20Apr.pdf.

U.S. Department of Veterans Affairs. (n.d.). *Veterans health administration*. Retrieved September 26, 2019 from https://www.va.gov/health/.

U.S. Department of Veterans Affairs, VHA Freedom of Information Act Office. (2014). *Budgetary obligation for fiscal year 2007 and 2014: Home Improvements and Structural Alterations Benefit Program*. (Freedom of Information Act request No. 15-00943-F). Unpublished raw data.

U.S. Department of Veterans Affairs, VHA Freedom of Information Act Office. (2019). *Budgetary obligation for fiscal year 2015 - 2017: Home Improvements and Structural Alterations Benefit Program*. (Freedom of Information Act request No. VHA-19-11564-F). Unpublished raw data.

U.S. Environmental Protection Agency. (2011). *Supporting sustainable rural communities*. Retrieved on June 23, 2019 from https://www.epa.gov/sites/production/files/documents/2011_11_supporting-sustainable-rural-communities.pdf.

U.S. Government Accountability Office. (2009). *Veterans Affairs: Implementation of temporary residence adaptation grants. (GAO Publication No. 09-637R.T)*. Retrieved from http://www.gao.gov/assets/100/96150.pdf.
U.S. Government Accountability Office. (2010). *Opportunities exist to improve potential recipients' awareness of the temporary residence adaptation grant. (GAO Publication No. 10-786)*. Retrieved on June 20, 2019 from https://www.gao.gov/new.items/d10786.pdf.
VHA Service Support Center. (2019). *Current enrollment cube (vssc.med.va.gov. Note: Behind the VA firewall)*. Retrieved on August 06, 2019. Unpublished raw data.

Supplementary reading recommendations

AARP. (2010). *Home and community preferences of the 45+ population*. Retrieved August 5, 2019 from http://assets.aarp.org/rgcenter/general/home-community-services-10.pdf.
Braubach, M., Jacobs, D. E., & Ormandy, D. (2011). *Environmental burden of disease associated with inadequate housing: Methods for quantifying health impacts of selected housing risks in the WHO European region. (Summary report, World Health Organization)*. Retrieved August 24, 2019 from http://www.euro.who.int/en/publications/abstracts/environmental-burden-of-disease-associated-with-inadequate-housing.-summary-report.
Congressional Research Service. (2013). *VA housing: Guaranteed loans, direct loans, and specially adapted housing grants*. Retrieved May 23, 2019 from https://www.fas.org/sgp/crs/misc/R42504.pdf.
DeJong, G. (1979). Independent living: From social movement to analytic paradigm. *Archives of Physical Medicine and Rehabilitation*, *60*, 435–446.
Kochera, A., Straight, A., & Guterbock, T. (2005, May). *Beyond 50.05: A report to the nation on livable communities: Creating environments for successful aging*. (AARP Research Report). Retrieved July 8, 2019 from http://assets.aarp.org/rgcenter/il/beyond_50_communities.pdf.
NIH National Institute on Aging. (n.d.). *Fall-proofing your home*. Retrieved May 5, 2019 from http://nihseniorhealth.gov/falls/causesandriskfactors/01.html.
Serlin, D. (2015). Constructing autonomy: Smart homes for disabled veterans and the politics of normative citizenship. *Critical Military Studies*, *1*, 38–46. https://doi.org/10.1080/23337486.2015.1005392.
U.S. Department of Veterans Affairs. (2014). Home Improvements and Structural Alterations (HISA) benefits program. Final rule. *Federal Register*, *79*, 71658.
U.S. Department of Veterans Affairs. (2015). *VA home loans*. Retrieved October 12, 2019 from https://www.benefits.va.gov/homeloans/adaptedhousing.asp.
University of Washington, Disabilities, Opportunities, Internetworking, and Technology. (2019, April 30). *What is the difference between accessible, usable, and universal design?* Retrieved April 15, 2019 from http://www.washington.edu/doit/what-difference-between-accessible-usable-and-universal-design.

Definitions

Accessibility There are two definitions presented in this work. An accessible environment can be approached, used, or experienced independently by individuals with diverse disabilities or abilities. Additionally, the word access is used to depict the ability without programmatic, environmental, or geographical barriers to receive, obtain, or seek heath care services.

Baby boomer generation Population segment of those born between the years of 1946 and 1964.

Disability The impact stemming from a physiological condition such as loss of hearing, limb, or mental acuity, which influences ones' ability to provide self-care (e.g., dressing, walking stairs)

Vendorization process The formal processes that a HISA HM awardees complete to have HISA funding disbursement after the prescribed HM is completed.

Visitability A measure of a home that indicates if the location infrastructure (e.g., entryway, interior doorways, bathroom facilities) is easily accessible for people with disabilities.

Infectious ecology: A new dimension in understanding the phenomenon of infection

Dmitry Nikolaenko[a], *Beth Ann Fiedler*[b]
[a]Environmental Epidemiology, Kiev, Ukraine, [b]Independent Research Analyst, Jacksonville, FL, United States

Abstract

All anthropogenic activities coupled with the natural dynamic ecological changes on Earth present new challenges to understanding the old problem of the manifestation of environmental pathogens. This chapter provides an overview of infectious ecology from the perspective of infection as a discrete phenomenon within a natural microorganism community that is disrupted by external forces. Various cases demonstrate the measured impact of the disruption of these forces, such as the sudden death of specific species, and illustrate the natural microbial dominance and their threat to public health by germinating, incubating, and/or transferring deadly infection. The new dimension generates awareness and a sense of urgency in understanding the factors that contribute to the activation of deadly pathogens. This is the basis for recommending preventive measures in the study of infectious ecology (1) by understanding the contribution of human disease and other triggers to harmful microbial disruption, and (2) suppressing activated infectious processes.

Keywords: Ecobiochemistry, Ecology, Environmental science, Emerging diseases, Zoonotic, Geography

Chapter outline

Three Facets of Public Health and Paths to Improvements. https://doi.org/10.1016/B978-0-12-819008-1.00015-8

Mankind's losing battle with nature

Anthropogenic activity and the dynamic and natural ecological changes occurring on planet Earth bring new challenges to understanding the manifestation of environmental problems associated with the outcomes of past actions. The development of science, sometimes considered a contributing factor in environmental decay, plays a unique role in emerging problems that wreak havoc on public health. But, environmental problems are not new to mankind. Instead, infectious ecology exposes that Planet Earth is not a "planet of human domination," but comprised of natural dominants that include microorganisms and insects eliciting practical threats associated with the transfer of infection. This domination becomes all too clear when human activity attempts to transform the natural biospheres by activation of the pathogenic properties of microorganisms. This awareness marks the beginning of their measurement and the threats associated with them that provide the basis for advancement in ecology and emergence of the study of infectious ecology.

The scientific community also fights with two diverse perspectives on infectious ecology just as man continues the losing battle against nature. In generic terms, the infection phenomenon is like a harmless rainbow that briefly appears and then quickly disappears, but later becomes an infectious rainbow with unforeseen massive and lethal consequences to certain biological species. Earth has experienced six massive extinction of species (Briggs & Crowther, 2001; Pimm et al., 2014; Sahney & Benton, 2008) demonstrating the plausibility of certain susceptibility even if not linked or investigated in relation to infectious factors. However, consistent with species susceptibility is the unexpected appearance of infectious disease with the exception of various strains of seasonal influenza. Former Soviet Union mathematicians and epidemiologists in the 1970s easily generated forecast models of this type of anthropocentric epidemic (Baroyan, Rvachev, & Ivannikov, 1977; Ivannikov, 1987). But there is a lack of information on infectious diseases associated with humans and their livestock in scientific, medical, and veterinary communities. This deficit in the collection of data and monitoring prevents forecasting infectious disease without which forecasting remains theoretical and not a practical tool in the important area of disease progression.

The minimal amount of monitoring that occurs is often imperfect, lost, undistributed, or thought to be insignificant to the progression of a disease. Attention to infectious events varies from the death of a number of rodents from the plague that remains relatively ignored to the deafening appearance of the Ebola manifestation in West Africa interpreted as a monumental threat to humanity. The infection seemingly appeared out of nowhere and continues to baffle many in the scientific community. However, some scientists could link the infectious catastrophe to the long-term mass death of apes in Africa precisely from Ebola (Genton et al., 2012, 2017). The "new" phenomenon of nature actually appeared 20 years earlier in the primate population. But the answer to why this is happening remains unclear. So, too, does the fundamental question of the nature of infection.

There are also new questions surrounding the role, if any, of climate change and the accumulation of information on the manifestation of infectious activities that must be addressed in the short-term perspective of *Homo sapiens* in the next few decades. In addition to Ebola, there are new and extraordinary anthrax outbreaks in the Siberian tundra (Bogdanov & Golovatin, 2017; Greenpeace, Russian Office, 2016; ZNAK News Service, 2016). Also, nearly half the population of the saiga antelope died from a bacterial blood infection in Kazakhstan during a period of 2 weeks (National Geographic Staff, 2018; Nicholls, 2016; Saiga Conservation Alliance, 2016). The revelation of the cause of death of the 200,000 saiga only serves to bring about more questions about the nature of catastrophic infectious events. The most significant question is how to prevent some catastrophic events that have short-term implications to the local ecosystem and long-term impact on the biosphere as a whole.

This chapter views the two major competing perspectives of infection and infectious disease and then details the contributions of three prevailing theorists to provide the impetus to break the stalemate. A case for infectious disease as a new paradigm is presented. Then anthrax is presented as an example of a disease event to illuminate key factors that distinguish the perspective of infectious ecology. This extensive background encompasses an overview of work from a number of manuscripts written in languages other than English. Thus, this paper incorporates some of these concepts from Eastern Europe and the United Kingdom into Western literature. A significant objective then is to further embrace the premise of collaborative research on the study of infection across multiple scientific disciplines. Ultimately, we demonstrate the potential application of the new paradigm of Infectious Ecology in the role of prevention instead of merely detection or validation after deadly disruption. The conclusions highlight several key aspects of why infectious ecology must step in beyond the identification of a hazardous event normally conducted through epidemiological examination of a hazardous natural event. First, the need for academic space to accommodate the dynamic objects of research; second, inclusion of the expert community on the discrete activation of pathogens toward understanding of the dynamics of trace elements that contribute to events; third, enhancing the taxonomy of the natural geography of microorganisms to inform research; and fourth, establishing the fundamental alternative of infectious ecology in understanding the phenomenon of infection.

The cold war over scientific approaches to infectious ecology

Science approaches the nature of infection and infectious diseases using two distinct methods. First being the long held perspective that infection is a continual or constant phenomenon; and second, that infectious disease is discrete, that is, the vital activity of certain natural microbial communities is an optional property that remains inert until acted upon. The second is a relative newcomer and the focus of more recent scientific direction (Nikolaenko, 2017a). We relay some significant differences in these approaches including their definition, application, and contribution to infectious ecology. Then we discuss prevailing theorists in the study of the pathogenic properties of microorganisms in view of the perspective of infectious ecology.

Defining key concepts

The continual approach to infectious ecology, based on an epidemiological approach, believes that infection is characterized by (1) a constant presence in nature, (2) identification as a contagion, and (3) is associated with numerous biological species-reservoirs. According to this established train of thought, mass manifestation of the pathogenic properties of microorganisms (e.g., epidemics, epizootics) is only the result of a certain set of circumstances. An estimation of the problem of people and their livestock in this perspective suggests that man cannot control nature, and thus, has no control over the perceived evil forces of nature that may be in play.

Traditionally, the concept of the permanent presence of infection in nature is associated with E.N. Pavlovsky (Pavlovsky, 1964). In canonical form, the concept of landscape epidemiology was outlined in the late 1930s, but developed into a systematic method in the 1960s which is the accepted form today. The concept has not received a consistent theoretical development and perhaps this is not accidental because the fundamental idea associated with the presence of species-reservoirs in nature and the permanent presence of infection inevitably leads to the fact that the fundamental thesis is categorically unconfirmed. In fact, there is substantial information refuting the basic premise that underlines this approach as a convenient hypothesis to create the binding manifestations of the pathogenic properties of microorganisms to the territories or what is known as landscape epidemiology. Unfortunately, the approach did not include the input of geographical science nor would scientists following these conventions consider the important theoretical and methodological innovations in Earth sciences such as ecotones, geosystems, and many other innovations. This concept invalidates the association with infectious ecology.

The inconsistency of the species-reservoirs concept remains prominent in many scientific circles, yet there is tangential evidence of the decline of this position. For example, the Federal State Health Agency, Russian State Anti-Plague Research Institute (2010) provides a history of the working group of I.S. Soldatkin in Microbe during the late 1970s and early 1980s. Microbe remains a leading post-Soviet scientific institution related to the study of dangerous pathogens whose scientists have conducted multiple studies with prominent publications building evidence refuting the concept of species-reservoirs (Rudenchik & Soldatkin, 1994; Rudenchik, Soldatkin, Lubkova, & Lobanov, 1983; Soldatkin & Rudenchik, 1971, 1988). The studies of the Soldatkin group radically changed the research strategy of pathogens in nature by revealing that, while there are quite unstable infectious chains, there is no infinite circulation of contagion. Thus, there is no need to use the concept of species-reservoirs because they act only as the first victim of the natural process associated with the discrete activation of the pathogen. But, these findings were in direct opposition to a large number of Soviet experts who did not repress the work but certainly the work was discontinued.

The new direction of infectious ecology characterizes infection as (1) a discrete phenomenon, (2) not isolated but rather part of a natural microbial community, and (3) maintaining the pathogenic option of microorganisms to respond to external and varying signals. In other words, infections can lead microorganisms to some specific state in which pathogenic properties are manifested. The commonality is that in the process of their manifestation the geochemical characteristics of the natural environment of a microorganism are always very significantly changed because the geochemical characteristics cause stress in the natural microbial communities. The consequence of stress may be the reaction associated with the manifestation of a variety of pathogenic properties of some microorganisms. For example, the role of humans in the process of generating geochemical stress can be extremely significant. In particular, the use of herbicides may be a signal to activate the pathogenic properties of a large number of microorganisms. There is reason to believe that a considerable number of cases of infectious activity are associated precisely with anthropogenic activities in relation to agriculture and not the erroneous association with biological weapons. Nevertheless, old science harkening to the Soviet Union Marxist-Leninist period would continue to place Pavlovsky in esteem for political reasons despite the growing evidence that supports the discrete property of infections.

However in the expert community related to bacteriology, epidemiology, and veterinary medicine, the first point of view is considered natural and allegedly without an alternative; the second position is marginal despite the underlying premise that supports the understanding of infection. Unfortunately, the lack of scientific consideration for the second position, with some prominent exception, creates a divide within the scientific community equal to the state of the Soviet Union-United States Cold War. Only this time the difference divides "comrades."

Unexpected light from the Microbe and the British

Though most of the Soldatkin Group at the Microbe was silent after the initial burst of studies debunking the prominent concept of the permanent presence of infection, Rothschild emerged with an ecological concept (Rothschild, 2011a, 2011b). His manuscripts display one of the first clear positions showing that infection has a discrete nature. While the most interesting results are associated with the manifestations of the plague in vivo, few experts could respond to this concept because of the categorical contradiction to the long-standing belief that infection is a constant phenomenon. Others acknowledged the breakthrough, but remain skeptical of the early position in favor of a more advanced theory of the discrete manifestation of the pathogenic properties of microorganisms that began with a new round of research in 2010s (Nikolaenko, 2016a, 2016b).

Accordingly, there were new opportunities for scientific understanding of the discrete nature of infections. But the dissemination of the publications was difficult to achieve for two main reasons, (1) the limitation of publication in the Russian language, and (2) publication in a limited number of scientific journals. While this can be interpreted as a certain shortcoming of the work, there is no reason to believe that the open publication of new fundamental results in commonly understood English language would have radically changed the outcomes. Nevertheless, consistent presentation of the theoretical concept continues to uncharacteristically befuddle the expert scientific community on the discrete nature of infection with few exceptions.

Enter Mark Purdey, a British scientific outsider who was devoid of the inspiring influence of Marxism-Leninism, who independently developed the concept of the geochemical factor in the cause of activation of pathogens (Bounias & Purdey, 2002; Purdey, 1994, 1996a, 1996b, 1998, 2000, 2001, 2003, 2004a, 2004b, 2004c, 2005a, 2005b, 2006b), while Rothschild was formulating his ecological paper in the midst of Russian naysayers. Within his framework, Purdey expressed the well-founded hypothesis that trace elements play an important role in the occurrence of prion diseases. Prion disease manifests itself in humans and domestic animals and embodies a large group of diseases. The most recognizable prion disease is mad cow or Creuzfeldt-Jacob disease (CDC, 2018; Meggendorfer, 1930), also known by other pseudonyms such as pseudo sclerosis spastic, syndrome of cortico-striospinal degeneration, and transmissible spongiform encephalopathy (TSE). However, some factors, such as the premature death of Purdey and some incorrect verification of his approach made based on the use of a Geographic Information System (GIS) (Chihota, Gravenor, & Baylis, 2004), undoubtedly hampered the already conservative response of the Western scientific community.

Despite the continued separation between the two fundamentally different points of view on the nature of the infection and the reigning status that infection is understood as a permanent phenomenon associated with species-reservoirs, the alternative perspective regarding the discrete nature of the infection associated with geochemical signals continues to find other light in publications, even if those manuscripts are only met with considerable silence (Nikolaenko, 2018g). The reason is that the idea still strongly contradicts the dominant epidemiology paradigm.

Still, the possibility exists for the conduct of decisive experiments in the spirit of the I.S. Soldatkin working group of Microbe, E.V. Rothschild, and Mark Purdey to prove that the manifestation of infection is precisely a discrete nature. There are later studies that have concurred with the seminal work (Nikolaenko, 2010a, 2011b, 2012, 2014a).

Can prevailing theorists break the stalemate?

There are four prevailing theorists in the study of the pathogenic properties of microorganisms, each uniquely evolving their perspectives from the standpoint of medicine (Weinberg), veterinary medicine (Purdey), ecology (Rothschild), or infectious ecology (Nikolaenko). There are obvious differences in terminology. However, there are consistent results among the fundamental characteristics from the point of view of microorganisms and their natural communities including (1) evidence indicating that the phenomenon of infection is discrete, and (2) microorganisms react to a signal from the external environment that is associated with geochemical and/or geophysical characteristics. The fundamental nature of the signal does not change; nonetheless, there may be a variety of manifestations. Thus, this section overviews the contribution of each perspective as the impetus for centralizing research that overcomes the obvious limits of the dominant paradigm that infection is a continual phenomenon in favor of the new findings that consider infectious ecology as discrete.

Eugene D. Weinberg and microorganism activation by metals

Since the early 1960s, Weinberg has contributed more than 100 papers in medicine, demonstrating that metals play an extremely important role in the manifestation of infectious diseases of warm-blooded organisms (Weinberg, 1974, 1978, 2009). The basic idea is that infection is a discrete phenomenon and depends on the environment of the microorganism. In the medical expert community, Weinberg's studies are rated extremely high in unique research methodology, but they are scarcely utilized in theoretical and methodological understanding of the phenomenon of infection. Thus, the field of epidemiology knows very little about these works, despite their prominence in medicine and his focus on warm-blooded organisms. The later focus also limits the generalization to the phenomenon of infection.

Mark Purdey and the epiphany of the properties of infection

The previously listed and legendary publications of Mark Purdey demonstrate his concentration of work from about 1990 to a few years past the Millennium (Environmental Epidemiology, 2019). His work resonates with the British public because of his response to a TSE emergency (McAlister, 2005; World Health Organization (WHO), n.d.) that arose in the United Kingdom at that time. Among Western experts, studies on transmissible infections relating to the use of herbicides are taboo, but this very problem was brought to the fore when herbicides caused this massive TSE outbreak that was predicated by Purdey's thesis relating to infections as a property.

Purdey was initially shunned by the scientific community for revealing that the most common factor in the manifestation of pathogenicity of microorganisms in the soil is a change in the dynamics of the microelement composition (Purdey, 1994). The UK TSE outbreak was a typical example of the introduction of pesticides into the soil, thus demonstrating that this process can be the basis for the massive manifestation of the pathogenic properties of a number of microorganisms. But the link between a seemingly trivial concept and the progression of disease stirred up such animosity in the scientific community because this revelation was the basis for the fundamentally new and scientific direction of infectious ecology.

Purdey's approach radically differs from prevailing scientists such as Rusev (2011, 2013) who sought to understand infectious disease through the lens of the factors that affect the population size of certain species of animals. Specifically, Rusev conducted research on hares (*Lepus europaeus*) in the Ukraine and the prominence of tularemia epizootic. In unique opposition, Purdey's work began to actively interfere with the theoretical tenets of epidemiology and veterinary medicine, suggesting that the massive use of pesticides led to tularemia that, in turn, led to the mass death of hares. From a specific outbreak of morbidity, he turned to the theory of epidemiology and showed that such an event categorically contradicts the environmentalist method. Further, Purdey's earlier work brought attention to the role pesticides played in the activation of microorganisms, such as tularemia, causing a rapid decline in hares in the Ukraine some years later.

The UK TSE outbreak informs Purdey's move away from the traditional epidemiological approach to the nature of infection as a tragic curiosity toward a fundamentally new theoretical explanation of the phenomenon of a number of infectious diseases through his own work and by drawing attention to other work toward the advancement of science. In his last work (Purdey, 2006a), he advocates research conducted on nanogeoscience (NGS) by the working group of Professor Vitaly Vodyanoy (Vodyanoy, 2010, 2012; Vodyanoy, Daniels, et al., 2016; Vodyanoy, Pustovyy, Globa, & Sorokulova, 2015, 2016) and his colleagues. This is an important contribution to the concept of inclusion of scientific innovation across multiple disciplines toward the study of infection as a discrete phenomenon. Unfortunately, scientists attempting to overcome traditional barriers by resurrecting and expanding Purdey's concepts to a new level were met with indifference in 2018 (Bogachevskay & Tymoshenko, 2018; Korolev & Kovalenko, 2018; Nikolaenko, 2018i; Tymoshenko & Korolev, 2018; Tymoshenko, Korolev, Bogachevskay, & Kovalenko, 2018), despite the likely potential impact of this approach in epidemiology, veterinary medicine, and other research.

Eugene V. Rothschild's concepts of microelements and their impact on infection

Rothschild is widely known in the post-Soviet Union era as an expert on the plague (Rothschild, 1973, 1978). He is categorically not known as the creator of the ecological concept of infectious diseases. But for more than a quarter of a century, his research focus on the role of the microelement composition of the soil in the manifestation of the pathogenic properties of microorganisms does contribute to the understanding of infectious ecology. The microelement concept emanates from the traditional understanding of the nature of infections from the Soldatkin Group at Microbe. Despite radical rethinking of the essence of the manifestation of the pathogenicity of microorganisms that points toward a clear understanding to the new direction in which infection is a discrete property, this concept has weak points in (1) terms of methodology and (2) neglecting to incorporate scientific innovations into the premise. Therefore, Rothschild understandably is not linked to the creation of the ecological concept of infectious diseases, nor does the concept stand up to scientific criticism. However, there is a case for the inclusion of the concept of the microelement concept as a stressor in the discrete activation of pathogens in the proposition of infectious ecology.

The S-Theory of Dmitry Nikolaenko

Nikolaenko developed the premise for the S-Theory in 2010, reasoning that microorganisms had a natural geography. In S-Theory, the proposition that infection is a discrete property of the ecological system of microorganisms is defined as an epigeosystem (EGS). Specific changes in the ecological organization of microorganisms can lead to the manifestation of pathogenic properties which is called an infectious disease. The range of such manifestations is large and their specificity is associated with specific microorganisms and their EGS.

The foundation of S-Theory and a priori analysis, including successful microgeography and nanocartography of the pathogenic properties of microorganisms stated (Nikolaenko, 2010a, 2011a, 2011b, 2011c), led to the development of the Advanced Space-Time Algorithm of Site Detection known as the ASTA methodology. The ASTA methodology, while still evolving, is a rational search for places that are most likely to manifest the pathogenic properties of microorganisms in natural conditions and is the methodological standard on the basis of which one can describe the natural geography of microorganisms. Various applications of the S-Theory have led to the introduction of new terminology and taxonomic units of geostationary research for the discrete activation of pathogens in addition to locating hot spots (Nikolaenko, 2018c, 2018h). Testing has revealed Francisella tularensis, *Bacillus anthracis*, and a massive array of data on anthrax in the Russian Empire that has been systematized enabling the important process of reconstruction of their natural geography. The achievements of digital cartography and nanogeoscience are fully taken into account and ASTA testing continues to achieve a high level of accuracy in determining these locations.

While there is a considerable amount of research done by ecologists on manifestations of infectious diseases, such as Rusev previously noted, such works have never claimed revision of the theoretical dogma of epidemiology. Outputs emphasize a situation report in relation to a certain species in which the massive use of pesticides led to an unprecedented activation of the manifestation of the pathogenic properties of Francisella tularensis. In other words, the issue is considered at a general level. On the contrary, modern methods allow a purposeful attempt to study the relationship of the geochemical impact and the activation of the pathogenic properties of microorganisms which is the root cause of the impact on warm-blooded animals (Nikolaenko, 2017b).

Another important contribution from Nikolaenko is the infusion of the concept of the microelement standard of a microorganism and natural community (Nikolaenko, 2010a, 2018b, 2018e, 2018j). We a priori allow the thesis that the microelement standard of a microorganism with pathogenic properties is understood as the comfortable structure of microelements characteristic of a strictly defined ecological organization of the microorganism. Noteworthy is that this does not lead to the occurrence of adaptive changes and does not cause the formation of epigeosystems (EGS) because the virulence of the manifestation of pathogenic properties depends on the microelement stress and the specific imbalance in EGS. This is an important area requiring further analysis which should refine the microelement definition across scientific disciplines using quantitative indicators. The establishment of these indicators will provide a natural set of baseline characteristics from which to measure deviation (microorganism environmental crisis) and identify the ontological basis for discrete activation of pathogens when changes occur during monitoring that may lead to infection.

To summarize, this section highlights four theorists from Eastern Europe and the United Kingdom with a focus on developing a new fundamental point of view on the nature of infections. However, noteworthy is that there are a number of specific approaches to various infectious diseases in which the authors, without accented theoretical reflection, reach the similar conclusion of the discrete nature of infection.

Explaining disease from the standpoint of infectious ecology: Anthrax

The modern study of anthrax began in 2010 on the basis of the cognitive standard of infectious ecology. Several studies accessed a massive array of data in the mid-19th century which was well-preserved revealing three historical manifestations of anthrax on territory of the Russian Empire in the 1870s and later in the Soviet Union (Nikolaenko, 2014b). Though the collection of a vast amount of data began in the early 19th century, the data from 1870 until the start of World War I in 1914 facilitated a detailed reconstruction of the natural geography of *B. anthracis*. The more recent data is amenable to GIS processing because of increased scientific rigor in the collection process that is important to paleoepidemiology.

A clear definition of the three versions of the manifestation of anthrax is important for a detailed study of the natural geography and ecology of *B. anthracis*. The most accurate knowledge of the first version leads to the capacity to correctly determine the place of geostationary research on the geography and ecology of this microorganism.

The research case of anthrax in the Russian Empire demonstrates important components and primary principles of the application of infectious ecology to the study of disease. For example, viewing the natural geography of pathogens sheds light on the progression of agitated microorganisms and the development of infection. Incorporating innovations from modern science (e.g., ASTA, nanocartography, and EGS) was a key element in unraveling data to improve understanding of the transition of microorganisms to a harmful agent. This is a strong indication that the study of the pathogenic properties of microorganisms in their natural environment can formulate a greater understanding of infection and disease. This is in opposition to current methods in which investigations rely on revisiting animal remains transferred to mass graves. Investigation of relocated deceased animals may also obscure causation in addition to the lack of understanding of natural microorganisms, impact of migration patterns, food supply, and other attributes at the point of origin of actual cause of death.

Natural geography and ecology of Bacillus anthracis

The study of the natural geography of microorganisms in infectious ecology is an important scientific task and distinctly separate area of research. The *B. anthracis* pathogen has a vast geography having manifested in many areas of the world, but this does not imply a random distribution. Instead, there are approximately 10 types of soil environments at depths from 100 to 300cm from the surface with identifiable characteristics that are associated with the natural geography and ecology of *B. anthracis*.

The destabilization of the ecology of these soil environments leads to the manifestations of the disease which is commonly referred as "anthrax." Thus, the history of the manifestation of anthrax in the 19th century gives many reasons to assert that the ecosystem of *B. anthracis* was destroyed by anthropogenic activity. A natural consequence was the unprecedented outbreak of infectious activity from this particular period of time. Noteworthy is that the earliest massive manifestation of anthrax in the 1870s coincided with the start of experiments with warm-blooded vaccinations. A sharp decrease in the incidence of anthrax at a later time can be explained apart from successful vaccinations that are out of the scope of this chapter.

Paleoepidemiology Bacillus anthracis

Paleoepidemiology, a systematic reconstruction of the geography and ecology of pathogens, is a new direction in science applied to the study of *B. anthracis* because of the volume of available statistics on the manifestation of anthrax. The paleoepidemiology process focuses on three key problems. First, identification of typology of soil environments with which *B. anthracis* is associated; second, typological analysis of the discrete activation of the pathogen *B. anthracis* focusing on the study of signals that lead to the manifestations of the pathogenic properties of the microorganism; and third, GIS analysis of an array of information related to *B. anthracis*. The paleoepidemiology of *B. anthracis* is extremely important for a fundamental understanding of the nature of the activation of the pathogenic properties of microorganisms.

Data demonstrate that the second appearance of anthrax began during the period between the period of time spanning the first incident and the beginning of WWI. Concurrently, systematic vaccinations against anthrax were initiated, but dramatically altered the natural geography of this pathogen. Thus, the manifestations of the disease could be significantly different depending on vaccinations and their ambiguous effectiveness. Nevertheless, great successes have been achieved in the confrontation of anthrax. Vaccinations decrease the number of diseases, but the geography of their manifestation begins to be clearly confused.

The result of the systematic implementation of the reconstruction of the natural geography of *B. anthracis* was a new scientific direction defined as paleoepidemiology. Paleoepidemiology provides a substantially new vision of the infectious activity of nature by introducing new scientific information on the ecology of microorganisms. Nikolaenko produced 14 volumes of data from 2013 to 2017 on the paleoepidemiology of anthrax in the Russian Empire (https://www.researchgate.net/profile/Dmitry_Nikolaenko2/publications). The ASTA methodology

(EGS/*B. anthracis*) (Nikolaenko, 2014b, 2018f, 2018h) is used to carry out the paleoepidemiological studies conducted in the area of the Novo-Ladoga canal that measures 250–400cm deep. The methodology determines locations with an accuracy of up to 30–40m in terms of the environmental studies of this pathogen.

The location of the study site, under investigation for about 30 years, is near the Novo-Ladoga canal on the territory of the Russian Federation and contributes to the success of the project (FBC, 2004–2019) because the canal, designed and built from 1860 to 1890, has become an unprecedented infectious catastrophe. The undesirable reputation stems from likely the longest and most powerful activation of *B. anthracis* ever recorded in the world over the entire observation period. The large-scale studies at the site have also been the impetus for growth of Russian imperial epidemiology and veterinary medicine.

Finally, we considered if the reconstruction of the events during canal construction affected the natural geography of *B. anthracis*. While the process does invade the territory, the pathogenic ecosystem does not die and is not greatly altered. The ASTA process takes soil samples from different depths so that the damage to the pathogen's natural ecology is minimal. This is confirmed on the basis of a comparison of samples taken from the same sites over a 2-year period. Therefore, we can accurately conclude that *B. anthracis* are ruined by drastic changes in geochemical characteristics of an ecological nature resulting in or the surprising outbreak of infectious activity.

Anthrax vaccine development and novel metal science

Evaluation of the impact of vaccinations on the manifestation of the pathogenic properties of *B. anthracis* is done only in terms of the study of the geography and ecology of the pathogen. Until recently, the geography of the pathogen *B. anthracis* was determined only by the manifestation of anthrax primarily in humans and livestock because warm-blooded creatures acted as sensors or indicators of the natural process. Before the advent of vaccinations, this gave relatively correct data regarding the geography of the pathogen. The massive introduction of vaccinations in humans and domestic animals has led to the fact that the vulnerability of warm-blooded animals to the manifestation of anthrax has decreased significantly due to effective vaccines. However, the application of the ASTA methodology is productive without waiting for the manifestation of disease by studying the natural geography and ecology of *B. anthracis*.

Meanwhile, preventive vaccinations saw indicators of anthrax drop dramatically to such an extent that *B. anthracis* can be considered an endangered species. Extinction is common to nature and there is no indication to believe that this condition would not apply to microorganisms. Still, the third and random manifestation of anthrax is a period without end, beginning about the time of WWI that is generally linked to land development (Nikolaenko, 2018a).

As a rule, environmental studies associated with complex relationships and long-term effects are recorded upon emergence of a complex problem. The infectious ecology probably has a similar story. The idea of an infectious ecology itself also implies a mandatory line of research related to preventive measures. But, prevention implemented from the standpoint of infectious ecology does not require consideration only through the prism of practical prevention, such as focusing on antibiotics or other pharmaceutical development, in response to microorganism activation (Nikolaenko, 2017a).

In an example of experimental work using $N=77$ relatively plague-resistant noon gerbils (*Meriones meridianus*) reported by Rothschild (2011b), variants of mineral additives were received daily with food for 16–30 days: (a) 2 mg of copper; (b) 2.1 mg of iron + 2 mg of copper; (c) 1.8 mg of iron + 4 mg of manganese (approximate animal weight is 100 g). Iron and copper were used in the form of sulfates. Then the experimental ($n=44$) and control ($n=33$) groups were infected with a pathogenic strain of the plague microbe in a moderate dose (Ld50). Within 20 days after infection, 2/44 animals in the experimental group died, but the infectious agent was not found in the sowing of organs of dead animals. The remaining 42 animals were healthy. Two-thirds (70%) of the animals in the control group of $n=33$ animals that did not receive mineral additives or cobalt and nickel additives died during the same period. After the experiment was completed, antibodies to the plague microbe were detected in the blood of all experimental animals. (Experimental notation: the protective effect was manifested only in anomalous conditions: with a shortage of all vital chemical elements in the natural environment (which is assumed to be indirect evidence).) Nevertheless, viewing how trace elements may safely overcome the pathogenic properties of Yersinia pestis, otherwise known as the plague, is an important principle in preventing pathogenic development having practical application to the prevention of warm-blooded disease (Mezentsev, Rothschild, Medzykhovsky, & Grazdanov, 2000; Nikolaenko, 2017a; Rothschild, 2011b) without the complex and often unknown long-term effects of pharmaceutical interventions.

The case for infectious ecology and a new scientific paradigm

Infectious Ecology is the direction of scientific research on the interaction of living organisms, their communities, their environment, and the infectious reaction to systemic changes associated with certain microorganisms. In this section, we view the basis of infectious ecology taking into consideration the known conceptual differences previously discussed. Then, we view the direction of scientific research and the important influence of multiple scientific communities in two schemes depicting these relationships.

At the base of infectious ecology is a new paradigm resting on two primary requirements.

- First, the introduction of a clear distinction between the concepts of infection, infectious process, and infectious disease which will enable the registration of a much larger amount of information on the infectious reality; and
- second, microorganisms have their own standard of ecological organization that does not adhere to the human ecology frame of reference. A new system of taxonomy will emerge to describe the ecological organization of microorganisms.

Together, these critical items provide the basis for collecting, monitoring, and analyzing data for the periodic occurrence of imbalance in trace elements when the microorganism is stressed. From these changes, the foundation of infectious ecology as a new scientific paradigm can build a distinct knowledge platform.

Infectious ecology as a direction of scientific research is the:

- study of the interaction of and connection to microorganisms with their environment because derivation of information during this state is critical in the explanation of the events of an infectious reality;
- study of the dynamics and natural geography of microbial communities, not single microorganisms, representing a systemic manifestation of the natural environment of microorganisms because the taxonomy of natural geography of microbial communities is the basis for understanding the manifestation of the phenomenon of pathogenicity in nature (Nikolaenko, 2010b);
- study of adaptation of microorganisms to the environment because adaptation is a resultant state of a complex system of changes in the relationship of a microorganism, the microbial community, and their environment in the soil. The concept of adaptation in relation to natural microbial communities is an essential task in the study of infectious ecology.

The study of infectious ecology requires the study of manifestations of complex relationships of microorganisms and their environment. Thus, there are a large number of points of contact with existing scientific disciplines. The following demonstrates two high-level schemes within the sciences in which infectious ecology can develop (Fig. 15.1).

Scheme 1 represents a starting point in the process of investigating specific cases of pathogenicity in the natural environment recognizing the value and contribution of other scientific communities. This issue should be treated purely pragmatically. However, the utilization of the full array of interdisciplinary connections can maximize the effective study of the pathogenicity of microorganisms. Of course, some scientific communities are more amenable to the collaborative process of infectious diseases than others (Nikolaenko, 2018d).

Specifically, the modern science research method of Nanogeoscience (NGS) emerging around the Millennium offers tremendous support to the development of infectious ecology. In contrast, epidemiologists continue to ignore the potential in the study of the manifestation of pathogenic properties of microorganisms by continuing to focus on locating species-reservoirs and the habitual neutralization of the consequences of this outbreak as recently as the investigation of the unprecedented Ebola epidemic (CDC, n.d.). Meanwhile, NGS made an important contribution to infectious ecology by providing a methodology to explore the natural environment of the microorganism. Prior to this new and fundamental thesis, the ability to systematically and thoroughly investigate the natural geography of microorganisms was impossible. For example, the manifestation of Anthrax in the territory of the Russian Empire (Nikolaenko, 2014b, 2018a, 2018f) is a study of the natural geography of *B. anthracis* and an example of the capacity to only conduct fragmentary analysis on some pathogens.

But the natural geography of microorganisms, as a new scientific direction, can be developed beginning with the influx of new terminology and geomonitoring scale especially created for infectious ecology. For example, the EGS dictates that work

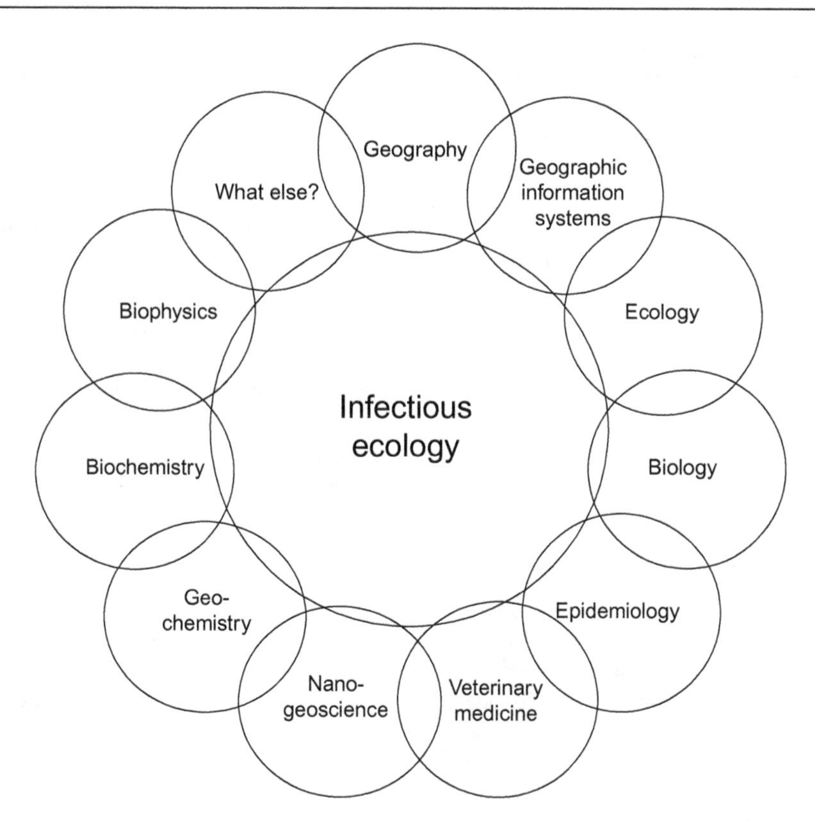

Fig. 15.1 Scheme 1: The place of infectious ecology in the system of sciences.
Credit: Created by Dmitry Nikolaenko.

begins with "one square meter" and gradually reduces scale to "one cubic centimeter" at a certain predetermined depth. This is a new standard of geomonitoring for manifestations of pathogenicity in nature which will allow scientists to properly comprehend the natural environment of microorganisms. (On the other hand, correlation of manifestations of anthrax to areas of 20–40 km^2 in landscape epidemiology is considered an acceptable scientific result or research scale. But, the epidemiological approach differs in other ways besides scale. In their method, vague binding manifestations of anthrax over a large area are taken into consideration, whereas the natural geography of microorganisms in the newest method targets a smaller scale and also studies ecotones—contact zones of various landscapes.)

The specificity of the object of the study of infectious ecology can be defined as a set of a number of subject (attribute) components designated here as Scheme 2 (Table 15.1). The relationship between the designated attribute components of the object of infectious ecology is far from unequivocal due to a large number of variations that experts must prepare to successfully navigate. Nature plays "infectious jazz" with quite flexible patterns, instead of the clarity of clearly written classical musical score which is always played the same way and at the same time. Variability is a reasonable expectation given the complexity of connections.

Table 15.1 Scheme 2: Attribute components of the object of study in infectious ecology.

Discrete activation of pathogens and infectious processes	1. Microorganism 2. Microbial community 3. Microbial environment 4. Adaptation of microorganisms and communities to change 5. Infection as new and optional information of nature 6. Infection process as an optional scenario of adaptation of microbial communities 7. Infectious disease of individuals of certain species as an environmental consequence

The advantages of infectious ecology are as follows:

- represents an opportunity to move away from anthropocentrism experts in epidemiology and veterinary medicine to expand their fixation on man and livestock toward a vast array of natural information on infections;
- a radical advancement in understanding the ecology of microorganisms by collecting data through geostationary studies associated with precisely defined types of natural microbial communities. The quality of this empirical information becomes unprecedented with the inception of a clear distinction between the concepts of infection, infectious process, and infectious disease in conjunction with microorganism taxonomy.

The success of the transition to an infectious ecology approach to infectious diseases is predicated on first understanding the microelement dynamics of the soil and this influence the manifestation of the pathogenic properties of microorganisms. This previously prohibitive path blocked by a vast number of restrictions is now uniquely positioned to achieve these milestones in the development of infectious ecology and understanding disease progression.

Discussion: Prevention and infectious ecology

Distinguishing the familiar medical concept of prevention or preventative medicine in relation to infectious disease can clarify the marked differences in the epidemiological approach vs the infectious ecology approach to understanding disease. Two factors separate these methodological perspectives that lead to identifying the cause of disease: (1) timing, and (2) perspective on prevention. The epidemiological approach is generally reactive when a pandemic or infectious disease is suspected, whereas the infectious ecology methodology is proactive/preventive targeting the identification of microorganisms and their natural environment to prevent disturbances that can lead to activation, and thus, the development of harmful pathogens.

While the epidemiological approach has led to the development of powerful vaccines that have been helpful in some ways to the general population, vaccines and antibiotics have not lived up to their perception as the panacea for all ills. These medical solutions have over time produced numerous and painful problems resulting from their active use because microorganisms and their natural communities have proven to be adaptive. The Russian Anthrax field investigation and antibiotic resistance point to the development of some vaccines that can be further damaging to the microorganism and compel changes in the natural geography of the microorganism that becomes increasingly difficult to abate. Further, inconsistencies within the medical community and general public regarding the safety and efficacy of vaccines and pharmaceutical solutions to infectious disease make a compelling case for supplemental and/or alternative approaches that address infectious disease.

Other facts gleaned from experimental research on warm-blooded animals, such as the noon gerbils noted previously, help to better explain why some rodent territories have not been affected by the plague while diseases have been recorded in the immediate vicinity. The composition of the chemical elements in blood changes at the peak of the infectious process when the content of iron and zinc is reduced by 1.5–3 times, but the concentration of manganese, copper, tungsten, molybdenum antagonists increases by 3–10 times. Such changes can be assessed as a natural protective reaction of a warm-blooded organism against infection. In this way, the reduction of molybdenum in the internal environment of animals provides such protection. Therefore, an informed conclusion is that the effect of protection against infection is likely due to the deficiency of molybdenum in animals. Manganese, iron, and copper are molybdenum antagonists. These metals in high doses, as well as sulfur in the composition of sulfates, could contribute to the removal of molybdenum from a warm-blooded organism. At the same time, under conditions of low content in the feed, the concentration of molybdenum in the organism of animals could decrease below the threshold level necessary for the normal metabolism of pathogenic bacteria, in particular, for the synthesis of pathogenicity factors.

The fact that the animals themselves probably did not suffer from a deficiency of molybdenum, and, in any case, remained alive for a long time, finds a plausible explanation. The need for molybdenum is significantly higher in living beings that are in the lower stages of evolution, such as bacteria and fungi, compared to mammals that are well-established. This, in principle, brings forth the option to solve the problem of the formation in a warm-blooded organism of a low concentration of molybdenum, safe for humans and animals, but blocking the manifestation of pathogenicity by microorganisms (Nikolaenko, 2017a).

This experimental work, performed by colleagues in frankly unfavorable conditions, was done a long time ago and seemingly lost to science. The reason could be attributed to the confrontation between the worlds of capitalism and socialism, but, instead, the plausible cause is attributed to the severe domination of the concept of species-reservoirs that heretofore has hindered alternative approaches to address infectious disease. However, objective recreation of these early experiments, not limited to laboratory tests and incorporating research advancements from other disciplines, should confirm the rigor of the science and reintroduce these novel solutions to the medical community as a viable alternative to typical preventative interventions.

Recommendations

Of great importance is the fundamental understanding that infectious ecology brings new concepts, definitions, and taxonomy to the scientific community. This position does not necessarily build on existing epidemiology and veterinary medicine because of operational standards based on new concepts. Instead, there is a clear expectation that the theoretical and methodological standard for formulating questions must change, too. The following summarizes key elements in the development of infectious ecology as a new scientific paradigm.

Dynamic object of research

1. The phenomenon of infection is variable in nature and there isn't any reason to believe that this natural object will change in the visible future.
2. The changing nature of the phenomenon of infection leads to the fact that the patterns of infectious activity, which appear in certain time intervals, can dramatically change their characteristics or disappear altogether. While why this happens is not presently clear because there appear to be no apparent laws in this object, there is a dominance of short-term patterns (rules) having clearly defined environmental characteristics of the manifestation.
3. The changing nature of the phenomenon of infection makes it extremely difficult to study because of the unique information that is undefined from the standpoint of the dominant paradigm. This object requires a radical change in perspective. The abundance of information that is generated as a result of the manifestation of the changeable nature of the infection can only be correctly understood within the framework of infectious ecology because of the focus on the discrete nature of the infection.
4. Promotion in the study of the phenomenon of infection is not only of a scientific nature. An important role in this process is played by the change of the infectious reality itself with new manifestations demonstrating that the object of study is actively changing. The study of infectious novelties is essential for advancing scientific knowledge that can be achieved only on the basis of new scientific points of view.

Dynamic object and expert community

5. The phenomenon of infection, as an object of study, can be viewed from significantly different points of view (Scheme 1). Consequently, study of this phenomenon can be divided into an infinitely large number of research subjects. In strict accordance with the specifics of the subject of the study, as well as the paradigm on the basis of which it is revealed, the results will be specific. However, this new path anticipates correcting the previous loss of information that was (1) characteristic of the stagnant cognitive process used in the epidemiology and veterinary medicine approach, and (2) may consider these previously dominant methodological approaches incorrect (nonscientific).
6. There are scientific innovations, such as NGS, that can be extremely useful in the study of the phenomenon of infection. But NGS in epidemiology and veterinary medicine is extremely

limited since introduced about 15 years ago. Now, the infectious ecology approach brings forth the notion of a rapid advance in understanding the nature of the infection precisely on the basis of the NGS by changing the fundamental expert position.

7. The contribution of other sciences, such as bacteriology, does not necessarily have to be associated only with epidemiology since this link limits the collection of data and is therefore ineffective. Much more interesting and effective may be the connection of bacteriology with infectious ecology because of the focus of the complex relationships arising in the process of discrete activation of pathogens. Scientific collaboration on this path can output a detailed study of the environment containing key information from bacteriology.
8. Infectious ecology is the direction of research and should not be associated only with a particular theory or methodology. There may be significantly different versions of the realization of the potential of infectious ecology from different scientific communities. The object of study is too complex and multifaceted to continue to beat one path when nature plays infectious jazz and its theoretical understanding cannot be developed from only one expert position.

Trace element dynamics of the soil and the discrete activation of pathogens

9. In the process of discrete activation of pathogenic properties of microorganisms, microelement gradients in the natural environment of microorganisms play an extremely important role. The microelement dynamic of the natural environment of microorganisms is the main factor leading to the manifestations of the pathogenic properties of microorganisms. This can be both natural and anthropogenic gradients.
10. The cases of discrete activation of pathogens associated with natural microelement gradients are probably quite widespread. In favor of this conclusion, there are numerous scientifically registered cases of a single infection of animals or the infection of a small number of animals. Such cases have been repeatedly described for tularemia and plague.
11. The natural causes of the formation of minor microelement characteristics can be very different. An example would be the case of black alder roots because they are characterized by a natural infection with Frankia Alni causing changes in trace elements in a small area. These changes formed a special ecotone resulting in the increase of risks of tularemia infection of rodents, but such cases often give infection to a single individual or a small group of animals.
12. The similar nature of discrete activation is associated with anthropogenic changes in which the scale of the infectious effects will be significantly different. Depending on the specificity of the microorganism with pathogenic properties and the anthropogenic impact, manifestations can be superior in terms of order of magnitude in comparison to what takes place in the natural environment.
13. Developing a conceptual and terminological system is necessary to study the microelement dynamics of the natural environment of microorganisms whose absence leads to a critical loss of information. A specific example is the concept of ecotones that enable scientists to correctly describe contact zones. The idea of an ecotone should be developed specifically for the case of natural microbial communities because they are the basis for the occurrence of uncharacteristic changes in the microelement dynamics of the natural environment of microorganisms.
14. The specific version of the manifestation of the virulence of pathogenic microorganisms can be interpreted as a function of the microelement gradient of the ecotone because this

causes an adaptive response of the microorganism to the ecological system. While virulence is a variable characteristic in nature, addressing the environmental mechanism of appearance instead of the uniqueness of the characteristics of virulence is important.

15. The process of discrete activation of pathogenic properties of microorganisms can be studied in detail. This requires geostationary monitoring and empirical information recording systems. There is a methodology that allows geostationary research at a fundamentally new level that is developed on the basis of S-Theory. Further development of S-Theory is expected from the research path presented in this paper.

Conclusion

In this chapter, we introduce a new theoretical and methodological basis for the study of infectious disease independent of the need to generate vaccines that have proven capacities to adapt and cause further harm. Infectious ecology represents a new standard of prevention that focuses on the prevention of the release of pathogens based on the capacity to locate likely sources of dangerous microorganisms. In this way, infectious ecology acts as a preventive measure for the release of contaminants by providing informed options to handle areas of potential contaminants with extreme caution or simply opt not to disturb areas with predictable hazards. Further, understanding the natural geography of microorganisms in their natural community provides the basis for finding alternative measures to reduce the risk to population health by providing alternative to pharmaceutical options and the preventive role of metals in the chemical composition of warm-blooded beings. The discussion and recommendations highlight several key aspects of why and how the research methodology of infectious ecology must step in to (1) expand knowledge of microorganisms, and (2) incorporate preventive strategies that suppress the activation of infectious disease through this fundamental understanding.

References

Baroyan, O. V., Rvachev, L. A., & Ivannikov, Y. G. (1977). *Modeling and forecasting influenza epidemics for the territory of the USSR/Academy of Medical Sciences of the USSR, Institute of Epidemiology and Microbiology*. 546 pp Moscow: N.F. Gamalei (in Russian).

Bogachevskay, E., & Tymoshenko, A. (2018). Dynamics of infectious reality and standards of expert reflection (the case of Mark Purdey). *Environmental Epidemiology, 12*(4), 100–110. (in Russian). 10.13140/RG.2.2.27538.96961.

Bogdanov, V. D., & Golovatin, M. G. (2017). Anthrax on Yamal: An ecological view of traditional reindeer herding. *Ecology, 2*, 1–6. https://doi.org/10.7868/S0367059717020056.

Bounias, M., & Purdey, M. (2002). Transmissible spongiform encephalopathies: A family of etiologically complex diseases—A review. *The Science of the Total Environment, 297*(I), 1–19. https://doi.org/10.1016/S0048-9697(02)00140-7.

Briggs, D., & Crowther, P. R. (Eds.), (2001). *Palaeobiology II* (1st ed.): Wiley-Blackwell.

Chihota, C. M., Gravenor, M. B., & Baylis, M. (2004). Investigation of trace elements in soil as risk factors in the epidemiology of scrapie. *The Veterinary Record, 154*(26), 809–813.

Environmental Epidemiology (2019). *Personalities: Mark Purdey (collected works)*. Retrieved July 6, 2019 from http://www.e-epidemiology.com/Mark-Purdey/.

FBC (2004–2019). *Administration: Volgo-Balt*. Retrieved April 16, 2019 from https://www.volgo-balt.ru/page/180.

Federal State Health Agency (2010). Russian State Anti-Plague Research Institute. *Microbe*, Retrieved April 11, 2019 from http://microbe.ru (in Russian).

Genton, C., Cristescu, R., Gatti, S., Levréro, F., Bigot, E., et al. (2012). Recovery potential of a western lowland gorilla population following a major ebola outbreak: Results from a ten year study. *PLoS One*, *7*(5), e37106. https://doi.org/10.1371/journal.pone.0037106.

Genton, C., Cristescu, R., Gatti, S., Levréro, F., Bigot, E., et al. (2017). Using demographic characteristics of populations to detect spatial fragmentation following suspected ebola outbreaks in great apes. *American Journal of Physical Anthropology*, *164*(1), 3–10. https://doi.org/10.1002/ajpa.23275.

Greenpeace, Russian Office (2016, August 3). *In the dissemination of anthrax on Yamal is guilty of climate change*. Retrieved April 11, 2018 from http://www.greenpeace.org/russia/ru/news/2016/03-08-2016_sibirskaya_yazva_climat/.

Ivannikov, Y. G. (1987). Results and perspectives of the mathematical forecasting of influenza epidemic. In: Y. G. Ivannikov (Ed.) *Vol. 2. Proceedings of the World Congress of Bernoulli Society* (pp. 543–546). Netherlands: VNU Science Press.

Korolev, A., & Kovalenko, M. (2018). Let's say Mark Purdey is still alive. *Environmental Epidemiology*, *12*(4), 111–128. (in Russian) 10.13140/RG.2.2.17380.88966.

McAlister, V. (2005). Sacred disease of our times: Failure of the infectious disease model of spongiform encephalopathy. *Clinical and Investigative Medicine*, *28*(3), 101–104. 16021982.

Meggendorfer, F. (1930). Klinische und genealogische Beobachtungen bei einem Fall von spastischer Pseudokosklerose Jakobs. *Zeitschrift für die Gesamte Neurologie und Psychiatrie*, *128*, 337–341. https://doi.org/10.1007/bf02864269 (in German).

Mezentsev, V. M., Rothschild, E. V., Medzykhovsky, G. A., & Grazdanov, A. K. (2000). The influence of trace elements on the infectious process in the plague in the experiment. *Journal of Microbiology, Epidemiology and Immunobiology*, *1*, 41–45 (in Russian).

National Geographic Staff (2018, January 29). *200,000 endangered antelope died. Now we know why*. Retrieved April 11, 2019 from https://news.nationalgeographic.com/2018/01/saiga-antelope-killed-bacteria-2015-mass-die-off-central-asia-spd/.

Nicholls, H. (2016, April 14). *Mass deaths of saiga antelope in Kazakhstan caused by bacteria*. Retrieved April 11, 2019 from https://www.theguardian.com/science/animal-magic/2016/apr/14/mass-death-saiga-antelope-kazakhstan-bacterial-infection.

Nikolaenko, D. (2010a). Geography of invisible world and its GIS-explanation, report. *Environmental Epidemiology*, *4*(2), 165–178. https://doi.org/10.13140/RG.2.2.20706.02243.

Nikolaenko, D. (2010b). Geoinformation modeling of the manifestation pathogenic properties of microorganisms and the hypothesis of infection as a property of EGS. *Environmental Epidemiology*, *4*(1), 102–105 (in Russian).

Nikolaenko, D. (2011a). Nanocartography and its potentials regarding infectious disease research. *Meta-Carto-Semiotics*, *4*, 44–46.

Nikolaenko, D. (2011b). Polygon for epidemiological research. Part 1. *Environmental Epidemiology*, *5*(3), 380–412 (in Russian).

Nikolaenko, D. (2011c). Theory of the infectious "sandwich" *Environmental Epidemiology*, *5* (5), 872–908 (in Russian).

Nikolaenko, D. (2012). The theory of an infectious sandwich, Version 1.2. *Environmental Epidemiology*, *2*, 193–246 (in Russian).

Nikolaenko, D. (2014a). Environmental epidemiology: New standard quantity & quality of empirical information, Part 2. *Environmental Epidemiology*, *8*(1), 136–148.

Nikolaenko, D. (2014b). Introduction to the information system "Anthrax in the Russian Empire" (version 1.0). *Environmental Epidemiology*, *8*(1), 51–112.

Nikolaenko, D. (2016a). Preface to the article by E.V. Rothschild "Ecological concept in the science of infections," 2nd Ed. *Environmental Epidemiology*, *10*(1), 14–18 (in Russian).

Nikolaenko, D. (2016b). Professor V.L. Adamovich and new paradigm of Environmental Epidemiology. *Environmental Epidemiology*, *10*(5–6), 3–190 (in Russian).

Nikolaenko, D. (2017a). Ecological dynamics of discrete activation pathogens: The cases of *Yersinia pestis*, *Francisella tularensis*, Filoviridae, report. *Environmental Epidemiology*, *11*(3), 84–126 (in Russian).

Nikolaenko, D. (2017b). *Infectious ecology as a new direction of scientific research. Ecology, environmental protection and balanced environmental management: Education – science – production. Kharkiv: KKNU, 161–163.* (in Russian).

Nikolaenko, D. (2018a). Anthrax in the Russian Empire: Eyes wide shut. *Environmental Epidemiology*, *12*(2–3), 226–314. (in Russian) 10.13140/RG.2.2.18456.90887.

Nikolaenko, D. (2018b). Bacteriology and infectious ecology: New perspectives of research of the pathogenic properties. *Bacteriology*, *3*(3), 68–77. (in Russian) 10.20953/2500-1027-2018-3-68-77.

Nikolaenko, D. (2018c). Dimensions of cartosemantics and use GIS in infectious ecology. *Environmental Epidemiology*, *12*(2-3), 137–141. (in Russian) 10.13140/RG.2.2.13567.51369.

Nikolaenko, D. (2018d). Infectious ecology research object and its differences from epidemiology. *Environmental Epidemiology*, *12*(4), 37–51. (in Russian) 10.13140/RG.2.2.22957.18408.

Nikolaenko, D. (2018e). Multidimensional system of geocoding information on infectious processes. *Environmental Epidemiology*, *12*(2–3), 85–111. (in Russian) 10.13140/RG.2.2.17321.08806.

Nikolaenko, D. (2018f). Paleoepidemiology of infectious diseases as a new scientific direction. *Environmental Epidemiology*, *12*(2–3), 198–225. (in Russian) 10.13140/RG.2.2.33921.35686.

Nikolaenko, D. (2018g). S-Theory: A new look at infections. *Environmental Epidemiology*, *12* (2–3), 47–70. (in Russian) 10.13140/RG.2.2.11973.35043.

Nikolaenko, D. (2018h). Spatial-temporal loci as a method of processing infectious information. *Environmental Epidemiology*, *12*(2–3), 141–161. (in Russian) 10.13140/RG.2.2.10002.35528.

Nikolaenko, D. (2018i). The history of infectious ecology: Mark Purdey. *Environmental Epidemiology*, *12*(4), 52–82. (in Russian) 10.13140/RG.2.2.25226.36802.

Nikolaenko, D. (2018j). The microelement norm and verification of the adaptation hypothesis of transformation PrPc into PrPSc. *Environmental Epidemiology*, *12*(4), 26–36. (in Russian) 10.13140/RG.2.2.15518.84806.

Pavlovsky, E. N. (1964). *The natural foci of vector-borne diseases in connection with the landscape epidemiology of zooanthroponosis.* Moscow; Leningrad, (in Russian).

Pimm, S. L., Jenkins, C. N., Abell, R., Brooks, T. M., Gittleman, J. L., et al. (2014). The biodiversity of species and their rates of extinction, distribution, and protection. *Science*, *344* (6187), 1246752. https://doi.org/10.1126/science.1246752.

Purdey, M. (1994). Are organophosphate pesticides involved in the causation of bovine spongiform encephalopathy (BSE)? Hypothesis based upon a literature review and limited trials on BSE cattle. *Journal of Nutritional & Environmental Medicine*, *4*, 43–81. https://doi.org/10.3109/13590849409034540.

Purdey, M. (1996a). The UK epidemic of BSE: Slow virus or chronic pesticide-initiated modification of the prion protein? Part 1: Mechanisms for a chemically induced pathogenesis/transmissibility. *Medical Hypotheses*, *46*(5), 429–443. https://doi.org/10.1016/S0306-9877(96)90022-5.

Purdey, M. (1996b). The UK epidemic of BSE: Slow virus or chronic pesticide-initiated modification of the prion protein? Part 2: An epidemiological perspective. *Medical Hypotheses*, *46*, 445–454. https://doi.org/10.1016/S0306-9877(96)90023-7.

Purdey, M. (1998). High-dose exposure to systemic phosmet insecticide modifies the phosphatidylinositol anchor on the prion protein: The origins of new variant transmissible spongiform encephalopathies? *Medical Hypotheses*, *50*(2), 91–111. https://doi.org/10.1016/S0306-9877(98)90194-3.

Purdey, M. (2000). Ecosystems supporting clusters of sporadic TSEs demonstrate excesses of the radical-generating divalent cation manganese and deficiencies of antioxidant co factors Cu, Se, Fe, Zn. Does a foreign cation substitution at prion protein's Cu domain initiate TSE? *Medical Hypotheses*, *54*, 278–306. https://doi.org/10.1054/mehy.1999.0836.

Purdey, M. (2001). Does an ultra violet photooxidation of the manganese-loaded/copper-depleted prion protein in the retina initiate the pathogenesis of TSE? *Medical Hypotheses*, *57*(1), 29–45. https://doi.org/10.1054/mehy.2001.1305.

Purdey, M. (2003). Does an infrasonic acoustic shock wave resonance of the manganese 3+ loaded/copper depleted prion protein initiate the pathogenesis of TSE? *Medical Hypotheses*, *60*(6), 797–820. https://doi.org/10.1016/S0306-9877(03)00007-0.

Purdey, M. (2004a). Chronic barium intoxication disrupts sulphated proteoglycan synthesis: A hypothesis for the origins of multiple sclerosis. *Medical Hypotheses*, *62*(5), 746–754. https://doi.org/10.1016/j.mehy.2003.12.034.

Purdey, M. (2004b). Elevated levels of ferrimagnetic metals in foodchains supporting the Guam cluster of neurodegeneration: Do metal nucleated crystal contaminants evoke magnetic fields that initiate the progressive pathogenesis of neurodegeneration? *Medical Hypotheses*, *63*(5), 793–809. https://doi.org/10.1016/j.mehy.2004.04.029.

Purdey, M. (2004c). Elevated silver, barium and strontium in antlers, vegetation and soils sourced from CWD cluster areas: Do Ag/Ba/Sr piezoelectric crystals represent the transmissible pathogenic agent in TSEs? *Medical Hypotheses*, *63*(2), 211–225. https://doi.org/10.1016/j.mehy.2004.02.041.

Purdey, M. (2005a). Erratum to "Elevated levels of ferrimagnetic metals in foodchains supporting the Guam cluster of neurodegeneration: Do metal nucleated crystal contaminants evoke magnetic fields that initiate the progressive pathogenesis of neurodegeneration?" [Medical Hypotheses, 65 (2004) 793–809]. *Medical Hypotheses*, *65*(6), 1207. https://doi.org/10.1016/j.mehy.2004.11.002.

Purdey, M. (2005b). Metal microcrystal pollutants: The heat resistant, transmissible nucleating agents that initiate the pathogenesis of TSEs? *Medical Hypotheses*, *65*, 448–477. https://doi.org/10.1016/j.mehy.2005.03.018.

Purdey, M. (2006a). Anti-lactoferrin toxicity and elevated iron: The environmental prerequisites which activate susceptibility to tuberculosis infection? *Medical Hypotheses*, *66*(3), 513–517. https://doi.org/10.1016/j.mehy.2005.09.027.

Purdey, M. (2006b). Auburn university research substantiates the hypothesis that metal microcrystal nucleators initiate the pathogenesis of TSEs. *Medical Hypotheses*, *66*(1), 197–199. https://doi.org/10.1016/j.mehy.2005.07.025.

Rothschild, E. V. (1973). *The spatial structure of the natural plague focus and the methods for studying it (using the example of the northern subzone of the Aral-Caspian deserts)*. Professor dissertation. Saratov. All-Union Research Anti-Plague Institute "Microbe" (in Russian).

Rothschild, E. V. (1978). *The spatial structure of the natural focus of the plague and its methods of study*. Moscow, 192 pp. (in Russian).

Rothschild, E. V. (2011a). Ecological concept in the science of infections. *Environmental Epidemiology*, *5*, 755–770 (in Russian).

Rothschild, E. V. (2011b). Infections in nature. Dangerous illnesses through the eyes of a naturalist. *Environmental Epidemiology*, *5*(4), 434–740 (in Russian).

Rudenchik, Y. V., & Soldatkin, I. S. (1994). The history of the fight against the natural foci of the plague. Domestic lessons. Interesting essays on the activities and figures of the anti-plague system in Russia and the Soviet Union. Moscow *Informa*, *2*, 61–85 (in Russian).

Rudenchik, Y. V., Soldatkin, I. S., Lubkova, I. V., & Lobanov, K. N. (1983). *Evaluation of the relationship of plague epizootic with the number of carriers and carriers in natural foci. Epidemiology and prevention of plague and cholera*: (pp. 3–11). Saratov, (in Russian).

Rusev, I. T. (2011). DDT pesticide as a provoking factor of activation of the parasitic tularemia ecosystem on Biryuchiy island. *Ecosystems, Their Optimization and Protection*, *4*(I), 144–156 (in Russian).

Rusev, I. T. (2013). *Ecosystems of the North-Western Black Sea region as the basis for the formation of dangerous faunistic complexes and their structural and functional organization*. Thesis for the degree of Doctor of Biological Sciences. Odessa, 393 pp. (in Ukrainian).

Sahney, S., & Benton, M. J. (2008). Recover from the most profound mass extinction of all time. *Proceedings of the Royal Society B: Biological Sciences*, *275*(1636), 759–765. https://doi.org/10.1098/rspb.2007.1370.

Saiga Conservation Alliance (2016, February 2). *Why did more than 200,000 saiga antelope die in less than two weeks?* Retrieved April 12, 2019 from http://saiga-conservation.org/2016/02/02/why-did-more-than-200000-saiga-antelope-die-in-less-than-two-weeks/.

Soldatkin, I. S., & Rudenchik, Y. V. (1971). Some questions of plague enzootic as a form of the existence of a self-regulating rodent-flea-pathogen system. *Fauna and Ecology of Rodents*, *10*, 5–29 Moscow: Moscow State University, (in Russian).

Soldatkin, I. S., & Rudenchik, Y. V. (1988). Epizootic process in natural foci of the plague (revision of the concept). In *Ecology of causative agents of sapronosis* (pp. 117–131). Moscow, (in Russian).

Tymoshenko, A., & Korolev, A. (2018). Mark Purdey and the post-Soviet scientific community: Why are scientific innovations so badly perceived? *Environmental Epidemiology*, *12*(4), 83–91. (in Russian) 10.13140/RG.2.2.13328.30727.

Tymoshenko, A., Korolev, A., Bogachevskay, E., & Kovalenko, M. (2018). Preliminary results of discussion of the topic "Mark Purdey and the perception of his scientific ideas" *Environmental Epidemiology*, *12*(4), 92–99. (in Russian) 10.13140/RG.2.2.22110.13123.

United States Center for Disease Control (CDC) (2018, October 2). *Creutzfeldt-Jakob Disease (CJD)*. Retrieved April 12, 2019 from https://www.cdc.gov/prions/cjd/index.html.

United States Center for Disease Control (CDC). (n.d.). 2014-2016 ebola outbreak in West Africa. [Last Reviewed March 8, 2019]. Retrieved April 11, 2019 from https://www.cdc.gov/vhf/ebola/history/2014-2016-outbreak/index.html.

Vodyanoy, V. (2010). Zinc nanoparticles interact with olfactory receptor neurons. *Biometals*, *23*, 1097–1103.

Vodyanoy, V. (2012). Characterization of primo-nodes and vessels by high resolution light microscopy. In K.-S. Soh, K. A. Kang, & D. K. Harrison (Eds.), *The primo vascular system: Its role in cancer and regeneration* (pp. 83–94). Springer Science.

Vodyanoy, V., Daniels, Y., Pustovyy, O., MacCrehan, W. A., Muramoto, S., & Stan, G. (2016). Engineered metal nanoparticles in the sub-nanomolar levels kill cancer cells. *International Journal of Nanomedicine*, *11*, 1567–1576. https://doi.org/10.2147/IJN.S101463.

Vodyanoy, V., Pustovyy, O., Globa, L., & Sorokulova, I. (2015). Primo-vascular system as presented by Bong Han Kim. *Evidence-based Complementary and Alternative Medicine*. https://doi.org/10.1155/2015/361974. 17 pp.

Vodyanoy, V., Pustovyy, O., Globa, L., & Sorokulova, I. (2016). *Evaluation of a new vasculature by high resolution light microscopy: Primo vessel and node*. https://arxiv.org/abs/1608.04276v1.

Weinberg, E. (1974). Iron and susceptibility to infectious disease. *Science*, *184*(4140), 952–956. https://doi.org/10.1126/science.184.4140.952.

Weinberg, E. (1978). Iron and infection. *Microbiological Reviews*, *42*(1), 45–66.

Weinberg, E. (2009). Iron availability and infection. *Biochimica et Biophysica Acta (BBA) - General Subjects*, *1790*(7), 600–605. https://doi.org/10.1016/j.bbagen.2008.07.002.

World Health Organization (WHO). (n.d.). Transmissible spongiform encephalopathies (TSE). Retrieved April 14, 2018 from https://www.who.int/bloodproducts/tse/en/.

ZNAK News Service. (2016, July 31). *To combat anthrax in the Yamal from Yekaterinburg sent specvojska*. Retrieved April 11, 2019 from https://www.znak.com/2016-07-31/na_borbu_s_sibirskoy_yazvoy_na_yamal_iz_ekaterinburga_napravleny_specvoyska.

Supplemental reading recommendations

Kremer, R. J. (2019). Bioherbicides and nanotechnology: Current status and future trends. In O. Koul (Ed.), *Nano-biopesticides today and future perspectives* (1st ed., pp. 353–366): Academic Press. https://doi.org/10.1016/B978-0-12-815829-6.00015-2.

Monti, M., Guido, D., Montomoli, C., Sardu, C., Sanna, A., et al. (2016). Is geo-environmental exposure a risk factor for multiple sclerosis? A population-based cross-sectional study in South-Western Sardinia. *PLoS One*, *11*(9), e0163313. https://doi.org/10.1371/journal.pone.0163313.

Purdey, M. (2005c). The pathogenesis of Machado Joseph disease: A high manganese/low magnesium initiated CAG expansion mutation in susceptible genotypes? *Journal of the American College of Nutrition*, *23*, 715S–729S. https://doi.org/10.1080/07315724.2004.10719415.

Singletary, M., Hagerty, S., Globa, L., Pustovyy, O., Vodyanoy, V., & Sorokulova, I. (2017). *Differential microbial effects of non-oxidized zinc metal nanoparticles*. https://doi.org/10.13140/RG.2.2.27885.90089.

Valera, P., Zavattari, P., Albanese, S., Cicchella, D., Dinelli, E., et al. (2013). A correlation study between multiple sclerosis and type 1 diabetes incidences and geochemical data in Europe. *Environmental Geochemistry and Health*, *36*(1), 79–98. https://doi.org/10.1007/s10653-013-9520-4.

Weinberg, E., & Moon, J. (2009). Malaria and iron: History and review. *Drug Metabolism Reviews*, *41*(4), 644–662. https://doi.org/10.1080/03602530903178905.

Definitions

Contagion disease causative agent; some microorganism that can be transmitted from individual to individual and cause an infectious disease in a population.

Continual continuous, having a permanent presence in nature.

Discrete separated, intermittent, not having a permanent presence in nature.

Epidemic massive infectious disease of people with unusually high rates for a given population and region; assumed that they might be a rational opposition to the infectious process.

Epidemiology the study of the origin and distribution of infectious disease within large populations; branch of medicine for investigative study to control or limit whose methodology primarily relies on the location of patient zero and/or species-reservoir.

Epizootics massive infectious disease of individuals of various species (i.e., animals, human) with unusually high rates for a given population and region; basic assumption is that they cannot be a rational opposition to the infectious process.

Landscape epidemiology scientific concept linking the geography of manifestations of infectious diseases of humans and domestic animals with natural landscapes and species-reservoirs; disputed dominant concept in epidemiology and veterinary medicine formulated in the 1920s.

Preventive medicine fundamentally the medical practice of promoting screening, testing, vaccination, diet, physical activity, and patient adherence to prescribed methods to prevent the onset of diseases such as diabetes or diseases of the cardiovascular system.

Prion diseases prions are a special class of infectious agents represented by abnormal tertiary proteins that do not contain nucleic acids which fundamentally challenges scientific understanding; the dominant understanding of the phenomenon of infection and infectious disease is clearly not consistent with the reality of what is now commonly defined in this category.

Species group of organisms in which any two individuals of the appropriate sexes or mating types can produce fertile offspring.

Species-Reservoir a disputed hypothesis in which pathogens of infectious diseases are ensured by the continuity of their existence due to certain species of animals.

Ukrainian energy sector in transformation

Ivan G. Savchuk[a], *Beth Ann Fiedler*[b]
[a]Institute of Geography of National Academy of Sciences, Kyiv, Ukraine,
[b]Independent Research Analyst, Jacksonville, FL, United States

Abstract

Ukrainian government faces several challenges in the energy sector including the (1) long-term problem of hazardous emissions from aging Soviet-built thermal and nuclear power plants that are unable to meet European safety standards; (2) lack of access to coal to support electricity production due to mine locations outside of their jurisdiction; and (3) slow growth of green energy which presently lacks the capacity to compensate for losses in electrical energy if dominant producers are closed. Consequently, this study utilizes a research methodology of comparative-geographical, descriptive geographic, and mathematical statistics to reveal declining population health in relation to current methods of energy production. Results indicate that initiating a formal plan to reduce dependence on traditional forms of electrical production and expediting transition to green energy can reduce morbidity and mortality of the population living directly in the zone of emissions and improve safety and the quality of life for the national population.

Keywords: Energy sector, Thermal power plants, Green energy, Population health.

Chapter outline

Three Facets of Public Health and Paths to Improvements. https://doi.org/10.1016/B978-0-12-819008-1.00016-X

Energy production and poor health

Dependence on dominant methods of energy production, such as thermal and nuclear power plants, belonging to infrastructure development following regulations and guidelines under the Union of the Soviet Socialists Republics (USSR) presently impacts national population health in Ukraine. Children are particularly susceptible to respiratory diseases in Oblasts (regions) of the country where thermal power plant (TPP) operations release contaminants into the air. This study utilizes a research methodology of comparative-geographics, descriptive geographics, and mathematical statistics to capture the link between energy production and poor health outcomes in Ukraine. Results demonstrate how the existing system of electricity production affects regional peculiarities in the spread of diseases. Establishing this link is the first step in developing viable energy alternatives that are less harmful to the general public.

A number of energy studies focus on the need to increase the share of renewable energy sources for total electricity production to combat environmental and public health concerns resulting from harmful substances emitted into the atmosphere and/or discharged into bodies of water (Costantini, Crespia, Martinic, & Pennacchio, 2015; del Rio & Bleda, 2012; Fouquet, 2012; Fouquet & Pearson, 2012; Grubler, 2012; Peters, Schneider, Griesshaber, & Hoffmann, 2012). Heretofore, there are a number of local examples (Bazelyuk, 2015; Gukalova & Poklyatskyi, 2014; Nikolayenko, 2017; Shovkun & Kostiuk, 2017; Shovkun & Myron, 2016; Yerem & Yerem, 2016). However, the need to highlight the public health impact of energy producing facilities is evident by the notable absence of general works on energy dependence and the problem of harmful emissions from power facilities. This is especially important in the context of energy transformations where priority development of certain types of power generating facilities must include baseline information on current population health and the outline of health outcome improvements as safer methods of energy production are introduced (Fig. 16.1).

From the point of view of medical geography, the focus is on the global spread of infectious diseases and in contrast to the concepts of the geography of health (Brown, McLafferty, & Moon, 2010). Though special attention is paid to the state role in overcoming the problems of dangerous diseases and general population health, the issue of the impact of energy changes on health and morbidity of the population has not been the subject of special geographic studies. Exceptions include studies on the effects of the operation of the nuclear power plants on the development of oncological diseases and cardiovascular disease in the United Kingdom (Committee on Medical Aspects of Radiation in the Environment (COMARE), 2019) and in Germany where leukemia was prevalent in children living near nuclear power plants (NPPs) (Ghirga,

Fig. 16.1 The administrative-territorial division of Ukraine.
Created by Ivan Savchuk. For further information see Eckert (2017); Eckert (2018); Eckert and Lambroschini (2017).

2010). However, other studies on the relationship between childhood cancer and NPPs in Switzerland where $N=2925$ (Spycher et al., 2011) and in Finland where $N<20$ (Heinävaara et al., 2010) conclude no statistical relationship. Nevertheless, the link between particulate matter and other emission pollutants to poor respiratory health and cardiovascular disease is linked to coal and oil-burning power plants such as TPPs (American Lung Association, 2019; US Environmental Protection Agency, 2016). In 2017, the European Union approved new standards for coal burning power plant emissions that are expected to reduce "20,000 premature deaths every year from coal plants alone," (Radosavljevic, 2017, *para* 7).

The World Bank (2017) reports that the percentage of national cause of death in 2016 for noncommunicable diseases for Ukraine is 91.1%, while neighboring countries, such as Hungary 93.8% and Poland 90.3%, demonstrate higher percentages than Europe and Central Asia (88.8%), France (87.6%), United States (88.3%), and the Russian Federation (87.4%). Cancer, diabetes, and cardiovascular disease are noncommunicable diseases but estimates include hereditary defects and diseases affecting digestive, skin, and musculoskeletal tissue.

Previously, researchers from the independent nation of Ukraine did not openly conduct or publish studies on the impact of energy production in relation to the morbidity of regional populations. However, some earlier research establishes a direct relationship between the ecological state and the morbidity of many diseases. "As a result of the migration of harmful substances in the environment, their concentration increase is observed not only in places of concentration of sources of pollution, but also at a considerable distance from them, i.e., in our time, almost all of the population is exposed to harmful substances," (Fashchevskiy, Paliy, Nemchenko, & Starostenko, 1992, p. 71). Fashchevskiy et al. also note an 80% rise in malignant tumors as a result of atmospheric pollution (1992, p. 84). Scant literature appears thereafter. Shevchenko (1994) discusses the natural and other factors influencing Ukrainian health through medical geography mapping of the area during the initial rise of the independent nation. While this scientist clearly establishes that there is a direct link between the ecological status and the general morbidity of the population, the question of the contribution of power plants to poor health was not examined.

However, a series of maps of morbidity distinguishing three groups of diseases depending on the ecological state did emerge. First, Rudenko, Lipinskyi, and Bondar (1993, pp. 49–50) introduced indicator pathology for oncological diseases, allergies, etc.; second, Baranovskiy et al. (2006, pp. 91, 108–116) introduced ecologically dependent pathologies for chronic bronchitis and pneumonia in children; and third, Baranovskiy et al. (2006, p. 106) introduced a moderate degree of dependence for chronic bronchitis and pneumonia in adults. While the present state of population health was analyzed more recently in separate articles (Batychenko, 2013; Gukalova & Poklyatskyi, 2014; Santalova, 2018; Shovkun & Myron, 2016; Zapadniuk, 2017), the question of the impact of electricity production on the morbidity in Ukraine has not been disclosed. This allows the assertion that this study is the first in Ukraine to view the role of this type of economic activity in relation to the health of the population.

The topicality of the study of modern transformations in the energy sector of Ukraine is largely due to two factors: 1) the adaptation of the national economy to the conditions for the simultaneous functioning of European integration, and 2) the implementation of European Union (EU) norms and rules in the field of transformation of the energy sector of the economy (EUR-Lex, n.d.). In particular, EU Regulation 2017/1369 is important to the transformation of electric power because the framework establishes a unified climate and energy policy intended to reduce demand that, in turn, should prompt positive changes in population morbidity and mortality. Therefore, authenticating the existence of a link between the level of pollution caused by the operation of energy enterprises and the consequential prevalence of morbidity of diseases is salient. Thus, the analysis of the relationship between the release of harmful substances into the air and these types of diseases will validate the negative impact of thermal energy on the health of Ukraine population. To accomplish this task, we must make some initial determinations in the main trends in the development of various types of power production and identify whether there is a relationship between the amount of emissions of harmful substances from the functioning of the energy sector, certain types of diseases, and the emergence and spread of disease which is largely due to the level of air pollution.

This chapter presents a general profile of the methods of energy production and the national energy security interests of power production presently available in Ukraine. The research methodology will introduce the various methods of data analysis to provide baseline indications of capacity and infrastructure. Then, development of the electric power industry in Ukraine is analyzed followed by the determination of the impact on regional health as a consequence of the volume of atmospheric emissions of harmful gases and the morbidity of illnesses that arose as a result of pollutants. This analysis establishes relevant information to provide baseline geographical health data in relation to TPPs which support Ukrainian transformation to green energy.

Thermal power plants, nuclear power plants, and limited alternatives

The structure of primary energy production and placement of generating capacities have a direct impact on the regional economy in Ukraine. Energy production is the main industry in some regions and specific cities, such as Burshtyn in Ivano-Frankivsk Oblast, serve as primary exporters (Savchuk, 2013, pp. 44–58). Therefore, the energy sector is important for national economic development. However, the existing system of electricity production is based on power facilities that were put into operation during the period of time prior to the 1991 dissolution of the USSR. The dissolution allowed countries like Ukraine to be established as an independent, self-governing nation giving them clear opportunity to strive for economic advances and improve upon the traditional types of power plants that presently dominate production (e.g., TPPs, NPPs) built under defunct Soviet standards.

Traditional types of power plants

The ecological situation and the state of health of the population in the area of radioactive contamination due to the accident at the Chernobylska NPP in 1986 are notable. But, the inability of NPPs to meet European safety standards (Council of Europe, 2018) represents the potential for significant population impact in terms of national energy security. A concurrent problem is that these behomeths produce more than half of all the energy in the country, but their lifespan ends in about 10 years (2030s). But the Chernobylska NPP accident is not the most significant cause of morbidity and mortality in Ukraine.

TPPs are becoming increasingly problematic in terms of their visible impact on population health and quality of life due to significant amounts of harmful substances, dust, and greenhouse gases pumped into the atmosphere. TPPs are the main sources of industrial air pollution in Ukraine (Ministry of Environmental Protection of Ukraine, 2018) because they are fueled by extracted coal which are major air pollutants in various Oblasts. The population living directly in the TPP zone of emissions has various respiratory health diseases, such as asthma, and medical conditions contributing to significant decline in health and death.

The dissolution of the Soviet Union in the modern era has also given rise to a new problem. TPPs require coal extracted from mines for normal operation, but the mines are located in the territory which is not currently controlled by the Government of Ukraine.

Despite the problems associated with electricity production, there are a limited number of alternatives. Converting facilities to utilize other types of fuel for electricity production appears to be the logical path. Other forms of power production taken into consideration include Hydroelectric Station (HESs) and Press Hydroelectric Stations (PHESs). However, water power utilization is at 90% in Ukraine (State Agency on Energy Efficiency and Energy Saving of Ukraine, 2009) and the additional 10% will be insufficient to make any dent in energy needs. Thus, these are no realistic options under the existing electricity production system. Neither is closure due to the significant reliance on the present system for domestic and export production.

Reducing the country's dependence on traditional forms of electricity production requires further development and expansion of green energy. But the establishment of new power generating capacities based on renewable wind and solar power sources is still insufficient. Together, these green energy alternatives produce less than 1.6% of all electricity in Ukraine (State Statistic Service of Ukraine, 2017). Further, development is growing too slowly to assume the energy production and export needs of the country and surrounding nations, while the state of health of the Ukrainian population continues to decline from pollutants attributed to operating NPPs and TPPs. But positive changes in the structure of energy production, such as incorporating green energy solutions, will undoubtedly lead to less atmospheric pollution and thus decrease the negative impact on population health linked to traditional methods.

Economic interests and raw materials in electricity production

The National Power Company Ukrenergo State Enterprise controls the network of transmission lines with a voltage of 220kV or more and performs the transport of electric current within the United Energy System of Ukraine. Final consumers receive electricity from 26 regional energy authorities (Oblenergos) who buy power from the Energorynok State Enterprise, as well as provide transport services to independent electricity suppliers. Oblenergos are the owners of local power networks with transmission voltage of 0.4–110kV and related equipment. Kyivenergo public joint-stock company serves the capital of the state. In 2014, DTEK Krymenergo public joint-stock company and EK Sevastopolenergo public joint-stock company were disconnected from the United Energy System of Ukraine.

The main sources of electricity generation in Ukraine are raw mineral resources. Ukraine has been producing large-scale coal for over 100years which has depleted most of the available large deposits. Similarly, most of the oil and natural gas was extracted in the 1970s when large deposits in the eastern state were put into operation. Currently, reserves comprising small and medium deposits of oil and natural gas are in operation, but the volume of their extraction does not meet the needs of the energy sector of Ukraine. Therefore, Ukraine must import a large amount of energy from neighboring countries at least until the development of biofuel processing technologies and their relevant resources have been established. Biofuel development is in the early stages in Ukraine, but the priority in agricultural development may be the foundation for hope that biofuel production will increase.

The use of renewable resources is clearly limited in Ukraine in comparison to the structure of energy production by type of energy sources in other countries (Table 16.1). The Ukrainian power system is based on only two types of energy, nuclear and coal, whereas in countries like India and the United States that have a different level of economic development, renewable energy sources account for a much larger share in the production of electric current.

Hydropower

The hydropower of Ukraine consists of nine large power plants that are under the control of the Ukrhydroenergo State Enterprise. Two hydro type stations are prominent in energy production. First, the Dnipro cascade of the Dniprovska HES on the Dnieper river in the Zaporizhia Oblast with the largest capacity (1539.8MW or 1/3 of 4610MW total of all hydroelectric plants); second, the Dnisterska PHES which is the largest operating station of this type in Europe with and operating capacity of 972MW producing 418.3 million kWh of electric current in 2017.

But, the national volume of power production at the hydroelectric power station is relatively small and depends heavily on the water level near the dam and reservoir reserves. However, the PHESs produce the required amount to cover daily fluctuations in the structure of electricity consumption during peak hours which is a significant role in the production of electric current in Ukraine. Continued construction of the Kanivska PHES would ensure coverage of the electric power requirements in

Table 16.1 Electricity production in the first five countries of the indicator and in Ukraine (2014), billions kWh (International Energy Agency (IEA), 2018).

Type of energy source	China	United States	India	Russia	Japan	Ukraine
Coal	4115.2	1712.6	966.5	158.3	348.8	70.5
Oil	9.5	39.9	22.7	10.7	116.4	0.2
Natural gas	114.5	1161.3	62.9	533.5	420.8	12.7
Biofuels	44.4	62.4	23.9	0.0	28.9	0.1
Household waste	13.0	19.4	1.5	3.1	6.6	–
Nuclear energy	132.5	830.6	36.1	180.8	–	88.4
Hydraulic energy	1064.5	281.5	131.6	177.1	86.9	9.3
Geothermal energy	0.1	18.7	–	0.4	2.6	–
Solar photocells	29.2	21.9	4.9	0.2	24.5	0.4
Solar thermocouples	0.0	2.7	–	–	–	–
Wind energy	156.1	183.9	37.2	0.1	5.0	1.1
Energy of tides and waves	0.0	–	–	–	–	–
Other sources	–	4.3	–	–	–	–
Total	5678.9	4339.2	1287.4	1064.2	1040.7	182.0

Credit: Created by Ivan Savchuk.

Cherkaska and Poltavska Oblasts at peak hours without the commissioning of reserve capacities at existing TPPs in these regions of Ukraine.

Hydropower in Ukraine still faces limitations due to scarcity in water resources and diminishing capacity during long periods of drought experienced in most of the country. In 2015, drought severely decreased the water content of Ukraine rivers leading to reduced capacity of the HESs and PHESs. Production in 2015 was 6.8 billion kWh compared to 9.1 billion kWh in 2014. The erratic nature of consistent water resources eliminates this method as a viable alternative to replacing thermal power in Ukraine. Accessing untapped water resources (e.g., mountainous shallow rivers of the Carpathians; restoration of individual small hydroelectric power plants placed on the plain rivers of Ukraine in the 2000s) are unable to substantially increase the volume of hydropower electricity production. Thus, hydropower on a small scale for rural customers may only support local needs. In 2016, 102 small Hydropower Plants (HPPs) with a total capacity of almost 80 MW produced only 0.2 billion kWh to customers who may otherwise have been without access to power.

Nontraditional renewable energy sources in Ukraine

The beginning of the 21st century brought experimental solar and wind power facilities into operation. But the use of nontraditional renewable energy sources there is rather insignificant even with a policy boost. The adoption of the legislation on the green tariff obligated Oblenergos to buy the wind (total capacity is 426.1 MW) and solar power station electric current (total capacity is 431.7 MW) at a higher cost, but provided the revenue necessary to commission new relevant power generating capacities. Two factors were decisive in determining the location of these facilities in southern Ukraine: (1) favorable natural conditions, and (2) insignificant population density in most of the coastal regions of Ukraine.

In 2017, 119 nontraditional renewable energy industrial enterprises that were mainly privately owned operated in the state. System Capital Management (SCM), an industry leader in power generating facilities in 2017, produced almost 40% of the total electric current on wind and solar power stations in Ukraine. However, Ukraine wind and solar power do not play a significant role in the production of electric current due to their lack of capacity. But the electricity produced by wind power in 2017 provided 1.1 billion kWh of electricity to the United Energy System of Ukraine to meet the needs of 2.8 million households in rural areas of the state suggesting that further development is advisable. Accordingly, wind and solar plants may continue to be developed in areas with a small population density to provide power to remote rural settlements. Further, this method will also allow reduction in the length of distribution networks and the consequent loss of electric current.

In Ukraine, the production of electric current with solar panels in housing stock is very small. There are only 204 contracts for the purchase and sale of electric energy produced from solar energy by generating installations of a private household. The high cost of the equipment, the lack of state support, the complicated procedure for connecting to the existing distribution network, and expensive loans for the installation of solar panels do not promote an opportunity for their widespread use in the general population, especially in rural areas. However, they are mainly established by owners of private hotels and large enterprises in urban areas. For example, the Ukrtransgaz Joint Stock Company (n.d.) placed a solar power plant with a capacity of 300 kW on the roof of the production base of Ukrgaztekhzviazok branch office of Ukrtransgas JSC in Boiarka, Kyivska Oblast.

In addition to items already noted, there are several negative aspects of the operation of nontraditional energy power plants that must be taken under consideration. These include noise pollution from normal operation (Bratley, 2019) and regional climate warming or dryness that could change biota and a subsequent decrease in the comfort of life for the population of the corresponding rural settlements (Harvey, 2015; Temple, 2018). These aspects give rise to the need to continue the search for alternative methods of energy production such as bioenergy.

Bioenergy is at large untapped reserve in the production of electric current in the countryside of Ukraine. Industrial plants for the production of electricity from biogas and biomass were built in the 2000s, but their capacity was only 52.4 MW in 2017. While corresponding power generating capacities cannot be of great industrial importance, they are quite suitable for ensuring an important source of electric current in the countryside if used to consume anthropogenic pollution such as significant volumes of unused raw materials, agricultural waste, and animal by-products. The EU member states are well into developing electricity production based on domestic waste recycling, but there are no comparable industrial enterprises in Ukraine. Instead, large incinerators in Kyiv and Kharkiv release significant volumes of harmful substances and stench into the atmosphere, making residents subject to falling ash, nonburned micro- and nanoparticles with carcinogenic properties, and unpleasant odors. Moving toward cleaner bioenergy conversion away from densely populated areas is a logical progression for public safety and health.

Current limitations and comparative cost of nontraditional renewable energy in Ukraine point to opportunities for renewable energy investments in rural settlements where extending traditional types of energy would be prohibitive. Development of the water power plant in the Carpathians is one example where the renewable energy project will save on costly infrastructure for the transfer of electric power to remote individual private estates, hotels, and restaurants facilities. Another example of the cost benefits associated with a renewable energy option is the construction of geothermal power stations in Zakarpatska Oblast that utilize thermal energy generated and stored by the Earth. For example, the Mutnovsko Geothermal energy plant draws heat from living volcanoes, while solar power plant development in southern Ukraine is favorable due to the longest duration of sunny days and minor atmospheric pollution.

Research methodology and data analysis

The basis of our analysis is to determine the dependence of the morbidity rate on population distribution in conjunction with the geographic locations of existing NPPs and TPPs based on open statistical information and materials published in official documents and information of the Ministry of Energy and Coal Mining of Ukraine and electropower companies. Official data on morbidity is provided only in the context of the regions and the capital of the state. Note that there is a lack of accessible data by the Ministry of Health of Ukraine to supplement the analysis; therefore, the level of analysis is regional with general implications for national population health due to the widespread impact of energy production. We selected only those types of morbidity that arise as a result of the negative impact of the functioning of energy companies because Ukrainian population morbidity indicators are presented in official open sources. These include various diseases of the respiratory organs that arise as a result of harmful emissions of the TPPs and thermal heat power plants (THPPs) that pollute the air.

Our research on the impact of changes in the energy sector structure of Ukraine on the health of the population is constructed as follows. First, the current state and problems of energy development are provided in an overview of structure, infrastructure, and the efficiencies of power production in dominant power producers. This includes an overview of the characteristics and complex policy surrounding NPPs and TPPs. Then we utilize geospatial modeling and comparative-geographical analysis of the natural, social, and economic factors that impact population morbidity and mortality in relation to power plant locations. This procedure is accomplished by documenting concentrations of disease that are most likely caused by the operation of power plants. The application of the mathematical-statistical and cartographic method for detecting dependencies between different quality phenomena, such as the production of electric current and morbidity caused by harmful emission into the atmosphere, yields two key points of information. First, a geographical description revealing the peculiarities of such influence; second, the ability to identify causal relationships between the production of electric current, the type of fuel, the type of environmental pollution, and population morbidity. Recommendations for introducing possible changes in the development of Ukraine's energy sector to improve the health of the population are summarized.

Structure, infrastructure, and efficiencies of power production

Several factors impact power production efficiency that can lead to unstable energy security in Ukraine. The best way to portray the current power development and energy sector capacity is to compare Ukrainian output from various types of power plants to the global leader France. Then we review the development of new power transmission lines with higher voltages and safety measures, provide further detail of the characteristics of dominant power producers, and demonstrate the complex geopolitical policy surrounding economic obligations and internal struggle. Finally, we discuss how these items threaten energy security and poor population health as a consequence of limitations of coal supply, policy, and barriers to a variety of raw materials.

Ukraine/France power development and energy sector capacity comparison

The production of electricity has been well-developed in Ukraine beginning with Soviet era infrastructure prior to independence. Energy production was mainly based on the availability of local mineral fuel resources. However, France is the prominent world leader in the overall structure of power production (Table 16.2) because they can generate large volumes of cheap electricity from NPPs. For the years spanning 2000–15, Ukraine had a 6.65% increase in overall energy production, while France had a 4.26% increase, but French NPP, which had a net percentage change increase of 5.18 from 2000 to 2015, is a model producer operating at high efficiency.

In 2015, NPPs in Ukraine produced the largest volume of electricity per unit of voltage power (Tables 16.2 and 16.3). Ukrainian NPPs had a net increase of 14.36% over the period of 2000–15. But, NPP production also took a 0.90% decrease from 2010 to 2015, offsetting the 15.39% increase in production from 2000 to 2010.

A comparison of the development of energy sector capacity in Ukraine and France demonstrates that the French NPP production is optimal and experiencing a need to slightly decline production of 0.19% from 2000 to 2010 and maintain that level in 2015. The net change in NPP production from 2000 to 2015 increased by 16.95% in Ukraine, but France was able to reduce their reliance on thermal power by 16.04% from 2000 to 2015. By comparison, TPP production in Ukraine had a net

Table 16.2 Development of power production in Ukraine and France, billions kWh (International Energy Agency (IEA), 2018).

Type of power plant	2000		2010		2015	
	Ukraine	France	Ukraine	France	Ukraine	France
Nuclear	77.3	415.0	89.2	425.4	88.4	436.5
Hydroelectrical	11.5	72.0	13.2	67.7	9.3	67.7
Solar and wind	0.0	8.1	0.1	10.5	1.6	23.1
Thermal	82.6	53.0	86.5	62.8	83.5	35.7
All	171.4	540.0	188.8	569.6	182.8	563.0

Credit: Created by Ivan Savchuk.

Table 16.3 Development of energy sector capacity in Ukraine and France, millions kW (International Energy Agency (IEA), 2018).

Type of power plant	2000		2010		2015	
	Ukraine	France	Ukraine	France	Ukraine	France
Nuclear	11.8	63.2	13.8	63.1	13.8	63.1
Hydroelectrical	4.7	25.4	5.5	25.4	5.9	25.4
Solar and wind	0.0	...	0.1	6.4	0.8	16.5
Thermal	36.3	26.8	35.2	27.4	35.3	22.5
All	52.9	115.3	54.6	123.8	55.8	129.1

Credit: Created by Ivan Savchuk.

decrease of 2.75% from 2000 to 2015. There is an obvious deficit in Ukrainian capacity for green energy (e.g., solar, wind) despite growth.

Transmission line upgrades and safety

Transmission lines play a role in electricity distribution after domestic power is produced by NPPs and TPPs. But there are several existing hurdles to achieving distribution efficiency in this area. The foremost of these are the age of existing power lines (40+ years old), replacing 100% of all 800 kV and 2/3 of 220–330 kV lines, a variety of nonstandardized voltages, and their role in connecting Ukraine into a single network with some of the largest power plants in Belarus, Moldova, Poland, Russia, Slovakia, and Hungary.

Work is underway to upgrade and expand high-voltage power lines in order to fully utilize energy generating capacities. For example, in 2015, the laying of a 750 kV transmission line from Rivnenska NPP to the 750 kV Kyiv substation (Makarivskyi district of Kyivska Oblast) has been completed to increase the supply of electric power to the capital of Ukraine and surrounding region (Ukrenergo State Enterprise, 2018a).

These improvements provide the electric power needs of the metropolitan region of the capital of the country and reduce losses in transmission of electric current as the volume of transmission is increased when transmission line voltage increases. Unfortunately, larger voltages pose a sharply increasing negative impact on the state of health of the population and so a safety perimeter, distance between the device and potential interaction, must be increased. The technical requirements and safety rules provide for different areas of the protection zone for transmission lines of different voltage. This is due to the negative influence of power generating and power transmission devices on human health. In accordance with the latest version of the *Rules for the Protection of Electric Networks* (approved by the Cabinet of Ministers of Ukraine on 1997.03.04, No. 209), significant restrictions were imposed on the economic activity within the security zones, the size of which directly depends on the voltage level of the corresponding electrical equipment. For example, high power international transmission lines at about 750kW require a 40m security zone perimeter of the device/line, while the smallest transmission lines ($\leq$1kV) require a 2m security zone perimeter (Cabinet of Ministers of Ukraine, 1997). Transformer substations, distribution points, and devices require a 3m security zone perimeter.

Lines of 800, 750, 500, and 400kV voltage are designed for transmission of electric current between regions of the state and in international deliveries (Ukrenergo State Enterprise, 2019). The first transmission line of 800kV in the world connected Volga Hydroelectric Station (Volzhsky, Volgograd Oblast, Russia) with the substation Mykhailivka (Luhanska Oblast) in 1962 (Mavili Elektronik Ticaret A.Ş., 2018). In 1981, a 750V transmission line began to transmit Ukrainian electric current for export at a distance of up to 1500km (932mile) between Donbaska-Vinnytska-Western Ukraine to the city of Albertirsa in Hungary (Ukrenergo State Enterprise, 2018b) (Fig. 16.2).

An electric current of 220V is commonly used in Ukrainian household electric appliances. These power transmission lines account for almost 3/4 of the entire length of the air power lines in Ukraine. Distribution substations, located near major cities and some border areas, control the flow of electrical current. For example, the Kyiv energy ring consists of five 330kV substations which regulate the flows of electric current to the state capital. The Mukachivska international substation (high-voltage transmission lines of 750kV) distributes the electric current from Ukraine to the adjacent countries within the framework of the electricity network at *Burshtynska TPP Island* (650MW capacity) that is part of the EU energy system. Among thermal energy enterprises in Ukraine, *Burshtynska TPP Island* operates exclusively for export in the mode of connection to the power networks of Slovakia, Romania, and Hungary. There is an increasing dilemma between the increase in such exports and the deterioration of the health of the population living in the Burshtynska TPP exposure zone and along the path of the high-voltage transmission lines.

Nuclear power plant units in Ukraine were built to reach an operating capacity of 1 million kW which allowed for growing demand in national current supply and planned export to Poland, Romania, Slovakia, and Hungary. The Khmelnytska NPPs were built in the inland regions of Ukraine under the conditions of the intergovernmental agreements to maintain the energy supply through a major transmission line (750kV)

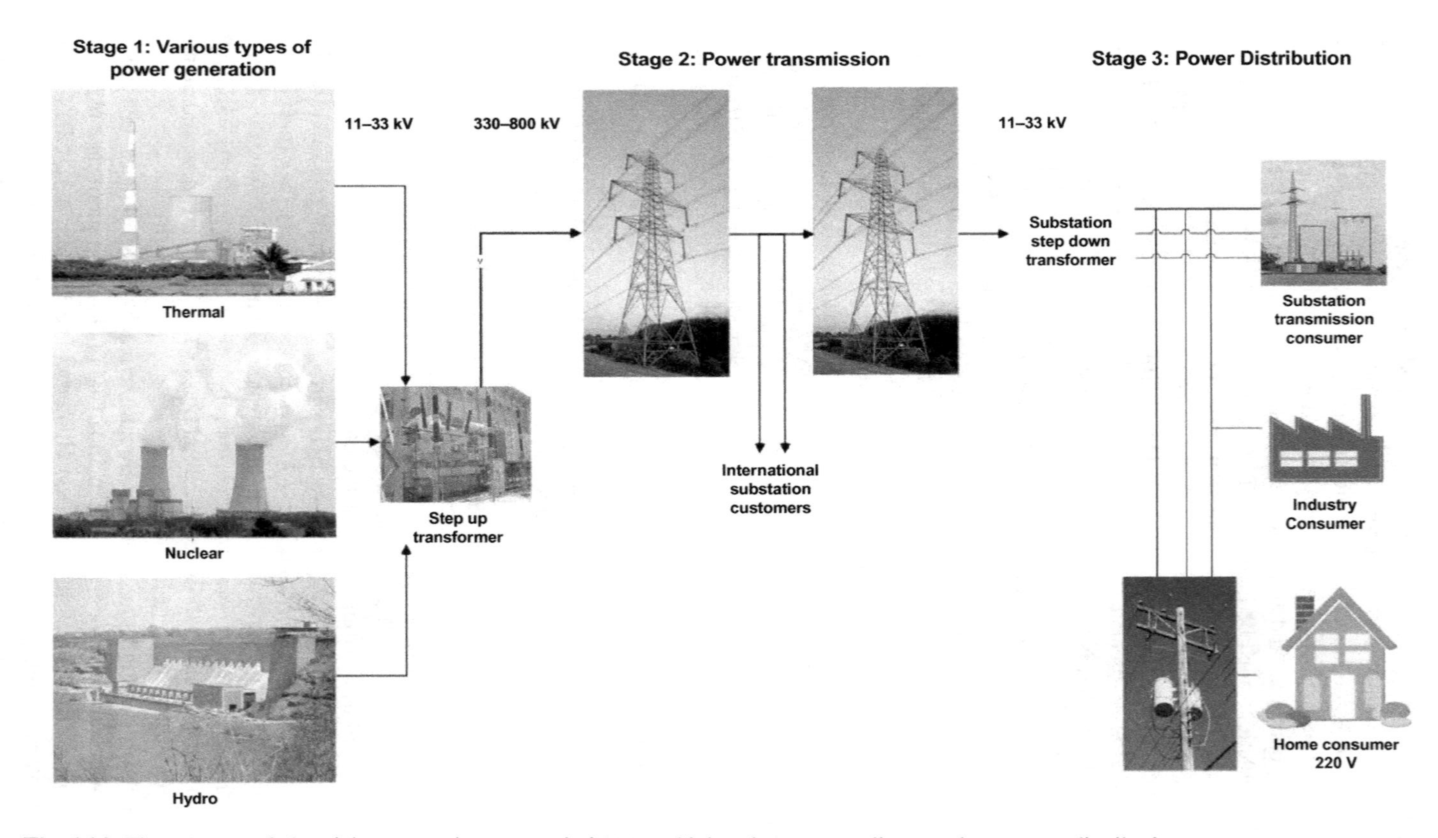

Fig. 16.2 Three stages of electricity generation, transmission over high-voltage power lines, and consumer distribution. Created by Beth Ann Fiedler.

built from them heading in a westerly direction. For the past 20 years, these NPPs have supplied half of the electricity production to the energy systems of these countries.

Employees and family members predominately live in their own satellite-town near the NPP where they work. Such cities, such as the city of Enerhodar in the Zaporizka Oblast, have become a center pole of socioeconomic growth for the population of adjacent territories. The largest power generating capacities in the country are concentrated here including the Zaporizhska NPP (6 million kW) which is the most powerful in Europe and the Zaporizhska TPP which is the most powerful in Ukraine (3.6 million kW). In 2017, they produced 1/5 of the total electric current in the state. This level of concentration in nuclear and thermal energy capacity is the only example of these conditions in one city and does not occur in any other country in Europe. However, the impact on human health and biota is apparent from the simultaneous influence of atmospheric emissions from the NPP and TPP and the discharges of waste water into the cooling ponds.

The South Ukraine power hub, including Yuzhno-Ukrainska NPP (3000 MW), Tashlykska PHEs (300 MW), and Oleksandrivska HPP (11.5 MW), has a special place in the energy sector of modern Ukraine because they operate in a single complex. These diverse energy generating systems are part of the National Nuclear Energy Company Energoatom State Enterprise which provide guaranteed supplies of fresh water to NPP power units, while surplus electric current production at night is used to fill the water storage tank of PHEs. Thus, the problem of the lack of regulatory capacities in the south of Ukraine was solved, the efficiency of electric current production was improved, and supplies of fresh water were guaranteed. But the natural evaporation of water from the surface of the reservoir, as well as emissions to the atmosphere of rare gases, is a consequence of the normal operation of the NPPs and HPPs. Thus, there is an increase in humidity and simultaneous impact on the atmosphere and hydrosphere due to harmful emissions in the area adjacent to the station and corresponding changes in the health of people living in the city of Yuzhno-Ukrainsk and adjoining rural settlements.

NPP characteristics and complex international policy

The closure of the Chernobylska NPP in 2000 prompted the development of the Rivnenska and Khmelnytska NPPs in 2004, the first built since the Soviet era, to compensate for energy production losses. Still, elimination of the consequences of the Chernobyl catastrophe of 1986 at this station will require considerable and long-term funding due to the persistent maintenance that prevents the release of radioactive micro- and nanoparticle dust in the center of the third power unit of this station. Chernobyl, the largest man-made disaster in Europe, refuted the myth about the safety and exceptional low cost of electric power at NPPs. Various international organizations funded construction of a new sarcophagus completed in 2017 over the destroyed fourth power unit of the Chernobylska NPP to help prevent deadly contaminants from being released in the third unit.

Today, all nuclear energy companies are separate structural subdivisions of the National Nuclear Energy Company Energoatom State Enterprise to reduce the likelihood of another NPP disaster. An important problem in regulating the daily operations of Ukraine's nuclear power industry is the lack of a national closed cycle from nuclear fuel production to disposal. Russia supplies relevant fuel cells produced from domestic uranium raw materials and now removes radioactive waste from Ukrainian nuclear power plants to the Chernobylska NPP exclusion zone as a permanent repository of spent nuclear fuel. The exclusion zone construction will allow fuel cells to be purchased from the United States, but there are looming and persistent questions about the safe operation of this nuclear power facility there due to several natural conditions in this part of Ukraine. For example, Polissia has high humidity, close proximity of groundwater to the surface, bogging of a large part of the area, the presence of active erosion, and karst processes that can significantly compromise the normal storage of spent nuclear fuel.

Nevertheless, nuclear power will remain the basis of the entire energy system of Ukraine for the coming years. Four operating NPPs, consisting of 15 power units with a total capacity of 13,835 MW, provide the basic foundation for the country's power network. They demonstrate their status as the largest level of operating capacity utilization at 73.0% in 2016. However, the lifespan of the relevant units is expiring from 2030 to 2035, but corresponding repairs and modernization should extend the present life expectancy to about 25 years. Cost to maintain and upgrade is a formidable barrier, but so is the cost of decommissioning that is also without the benefit of energy production.

Other problems plague the NPPs. For example, an urgent resolution to address the lack of exhaust technology for dismantling and utilization of the reactor zone is required. Further, the prolongation of their operation period is a controversial issue in view of the existing International Atomic Energy Agency (IAEA) standards for the safe operation of nuclear reactors. Their gradual decommissioning will lead to serious problems with the provision of electric current requirements, increase in energy cost, and funding sources for dismantling the equipment and the safe operation of the relevant NPP power units. However, the costly German experience of retrofitting Soviet-built NPPs to steam and gas thermal power plants (World Nuclear Association, 2019), now under the control of German government-owned Energiewerke Nord GmbH (EWN), serves as a salient example of the obstacles facing changes in Ukraine energy production. The absence of large supplies of natural gas deposits in Ukraine also makes retrofitting applicable only if natural gas can be purchased abroad at reasonable prices that do not significantly increase the cost of generated electricity. But this direction is not stable due to the unpredictable nature of price fluctuations for imported natural gas. Therefore, logic dictates the development of measures for modernization of existing NPPs in Ukraine and the gradual increase of capacities at the TPPs as reserve producers.

TPP problematic supply chain, trade policy, and competing energy production

Utility ownership, access to energy grade coal, and economic reliance on hazardous methods of energy production surround the method of TPP electricity generation that is dependent on scarce minerals and specific types of coal. Add into the mix the external geopolitical disruption of Russian occupation, annexed Crimea, and internal division over raw materials and we enter into the complex world of Ukraine energy, especially TPP-generated electricity. A seemingly less obvious problem is the growing concern for public health in densely populated settlements and large concentrations of industrial production where TPPs are located due to their consumer orientation.

Supply chain barriers

Domestic generating facilities of thermal power in Ukraine vary in operating capacity and may use various types of mineral fuel which can produce electric current from a consistent supply of domestic coal. Each Ukrainian TPP is focused on the consumption of coal of certain energy brands and cannot, without reducing production efficiency and growth in fuel consumption, operate using other types of fuel. For example, Burshtynska TPP works exclusively on the coal of the Lviv-Volyn Basin. This leads to significant pollution of the environment because this brand causes harmful nitrogen and sulfur compound emissions and generates volumes of coal ash dumps containing toxic substances that become airborne. Thus, large Ukrainian thermal power plants are among the foremost stationary sources of atmospheric air pollution on a national scale, leading to an increase in diseases such as asthma, acute respiratory illnesses, and allergies that can be fatal to the population living in the zone of their influence.

Inconsistency of the brand of domestic coal utilized at the TPP facilities, which has been the source of internal competition, have led certain areas in eastern Ukraine to prevent the regular supply of coal from the mines of Donbas. The crisis reached a peak in 2014 when a significant drop in production occurred for a period of several weeks when almost half of the generation capacity of coal TPPs in the country had been operating using Donetsk anthracite coal resulting in operational shutdown of large TPPs, such as Trypilska TPP in the city of Ukrainka in the Kyivska Oblast. Further, the disruption in domestic coal supply chains led to the purchase of energy coal from Australia and South Africa which sharply increased harmful atmospheric emissions (Coal-and in the African Coal!, 2019). Unfortunately, the systems of purification and combustion of coal at the TPP and THPP were not adjusted to the physical and chemical characteristics of imported coal which led to increased volumes of waste and cost of electric current.

The recalibration of TPPs and THPPs to work on imported coal is also problematic in view of the fact that the global market allows for energy coal by negotiating fixed prices and volume for long-term contracts for a minimum of 5 years. But Ukrainian companies were unable to break into the global supply chain and their attempt to make large-scale purchases of steam coal in the free world market only led to the purchase of various small batches of coal. Thus, the normal functioning of the TPPs and THPPs of Ukraine is possible only in the context of a purposeful state policy on the purchase of large guaranteed supplies of energy grade coal from specific mines or coal cuts that have physical and chemical properties similar to those brands of Donetsk coal that were previously supplied to them.

Kyiv's economic dependency on electricity production

Ukraine's 26 THPPs provide hot water and heat through a centralized network. Kyiv contains the largest THPP-5 and the largest number of employees work in the production of electricity and by-products there. Thus, the deindustrialization of the largest industrial center in the capital has created a phenomenon of disproportionate importance of energy to economic development. Further, the shrinking industrial base has not diminished normal operation of the THPPs' large areas under the sanitary zone and warehouses of energy coal continue to be allocated. Meanwhile, high pipes, which emit harmful substances into the atmosphere, lead to the spread of hazardous levels of emissions to a significant area in and outside the city, leading to regional deterioration of air quality. The operation of their own low-power industrial generating facilities causes the combination of emissions from the metropolitan THPPs and local boiler houses. The consequences of a distressed ecological situation present environmental conditions from electricity production, leading to an increase in the number of patients with asthma, allergies, and acute respiratory diseases among the population in the suburban zone of Kyiv.

Utility generation ownership, natural gas, and gas turbine alternatives

Though some small energy enterprise businesses exist, the main generating capacities belong to the Public Joint Stock Company (PJSC) System Capital Management (SCM) which controls DTEK Dniproenergo PJSC, DTEK Zakhidenergo PJSC, and DTEK Vostokenergo LLC. DTEK was formed in 2005 and is Ukraine's largest energy group. In 2016, SCM produced more than half of the total TPPs' electric current of Ukraine. However, the largest public utility enterprise in the state is Kyivenergo PJSC which operates municipal THPP with a capacity of 1200 MW and an electricity distribution network in the capital of Ukraine. Relevant THPPs mainly produce electricity, hot water, and heat by burning natural gas. Natural gas greatly improves the state of air quality, reduces environmental pollution, and significantly reduces the volume of transport associated with the coal supply chain.

The political confrontation with Russia since 2014 led to a reorientation of Ukraine to import gas from Norway and other European countries, leading to higher costs associated with natural gas, especially when compared to the costs of Ukrainian coal. In view of this, the reequipping of TPPs to use imported natural gas was suspended. Accordingly, Ukrainian energy cannot develop in the same manner as in the EU member states which provides for the transition of most TPPs to the use of natural gas as the main type of fuel. Now the majority of natural gas is officially bought from Ukrainian companies involved in mining within the country. According to the laws of Ukraine, companies that produce natural gas in the country must sell the product to consumers in Ukraine, but the natural gas produced in Ukraine is insufficient for the needs of the population and industry. The state gives priority to deliveries of natural gas to the population, while the electric power industry must purchase natural gas abroad. Therefore, the cost of selling electricity to domestic consumers is rising.

In Ukraine, industrial power generating capacities based on secondary heat are commissioned at enterprises of different types of economic activity. For example, a gas turbine power plant with a capacity of 300 MW operates on the heat released during metallurgical production by Alchevsk Iron and Steel Works PJSC in Alchevsk in the Luhanska Oblast. The electric current produced here also supplies the needs of the power plant. This is a rather promising direction of reducing Ukraine's dependence on imported energy. But gas turbine plants have limited use because the relevant large industrial enterprises of ferrous metallurgy or chemistry are located only in a small number of cities.

Recovering coal mine methane

Since 2006, Ukraine boasts of one of the largest cogeneration power stations in the world for the utilization of coal mine methane for the production of electric current and the supply of heat at the O.F. Zasiadko mine (Donetsk) and at Chervonoarmiiska-Zakhidna № 1 mine (Udachne town within the city of Pokrovsk, Donetska Oblast). These locations comprise one of the largest reserves of methane in coal seams for the production of electric current, but precaution must be maintained by controlling the methane outflow to significantly reduce the probability of mine explosions reducing injury and fatality to miners. Coal mine methane to produce electric current can be extracted from operational coal mines and those that have been closed. Activating closed mines for methane extraction can simultaneously increase the employment rate of the local population, reduce the import of natural gas, and reduce atmospheric emissions because the methane capture process is conducted underground where cogeneration power stations are located.

Geospatial modeling findings and interpretation

Power engineering is a primary industry for Ukraine, but is also a source of energy insecurity and pollution that impacts public health. The difficulty in balancing consumer and industrial energy needs, environmental impact, and the cost of upgrading aging Soviet era power plants to meet current energy security standards requires consideration for planning, development, and international collaboration. Establishing new methods of electricity production, which is presently one of the largest atmospheric pollutants, is important to addressing chronic respiratory diseases generally and specifically for children. We generate a snapshot of the Ukrainian energy security status drawing from a number of resources to demonstrate the energy production environmental impact on public health.

Unstable energy security

After analyzing the existing structure of electric power production in Ukraine, we may conclude that the predominance of old thermal and nuclear power plants is a threat to the security of the country and has a noticeable negative impact on the health of the population. The regions of Ukraine where electricity is produced demonstrate the largest amount of atmospheric pollutant emissions per unit of electric current produced (Table 16.4). At the same time, there is an uneven distribution in atmospheric emissions of harmful substances due to the functioning of the energy sector. The capital

Table 16.4 Volume of air pollution emissions resulting from the supply of electricity, gas, steam, and air-conditioned air per unit of electricity produced in the regions of Ukraine (2017).

Regions (capital of region) of Ukraine	Amount of emissions (t/kWh)[a–c]	Population[a,d]
Central area		
City of Kyiv (Kiev)	117.0	2,906,569
Cherkaska Oblast (Cherkasy)	3,537,200.0	1,220,363
Chernihivska Oblast (Chernihiv)	77,461.8	1,020,078
Kirovohradska Oblast (Kropivnytskiy)	5607.6	956,250
Kyivska Oblast (Kyiv)	87,403.0	1,754,284
Poltavska Oblast (Poltava)	137,760.2	1,413,829
Sumska Oblast (Sumy)	76,459.6	1,094,284
Zhytomyrska Oblast (Zhytomyr)	...	1,231,239
Vinnytska Oblast (Vinnytsia)	57,116.1	1,575,808
East area		
Donetsk Oblast (Donetsk)	...	4,200,461
Kharkivska Oblast (Kharkiv)	149.0	2,694,007
Luhanska Oblast (Luhans'k)	61,444.9	2,167,802
South area		
Autonomous Republic of Crimea (Simferopol)[e]	...	...
City of Sevastopol' (Sevastopol)[e]	...	...
Dnipropetrovska Oblast (Dnipro)	469,533.3	3,231,140
Khersonska Oblast (Kherson)	308,758.4	1,046,981
Mykolaivska Oblast (Mykolaiv)	5,613,603.4	1,141,324
Odeska Oblast (Odessa)	33,170.3	2,383,025
Zaporizhska Oblast (Zaporizhia)	419,705.5	1,723,171
West area		
Chernivetska Oblast (Chernivtsi)	17,258.2	906,701
Ivano-Frankivska Oblast (Ivano-Frankivs'k)	51,087.4	1,377,496
Khmelnytska Oblast (Khmelnytskyi)	20,576,718.0	1,274,409
Lvivska Oblast (L'viv)	64,346.7	2,529,608
Rivnenska Oblast (Rivne)	33,368,987.6	1,160,647
Ternopilska Oblast (Ternopil)	58,571.4	1,052,312
Volynska Oblast (Lutsk)	309,333.3	1,038,457
Zakarpatska Oblast (Uzhorod)	55,096.8	1,258,155

[a] Calculated according to source: State Statistic Service of Ukraine (2016), Ministry of Environment and Natural Resources of Ukraine (2017).
[b] Data on volumes of pollutant emissions into the air in Donetska Oblast and data on the volume of electric power production in Zhytomyrska Oblast do not officially publish.
[c] t/kWh = t/1000 W.
[d] Population numbers are 2018 estimates (State Statistic Service of Ukraine, 2018).
[e] The population of city of Sevastopol and Autonomous Republic of Crimea does not officially publish.

city of Kyiv has the largest concentration of THPPs and emissions of 35.2 million tons in 2017. Statistical information and mathematical calculation indicate that the capital emits 63.8% of the total emissions of harmful substances in the country. This has a significant negative impact on the inhabitants of the most populous city of Ukraine.

Energy production environmental impact on public health

Fig. 16.3 establishes the environmental impact of energy production leading to morbidity from respiratory system diseases confirming that traditional types of power plants emitting pollutants have a detrimental impact on population health in all regions where the largest power companies are located (Table 16.5). The overall morbidity from respiratory diseases per 1000 children in these energy producing locations was higher than the average for Ukraine. This finding directly points to the connection between the operation of large power plants and the morbidity of the population. Furthermore, the high concentration and repeat events of children with respiratory diseases in the city of Kyiv (1082.24/1000 children), the Dnipropetrovska Oblast (1051.79/1000 children), and the Kyiv Oblast (1251.93/1000 children) reported by the Ministry of Health of Ukraine (2016) represents a significant indicator since the population in Kyiv in 2016 comprised only 497,272 children (under the age of 15).

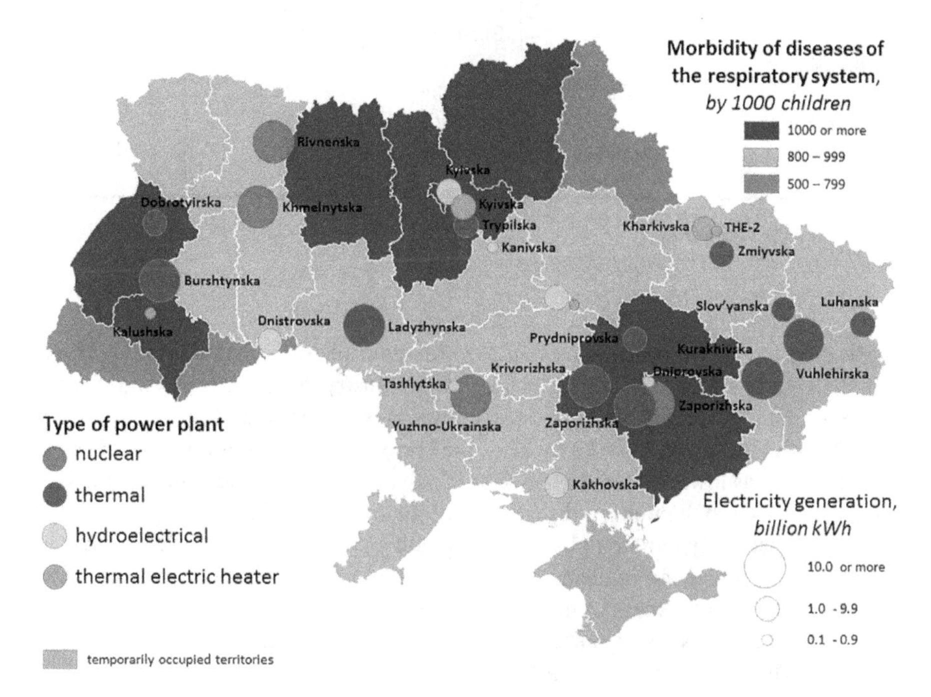

Fig. 16.3 Production of electrical energy in Ukrainian cities and morbidity of respiratory system diseases (compiled from Ministry of Energy and Coal Mining of Ukraine, 2016; Ministry of Health of Ukraine, 2016).
Created by Ivan Savchuk.

Table 16.5 The largest enterprises of the electric power industry by volumes of pollutants' emissions (Ministry of Energy and Coal Mining of Ukraine, 2016; Ministry of Health of Ukraine, 2016).

Oblast	**Power plant**	**City/town**	**Volume of emissions (thousands tons per year)**	**Volume of energy production (billions kWh)**
Vinnytska	Ladyzhynska TPP	Ladyzhyn	94.2	5.2
Dnipropetrovska	Prydniprovska TPP	Dnipro	24.0	1.3
Donetska	Slovianska TPP	Mykolaivka	19.5	2.3
	Kurakhivska TPP	Kurakhove	113.7	6.8
	Vuhlehirska TPP	Svitlodarsk	96.1	4.1
Zaporizka	Zaporizhska TPP	Enerhodar	105.2	6.3
Ivano-Frankivska	Burshtynska TPP	Burshtyn	31.9	8.8
Kyivska	Trypilska TPP	Ukrainka	23.3	0.8
Luhanska	Luhanska TPP	Shchastia	45.7	2.7
Lvivska	Dobrotvirska TPP	Dobrotvir	49.8	2.6

Credit: Created by Ivan Savchuk.

While Kyiv has a high concentration of harmful atmospheric emissions, the contribution of all major enterprises of the electric power industry that operate by burning coal to overall poor air quality leads to chronic respiratory conditions (Fig. 16.3).

Only in Kyivska and Lvivska Oblasts, the corresponding morbidity rate for bronchial asthma was lower than the average in Ukraine. A viable explanation for this variance can be due to the uneven distribution of the population in these regions where there are respectively remote Polissia and Carpathian parts from the power stations in which the atmospheric air is cleaner because of high forest distribution. In other areas, we have a significant excess of the morbidity of children that is consistent with high levels of harmful emissions from the largest power companies in Ukraine (Tables 16.5 and 16.6). The existence of large ash dumps near TPPs where the slag and coal dust ash are stored greatly impacts the ecological situation in the areas and settlements indicated in the same tables. Slag and coal dust are both carcinogens. The ash dump of the Burshtyn TPP operates solely for the supply of electric current for export. In 2017, this facility produced 133.5 thousand tons of slag and 503.9 thousand tons of dust were

Table 16.6 The level of morbidity from children's respiratory diseases and bronchial asthma in Oblasts of Ukraine (per 1000 children), where the largest (in terms of emissions of pollutants into the air) enterprises of the electric power industry operate (Ministry of Health of Ukraine, 2016).

Oblast	Respiratory diseases	Bronchial asthma
Total	**907.66**	**0.62**
Vinnytska	977.77	1.03
Dnipropetrovska	1051.79	0.85
Donetska	921.38	0.84
Zaporizka	1105.20	1.12
Ivano-Frankivska	1066.53	0.64
Kyivska	1251.93	0.43
Luhanska	1027.72	0.67
Lvivska	1037.88	0.47

Credit: Created by Ivan Savchuk.

accumulated. Both products cannot be disposed of or recycled according to present technologies in Ukraine. Thus, the further operation of this power plant will continue to negatively impact the health of citizens in the densely populated rural regions of Western Ukraine.

Discussion and recommendations

The transformation of Ukraine's energy sector faces significant challenges broadly based on the lack of logistic supply chain capacity for required minerals and raw materials, minimal purchasing power in international trade, and shortage of specific industrial development that could reposition the nation in energy production. Foreign policy challenges and development encompass these three items and extend to the delayed implementation of EU norms and rules in the functioning of Ukraine energy sector. While the state's role in regulating the market for electric current sales is high, such as the introduction of the green tariff that increased investments in renewable energy, they are not at the level sufficient for the EU.

Historically, these and other obstacles have been the impetus to preserve the dominant role of the NPPs and TPPs in the production of electric current in facilities built according to standards and norms that are significantly different from the EU. Thus, state-directed policy has been minimal and European integration in the energy sector has been almost impossible without significant changes in the production infrastructure. The additional constraints derived from the complex combination of state-owned enterprise and private companies who control the sale of electricity warrant intricate negotiations between and among their owners, the state, and representatives from countries who benefit from Ukrainian energy exports. Still, Ukraine and several nations hold a promising position based on their existing interdependencies and intergovernmental agreements from which to build upon inclusive of private enterprise.

We have presented the pros/cons of the traditional sources in the ongoing debate between economic gains and population health. Further, even forms of renewable energy pose a danger to the regional environments. While we know that some methods of energy production are less hazardous to the environment, and thus, human health, these options are fraught with the reality that the existing system of management and investments in power engineering may only be possible at the expense of the development of the TPP. On one hand, this would reduce the health hazards associated with the TPP. On the other hand, renewable energy and alternative methods will fall short of demand. Following the same logic, the level of NPP energy production cannot simply come to a resounding stop. The additional environmental degradation and local climate change as a result of some renewable energy production options, coupled with less than efficient capacity, require a new but not necessarily radically innovative direction.

What may be lost in this presentation of information is that Ukraine is the second country in the world after France in terms of the share of electric power produced by NPPs. This may be the most important piece of information for the state to draw attention to NPP modernization and developing an industrial base to support the transition to green energy during the extended life span of existing power plants. The proposition includes the recruitment of specialized industries to Ukraine (e.g., ferrous metallurgy, chemistry) and contracting/gaining experience in purchasing commodities (e.g., volume trading in natural gas, energy grade coal) to hedge known fluctuations in pricing. Finally, the formation of an intergovernmental Eastern European Energy Consortium comprised of Ukraine and those countries reliant on Ukraine's energy production to increase access to high-quality raw materials at the best contracted pricing.

Forming a multinational consortium, developing diplomatic relations with international suppliers, initiating long-term contract negotiations for raw materials, and generating interest in development are important government functions that must be initiated to improve the quality of life and population health. This process also introduces emerging employment opportunities to offset changes in energy production. Without an intentional plan, Ukraine power production may not be able to meet customer demand, especially if funds to retrofit aging NPPs are not obtained before they experience their end of life in the 2030s. Otherwise, the high environmental impact of electric power production from TPPs and NPPs revealed in the analysis will continue unless there is an expedited move to alternative energy production that can markedly improve air quality and population health.

In the interim, gaining access to energy grade materials and developing natural gas and gas turbine plants will incrementally help to reduce emissions and improve air quality. For example, Kyiv experienced high levels of emissions and morbidity in association with TPPs, but the volume of emissions of harmful substances per kW of produced electricity is the smallest in the state (117 tons) due to the use of natural gas as the fuel for energy enterprises of the Ukrainian capital. Thus, the city serves as an example of the need for energy transformation of coal burning electric power production to natural gas and other renewable energy sources to significantly reduce the amount of emissions per kW and change the structure of harmful emissions to those that are more environmentally friendly.

Finally, expediting the transition to green energy and other energy reforms is critical to the Ukraine economic interests in energy production, emission reduction, and ultimately reduction in morbidity of the population in the contaminated areas, particularly in those cities where TPPs are located. Two major obstacles to this approach are (1) the absence of strategy to implement the requirements of the EU energy legislation, and (2) the sharp deterioration of relations with Russia as a result of the annexation of the Crimea and the undeclared war in the Eastern Ukraine in 2014 (Kuzio & D'Anieri, 2018). These items have a significant and ongoing impact on the country's development. Drawing attention in this format is one way to attract reasonable input and interest in mutually beneficial solutions for Ukraine, Russia, export nations, and the international community. Of utmost concern is the looming end of life expiration of existing NPP units in 2030 and the ongoing concerns resulting from the Chernobyl meltdown. Drawing business enterprise to the state will require resolution with Russia regarding occupation of certain areas and likely require intervention from various nations and international organizations.

In summary, the introduction of EU norms in the energy sector of Ukraine can establish the necessary framework, guidelines, and standards to move forward with European integration. Ukraine milestones based on EU norms, such as achieving 25% production of the electric current from renewable energy sources in 2025, are possible due to the modernization of the TPPs already successfully navigated by the French. With this objective, coupled with an increase in production at the wind and solar farms in remote regions, Ukraine can build long-lasting positive effects toward balancing the present battle of meeting economic commitments against growing concern for public health. Clearly, reducing population morbidity related to TPP energy production is possible by reducing the share of coal-fired power stations by transitioning them to receive natural gas as the main type of fuel. Thus, upgrading existing power generating capacities and purification systems at existing TPPs must be scheduled to permit sufficient energy production while systems are offline. These and other planned internal measures by Ukraine can emulate the French model and coordinate and establish a strategic plan given the EU framework beginning with development of a set of measures aimed at increasing the safety of the NPP and extending their service life after 2030; gradual retrofit of TPPs to burn natural gas and increased understanding of how these actions can reduce atmospheric emissions of harmful substances will, in turn, reduce the incidence of the diseases of the respiratory system.

Conclusion

In this chapter, we introduced a methodology of mathematical and statistical analysis to produce a geospatial model of the impact of thermal and nuclear power plants on population health. Traditional power plants developed under Soviet era specifications came under inspection to reveal several problems that threaten the energy security of the nation, their deadly impact on respiratory health, and insufficient capacity to meet power needs in a manner that is more conducive to public health. Aging infrastructure, lack of government purchasing power, and inherent problems with alternative energy are some topics under discussion. A primary recommendation is the development of a strategic development plan in conjunction with the EU energy legislation and the French model of energy production in order to integrate technologies for an improved future.

References

American Lung Association. (2019). *Reducing air pollution from power plants: The success of the mercury and air toxics standards.* Retrieved June 9, 2019 from https://www.lung.org/about-us/blog/2019/03/reducing-air-pollution.html.

Baranovskiy, V. A., et al. (2006). *Ukraine. Ecological-geographical atlas.* Kyiv [Барановський В.А. та ін. Україна. Еколого-географічний атлас. (2006) Київ. 220 с.] (in Ukrainian).

Batychenko, S. (2013). Human-geographical studies of socio-economic factors affecting population morbidity of Ukraine. *Ekonomichna ta Sotsialna Geografiya, 67*, 82–88 (in Ukrainian).

Bazelyuk, J. (2015). Features of the medical-geographical zoning of the territory (on the example of Volyn region). *Siences magasin of Lesia Ukrainka Esteuropean national university. Serie Geography, 15*, 35–41 (in Ukrainian).

Bratley, J. (2019, March 22). *The impacts of wind turbines.* Clean Energy Ideas. Retrieved June 12, 2019 from https://www.clean-energy-ideas.com/wind/wind-turbines/the-impacts-of-wind-turbines/.

Brown, T., McLafferty, S., & Moon, G. (Eds.), (2010). *A companion to health and medical geography.* UK: Wiley-Blackwell.

Cabinet of Ministers of Ukraine (1997). *Rules of the security at electric networks (Resolution of 04.03.1997 No. 209).* Retrieved May 20, 2019 from https://zakon.rada.gov.ua/laws/show/209-97-%D0%BF (in Ukrainian).

Coal-and in the African Coal! (2019). *ВУГІЛЛЯ – І В АФРИЦІ ВУГІЛЛЯ!*. Official site of the city of Ukrainka (in Ukrainian).

Committee on Medical Aspects of Radiation in the Environment (COMARE). (2019). *Current COMARE Work Programme (April 2019 – March 2020).* Retrieved May 7, 2019 from https://www.gov.uk/government/groups/committee-on-medical-aspects-of-radiation-in-the-environment-comare.

Costantini, V., Crespia, F., Martinic, C., & Pennacchio, L. (2015). Demand-pull and technology-push public support for eco-innovation: The case of the biofuels sector. *Research Policy, 44*, 577–595.

Council of Europe. (2018). *Nuclear safety and security in Europe. Resolution 2241.* Retrieved May 7, 2019 from http://assembly.coe.int/nw/xml/XRef/Xref-XML2HTML-EN.asp?fileid=25175&lang=en.

del Rio, P., & Bleda, M. (2012). Comparing the innovation effects of support schemes for renewable electricity technologies: A function of innovation approach. *Energy Policy, 50*, 272–282.

Eckert D., Ukraine or the uncertain limits of an European state, [L'Ukraine ou les contours incertains d'un Etat européen], L'Espace Politique 33, 2018, https://doi.org/10.4000/espacepolitique.4411, [En ligne], 2017-3, mis en ligne le 31 janvier 2018, Retrieved May 7, 2019 from. https://journals.openedition.org/espacepolitique/4411, (in French).

Eckert D., Ukraine: les deux temps de l'incorporation à l'Union soviétique (version en ukrainien), Retrieved May 7, 2019 from. https://hal.archives-ouvertes.fr/medihal-02018890, 2019, (in French).

Eckert D. and Lambroschini S., La ligne de démarcation entre séparatistes du Donbass et reste de l'Ukraine, Retrieved May 7, 2019 from. http://mappemonde.mgm.fr/119lieu1/, 2017.

EUR-Lex. (n.d.). European Union energy law. Retrieved May 7, 2019 from https://eur-lex.europa.eu/search.html?CC_1_CODED=12&qid=1557753305740&DTS_DOM=ALL&type=advanced&lang=en&SUBDOM_INIT=ALL_ALL&DTS_SUBDOM=ALL_ALL.

Fashchevskiy, N. I., Paliy, T. M., Nemchenko, M. P., & Starostenko, A. G. (1992). *Territorial organization of vital activity of the population (Kyiv).* [Фащевский Н.И., Палий Т.М., Немченко М.П., Старостенко А.Г. (1992) Территориальная организация жизнедеятельности населения. Киев] (in Russian).

Fouquet, R. (2012). The demand for environmental quality in driving transitions to low-polluting sources. *Energy Policy*, *50*, 130–141.

Fouquet, R., & Pearson, P. J. G. (2012). Past and prospective energy transitions: Insights from history. *Energy Policy*, *50*, 1–7.

Ghirga, G. (2010). Cancer in children residing near nuclear power plants: An open question. *Italian Journal of Pediatrics*, *36*, 60. https://doi.org/10.1186/1824-7288-36-60.

Grubler, A. (2012). Energy transitions research: Insights and cautionary tales. *Energy Policy*, *50*, 8–16.

Gukalova, I., & Poklyatskyi, S. (2014). Monitoring of health and demographic indicators in the context of the environmental situation (on example of Kyiv Dnieper region). *Ekonomichna ta Sotsialna Geografiya*, *70*, 81–91. Retrieved May 7, 2019 from 10.17721/2413-7154/2014.70.81-91. (in Ukrainian).

Harvey, C. (2015, November 2). Surprising study finds that solar energy can also cause climate change (a little). *Washington Post*. https://www.washingtonpost.com/news/energy-environment/wp/2015/11/02/surprising-study-finds-that-solar-energy-can-also-cause-climate-change-a-little/?noredirect=on&utm_term=.0c47d97b6e54.

Heinävaara, S., Toikkanen, S., Pasanen, K., Verkasalo, P. K., Kurttio, P., & Auvinen, A. (2010). Cancer incidence in the vicinity of Finnish nuclear power plants: An emphasis on childhood leukemia. *Cancer Causes & Control: CCC*, *21*(4), 587–595. https://doi.org/10.1007/s10552-009-9488-7.

International Energy Agency (IEA). (2018). *World energy balances: Overview. 2018 Edition.* Retrieved June 11, 2019 from https://www.iea.org/newsroom/events/statistics-world-energy-balances-2018-overview.html.

Kuzio, T., & D'Anieri, P. (2018, June 25). *Annexation and hybrid warfare in Crimea and Eastern Ukraine.* Retrieved June 15, 2019 from https://www.e-ir.info/2018/06/25/annexation-and-hybrid-warfare-in-crimea-and-eastern-ukraine/.

Mavili Elektronik Ticaret A.Ş. (2018). *Volga hydroelectric station.* Retrieved May 7, 2019 from https://www.mavili.com.tr/en/volga-hydroelectric-station.html.

Ministry of Energy and Coal Mining of Ukraine. (2016). *Electricity production by power companies and power plants of Ukraine for 12 months of 2016.* Retrieved May 7, 2019 from http://mpe.kmu.gov.ua/minugol/control/uk/publish/article?art_id=245183779&cat_id=245183225.

Ministry of Environment and Natural Resources of Ukraine. (2017). *Regionals reports on the state of the environment in 2016.* Retrieved May 20, 2019 from https://menr.gov.ua/news/31778.html.

Ministry of Environmental Protection of Ukraine. (2018). *Ministry of Environmental Protection has prepared the rating "TOP-100 largest pollutant enterprises" for 2017*. Retrieved May 7, 2019 from https://menr.gov.ua/news/32941.html. (in Ukrainian).

Ministry of Health of Ukraine. (2016). *Center of medical statistics. [database]* Retrieved May 7, 2019 from http://medstat.gov.ua/ukr/statdan.html.

Nikolayenko, D. (2017). Microgeography and ecology of *Riparia riparia* nesting on banks of the Trubezh River: A pilot epidemiological project. *Environmental Epidemiology*, *1*, 90–116 (in Russian).

Peters, M., Schneider, M., Griesshaber, T., & Hoffmann, V. H. (2012). The impact of technology-push and demand-pull policies on technical change—Does the locus of policies matter? *Energy Policy*, *50*, 1296–1308.

Radosavljevic, Z. (2017, August 17). *New EU rules for lower power plant emissions take effect*. Retrieved June 9, 2019 from https://www.euractiv.com/section/air-pollution/news/new-eu-rules-for-lower-power-plant-emissions-take-effect.

Rudenko, L. H., Lipinskyi, V. M., & Bondar, A. L. (Eds.), (1993). *Ukraine. Natural environment and people*. Map series. Kyiv. [Україна. Природне середовище та людина. Серія карт. Відп. ред. Л.Г. Руденко, В.М. Ліпінський, А.Л. Бондар. Київ, 1993. 55 с.] (in Ukrainian).

Santalova, S. O. (2018). Preconditions of the especially dangerous population diseases' emergence, development and spreading in Ukraine. *Ukrainian Geographical Journal*, *4*, 41–48. https://doi.org/10.15407/ugz2018.04.041 (in Ukrainian).

Savchuk, I. (2013). The main trends in the cities development in Ukraine. In *Urban territories changes in Ukraine*. Kyiv [Савчук И. Главные тенденции в развитии городов Украины в *Изменения городского пространства в Украине*. Под ред. Л.Г. Руденко. Киев, 2013. 160 с.] (in Russian).

Shevchenko, V. A. (1994). *Medico-geographical mapping of the territory of Ukraine*. Kiev. [Шевченко В.А. Медико-географическое картографирование территории Украины. Киев, 1994. 157 с.] (in Russian).

Shovkun, T., & Kostiuk, V. (2017). Spatial patterns of breast cancer morbidity (the case of Chernihiv Region). *Ekonomichna ta Sotsialna Geografiya*, *77*, 71–77. Retrieved May 7, 2019 from 10.17721/2413-7154/2017.77.71-77. (in Ukrainian).

Shovkun, T., & Myron, I. (2016). Geographical aspects of endocrine system morbidity of population in Chernihiv and Zhytomyr regions. *Ekonomichna ta Sotsialna Geografiya*, *76*, 44–49. Retrieved May 7, 2019 from 10.17721/2413-7154/2016.76.44-49 (in Ukrainian).

Spycher, B. D., Feller, M., Zwahlen, M., Röösli, M., von der Weid, N. X., et al. (2011). Childhood cancer and nuclear power plants in Switzerland: A census-based cohort study. *International Journal of Epidemiology*, *40*(5), 1247–1260. https://doi.org/10.1093/ije/dyr115.

State Agency on Energy Efficiency and Energy Saving of Ukraine. (2009). *Hydropower*. Retrieved May 7, 2019 http://saee.gov.ua/uk/ae/hydroenergy.

State Statistic Service of Ukraine. (2016). *Capacity of power plants and release of electricity, by region in 2016 in Energy supply and use. (Kyiv)*. Retrieved May 7, 2019 from http://ukrstat.gov.ua.

State Statistic Service of Ukraine. (2017). *Energy balance of Ukraine for*, 2017. Retrieved May 7, 2019 from http://ukrstat.gov.ua.

State Statistic Service of Ukraine. (2018). *The number of Ukrainian population (2018)*. Retrieved May 20, 2019 from http://ukrstat.gov.ua.

Temple, J. (2018, October 4). Wide-scale US wind power could cause significant warming. *MIT Technology Review*. Retrieved June 12, 2019 from https://www.technologyreview.com/s/612238/wide-scale-us-wind-power-could-cause-significant-warming/.

U.S. Environmental Protection Agency. (2016). *Air pollution and heart disease research.* Retrieved June 9, 2019 from https://www.epa.gov/sites/production/files/2016-01/documents/air_pollution_and_heart_disease_fact_sheet.pdf.

Ukrenergo State Enterprise. (2018, November 27a). *Ukrenergo enhances the reliability of the power supply of Kyiv region.* Retrieved May 7, 2019 from https://ua.energy/media/pres-tsentr/pres-relizy/ukrenergo-pidvyshhylo-nadijnist-elektropostachannya-kyyivshhyny.

Ukrenergo State Enterprise. (2018, Feruary 8b). *Ukrenergo will renovate one of the most powerful export transmission lines.* Retrieved May 7, 2019 from https://ua.energy/media-2/news/ukrenergo-will-renovate-one-powerful-export-transmission-lines.

Ukrenergo State Enterprise. (2019). *Technical report.* Retrieved May 20, 2019 from https://ua.energy/diyalnist/zvitnist/tehnichna/#1534493972464-f265f58b-260c3a1a-868f.

Ukrtransgaz Joint Stock Company. (n.d.). VIDNOVLYUVAL'NA ENERHETYKA (2018) Renewable Energy. [ВІДНОВЛЮВАЛЬНА ЕНЕРГЕТИКА] Retrieved May 7, 2019 from http://utg.ua/utg/business-info/vidnovliuvalna-energetyka (in Ukrainian).

World Bank. (2017). *Cause of death, by non-communicable diseases (% of total), Ukraine [Data World Health Organization World Health Statistics].* Retrieved June 10, 2019 from https://data.worldbank.org/indicator/SH.DTH.NCOM.ZS?locations=UA&view=map.

World Nuclear Association. (2019). *Nuclear power in Germany. [Last updated March 2019].* Retrieved June 15, 2019 from http://www.world-nuclear.org/information-library/country-profiles/countries-g-n/germany.aspx.

Yerem, T. V., & Yerem, K. V. (2016). Medico-geographical characteristic of residing conditions of the population in the Transcarpathian region by way of example of the total environmental contamination study. *Environment & Health*, *1*, 59–61 (in Ukrainian).

Zapadniuk, S. O. (2017). Trends of spreading of specially dangerous diseases of the population in Ukraine and some countries of Europe. *Ukrainian Geographical Journal*, *4*, 53–61. Retrieved May 7, 2019 from 10.15407/ugz2017.04.053. (in Ukrainian).

Supplementary reading recommendations

Baker, P. J., & Hoel, D. G. (2007). Meta-analysis of standardized incidence and mortality rates of childhood leukaemia in proximity to nuclear facilities. *European Journal of Cancer Care*, *16*(4), 355–363. https://doi.org/10.1111/j.1365- 2354.2007.00679.

Brody, M., Caldwell, J., & Golub, A. (2005). Developing risk-based priorities for reducing air pollution in urban settings. *Ukraine Journal of Toxicology and Environmental Health*, *68*(9), 356–357.

European Parliament, Council of the European Union, Regulation (EU) 2017/1369 of the European Parliament and of the Council of 4 July 2017 setting a framework for energy labelling and repealing Directive 2010/30/EU. Retrieved June 10, 2019 from. https://eur-lex.europa.eu/legal-content/EN/TXT/PDF/?uri=CELEX:32017R1369&rid=2, 2017.

International Atomic Energy Agency (IAEA). (2019). *Resources. [online database].* Retrieved June 15, 2019 from https://www.iaea.org.

Kounalakisa, M. E., & Theodoroub, P. (2019). A hydrothermal coordination model for electricity markets: Theory and practice in the case of the Greek electricity market regulatory framework. *Sustainable Energy Technologies and Assessments*, *34*, 77–86. https://doi.org/10.1016/j.seta.2019.04.012.

Page, S. E. (2006). Path dependence. *Quarterly Journal of Political Science*, *1*, 87–115.

Sevenster, M., Croezen, H., van Valkengoed, M., Markowska, A., & Dönszelmann, E. (2008). *External costs of coal Global estimate Delft*. Retrieved June 15, 2019 from https://www.cedelft.eu/publicatie/external_costs_of_coal/878?PHPSESSID=f1382192-38c72e8038a0a5694354af1d.

Tighe, C. (2010, April 23). Doubt cast over power plant's future. *Financial Times*. Retrieved June 15, 2019 from http://www.ft.com/cms/s/0/d7529c58-4e39-11df-b48d-00144feab49a,s01=1.html#axzz1JyBOJ2Gb.

Walker, E. W. (2017). *Eurasian geopolitics*. Retrieved May 1, 2019 from https://eurasiangeopolitics.com/ukraine-maps.

Wynn, G., & Coghe, P. (2017). *Europe's coal-fired power plants: Rough times ahead, Analysis of the impact of a new round of pollution controls*. Institute for Energy Economics and Financial Analysis (IEEFA). Retrieved June 15, 2019 from http://ieefa.org/wp-content/uploads/2017/05/Europe-Coal-Fired-Plants_Rough-Times-Ahead_May-2017.pdf.

Christer, E. (2010). *Emission ceilings may be further postponed*. Air Pollution and Climate Secretariat. Retrieved June 15, 2019 from http://airclim.org/acidnews/2010/an3-10.php#fourteen. Evaluation of the Member States' emission inventories 2004–2006 for LCPs under the LCP Directive (2001/80/EC), European Commission, 2008.

Definitions

Ecological dependence natural factors that affect the living conditions of people and their health.

Emission(s) a harmful by-product of producing energy, accessing raw materials, or other production/manufacturing that pollutes the air in the form of gas, light, or various particulate matter such as dust.

Energy security a complex network of measures to diversify the sources of supply of primary energy sources and the reliability of the functioning of the country's energy system.

Geography of health a science that studies the spatial patterns of the spread of morbidity or conditions that cause the emergence of certain diseases and measures to overcome them.

Indicator pathology indicator of the presence of the harmful substance/agent that occurs with the appearance or transmission of a pathogen.

Karst process natural weathering of topography, such as limestone and other carbonate bedrock, that forms unstable substructure caverns and shafts that divert waterways.

Medical geography a science that studies the spatial distribution of morbidity and health systems in different countries and regions.

Promoting healthcare access and addressing systemic health disparities in toxic environments

17

Tracy Wharton[a], *Andrés Cubillos Novella*[b], *Kenan Sualp*[c]
[a]School of Social Work, College of Medicine: Clinical Sciences, University of Central Florida, Orlando, FL, United States, [b]Institute of Public Health, Pontificia Javeriana University, Bogotá, Colombia, [c]College of Community Innovation and Education, University of Central Florida, Orlando, FL, United States

Abstract

Neighborhood-level factors and health disparities are often overlooked when conceptualizing the impact of the environment as either a protective or risk factor. Models of health and resource disparity inform our examination of the need to address health outcomes for marginalized populations in the absence of population mobility, broad exposure to toxins, and access barriers to resources. This chapter engages epidemiological data, a survey of available literature, and a case study of the City of Orlando to engage questions surrounding disparities of impact from toxic environments or physical impediments to resource access. A pattern of disparate exposure to toxic environments can be seen in data and mapping, perpetuating historical environmental injustice. Addressing environmental toxin exposure is a necessary step to addressing health disparities and inequities. Engagement with community stakeholders to identify strategies that benefit both business and community interests is a critical step.

Keywords: Environment, Disparity, Poverty, Toxin exposure, Orlando, Health

Chapter outline

Three Facets of Public Health and Paths to Improvements. https://doi.org/10.1016/B978-0-12-819008-1.00017-1

Environmental impact on well-being

Toxins in the environment, disparities among exposure, air quality, access to healthy food, transportation, and healthcare access all play critical roles in developing thriving communities. Geographic location is an established indicator of psychosocial and physical well-being impacting the economic health and quality of life of populations. Research on this issue is not new, although the linkages between health and the environment are becoming more well-known in popular culture, with organizations such as the Robert Wood Johnson Foundation announcing that "Where you live or work, your age, your preexisting health conditions or chronic illnesses, as well as your race or income all influence what health harms from climate change you experience—and how strongly you feel them," (Painter & Gandhi, 2019, *para* 6). Those who are subjected to the lower end of disparities in these domains face a feedback loop where environmental exposures and contextual stressors call for improvements in lifestyle, such as exercise and medication compliance, while at the same time generating ever-increasing pressure on access to resources necessary to meet those needs. The problem under investigation is threefold in nature: (1) How do we conceptualize health and well-being in diverse populations with few options to escape poverty where they live?, (2) what is the importance of neighborhood-level factors and health disparities, and (3) how do we conceptualize the environment as a protective or risk factor for those who live there?

Globally, news articles and scientific works have highlighted the impact of climate change on economic inequality, where poorer nations clustered near the equator are faced with the most dire consequences of extreme heat, weather events, rising sea levels, and other neighborhood-level disparities. Nations around the world must attend to the consequences of increased air pollution, residual toxins in soil and water, and the impact of poverty and environmental inequality on well-being (Aizer, 2017; Clark, Millet, & Marshall, 2017; GBD 2017 Risk Factor Collaborators, 2018; Salazar, Clauson, Abel, & Clauson, 2019; Sengupta, 2019). It is sometimes easy to dismiss these news reports as events happening elsewhere to people in other parts of the world or who are not at all like "us"; but, environmental impact is not an abstract issue and disparities of exposure to a range of toxins is something that requires attention even in our own backyards.

The United States has a well-established history of economic injustice, disparities of healthcare access, and availability of housing for families of color in limited areas within cities that have set the course of development on a skewed path. Cities such as St. Louis, Detroit, and Orlando have grown within these frameworks leaving some areas of population stuck in the midst of "development." As construction of new highways or expansion of business areas has led to gentrification, historical areas of minority demographics have been cut off from important assets. For example, living in close proximity to a power plant, or in a neighborhood that is encircled by highways may limit access to parks, safe areas for outdoor activities, transportation lines, and healthy food options and expose residents to high levels of toxins in the air and water as a result of surrounding development (Interdisciplinary Environmental Clinic, 2019; Painter & Gandhi, 2019). When residents of an area are unable to escape the impacts of environmental change around them, the ripple effect can be seen in all areas of their lives from health outcomes to work productivity. For the working class families and others who may be home owners, development has changed neighborhood factors, such as highways overhead rather than trees to such an extent that selling property to relocate may not be a viable option. The consequences of these conditions continue to expose residents to unwanted environmental factors that prevent mobility, leaving them stuck.

We draw attention to literature that concurs with the problem of geographic-based environmental (in)justice across the United States and share case examples where failing infrastructure and intergenerational poverty can be closely linked with racial, ethnic, and healthcare disparities. The impact of place on health outcomes suggests that we must concurrently address environmental disparities in order to adequately address racial inequalities. Toxicity in landscapes across the United States that hinders residents from achieving economic advancement which would provide them with the resources needed to adequately address associated health conditions such as asthma, hypertension, and neurodevelopmental impacts supports this position.

Further, the Centers for Medicare and Medicaid Services (CMS) Accountable Health Communities Model highlight the core components of health outcomes: housing stability, food insecurity, transportation needs, utility access, and interpersonal safety (Billioux, Verlander, Anthony, & Alley, 2017; CMS, n.d.). This is consistent with the Organization for Economic Cooperation and Development (OECD) and Centers for Disease Control and Prevention (CDC) frameworks that emphasize education, health, and financial resources as central to well-being of individuals, families, and communities (CDC, 2017; OECD, 2012).

This chapter provides important background context that focuses on the relationship between environmental disparity, healthcare access, and concentrated neighborhood disadvantages, some exploration of how theory and social determinants of health have an impact on health outcomes, and a case study using data to build "heat maps" to demonstrate the adverse effects of environmental disparities in communities located in the western sectors of the City of Orlando: (1) Parramore, and (2) Pine Grove. These two communities incorporate about 80,000 of the 2.4 million residents in the greater Orlando metro area. The area has historically been populated by African-American families with surrounding areas including Spanish and Creole-speaking residents. We conceptualize health discrepancies and the relevance of neighborhood-level health using this area to highlight clear disparities in environmental exposure that play out within a racial and economic framework. We also discuss both risk and protective factors that emerge and posit how these could be addressed through collaborations with stakeholders. The presentation of "heat maps" of several sectors of the City of Orlando visually depict how racial distribution and poverty are highly correlated with health disparities and environmental toxin exposure. We conclude with some suggestions for moving forward to address the issues presented in the chapter. Some of these recommendations include incentivizing city planners and administrators to directly engage with communities and that city-level economic development plans should include an assessment of the impact of population health supporting residential well-being.

Conceptualizing health and well-being in the context of place

Access is essential when we talk about healthcare systems, but access is not the only aspect because healthcare is a complex topic to consider for intervention. Consideration of healthcare is conceptualized as access (the ability to obtain care that is beneficial), quality of care, and health indicators that impact on well-being outcomes and drive people to seek care in particular areas (Levesque, Harris, & Russell, 2013). While there are arguments regarding need, who deserves assistance, and morality surrounding health insurance and care access in the United States, research suggests that broad care coverage could reduce inequities, improve universal access to services, and improve quality of services (McCollum et al., 2019; Meyer, Luong, Mamerow, & Ward, 2013; Starfield, 2011).

Social determinants of health (SDOH) are important indicators of health and well-being (McGibbon, Etowa, & McPherson, 2008; Soeung, Grundy, Sokhom, Blanc, & Thor, 2012; WHO, n.d.). According to the World Health Organization (WHO), SDOH are the conditions into which people are born, grow, work, live, and age, as well as the wider set of forces and systems shaping the conditions of daily life (WHO, n.d.). If health systems are to be socially just and equitable, issues such as physical and financial access and cultural responsiveness must be part of the systemic design. One theoretical model which captures this concept particularly well is the Tanahashi Equity model (Henriksson, Fredriksson, Waiswa, Selling, & Peterson, 2017; see Fig. 17.1). This model defines five dimensions to consider regarding adequate healthcare for a population: availability, accessibility, acceptability, contact, and quality. These aspects are at the core of discussions related to distributional equity of healthcare services (McCollum et al., 2019). At each level of this model, several factors converge to influence who has access to services within the larger health system. The environment, in terms of physical and social dimensions, has a strong relationship

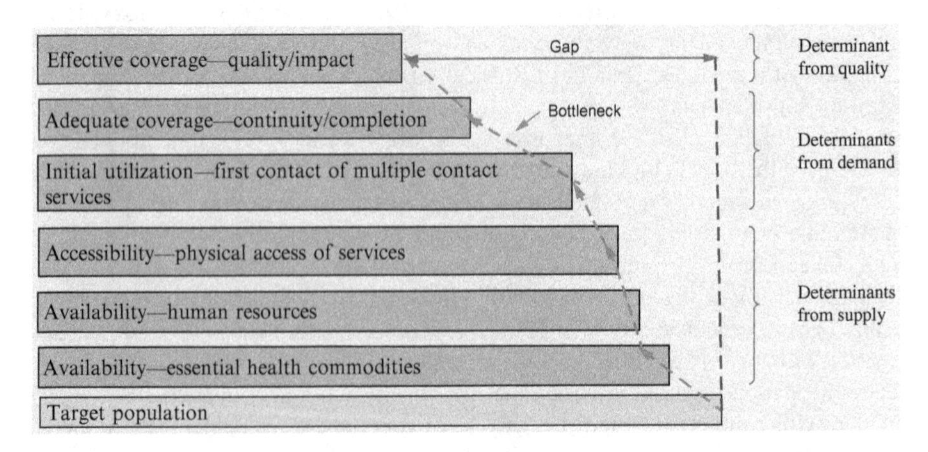

Fig. 17.1 Tanahashi model (Used by Permission, Henriksson et al., 2017).

with accessibility and subsequent use of health services. The National Research Council, related to the Institute of Medicine in the US, notes that many aspects of the physical and social environment can affect people's health and access to care (O'Connell & Sharkey, 2013), and the OECD identifies the critical nature of environmental context in health.

Woolf & Aron (2013, p.194):

> *The factors in the physical environment that are important to health include harmful substances, such as air pollution or proximity to toxic sites (the focus of classical environmental epidemiology); access to various health-related resources (eg, healthy or unhealthy foods, recreational resources, medical care); and community design and the "built environment" (eg, land use mix, street connectivity, transportation systems).*

The impact of various environmental and other conditions is understood within the framework of social determinants of health. The term "social determinants" often evokes factors such as the characteristics of health-related neighborhoods (e.g., walking areas, recreational areas, and accessibility of healthy foods), which can influence health-related behaviors. However, evidence continues to highlight socioeconomic factors such as income, wealth, and education as fundamental causes that influence health outcomes within populations (Braveman & Gottlieb, 2014; CDC, 2017).

According to the WHO (2013), health among individuals within geographic areas is influenced by both developmental and biological context; social networks, habits, customs and beliefs, education, and culture all play strong roles, along with social class and economic strength of an area. The Tanahashi model helps to explain this phenomenon, noting that supply, demand, and need are all linked within the context of a geographic area (Henriksson et al., 2017). The gap between the overall needs of a population and the availability and quality of services can create a bottleneck in provision of care that impacts the overall health and well-being of communities.

Culture and ethnicity are tied to health inequities. This is not only an issue of differences in access to or preference for care, but also language accommodation and systemic bias, outreach and accessibility to communities, and overlap with racial disparity. In Florida, issues of race, ethnicity, and language all play a part in discrimination and social injustice, yet data are difficult to parse on this point. With large areas that are demographically Latinx, bias related to perceptions of class, language use, and skin shade all may become conflated in this geographic area. Individuals who are identified as "White" and "Hispanic/Latinx" may face little discrimination if they have no discernible accent to their speech. Yet, a similar individual with a Spanish-language accent may face substantial discrimination despite the color of her skin. Such intrinsic bias can have insidious harms on the health status of people who are identified as a member of a minority group (Ouyang & Pinstrup-Andersen, 2012). Culture and language use, considered as SDOH, can have impact on how (or whether we are able to accurately) we measure inequality and health across cultural boundaries (Lauster & Tester, 2010).

As Rothenberg (1998) points out, many theories related to social identity posit primacy of one domain over the other, shifting between race, gender, ability, or socioeconomic status, for example. Application of theory, however, must consider the *intersectionality* of these domains and the complex nature of the human experience. Grounded in Western orientations of "difference," analysis often applies false dichotomies, such as White and non-White, or women and men, and privileges White, English-speaking, able-bodied, heterosexual, cisgendered, and Christian as the default, labeling anyone diverging from those categories as different (Delgado & Stefancic, 2001; Rothenberg, 1998). In reality, however, such terms fail to acknowledge the rich heterogeneity of humanity.

Social determinants of health, for example, may identify access to healthy foods as a central need, but analysis of available data may fail to take into consideration cultural dietary variances and preferences of those who live within a region. If the only food that is accessible in an area is food that no one wants to eat, that is equally as problematic as not having the option at all. Similarly, health care providers are generally defined as licensed medical doctors offering outpatient primary care or emergency services in an area. This does not, however, consider the potential presence of community health workers, nurse-led clinics, mobile clinics, or curanderas (community healers). Advantage and disadvantage work along multiple dimensions simultaneously, and such compounding of forces must be considered by policy makers and community leaders as solutions are designed.

The Accountable Health Communities Model being applied by the Centers for Medicare and Medicaid Services (CMS) attempts to address this issue by recognizing evidence that is emerging in relation to the inclusion of health-related social needs factors by enhancing clinical-community links. These linkages are purported to have a positive impact on health outcomes with the added bonus of cost reduction. "Unmet health-related social needs, such as food insecurity and inadequate or unstable housing, may increase the risk of developing chronic conditions, reduce an individual's ability to manage these conditions, increase health care costs, and lead to avoidable health care utilization," (CMS, n.d., *para* 2). The CMS model is focused on both including assessment of SDOH among those accessing healthcare and aligning community needs with local provision of care to build strategies to expand and support accessibility for those living in those communities. This approach recognizes the diversity and heterogeneity of geographic areas and attempts to align national and global priorities for health and well-being with the available resources in a local area, with attention to activating mechanisms to eliminate barriers for sustainable change.

Starfield (2007) considers that equity in health is related to how people are characterized socially, geographically, or demographically. According to Starfield, there are several pathways in which inequity is generated and maintained. First, the extent of inequities across the country varies according to the measure of health and the relative frequency of different type of illness; second, the level of geographic aggregation of data influences conclusions about the nature and extent of inequities; third, health services can contribute to reductions in inequity in health (Starfield, 2007). This highlights the challenges mentioned previously regarding language and cultural demographics, as well as issues of variance within large counties. If data are only available within a large geographic tract (e.g., Orange County in Florida), rather than at the census tract level, important variance within the demographics of that area is lost.

Maignant and Staccini (2018) expand on this idea and define the relationship between environment and health, considering two categories: structural and organizational. They posit that not only does the physical environment impact on the ability to work in a range of ways, but that considerations about pollutants necessarily take into account not only present-day harm, but also consideration of what kind of environment should be left for future generations. Such assessment of actions not only can prevent or reduce harm for those living in a space at the current time, but may serve to raise awareness of issues that may impact future generations and the impact of environmental deterioration on health (Maignant & Staccini, 2018). Some theorists propose that lifetime exposure to toxins in the environment as well as both psychological and socioeconomic stressors contributes to overall health independently of genetic disease transmission. This concept moves beyond the relationship linking a simple contaminant and a single associated effect, focusing on complex and multiple exposures, known as a cocktail or cumulative effect (Maignant & Staccini, 2018).

Theoretical linkages between populations and health outcomes are clear in that there is a solid link between barriers to healthcare and environment, culture, socioeconomic conditions, and available access to both providers and clinics. Health and well-being are complex constructs and the development of the environment has specific influences on demographic groups, impacting physical and environmental conditions such as safety, where the worst outcomes are usually related to areas with the largest resident minority groups.

Environment as a risk or protective factor

Environmental disparity may be more commonly associated in popular culture with Sub-Saharan terrain, but this problem is not limited to foreign land and the phenomenon can be found in many US cities, particularly in urban areas with a history of racial divide. In the United States, concerns about health and resource disparity have often been documented along racial lines with recognizable and persistent impact of racial disparities on health and well-being (Chavis, 1994; Rothenberg, 1998). The general term used to define this concept is *environmental racism.*

Sicotte (2008, p. 1140):

> *Environmentally poor or hazardous conditions suffered by a community due to the race or ethnicity of its residents. It is nested within an understanding of environmental inequity... (and) enclosed within the broadest term:* environmental injustice... *unhealthy and unfair environmental conditions when suffered by any type of community.*

A substantial body of work describes environmental racism in the United States, both in terms of policy creation that excludes communities of color (e.g., "redlining"—the practice of defining who was allowed to purchase homes in various areas) and targeting of communities (e.g., placement of unwanted features like garbage dumps or processing plants). This perspective includes a deliberate attempt to create disparities throughout history and the impact of such choices on the longitudinal development of present-day disparities (Chavis, 1994; Clark et al., 2017; Pulido, 2000; Salazar et al., 2019; Sicotte, 2008).

Geographic location can be a predictor of psychosocial and physical well-being that, in turn, impacts the economic health and quality of life for various populations. This speaks to an issue of *distributional justice*. While national law includes guarantees of "meaningful involvement and fair treatment regarding development, implementation, and enforcement of environmental laws, regulations, and policies," (US EPA, 2006), a history of uneven economic development in the country has deeply embedded issues of inequity in neighborhood quality, financial opportunity, and healthcare access (Fitzpatrick & LaGory, 2003; Salazar et al., 2019; Sicotte, 2008). Indeed, evidence suggests that burden of exposure to toxic waste has been "channeled into communities with less social and economic power," (Sicotte, 2008, p. 1142) as well as lower income urban areas consistently since the 1970s. Data show that income inequality has increased in every state in the United States since that time. Policy decisions, however, are often shaped by those who are at the top end of resource access and quality of environmental surroundings (Salazar et al., 2019).

Although racial inequality remains persistent across the country, in some areas of deep poverty, intergenerational struggles with healthcare access, low socioeconomic power, and poor social capital have flattened out some disparities across race, lowering all who live in the area to the same level of poverty. In lower central Alabama, for example, the area termed "the Black belt"—so named for the dark colored, iron-rich soil—is an area of intergenerational poverty, despite repeated initiatives to capitalize on the natural resources in the area (Wharton & Church, 2009). While racial gaps in income and home ownership have gotten smaller, low home values, ongoing pollution of soil, water, and air, isolation from healthcare, transportation, and access to healthy food have kept residents of this region in a cycle of chronic, intergenerational struggle (Wharton & Church, 2009).

Although this impact of poverty is strong, racial disparities persist and a consistent body of evidence addresses issues of spatial inequality and relationship to injustice (Interdisciplinary Environmental Clinic, 2019; Pulido, 2000; Salazar et al., 2019). Salazar et al. highlight findings that while differing factors may drive disparities of race and economy, "states where (toxin) exposure remains at the highest levels also exhibit greater poverty-based and, especially, race-based environmental inequality," (2019, p. 598). Literature is consistent regarding the consequences of toxin exposures, indicating that even after passage of legislation to ensure clean air and water, those living in areas with consistently high exposure rates showed higher rates of infant mortality and newborn intensive care admissions (Aizer, 2017), neurodevelopmental disorders (Hennig et al., 2018; Legot, London, Rosofsky, & Shandra, 2012), hypertension (Hennig et al., 2018), asthma (Di et al., 2017; Legot et al., 2012), and ADHD (Johnson, Riis, & Noble, 2016; Weissenberger et al., 2017). Additionally, evidence reliably notes that people of color resided in areas that were an order of magnitude more toxic than their White, non-Hispanic counterparts (Aizer, 2017;

Clark et al., 2017; Interdisciplinary Environmental Clinic, 2019; Legot et al., 2012; Weissenberger et al., 2017).

Across the country, disparity in who is exposed to environmental toxins and which neighborhoods are given to border highways, garbage dumps, and long-term construction sites is a crisis that is systemic, rather than isolated (Filippelli & Laidlaw, 2010; Hadler et al., 2018; Kim & Williams, 2016; Legot et al., 2012; Sicotte, 2008). Despite our standing as one of the wealthiest nations in the world, we have repeatedly learned of areas where there is a lack of clean and safe drinking water. Stories about Flint, Michigan, and the Navajo nation have highlighted the socioeconomic differences and low political power of those who are impacted by lack of water access. The Lakota Sioux nation and the fight at the Standing Rock reservation stand as a stark example of how choices are made regarding who should be put at risk for water contamination. Mapping for an oil pipeline was intentionally chosen to cross the water source for the Standing Rock Reservation, home to roughly 6200 people, rather than travel near the urban, and mainly White, population of Bismarck, ND, a city of 66,980 people, or take a longer route that would avoid population centers altogether. Environmental studies were split on the overall impact of one route versus another and legal documents indicate callous disregard by the contractors for safety regulations. Economic incentives to go the shortest route were valued by the designers more than the well-being of those living on the reservation.

Similarly in Flint, where lead in the pipes was poisoning children at rates well above safety levels, families who had the financial capacity to replace pipes and install filters on their properties were able to return to relative safety from lead exposure fairly quickly. On the other hand, the majority of this low income, postfactory town were forced to monitor toxin exposures, drink bottled water for years, and limit bathing, handwashing, or dishwashing. In addition to the obvious hazards from contaminants, noted decreases in the ability to support good hygiene factors such as handwashing can also contribute to spread of infectious disease. Thousands of children from this area will face the lifelong consequences of high levels of lead exposure, and the town will have to deal with higher rates of learning disabilities, behavioral disorders, and neurological disorders long after the water pipes are fixed. Following the same story line, the activist Majora Carter gave a TED talk describing the disparities in neighborhood exposures in the South Bronx in New York City (Carter, 2006). She described the placement of city waste dumps in poor neighborhoods and how only those in low socioeconomic brackets seemed to be exposed to the environmental toxins produced by a bustling city.

For families who own property, or those who live on ancestral lands, selling and moving may not be a simple option. Alongside the psychological connections to homes that have been inhabited for generations, economic pressures and the need to access transportation often keep people stuck in places that may not be optimal for psychosocial or physical well-being. Critically, evidence suggests that while economic changes, such as closing of factories, may lead to overall improvement in *environmental* conditions (fewer factories leading to less pollution), there is also a correlational impact on *political* power, as individuals living in those areas lose income, influence, and economic stability (Salazar et al., 2019). This change feeds into policy decision making in important ways, leading to uneven determinations of need and resource allocation for mitigation of toxic environmental conditions, as those who are most impacted become less able to influence policy development.

The picture, however, is not entirely bleak. There are a number of strengths and protective factors that can mitigate adverse impacts for those stuck in environmentally or economically disadvantaged areas. Resilience among residents of neighborhoods, access to education for children, and the right to organize for political influence are strengths-focused characteristics that should not be overlooked. Evidence suggests that neighborhood-level proximity to others who speak the same language or share cultural, racial, or ethnic experience may be protective against some of the stress produced by the environment and social context, such as maintenance of ethnic cultural traditions, stronger social network support for family caregiving, and lower levels of depression (Rote, Angel, & Markides, 2017). However, while there is a buffering effect related to language and demographic similarity (sometimes referred to as "the barrio advantage"), this is quickly offset by resource deprivation, imposition of environmental toxin exposure, and financial strain in neighborhoods that are both poor and majority inhabited by minority families (Di et al., 2017; Pinderhughes, 2017; Rote et al., 2017; The National Academy of Medicine, n.d.).

Substantial research links race, education, and wealth (Chetty et al., 2016; Shiels et al., 2019), with continued widening gaps in life expectancy among poor and wealthy in the United States and persistent racial disparities. Cities such as San Francisco and New York, rating relatively high on both wealth and education measures, presented an average of 5 years of additional life when compared to the somewhat lower indexed cities of Detroit or Gary, Indiana (Chetty et al., 2016). National statistics continue to evidence strong correlations between all-cause mortality and residence. For example, wealthier counties have longer life expectancy than poorer counties and this holds across all categories of ethnicity, race, and age. Differences are notable across racial groups, however, which is neither new nor surprising information in the United States, with mortality highest among people of color and lowest among White demographics (Shiels et al., 2019). Income and resource accessibility, however, are central factors in unpacking such disparities; recent analyses indicate that these factors may be flattening out mortality rates and impacting disparities related to obesity, physical inactivity, and access to healthcare across racial lines, particularly in smaller urban counties and rural areas of the country, even as disparities persist in larger urban areas (Salazar et al., 2019; Shiels et al., 2019).

Healthcare access as an environmental factor

Access to healthcare and insurance to pay for health-related services in the United States are both closely linked to the issues outlined so far. While insurance coverage across the nation has improved markedly since the roll-out of the Affordable Care Act in 2010 (Blewett, Planalp, & Alarcon, 2018; Kino & Kiwachi, 2018), there remain pockets of people who are uninsured or underinsured (Dreier, Mollenkopf, & Swanstrom, 2014). These often correspond to geography, where limited access or structural barriers of access (such as walkability or transportation) to healthcare providers and facilities to meet the needs of the population feed into the overall need for insurance among residents. Intergenerational patterns of chronic health conditions related to toxic environmental exposure characterize areas that are highly correlated to healthcare deserts where care is not available at all.

In addition to persistence of limitations to insurance access in vulnerable areas, concentrations of chronic illness and isolation may exacerbate disparities. Areas of toxin exposure may lead to higher rates of respiratory illnesses and limited access to grocery stores and safe areas to walk or play in neighborhoods may lead to higher rates of obesity, diabetes, and heart disease. Further, chronic exposure to soil and water toxins may lead to high rates of autoimmune disorders. However, health centers that can serve these needs are often not located parallel to the toxic sources and residents may face barriers such as transportation and insurance coverage required to access such resources. Spatial distance, transportation, and cost limitations for accessing services located outside of the immediate area can exacerbate conditions and disease rates. These factors create a gap between need and receiving services and illustrate the adverse effect of spatial and resource disadvantage on health outcomes, quality of life, and longevity (Dreier et al., 2014; The National Academy of Medicine, n.d.). Furthermore, the relationship between socioeconomic status and burden of such issues is strong (GBD 2017 Risk Factor Collaborators, 2018).

The importance of neighborhood-level factors and disparities

If air quality, neighborhood safety, access to healthcare, and water and soil safety are factors that influence the quality of health and well-being of residents, people residing in marginalized or vulnerable areas must be empowered to use existing evidence to generate action to address the social injustice of these disparities. With increasing numbers of "climate refugees" fleeing from rising sea levels and shifting weather events that put their physical homes in danger, and populations chronically exposed to toxins in the air, water, and soil, as well as proximity to waste dumps or factories that are co-located with playgrounds in neighborhoods, evaluating data and potential actions are critical approaches that we must take to mitigate the impact on both physical and mental health.

In the sunshine state of Florida, demographics tell stories of diverse populations spread out across very urban and very rural areas. Florida's history mirrors the rest of the nation in use of redlining tactics and segregationist development. Historical limitations on who was allowed to purchase homes or the value of mortgages allowed in certain areas have created pockets of inequity that have persisted for generations. Jargowsky (2015) describes the growth of suburban areas to accommodate White and wealthy flight from urban areas resulting in population loss and business decline, which, in turn, led to public housing development and low property values reinforcing spatial disparities. Two trends, including highway development through minority neighborhoods and gentrification in some previously minority areas, have led to forced migration among those who find themselves no longer able to afford property taxes. Additionally, influx of people fleeing natural disasters from the Caribbean and economic upset in other countries (e.g., Venezuela, Guatemala) have led to rapid increases in population density in urban areas in Florida where affordable housing and transportation options are slightly less scarce than more spatially disparate locations. The common trait among these groups before relocation is their residence in areas that are disproportionately exposed to toxins and other environmental stressors by comparison to wealthier areas of town that are less demographically heterogeneous.

People who live in dense urban areas have higher risks for physical and psychological impact as a result of environmental stressors. Craven (2018) illustrates this problem with evidence that even breathing is a risk in Orlando's poorest neighborhoods. Descriptions of children wearing masks to protect themselves from the pollution caused by two large highway construction projects alongside the neighborhood are a stark contrast to some other areas of town where trees are protected and green space abounds. For children in these urban neighborhoods, thousands of cars passing on the highways continuously release particulate matter that raises rates of respiratory distress and impacts development among even the healthiest children (Aizer, 2017; Clark et al., 2017; Filippelli & Laidlaw, 2010; Legot et al., 2012). Similar to the situations experienced by the Lakota nation with the routing of the oil pipeline, routing of roads in urban areas such as Orlando inherently values economic impacts and forces choices about whose neighborhoods will be subjected to such development.

The case of the City of Orlando

The urban areas of the City of Orlando located in central Florida provide an excellent case study. High impact business development, including a sports stadium and university campus, and highway expansion have spread across large tracts of land in an area that has been historically populated by African-American families. The areas known as Parramore and the nearby area of Pine Hills, just west of the major north-south highway artery of Interstate Four (I-4), are characterized by small homes that have been owned for decades, active neighborhood centers that serve area children and older adults, and a demographic of older adults who use the sidewalks in the area to run errands and visit neighbors. Although these areas have been relatively stable, recent years have seen an increase in crime rates as business sprawl has encroached on the neighborhoods and the construction has moved ever nearer, leading to the elimination of trees and other buffer zones from the traffic and construction. Similarly, in nearby urban areas, apartment blocks situated in high traffic areas with few sidewalks or open green spaces, such as parks, are mostly populated by Spanish or Creole-speaking families, many of whom work in the hospitality and agricultural industries that thrive across the region.

Using data from the Centers for Disease Control and Prevention's Environmental Public Health Tracking Network (CDC, 2019), the US Census (US Census Bureau, 2019), and data from the City of Orlando (City of Orlando, n.d.), we examined density of population by race and socioeconomic status, along with prevalence of obesity, asthma, chronic obstructive pulmonary disease (COPD), and toxin exposures in the air and water. Using these data, we map prevalence of some of these variables by census tract, but note that some toxin exposure is only available by county. Data such as these give us the information needed to produce maps that can quickly show the distribution of these issues across geographic areas using GIS mapping.

The Orlando metro area (i.e., Orlando, Kissimmee, and Sanford) comprises just over 16% of Orange County, FL and is located about halfway down the peninsula. The large urban area is surrounded by dense suburbs and accounts for about 420,000 of the 1.4 million estimated residents in Orange County in 2019 covering roughly 147/903 square miles in the county per the last assessment in 2010 (US Census Bureau, n.d. [database]). Although many imagine the state as a narrow peninsula, the drive from coast to coast is approximately 3 hours and Orlando is right in the middle. While the state benefits from weather patterns off the Atlantic Ocean and Gulf of Mexico, Orange county is somewhat less impacted by patterns that would disperse toxins in the air. The average air toxin rate is higher in that county than most of the rest of the state, and higher than many other urban areas, at 2.55 $\mu g/m^3$ of air toxins; consequently, data indicate that cancer risk related to air toxin exposure is higher in Orange County than nearly every other county in the state (CDC, 2019, National Environmental Public Health Tracking Network database). While it is outside of the strict limits of these data to interpret reasons for this, other research supports the conclusion that heavy traffic and construction have led to high levels of fine particulate matter in the air in the area (Clark et al., 2017; Craven, 2018; Di et al., 2017). This particulate matter is easily found in low-income neighborhoods in the areas west of I-4, historically an area with primarily African-American residents.

Fig. 17.2 shows the density of minority populations in the area. Imagine a north-south dividing line, right through the center in Orlando in this graphic, and a west-east dividing line that runs just below the name of the city. Those are the major highway arteries: I-4 running top to bottom, roughly along the historical train routes in the area that marked racial segregation, and Highway 408 running left to right. The crossing lines in the middle of the urban area are the literal dividing lines for the city. This is further emphasized in Fig. 17.3, which overlays racial distribution onto data identifying areas where greater than 35% of households in an area have incomes below the official poverty line. Although formal segregation has been relegated to the dustbin of history, the maps clearly depict ongoing divisions. Generated from CDC data (2019), the maps use "White, non-Hispanic" as the reference category and all other categories (including Latinx/Hispanic categories) are included in "non-White" to map population density.

Figs. 17.2 and 17.3 demonstrate that the areas west of I-4 are greater than 70% minority residence areas with a substantial portion of that density falling to the north of the Highway 408 line. Data calculated for these areas indicate that between 1/3 and 2/3 of adults living in these areas report that they have no leisure time physical activity, and 34% report having no health insurance (CDC, 2019). These statistics may be linked to the high rates of obesity and COPD that can be seen in the same area.

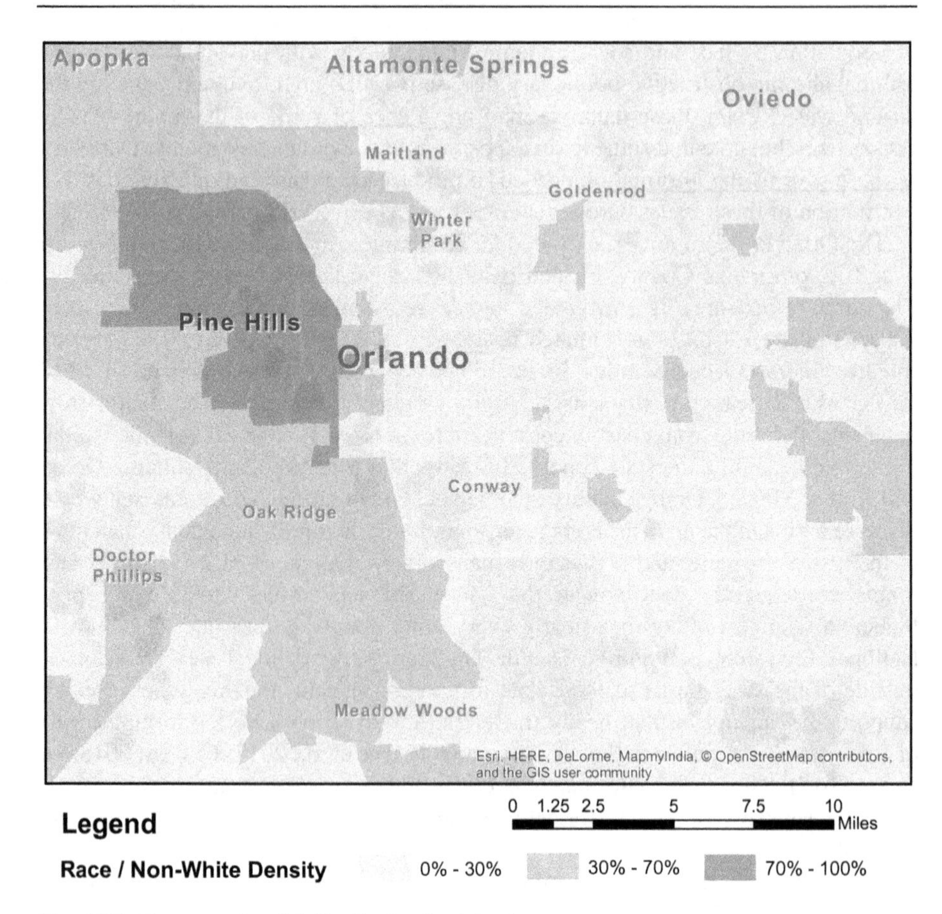

Fig. 17.2 Heatmap of racial distribution in Orlando metro area.

Significant development in these same areas, including a major highway expansion and several major building construction projects, correlates to high rates of asthma among adults in the area as shown in Fig. 17.4. Overall, the zones west of the highway are considered to be areas of concentrated disadvantage, where poverty, resource access, and racial disparities intersect (Sampson, 2013). While people living in these areas may possess health insurance as high as 66% self-reported in the raw data (CDC, 2019), there is lack of information to confirm whether local primary care locations accept those insurances or if co-payments for care are manageable. Additionally, although major medical centers are geographically proximal (i.e., within a few miles of the cross-point of the highways, to the southeast), walkability in this part of the city is low and parking for those with cars can be challenging. Structural barriers and lack of infrastructure, such as bus-stop shelters to protect from the daily rain and extreme heat, can impede the actual ability to use services which may be present in the overall environment.

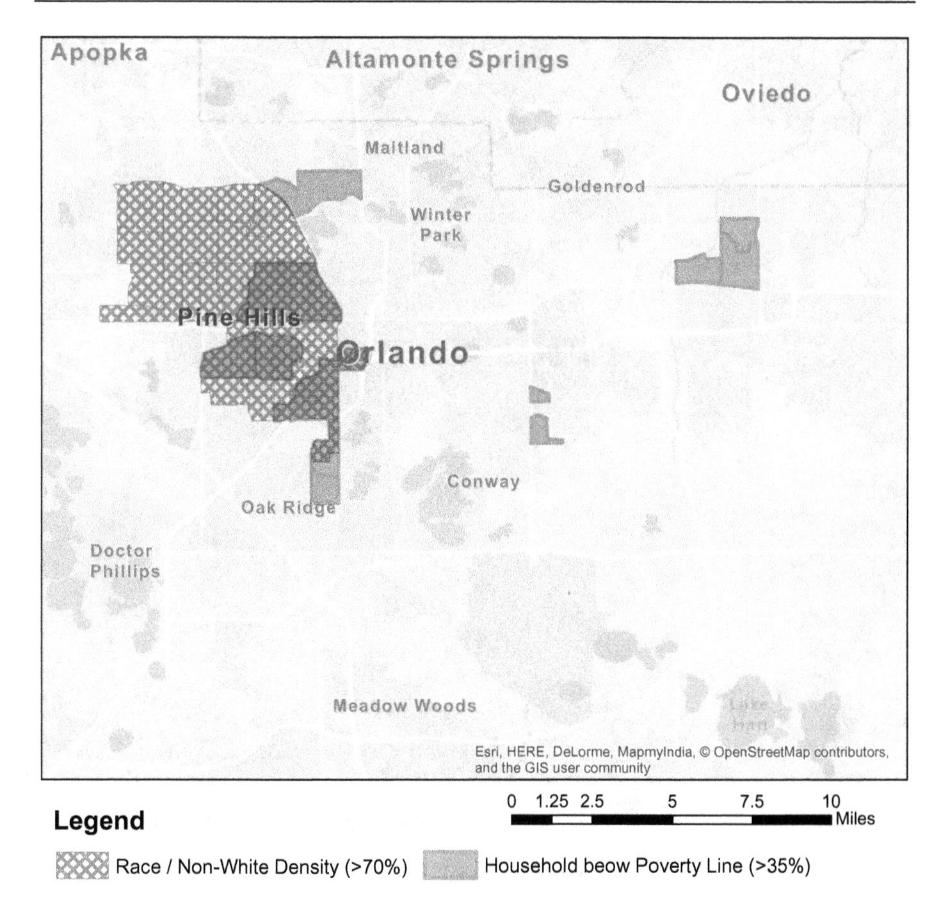

Fig. 17.3 Heatmap of race and poverty distribution in Orlando metro area.

Disease prevalence, correlation with racial distribution and poverty rates, and limited meaningful access to services are readily observable in the context of data for the sections of Orlando west of I-4. Although medical care may be geographically proximal to these areas, there are few grocery stores in the downtown area of the city (i.e., food desert). While bus service is available to several major retail centers, travel by this service is known to be slow, sometimes taking an entire afternoon to accomplish the task of food shopping for a family. More convenient food options in the downtown area include small shops that stock little produce or meat and are more likely to offer a range of less than optimal food choices, leaving families with an undesirable choice when facing nutritional decision making. Such structural barriers related to walkability and presence of sidewalks and lighting, bus shelters, limited parking and slow or limited public transportation, or overall personal safety for walking alone can prevent access to resources that may be geographically nearby, yet still effectively out of reach. Furthermore, while policy actors may point to bus

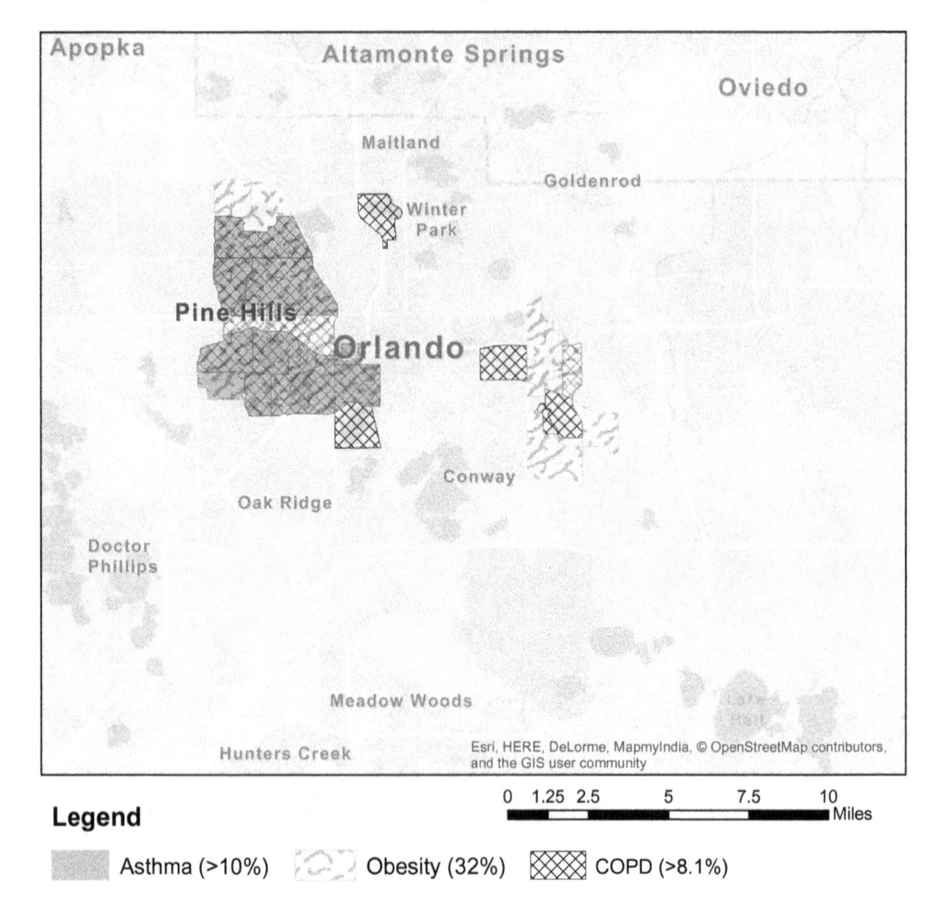

Fig. 17.4 Heatmap of concentrated disease areas.

transportation to resources, geographic proximity of hospitals, and the overall impact of economic development as a stimulus for the city, the operationalization of meaningful access to options for mitigating the toxic impacts of the environment around them remains slow to come for people in these sections of the metro area.

Recommendations and conclusions

While entrenched social inequity is a long-term challenge, there are some actions that can be embedded into governance and community action that may prove impactful in the short-term. For administrative officials, building rapport and relationships with diverse communities across regions is critical. Inequities by race and by poverty involve different phenomena, and strategies for addressing these in communities must be subsequently responsive and involve engaging the unique voices of those who are most impacted (Delgado & Stefancic, 2001; Salazar et al., 2019). Engagement directly with residents, with a genuine approach that includes both listening to concerns and asking for input, can provide solutions that benefit everyone and recognize the environmental and intersectional cultural patterns that are present in the most deeply impacted areas. Communities have an on-the-ground perspective that can benefit officials, not only in providing opportunities to improve the overall health and well-being of people living in the area, but in identifying areas where construction, urban planning, or economic activity might be improved.

Global data confirm that obesity, heart disease, and healthcare access are strongly correlated to race, ethnicity, and poverty, and that those at the bottom end of disparities face barriers of political power in policy making that could mitigate their situations (GBD 2017 Risk Factor Collaborators, 2018; Salazar et al., 2019). Recognition of the impact of political power on health and well-being of nearby residents must lead to mitigating approaches that include attention to environmental conditions that impact public health. Attention to walkability of neighborhoods into design, mitigation of toxin exposure in soil, water, and air, and intentional focus on addressing issues of environmental racism and economic injustice can help to improve both chronic disease rates and mental health stressors.

In this context, community organizers must consider diverse and creative paths to improvement in order to make sustainable impacts on intergenerational health disparities related to environmental stressors and socioeconomic barriers to services. Clearly, these barriers also include limitations to health insurance and medical care, persistent exposure to environmental toxins, and limited options for healthy food that are significant contributors to intergenerational transmission of poverty, making these important issues to address if sustainable change is to occur (Aizer, 2017; Rote et al., 2017). Organizing for political influence and advocacy, although never easy, is a critical point of intervention for marginalized communities. Policy is often made by those who are far from the communities being impacted, and pressure must be brought to educate policy makers about the broad impacts of decisions such as road mapping, distribution of businesses and construction projects across metro areas, and density of population relative to air pollution. This must go beyond meetings with gatekeepers to engage people who live in resource poor areas to problem solve neighborhood-level solutions. Additionally, political attention to environmental impact on well-being can help push development in a direction that supports residential well-being. For sustainable change to occur, top-down imposition of environmental planning must end.

Growing partnerships with universities and colleges can open opportunities for both interprofessional training of the healthcare workforce and doors to creative solutions for addressing healthcare access. Local mobile health clinics, community engagement for dissemination of health promotion and public health education, and broad implementation of risk screening tools related to health-related social needs (e.g., Billioux et al., 2017; National Academy of Medicine Accountable Health Communities screening tool) may offer paths toward improvement. With such strong evidence that food instability, transportation and utility assistance, and safe neighborhood residence are linked to health-related outcomes, practice guidelines are moving toward consistent recommendations to assess for these factors in every healthcare setting. Evidence is still unclear on whether screening tools that identify contextual risk can improve healthcare outcomes for individual patients, when applied across broad outpatient services. Concerns about inadequate training, perceptions of burden in practice, and issues of limited resources for referral are a persistent thread. Evidence suggests, however, that this approach, incorporating psychosocial determinants of health into primary outpatient care, shows promise, particularly when brief, evidence-based screening tools are employed (Billioux et al., 2017).

Similarly, while mobile or "pop-up" health clinics should not be a permanent solution to access for vulnerable populations, they can offer a measure of preventative care and management of chronic health issues that reduce impact on emergency and urgent care providers in a region. Most importantly, there must be broad acknowledgment among leadership at local, state, and national levels that (1) environmental stressors generate healthcare consequences which ripple out to economic impact on entire regions, and (2) longitudinal data that provide objective ability to assess change along lines of disparity are critical to advancement of research in this area. Such debate need not be between business hawks and environmental doves; this debate is more appropriately framed as both a humanitarian and social justice issue, as well as an economic imperative. Exposure to toxins and other environmental stressors will continue to lead to higher healthcare costs. Even if the United States had a ready solution to our health insurance dilemmas, the disparate impact of such stressors on the poor and uninsured ripples out to labor force and tax impacts for the broader population, making this an issue that should be at the top of our national priority lists.

References

Aizer, A. (2017). The role of children's health in the intergenerational transmission of economic status. *Child Development Perspectives*, *11*(3), 167–172. https://doi.org/10.1111/cdep.12231.

Billioux, A., Verlander, K., Anthony, S., & Alley, D. (2017). Standardized screening for health-related social needs in clinical settings: The Accountable Health Communities Screening Tool. *National Academy of Medicine: Perspectives*, *7*(5). https://doi.org/10.31478/201705b.

Blewett, L., Planalp, C., & Alarcon, G. (2018). Affordable care act impact in Kentucky: Increasing access, reducing disparities. *American Journal of Public Health*, *108*(7), 924–929.

Braveman, P., & Gottlieb, L. (2014). The social determinants of health: It's time to consider the causes of the causes. *Public Health Reports*, *129*(Suppl. 2), 19–31. https://doi.org/10.1177/00333549141291S206.

Carter, M. (2006). Greening the Ghetto. In *TED talks*. New York, NY. Available at: https://www.ted.com/talks/majora_carter_s_tale_of_urban_renewal.

Centers for Disease Control and Prevention (CDC). (2017). *Ten essential public health services and how they can include addressing social determinants of health inequities*. Washington, DC: Centers for Disease Control and Prevention. https://doi.org/10.1073/pnas.0702212104.

Centers for Disease Control and Prevention (CDC). (2019). *National environmental public health tracking network (raw data)*. Retrieved from www.cdc.gov/ephtracking.

Centers for Medicare and Medicaid Services (CMS). (n.d.). Accountable health communities model. Washington, DC: Centers for Medicare and Medicaid Services. Retrieved August 4, 2019, from https://innovation.cms.gov/initiatives/ahcm/.

Chavis, B. (1994). *Unequal protection: Environmental justice and communities of color*. San Francisco, CA: Sierra Club Books.

Chetty, R., Stepner, M., Abraham, S., Lin, S., Scuderi, B., et al. (2016). The association between income and life expectancy in the United States. *JAMA*, *315*(16), 1750–1766. https://doi.org/10.1001/jama.2016.4226.

City of Orlando. (n.d.). City of Orlando map library. Available at: https://www.orlando.gov/Our-Government/Records-and-Documents/Map-Library.

Clark, L., Millet, D., & Marshall, J. (2017). Changes in transportation-related air pollution exposures by race-ethnicity and socioeconomic status. *Environmental Health Perspectives*, *125*(9), 1–10. https://doi.org/10.1289/ehp959.

Craven, J. (2018, Jan 23). Even breathing is a risk in one of Orlando's poorest neighborhoods. *Huffpost*. Available at: https://www.huffpost.com/entry/florida-poor-black-neighborhood-air pollution_n_5a663a67e4b0e5630072746e.

Delgado, R., & Stefancic, J. (2001). *Critical race theory*. New York, NY: New York University Press.

Di, Q., Wang, Y., Zanobetti, A., Wang, Y., Koutrakis, P., et al. (2017). Air pollution and mortality in the Medicare population. *New England Journal of Medicine*, *376*(26), 2513–2522. https://doi.org/10.1056/NEJMoa1702747.

Dreier, P., Mollenkopf, J., & Swanstrom, T. (2014). *Place matters: Metropolitics for the twenty-first century*. Lawrence, KS: University Press of Kansas.

Filippelli, G., & Laidlaw, M. (2010). The elephant in the playground: Confronting lead-contaminated soils as an important source of lead burdens to urban populations. *Perspectives in Biology and Medicine*, *53*(1), 31–45. https://doi.org/10.1353/pbm.0.0136.

Fitzpatrick, K., & LaGory, M. (2003). "Placing" health in an urban sociology: Cities as mosaics of risk and protection. *City & Community*, *2*(1), 33–46.

Global Burden of Disease (GBD) 2017 Risk Factor Collaborators. (2018). Global, regional, and national comparative risk assessment of 84 behavioural, environmental and occupational, and metabolic risks or clusters of risks for 195 countries and territories, 1990-2017: A systematic analysis for the Global Burden of Disease Study. *The Lancet*, *392*, 1923–1994. https://doi.org/10.1016/S0140-6736(18)32225-6.

Hadler, J., Clogher, P., Huang, J., Libby, T., Cronquist, A., et al. (2018). The relationship between census tract poverty and shiga toxin—producing *E. coli* risk, analysis of Food-Net Data, 2010 – 2014. *Open Forum Infectious Diseases*, 1–7. https://doi.org/10.1093/ofid/ofy148.

Hennig, B., Petriello, M. C., Gamble, M. V., Surh, Y., Kresty, L. A., et al. (2018). The role of nutrition in influencing mechanisms involved in environmentally mediated diseases. *Reviews on Environmental Health*, *33*(1), 87–97. https://doi.org/10.1515/reveh-2017-0038.

Henriksson, D. K., Fredriksson, M., Waiswa, P., Selling, K., & Peterson, S. S. (2017). Bottleneck analysis at district level to illustrate gaps within the district health system in Uganda. *Global Health Action*, *10*(1). https://doi.org/10.1080/16549716.2017.1327256.

Interdisciplinary Environmental Clinic. (2019). *Environmental racism in St. Louis*. St. Louis, MO: Washington University School of Law.

Jargowsky, P. (2015). *Architecture of segregation: Civil unrest, the concentration of poverty, and public policy*. The Century Foundation.

Johnson, S. B., Riis, J. L., & Noble, K. G. (2016). State of the art review: Poverty and the developing brain. *Pediatrics*, *137*(4), e20153075. https://doi.org/10.1542/peds.2015-3075.

Kim, M. A., & Williams, K. A. (2016). Lead levels in landfill areas and childhood exposure: An integrative review. *Public Health Nursing*, *34*(1), 87–97. https://doi.org/10.1111/phn.12249.

Kino, S., & Kiwachi, I. (2018). The impact of ACA Medicaid expansion on socioeconomic inequality in healthcare services utilization. *PLoS One*, *13*(12), e0209935.

Lauster, N., & Tester, F. (2010). Culture as a problem in linking material inequality to health: On residential crowding in the Arctic. *Health and Place*, *16*(3), 523–530. https://doi.org/10.1016/j.healthplace.2009.12.010.

Legot, C., London, B., Rosofsky, A., & Shandra, J. (2012). Proximity to industrial toxins and childhood respiratory, developmental, and neurological diseases: Environmental ascription in East Baton Rouge Parish, Louisiana. *Population and Environment*, *33*, 333–346. https://doi.org/10.1007/s11111-011-0147-z.

Levesque, J., Harris, M., & Russell, G. (2013). Patient-centred access to health care: Conceptualising access at the interface of health systems and populations. *International Journal for Equity in Health*, *12*(18). https://doi.org/10.1186/1475-9276-12-18.

Maignant, G., & Staccini, P. (2018). From the territory to health territories. In *Statistical, mapping and digital approaches in healthcare*. New York: Elsevier Publishing. https://doi.org/10.1016/b978-1-78548-211-3.50001-0.

McCollum, R., Taegtmeyer, M., Otiso, L., Mireku, M., Muturi, N., Martineau, T., et al. (2019). Healthcare equity analysis: Applying the Tanahashi model of health service coverage to community health systems following devolution in Kenya. *International Journal for Equity in Health*, *18*(1). https://doi.org/10.1186/s12939-019-0967-5.

McGibbon, E., Etowa, J., & McPherson, C. (2008). Health-care access as a social determinant of health. *Canadian Nursing*, *7*(104), 22–27.

Meyer, S., Luong, T., Mamerow, L., & Ward, P. (2013). Inequities in access to healthcare: Analysis of national survey data across six Asia-Pacific countries. *BMC Health Services Research*, *13*, 238. https://doi.org/10.1186/1472-6963-13-238.

O'Connell, T., & Sharkey, A. (2013). *Reaching universal health coverage through district health system strengthening: Using a modified Tanahashi model sub-nationally to attain equitable and effective coverage*. New York, NY: UNICEF Health Section. Program Division working papers.

OECD. (2012). *OECD environmental outlook to 2050: Consequences of inaction*. Paris: OECD Publishing. https://doi.org/10.1787/9789264122246-en.

Ouyang, Y., & Pinstrup-Andersen, P. (2012). Health inequality between ethnic minority and Han populations in China. *World Development*, *40*(7), 1452–1468. https://doi.org/10.1016/j.worlddev.2012.03.016.

Painter, M., & Gandhi, P. (2019, Sept 30). *How communities are promoting health and responding to climate change*. Robert Wood Johnson Foundation Culture of Health Blog. Available at: https://www.rwjf.org/en/blog/2019/09/how-communities-are-promoting-health-and-responding-to-climate-change.html?rid=0034400001rm6qkAAA&et_cid=1845719.

Pinderhughes, H. (2017). The interplay of community trauma, diet, and physical activity. *NAM Perspectives*, *7*(8), 8–10. https://doi.org/10.31478/201708a.

Pulido, L. (2000). Rethinking environmental racism: White privilege and urban development in southern California. *Annals of the Association of American Geographers*, *90* (1), 12–40.

Rote, S. M., Angel, J. L., & Markides, K. (2017). Neighborhood context, dementia severity, and Mexican American caregiver well-being. *Journal of Aging and Health*, *29*(6), 1039–1055. https://doi.org/10.1177/0898264317707141.

Rothenberg, P. (1998). *Race, class, and gender in the United States: An integrated study* (4th ed.). New York, NY: Worth Publishers.

Salazar, D. J., Clauson, S., Abel, T. D., & Clauson, A. (2019). Race, income, and environmental inequality in the U.S. states, 1990 – 2014. *Social Science Quarterly*, *100*(3), 592–603. https://doi.org/10.1111/ssqu.12608.

Sampson, R. (2013). *Great American city: Chicago and the enduring neighborhood effect.* Chicago, IL: University of Chicago Press.

Sengupta, S. (2019, April 22). Global wealth gap would be smaller today without climate change, study finds. *The New York Times.*

Shiels, M. S., Berrington de González, A., Best, A. F., Chen, Y., Chernyavskiy, P., et al. (2019). Premature mortality from all causes and drug poisonings in the USA according to socioeconomic status and rurality: An analysis of death certificate data by county from 2000–15. *The Lancet Public Health*, *4*(2), e97–e106. https://doi.org/10.1016/S2468-2667(18)30208-1.

Sicotte, D. (2008). Dealing in toxins on the wrong side of the tracks: Lessons from a hazardous waste controversy in Phoenix. *Social Science Quarterly*, *89*(5), 1136–1152.

Soeung, S. C., Grundy, J., Sokhom, H., Blanc, D. C., & Thor, R. (2012). The social determinants of health and health service access: An in depth study in four poor communities in Phnom Penh Cambodia. *International Journal for Equity in Health*, *11*(1), 46. https://doi.org/10.1186/1475-9276-11-46.

Starfield, B. (2007). Pathways of influence on equity in health. *Social Science and Medicine*, *64*(7), 1355–1362. https://doi.org/10.1016/j.socscimed.2006.11.027.

Starfield, B. (2011). The hidden inequity in health care. *International Journal for Equity in Health*, *10*, 15. https://doi.org/10.1186/1475-9276-10-15.

National Academy of Medicine. (n.d.). Social determinants of health. Retrieved August 4, [illegible]9, from https://nam.edu/tag/social-determinants-of-health/.

US Census Bureau. (2019). *American Community Survey, 2011-2015*. Washington, DC. Available at: https://www.census.gov/programs-surveys/acs.

US Census Bureau. QuickFacts: Sanford city, Florida; Kissimmee city, Florida; Orange County, Florida; Orlando city, Florida. Retrieved February 21, 2020 from https://www.census.gov/quickfacts/fact/table/sanfordcityflorida,kissimmeecityflorida,orangecountyflorida,orlandocityflorida/PST045219, n.d.

US Environmental Protection Agency (US EPA). (2006). *Environmental justice. Frequently asked questions.* Retrieved from http://www.epa.gov/compliance/resources/faqs/ej/index.html.

Weissenberger, S., Ptacek, R., Klicperova-baker, M., Erman, A., Schonova, K., Raboch, J., et al. (2017). ADHD, lifestyles and comorbidities: A call for an holistic perspective—From medical to societal intervening factors. *Frontiers in Psychology*, *8*, 454–467. https://doi.org/10.3389/fpsyg.2017.00454.

Wharton, T., & Church, W. (2009). Consideration of one area of persistent poverty in the United States. *Social Development Issues*, *31*(1), 27–28.

Woolf, S., & Aron, L. (2013). *U.S. health in international perspective: Shorter lives, poorer health.* National Research Council (US); Institute of Medicine (US)Washington, DC: National Academies Press. Available at: www.ncbi.nlm.nih.gov/books/NBK154491.

World Health Organization (WHO). (2013). *Evaluating intersectoral processes for action on social determinants of health: Learning from key informants* [Discussion Paper Series on Social Determinants of Health No. 5]. Geneva: Social Determinants of Health.

World Health Organization (WHO). (n.d.). Social determinants of health. Retrieved August 4, 2019, from https://www.who.int/social_determinants/en/.

Supplementary reading recommendations

Association of State and Territorial Health Officials. (2019, May 20). *ASTHO briefs: Top 10 promising approaches to improving population health*. Available at: http://www.astho.org/.

Centers for Disease Control and Prevention (CDC). (2015). *Assessment & planning models, frameworks, & tools*. Available at: https://www.cdc.gov/publichealthgateway/cha/assessment.html.

Centers for Disease Control and Prevention (CDC). (2018a). *Ten essential public health services and how they can include addressing social determinants of health inequities*. Available at: https://www.cdc.gov/publichealthgateway/publichealthservices/pdf/Ten_Essential_Services_and_SDOH.pdf.

Centers for Disease Control and Prevention (CDC). (2018b). *The Public health system & the 10 essential public health services*. Available at: https://www.cdc.gov/publichealthgateway/publichealthservices/essentialhealthservices.html.

Magnan, S. (2017). Social determinants of health 101 for health care: Five plus five. *NAM Perspectives*. Discussion Paper, National Academy of Medicine, Washington, DC. https://doi.org/10.31478/201710c.

Mongeon, M., Levi, J., & Heinrich, J. (2017). Elements of accountable communities for health: A review of the literature. *NAM Perspectives*. Discussion Paper, National Academy of Medicine, Washington, DC. https://doi.org/10.31478/201711a.

National Academy of Medicine. (2018, Oct 1). *Social determinants of health*. Washington, DC: National Academies of Sciences. Available at: https://nam.edu/tag/social-determinants-of-health/.

US Public Health Service. (2017). *Inter-Professional Education Collaborative (IPEC) Award Webinar: Harnessing the strength of interprofessional teams*. Available at: https://www.ipecollaborative.org/archived-webinars.html.

World Health Organization. (2018). *Monitoring health for the Sustainable Development Goals (SDGs). Report by the Global Health Observatory*. Available at: https://www.who.int/gho/publications/world_health_statistics/2018/en/.

Definitions

Environmental Racism imposition of environmentally poor or dangerous conditions on communities that are identifiable by concentration of racial or ethnic minority groups.

Environmental (In)justice imposition of toxic or unequal environmental conditions on any community.

Health disparities a higher burden of illness, injury, disability, or mortality experienced by one group relative to another. Specifically, these differences are not attributable to variations in general healthcare need, patient preferences, or treatment recommendations.

Poverty line the official line over which a family is determined to have minimally adequate financial resources for survival. The poverty line was originally conceived by an economist at the US Social Security Administration in the 1960s. At the time, the conceptualization was based on a "market basket" approach—how much would a family need to adequately feed everyone and pay normal expenses related to home ownership and work? The poverty line has been controversial as expenses and distribution of household income has shifted in the United States. For this reason, many programs use a benchmark of 150%–300% of the poverty line as the standard of entrance for benefits. The official poverty line is not the same as the "cost of living" in an area. The *official poverty line* in 2019 for a family of 4 living in the contiguous lower 48 US states is $25,750; the *estimated cost of living* in the Orlando metro area, according to the United Way annual report in 2018, is approximately $55,000.

Redlining the historical practice of identifying areas where non-White families were prohibited from home ownership or unable to get a mortgage to purchase property, insurance, or other services such as retail outlets, student loans, or healthcare. These were said to be drawn in red lines on city maps. In the 1930s, federal programs created "security maps" to identify areas of high-risk lending for financial services, which often identified minority neighborhoods, leading to a long-standing phenomenon where banks would lend to low-income White families, but not to high-income Black families. This practice continued through the 1970s, when court challenges prompted legislative action to bring transparency and an end to the discrimination. The *Fair Housing Act of 1968* and the *Community Reinvestment Act of 1977* were supposed to address and end such practices. Court challenges that exposed redlining practices, however, have been heard as recently as 2015.

Social Determinants of Health (SDOH) the conditions into which we are born, grow and develop, live and work, as well as the wider set of forces and systems that shape the conditions of our daily lives. This is the "shorthand" for the things about our lives—where we live, what we eat, how we are treated in our neighborhoods, what kind of education we have access to, whether we have networks of support in our lives, etc. —that influence our health and well-being.

Index

Note: Page numbers followed by *f* indicate figures, *t* indicate tables and *ge* indicate glossary terms.

B

C

D

I

O

P

Q

R

S